Role of Entomopathogenic Bacteria in Insect Pest Management

This book highlights the important contribution of entomopathogenic bacteria in transforming the management of insect pests. It meticulously details their mechanisms, historical development, and crucial role in integrated pest management (IPM) strategies. With a focus on sustainable and environmentally friendly pest control, the book categorizes and explains various entomopathogenic bacteria, providing insights into their mode of action, advantages, and disadvantages. It also compiles essential information about insecticidal toxins produced by these bacteria and their historical context, spanning over a century of research. Additionally, the book sheds light on ongoing research, including genetic engineering, mutualistic relationships, and resistance management. Through its holistic approach, this book underscores the significance of entomopathogenic bacteria in achieving the UN Sustainable Development Goals and various ecosystem services. It offers a vital perspective on their ecological, economic, and policy implications, making it important for those seeking sustainable solutions in agriculture and pest management. This book is intended for professionals and researchers in the fields of entomology, agriculture, and pest management.

Role of Entomopathogenic Bacteria in Insect Pest Management

Swagata Thakur, Samrat Saha, Govindaraj Kamalam Dinesh,
and Archana Anokhe

CRC Press
Taylor & Francis Group
Boca Raton London New York

CRC Press is an imprint of the
Taylor & Francis Group, an **informa** business

Designed cover image: Shutterstock ID 1883590570

First edition published 2026
by CRC Press
2385 NW Executive Center Drive, Suite 320, Boca Raton FL 33431

and by CRC Press
4 Park Square, Milton Park, Abingdon, Oxon, OX14 4RN

CRC Press is an imprint of Taylor & Francis Group, LLC

ISBN: 978-1-032-73130-8 (hbk)
ISBN: 978-1-032-73131-5 (pbk)
ISBN: 978-1-003-42688-2 (ebk)

DOI: 10.1201/9781003426882

Typeset in Times
by Apex CoVantage, LLC

Dedicated to Those Who Work for the Well-Being of the Society

Contents

Foreword

The growing demand for sustainable agricultural practices has intensified the search for effective, environmentally friendly alternatives to conventional pest management strategies. Among these, entomopathogenic bacteria (EPB) have emerged as powerful biocontrol agents, offering an ecologically sound solution to insect pest management. It is my pleasure to introduce *Role of Entomopathogenic Bacteria in Insect Pest Management*, an invaluable contribution by Swagata Thakur, Samrat Saha, Govindaraj Kamalam Dinesh, and Archana Anokhe. This book provides a rigorous and well-structured exploration of the critical role of EPB in sustainable agriculture.

The global challenge of pest resistance to chemical pesticides, coupled with environmental degradation, necessitates a paradigm shift in pest management. This volume addresses these pressing concerns by comprehensively presenting the taxonomy, mechanisms of action, and practical applications of EPB. With 17 meticulously curated chapters, the book covers the historical background, classification, and ecological significance of these bacteria, while also delving into advanced topics such as genetic engineering, resistance management, and commercial formulation.

A key highlight of this work is its interdisciplinary approach, bridging foundational research with practical applications. The chapters not only examine the microbial mechanisms that drive pest suppression but also assess the economic and ecological impact of EPB on sustainable agriculture. Additionally, discussions on Bt crops, biosafety regulations, and policy frameworks ensure that the book remains relevant to researchers, policymakers, practitioners, and students alike.

This publication aligns with India's commitment to integrating sustainable practices into agricultural and environmental policies. By bringing together experts in microbiology, environmental science, and agricultural research, this volume serves as a crucial reference for advancing biological pest control solutions. I commend the editors and contributors for their dedicated efforts in compiling this comprehensive resource and extend my gratitude to CRC Press and Taylor & Francis Group for supporting the dissemination of this essential knowledge.

I am confident that this book will inspire further innovation in biological pest management and contribute to building a more sustainable future for global agriculture.

Prof Himanshu Pathak
Director General
International Crops Research Institute
for the Semi-arid Tropics (ICRISAT), Hyderabad

Foreword

I am pleased to introduce the Role of Entomopathogenic Bacteria in Insect pest management, a captivating perspective of Swagata Thakur, Samrat Saha, Govindaraj Kamalam Dinesh, and Archana Anokhe. This timely and thorough volume presents a ground-breaking approach to biological pest control, addressing one of agriculture's most pressing challenges: the management of insect pests by using sustainable, ecologically balanced practices. This book, published by CRC Press, Taylor & Francis Group, significantly contributes to integrated pest management and is translated directly into the ICAR. mission to advance science for sustainable agriculture.

Conventional chemical pesticides are effective but carry the twin burdens of environmental degradation and insect resistance; entomopathogenic bacteria (EPB) represent promising alternatives to conventional chemical pesticides. This book is an outstanding because it deals with these issues appropriately and, more importantly, how EPB helps to wean us off chemical controls and preserve eco-friendly agricultural practices.

This volume of 17 chapters provides readers with the scope of knowledge about entomopathogenic bacteria, from basics, history and historical insights to finer taxonomic aspects and biotechnological applications of EPB in controlling pests. Structurally, the book begins with principles and moves quickly into cases of specific bacterial taxa before a detailed discussion of genetic engineering, resistance management, and policy consideration. The complex dynamics of large-scale implementation of EPB results in key discussions on the genetic modification of EPB, the safety of Bt crops, and the criteria for commercialization. Finally, further weaving the regulatory framework and policy perspective, the book emphasizes using entomopathogenic bacteria as an alternative to genetically modified crops and. chemical insecticides.

I congratulate the authors for writing this reference in a way that should make a perfect resource for scientists, students, policymakers, and agricultural practitioners. In my view, this excellent book, 'Role of Entomopathogenic Bacteria in Insect Pest Management', will undoubtedly play a part in developing sustainable pest management and framing the direction for biological control in agriculture.

(Ch. Srinivasa Rao)
Director
ICAR - Indian Agricultural Research Institute, New Delhi

Preface

Entomopathogenic bacteria (EPB) as a powerful and efficient insect pest management strategy is gaining prominence due to rising concerns regarding pesticide resistance, environmental degradation, and the necessity of sustainable agricultural practices. These environment-friendly and effective naturally occurring microbial agents constitute an innovative path for integrated pest management. This book, *Role of Entomopathogenic Bacteria in Insect Pest Management*, brings together comprehensive research and insights into the history, classification, pathogenic potential, application, improvement, and future potential of EPBs as an effective tool for sustainable pest management. The authors, Swagata Thakur, Samrat Saha, Govindaraj Kamalam Dinesh, and Archana Anokhe, have extensively detailed their structure and plausible exploitation in pest management across different phyla.

The introductory chapters lay the foundation, with Chapter 1 exploring the essential question: 'Why entomopathogenic bacteria?' This is followed by a brief history of EPB and an in-depth discussion of the classification of EPB, providing readers with a clear context for the bacterial groups covered at the initial stage.

The subsequent chapters explore specific EPB taxa, which include thorough overviews of major Firmicutes members, *Bacillus thuringiensis*, *Lysinibacillus sphaericus*, *Paenibacillus popilliae*, and other lesser-known entomopathogenic bacteria, presenting a balanced perspective of how each genus specializes in biocontrol of pest or can be explored further. The other potential entomopathogenic bacterial phyla have been critically described in the chapters on Actinobacteria, Proteobacteria, and Tenericutes, which have diverse mechanisms of action and applications.

The book's later chapters deal with modern developments in EPB genetic engineering for improved pest control and critical appraisal of GM technology. The problem of insect resistance against EPB is also addressed, which is a crucial argument in supporting the retaining of biocontrol measures and questioning their efficacy.

The practical considerations in bringing EPB to market are discussed in the chapters on 'selection criteria and legislation procedures' and 'mass production and commercial formulation of important EPB'. Readers interested in microbial pest control agents' commercialization and regulatory aspects will find these sections more valuable.

Comprising an ecological, economic analysis and ecosystem services of EPB with perspectives on policy and future direction for using EPB, the concluding chapters summarize the work. The authors discuss EPB as an alternative to genetically modified crops and chemical insecticides and in a future where EPB agents will be a crucial part of sustainable agricultural ecosystems.

Role of Entomopathogenic Bacteria in Insect Pest Management is a timely and end-to-end resource for researchers, practitioners, students, and policymakers. This book illuminates the promise and practicality of EPB through an expertly curated sequence of topics, including detailed explorations of foundational and advanced biocontrol. We hope this work will spark additional research and development of microbial solutions towards developing sustainable, eco-friendly pest management.

The authors thank the many other researchers and practitioners whose work has facilitated the development of entomopathogenic bacteria as viable agents in integrated pest management. We hope this volume will be an exhaustive resource, a one-stop solution and a clarion call to prompt ongoing conservation of insect pest management and agricultural resiliency worldwide.

Acknowledgements

As a prelude to our thanksgiving, at first, we wish to thank the Almighty for giving us strength, courage, and confidence; after all, He is the 'greatest'.

We are extremely thankful to our family members for supporting us every moment. Our deepest sense of gratitude and indebtedness to our mentor, Dr. Vinay K. Kalia, Rtd Principal Scientist, Indian Council of Agricultural Research, Indian Agricultural Research Institute, New Delhi, for her guidance and unwavering support and encouragement throughout this journey. Ma'am, your belief in our vision made this possible.

We are also profoundly grateful to the Indian Council of Agricultural Research, Indian Agricultural Research Institute, New Delhi, where the idea of writing this book was conceived and whose resources and opportunities needed to bring this project to fruition were utilized.

We express our heartfelt gratitude to our mentors, Prof. Himanshu Pathak, Director General, International Crops Research Institute for the Semi-Arid Tropics, Hyderabad, and Prof. Cherukumalli Srinivasa Rao, Director, ICAR-Indian Agricultural Research Institute, New Delhi, for their sustained guidance and continuous support. We are particularly thankful for providing valuable foreword and wishes for this book.

Special thanks to Dr. Bhavik Sawhney, Senior Editor (Life Sciences), and CRC Press/Taylor & Francis Group for their expertise and dedication in refining this work. Your insight and professionalism have enriched the final product immensely.

Lastly, to our readers, researchers, and scholars, your enthusiasm and curiosity are the driving forces behind our work. Thank you for allowing us to share our ideas and stories with you.

Author Biographies

Swagata Thakur

Swagata Thakur is currently pursuing her PhD in plant pathology at the Indian Agricultural Research Institute. She has qualified for ICAR AIEEA-PG (Indian Council of Agricultural Research All India Entrance Examination for Agriculture—Post-Graduation) 2020 in plant science with All India Rank (AIR) 16 and availed of junior research fellowship for her MSc from the Indian Agricultural Research Institute. During her MSc tenure, she worked on phenotypic and molecular characterization of finger millet against blast disease caused by *Pyricularia grisea* and gained expertise in finger millet pathology and molecular biology. She has also qualified for ICAR AICE-JRF/SRF (All India Competitive Examinations Junior Research Fellowship, Senior Research Fellowship) 2022 in plant pathology with AIR 10 and is availing a senior research fellowship. She has also availed a DST-INSPIRE fellowship and cleared the ASRB-NET, UGC-NET, and DBT-BET examinations. For her PhD, she is working on the bakanae disease of rice and availing expertise in rice pathology and molecular biology. She has published two research articles, several popular articles/technical articles, and book chapters; attended and presented her research works at several national and international conferences; and received the best oral and poster presentation award. She can be contacted at sthakurubkv1998@gmail.com

Samrat Saha

Samrat Saha has completed his BSc in agriculture in 2020 and his MSc in agricultural entomology in 2022 from Uttar Banga Krishi Viswavidyalaya. He is currently pursuing his PhD in the Department of Entomology at Uttar Banga Krishi Viswavidyalaya. He has gained interest and expertise in the field of apiculture, bee ecology, behaviour, plant–pollinator interaction, and melissopalynology and has published several research articles, popular articles, and book chapters in different reputed journals, magazines, and publishers. He has qualified for ASRB-NET 2023. He has also participated in different national and international seminars and conferences and was awarded best presentation awards. He can be contacted at isamratsahamtb43@gmail.com

Dr Govindaraj Kamalam Dinesh

Dr. G K Dinesh (Govindaraj Kamalam Dinesh) serves as an Assistant Professor of Environmental Sciences at College of Agriculture, Central Agricultural University, Iroisemba, Imphal, India, under the Government of India. He earned his PhD in Environmental Science from ICAR-Indian Agricultural Research Institute, New Delhi, achieving a national-level 5th rank and receiving the IARI Ph.D. Senior Research Fellowship through a national entrance examination. Additionally, he is a recipient of the 'Young Advantage Fellow' award from the Institute of Biosciences, USA, and holds an honorary Research Fellow position at INTI International University, Malaysia.

Dr. Dinesh completed his Bachelor's degree in Agriculture at Annamalai University, Chidambaram, Tamil Nadu, India, and his Master's degree in (Agriculture) Environmental Science from Tamil Nadu Agricultural University, Coimbatore, Tamil Nadu, India. During his Master's program, he received a fellowship for academic excellence through the Student Junior Research Fellowship under the TNAU-Private Industrial Agency Scheme (Sakthi Sugars Pvt. Ltd.) and conducted a long-term impact assessment study on distillery effluent application on soil and Cassava and Jasmine crops. His Master's thesis focused on the ecology of birds and insects in organic and inorganic paddy ecosystems.

He has qualified for the ICAR and UGC National Eligibility Tests for Professorship multiple times. His PhD research on conservation agriculture and ecosystem services valuation was supported by the University Grants Commission, Government of India, through the UGC Junior Research Fellowship, and his work was recognized as significant by the Division of Environment Science, ICAR-Indian Agricultural Research Institute, India.

Dr. Dinesh's research expertise encompasses carbon sequestration, ecosystem services, agroecology, climate-smart agriculture, farming systems, sustainable agriculture, crop simulation modeling, GHG emissions, agrobiodiversity, bioremediation, economic ornithology, waste recycling, and pollution control. He has presented his research at numerous national and international conferences, earning accolades such as the 'Best Oral Paper Presentation Award', 'Best M.Sc. Thesis Award', and 'Young Researcher Award for Excellence in Research'.

He has authored over 35+ peer-reviewed publications, 15+ book chapters, 10+ books, and 70+ popular scientific articles in national and international journals and magazines. He has amassed over 250+ citations, an h-index of 9, and a cumulative impact factor exceeding 32. He is also acting as an Editor-in-chief for Enlighten Agriculture Scientific Magazine.

Beyond academia, Dr. Dinesh has translated numerous technical documents, including the Environmental Impact Assessment Draft by the Government of India. Since June 2024, he has served as a Subject Expert Member in the Book Selection Committee for Public Libraries under the Government of Tamil Nadu. He is also an expert member of various organizations, including the Food and Agricultural Organization, Rome, and has experience as a journalist and editorial member for a leading magazine. His mission is to bridge the gap between science and society, fostering an inclusive and dialogic classroom environment. Currently, he is writing and editing books for prominent publishers such as CRC Press (UK), Nova Science (USA), and Bentham Science Publishers (UAE). He can be contacted at gkdineshiari@gmail.com or gkdinesh@myyahoo.com

Dr Archana Anokhe

Dr. Archana Anokhe is now working at ICAR–Directorate of Weed Research, Jabalpur, India. She was also a visiting scholar at the Agricultural Research Organisation (ARO), in Israel. She has qualified for a junior research fellowship with All India Rank (AIR) 13, a senior research fellowship with AIR-4, and agricultural research services with AIR-8; she was awarded the Women Scientist Award in 2022, Young Scientist Award in 2021, best PhD thesis award in 2020, Best Student Award in 2019, best oral presentation award at the National Entomologist Meet in 2015, and the best poster presentation in the Indo-US International Conference in 2017. She was also awarded a National Talent scholarship and a UGC national fellowship during her graduation and PhD, respectively. Apart from PhD, she has also undergone different degree programmes, such as post-graduate diploma in education technology management and a master of arts in English. She has been involved in the teaching of courses like insect pathology, host plant resistance, advances in host plant resistance, pest of horticultural and plantation crops, insect ornithology, forest entomology, and medical entomology. She has published more than ten research papers in national and international journals and has also published more than 50 popular articles/blogs and technical bulletins and is a member of many institutional committees. She has also been interviewed at Doordarshan and All India Radio. She is the associate editor of the *Indian Journal of Entomology*, the *Indian Entomologist* magazine, and the *Agro-Titan* magazine. Her area of specialization is insect pathology and virus–vector relationship studies. She can be contacted at anokhearchana12@gmail.com

Abbreviations

®	Registered trademark
µm	Micrometer
Aam1	*Aedes aegypti* maltasc 1
ABC	ATP-binding cassette
ABCA2	ATP-Binding Cassette Subfamily A Member 2
ABCB1	ATP Binding Cassette Subfamily B Member 1
ABCC2	ATP Binding Cassette Subfamily C Member 2
ABCG1	ATP Binding Cassette Subfamily G Member 1
ADH	Arginine dihydrolase
ADP	Adenosine Diphosphate
afp	Antifeeding prophage
Agm1	Anopheles gambiae maltase 1
Agm3	Anopheles gambiae maltase 3
ags	Agropine
AIZ	*Bacillus thuringiensis* serovar *aizawai*
$AlCl_3$	Aluminium chloride
ALE	*Bacillus thuringiensis* serovar *alesti*
ALP	Alkaline phosphatase
AMA	*Bacillus thuringiensis* serovar *amagiensis*
AMPs	Antimicrobial peptides
AND	*Bacillus thuringiensis* serovar *andaluciensis*
ANK	Ankyrin
APN	Aminopeptidase-N
AprA	Alkaline protease
ARG	*Bacillus thuringiensis* serovar *argentinensis*
AST	*Bacillus thuringiensis* serovar *asturiensis*
ATP	Adenosine Triphosphate
AZO	*Bacillus thuringiensis* serovar *azorensis*
BAL	*Bacillus thuringiensis* serovar *balearica*
BASF	Badische Anilin- und Sodafabrik (Baden Aniline and Soda Factory)
BBMF	Brush border membrane fractions
BBMV	Brush Border Membrane Vesicles
BCA	Biological control agent
Bin/Btx	Binary toxins
BOL	*Bacillus thuringiensis* serovar *bolivia*
BoNT	Botulinum Neurotoxin
BPH	Brown plant hopper
BRA	*Bacillus thuringiensis* serovar *brasiliensis*
BRAI	Biotechnology Regulatory Authority of India
Bt	*Bacillus thuringiensis*
Bti	*Bacillus thuringiensis* var *israelensis*
Btk	*Bacillus thuringiensis* var *kurstaki*
Btt	*Bacillus thuringiensis* subsp. *tenebrionis*
C	Cytokinin
$CaCl_2$	Calcium chloride
CADR	Cadherin-like protein
CAGR	Compound annual growth rate
CAM	*Bacillus thuringiensis* serovar *cameroun*
CaMV	Cauliflower mosaic virus
CAN	*Bacillus thuringiensis* serovar *canadensis*
CCD	Colony collapse disorder
CDC	Cholesterol-dependent cytolysin
CDC	Centers for Disease Control and Prevention

CDC	Cholesterol-dependent cytolysins
CFU	Colony forming unit
CHA	*Bacillus thuringiensis* serovar *chanpaisis*
Chi	Chitinase
CHRD	Cysteine- and histidine-rich domain protein
CI	Cytoplasmic incompatibility
CIBRC	Central Insecticides Board and Registration Committee
COL	*Bacillus thuringiensis* serovar *colmeri*
COR	*Bacillus thuringiensis* serovar *coreanensis*
Cpb	Spore coat-parasporal body complex
CPB	Colorado potato beetle
Cpm1	*Culex pipiens* maltase 1
CPV	Cytoplasmic Polyhedrosis Virus
Cqm1	*Culex quinquefasciatus* maltase 1
CRIIGEN	Committee for Independent Information and Research on Genetic Engineering
Cry	Crystal toxin
CSPB	Canoe-shaped parasporal body
CTA	Caprylate-thallous agar
CTNBio	Brazilian National Biosafety Technical Commission
Cyt	Cytolytic toxin
DAK	*Bacillus thuringiensis* serovar *dakota*
DAR	*Bacillus thuringiensis* serovar *darmstadiensis*
DBM	Diamondback moth
DBT	Department of Biotechnology
DDT	Dichlorodiphenyltrichloroethane
DNA	Deoxyribonucleic acid
DNase	Deoxyribonuclease
dpi	Days post-infection
DPPQS	Directorate of Plant Protection, Quarantine and Storage
dsDNA	Double-stranded deoxyribonucleic acid
dsRNA	Double-stranded ribonucleic acid
EB	Elementary body
EC	Enterocyte
ECB	European corn borer
EE	Enteroendocrine
EFB	European Foulbrood
EFSA	European Food Safety Authority
EIL	Economic injury level
EMS	Ethyl methane-sulphonate
ENT	*Bacillus thuringiensis* serovar *entomocidus*
EPA	Environmental Protection Agency
EPB	Entomopathogenic Bacteria
EPF	Entomopathogenic Fungus
EPN	Entomopathogenic Nematode
EPSPS	5-enolpyruvylshikimate-3-phosphate synthase
ESA	Entomological Society of America
EU	European union
ExsC	Exosporium protein C
FDA	Food and Drug Administration
$FeSO_4$	Ferrous sulfate
FIN	*Bacillus thuringiensis* serovar *finitimus*
FSB	Fruit and shoot borer
FUK	*Bacillus thuringiensis* serovar *fukuokaensis*
g	Gram
G	Guanine
GABA	Gamma-aminobutyric acid
GAD	Glutamic acid decarboxylase

GAL	*Bacillus thuringiensis* serovar *galleriae*
GalNAc	N-acetylgalactosamine
GC	Goblet cell
GCR	Glycoconjugate
GCSC	Germany-China Scientific Cooperation
GDP	Gross domestic product
GE	Genetic engineering/Genetically engineered
GEAC	Genetic Engineering Approval Committee
GI	Gastrointestinal
GM	*Genetically modified*
GPI	Glycosylphosphatidyl-inositol
GRA	*Bacillus thuringiensis* serovar *graciosensis*
GRAS	Generally Regarded as Safe
GRDC	Grands Research and Development Corporation
GUI	*Bacillus thuringiensis* serovar *guiyangiensis*
GV	Granulosis Virus
GVA	Gross Value Addition
h/hrs	Hours
H_2S	Hydrogen sulphide
ha	Hectares
HDPE	High-density polyethylene
HIG	*Bacillus thuringiensis* serovar *higo*
HT	Herbicide-tolerant
HUA	*Bacillus thuringiensis* serovar *huazhongensis*
IB	Intermediate Body
IBE	*Bacillus thuringiensis* serovar *iberica*
ICPs	Insecticidal crystal proteins
IEBC	International Entomopathogenic Bacillus Centre
IJs	Infective juveniles
Imd	Immune deficiency
IND	*Bacillus thuringiensis* serovar *indiana*
IP	Insecticidal protein
IPM	Integrated Pest Management
IPPC	International Plant Protection Convention
IR	Insect-resistant
IR-HT	Insect-resistant and herbicide-tolerant
IRM	Insecticide Resistance Management
IRRI	International Rice Research Institute
ISC	Intestinal stem cell
ISP	Insecticidal secreted proteins
ISR	*Bacillus thuringiensis* serovar *israelensis*
ITS	Internal transcribed spacer
JAP	*Bacillus thuringiensis* serovar *japonensis*
JEG	*Bacillus thuringiensis* serovar jegathesan
JH	Juvenile Hormone
JHEs	Juvenile hormone esterases
JIN	*Bacillus thuringiensis* serovar *jinghongiensis*
JOR	*Bacillus thuringiensis* serovar *jordanica*
kb	Kilobase
kDa	Kilodaltons
KEN	*Bacillus thuringiensis* serovar *kenyae*
KIM	*Bacillus thuringiensis* serovar *kim*
km	Kilometers
KNO3	Potassium nitrate
KON	*Bacillus thuringiensis* serovar *konkukian*
KUM	*Bacillus thuringiensis* serovar *kumamotoensis*
KUR	*Bacillus thuringiensis* serovar *kurstaki*

KYU	*Bacillus thuringiensis* serovar *kyushuensis*
L	Liters
LA	Luria-Bertani agar
LB	Luria-Bertani broth
LC_{50}	Lethal concentration of the toxin causing 50% mortality
LD_{50}	Lethal Dose of the toxin causing 50% mortality
LDPE	Low density polyethylene
LEE	*Bacillus thuringiensis serovar leesis*
Leu	Leucine
LMC	Learning and Memory Capacity
LON	*Bacillus thuringiensis serovar londrina*
Ls	Lysinibacillus spharicus
LSF	Liquid state fermentation
LT_{50}	Lethal Time of the toxin causing 50% mortality
M	Molar
MAHYCO	Maharashtra Hybrid Seed Company Ltd.
MAL	*Bacillus thuringiensis* serovar *malaysiensis*
mART	mono-ADP-ribosyltransferase
mas	Mannopine
Mbp	Mega base pair
MBS	Minimal Basal Salt
Mcf1	Makes caterpillars floppy 1
MDCK	Madin and Darby canine kidney cells
MDF	Midgut differentiating factor
MED	*Bacillus thuringiensis* serovar *medellin*
MEX	*Bacillus thuringiensis* serovar *mexicanensis*
MGPS	Multiple gene pyramiding and silencing
$MgSO_4$	Magnesium sulfate
mi ha	Million hectares
min	Minutes
mL	Milliliters
MLOs	Mycoplasma-like organisms
mmt	Million metric ton
$MnSO_4$	Manganese sulfate
MoA	Mode of action
MOG	*Bacillus thuringiensis* serovar *mogi*
MON	*Bacillus thuringiensis* serovar *monterrey*
MOR	*Bacillus thuringiensis* serovar *morrisoni*
MPCA	Microbial pest control agent
Mre	Murein formation gene cluster E
MRL	Maximum residue limit
mRNA	Messenger ribonucleic acid
MSL	Male-specific lethal
MT	Metric tonne
MTX	Mosquitocidal toxin
MUJ	*Bacillus thuringiensis* serovar *muju*
MYPGP	Mueller-Hinton broth, yeast extract, potassium phosphate, glucose, and pyruvate agar
N	Normal
NaCl	Sodium chloride
NaOCl	Sodium hypochlorite
NAV	*Bacillus thuringiensis* serovar *navarrensis*
NBTA	Nutrient bromothymol blue-triphenyl tetrazolium chloride agar
NEO	*Bacillus thuringiensis* serovar *neoleonensis*
NIG	*Bacillus thuringiensis* serovar *nigeriensis*
NOV	*Bacillus thuringiensis* serovar *novosibirsk*
NptII	Neomycin phosphotransferase II
NPV	Nuclear Polyhedrosis Virus

NSKE	Neem Seed Kernel Extract
NVBDCP	National Vector Borne Disease Control Programme
ocs	Octopine
OST	*Bacillus thuringiensis* serovar *ostriniae*
OSW	*Bacillus thuringiensis* serovar *oswaldocruzi*
OTU	Ovarian tumour deubiquitinase
pADAP	Amber disease-associated plasmid
PAH	*Bacillus thuringiensis* serovar *pahangi*
PAK	*Bacillus thuringiensis* serovar *pakistani*
PAL	*Bacillus thuringiensis* serovar *palmanyolensis*
PBW	Pink bollworm
pcps	Pyrrolidone carboxyl peptidases
PCR	Polymerase chain reaction
PEG	Polyethylene glycol
PEMBA	Polymyxin pyruvate egg-yolk mannitol–bromothymol blue agar
PEPC	Phosphoenolpyruvate carboxylase
PFTs	Pore-forming Toxins
PG	Peptidoglycan
PGPR	Plant growth-promoting rhizobacterium
Phe	Phenylalanine
PIN	*Bacillus thuringiensis* serovar *pingluonsis*
PIPs	Plant-incorporated protectants
Pir	*Photorhabdus* insect-related proteins
PIR	*Bacillus thuringiensis* serovar *pirenaica*
PMP1	Paraclostridial mosquitocidal protein 1
PMR	Post-Mating Responses
POL	*Bacillus thuringiensis* serovar *poloniensis*
PON	*Bacillus thuringiensis* serovar *pondicheriensis*
PS	Parasporin
PTM	Potato tuber moth
PTP	Pollen-tube pathway
PUL	*Bacillus thuringiensis* serovar *pulsiensis*
PVA	Polyvinyl alcohol
PVC	*Photorhabdus* virulence cassettes
pvf	*Pseudomonas virulence factor*
PVY^O cp	Potato virus Y (ordinary strain) coat protein
R&D	Research and development
RB	Reticulate bodies
RCGM	Review Committee on Genetic Manipulation
rDNA	Ribosomal deoxyribonucleic acid
RH	Relative humidity
RIP	Ribosome-inactivating protein
RISC	Ribonucleic acid Induced Silencing Complex
RISC	RNA-induced silencing complex
RKN	Root-knot nematode
RNA	Ribonucleic acid
RNAi	Ribonucleic acid interference
RON	*Bacillus thuringiensis* serovar *rongseni*
ROS	*Bacillus thuringiensis* serovar *roskildiensis*
RR	Homozygous resistant insects
rRNA	Ribosomal Ribonucleic acid
RS	Heterozygous offspring
SCP	Spore-crystal preparations
SDGs	Sustainable development goals
SEM	Scanning electron microscope
SEO	*Bacillus thuringiensis* serovar *seoulensis*
sep	*Serratia entomophila* pathogenicity

SHA	*Bacillus thuringiensis* serovar *shandongiensis*
SIL	*Bacillus thuringiensis* serovar *silo*
SIN	*Bacillus thuringiensis* serovar *sinensis*
Sip	Secreted Insecticidal Proteins
siRNA	Small interfering ribonucleic acid
SlpC	Surface layer protein or S-layer protein
SNARE	Soluble N-ethylmaleimide-sensitive factor activating protein receptor
SOO	*Bacillus thuringiensis* serovar *sooncheon*
SOT	*Bacillus thuringiensis* serovar *sotto*
***Sp*AID**	*Spiroplasma poulsonii* AndroclDin
SROs	Sex ratio organisms
SS	Homozygous susceptible insects
SSF	Solid-state fermentation
ssRNA	Single-stranded ribonucleic acid
SUM	*Bacillus thuringiensis* serovar *sumiyoshiensis*
SYL	*Bacillus thuringiensis* serovar *sylvestriensis*
T1SS	Type I secretion system
T3SS	Type III secretion system
T6SS	Type VI secretion system
Tc	Insecticidal toxins complex
TCP	3,5,6-trichloro-2-pyridinol
T-DNA	Transferred DNA
THA	*Bacillus thuringiensis* serovar *thailandensis*
THO	*Bacillus thuringiensis* serovar *thompsoni*
Thu	Thuringiensin
THU	*Bacillus thuringiensis* serovar *thuringiensis*
Ti plasmid	Tumour-inducing plasmid
™	Trademark
TOC	*Bacillus thuringiensis* serovar *tochigiensis*
TOG	*Bacillus thuringiensis* serovar *toguchini*
TOH	*Bacillus thuringiensis* serovar *tohokuensis*
TOL	*Bacillus thuringiensis* serovar *tolworthi*
TOU	*Bacillus thuringiensis* serovar *toumanoffi*
TSE	Transmissible spongiform encephalopathy
U.S.	United States
UAV	Unmanned aerial vehicle
UniProtKB	Universal Protein Resource Knowledgebase
USGS	United States Geological Survey
UV	Ultra-violet
v/v	Volume by volume
VAZ	*Bacillus thuringiensis* serovar *vazensis*
Vip	Vegetative Insecticidal Proteins
VP	Voges-Proskauer test
w/v	Weight by volume
WCR	Western corn rootworm
WHO	World Health Organization
WP	Wettable Powder
WRA	*Bacillus thuringiensis* serovar *wratislaviensis*
XIA	*Bacillus thuringiensis* serovar *xiaguangiensis*
YOS	*Bacillus thuringiensis* serovar *yosoo*
YSB	Yellow stem borer
YUN	*Bacillus thuringiensis* serovar *yunnanensis*
ZHA	*Bacillus thuringiensis* serovar *zhaodongensis*
$ZnSO_4$	Zinc sulfate

1 Why Entomopathogenic Bacteria?

ABSTRACT

People are over-relying on chemical insecticides as they are cheap, are easy to use, and show quick visible impact against insect pests. However, the indiscriminate use of chemical insecticides possesses a great threat to the environment and human health, biodiversity, and beneficial organisms, and insects also develop resistance against those chemicals, making them completely ineffective. To protect the environment from these consequences, modern agriculture is moving towards the use of different biopesticides. Entomopathogenic bacteria are an excellent alternative to chemical insecticides that are eco-friendly; are specific to target organisms and safe to non-target ones, as well as humans; and possess less chance of resistance development.

1.1 INTRODUCTION

> 'The ultimate goal of farming is not the growing of crops, but the cultivation and perfection of human beings.'
>
> **Masanobu Fukuoka, proponent of 'natural farming'**

The population is the strength of a society, and overpopulation is a threat to the environment. Whenever the term 'population growth' arises, another term comes side by side, that is, 'agriculture'. Agriculture, along with its allied sectors, built the basic pillar of our society, with the farming community as the mason of that pillar. This sector is undoubtedly the largest livelihood provider that uplifts the socio-economic condition of rural people. According to World Bank 2023 data, the agricultural sector accounts for 4% of the world's gross domestic product (GDP), whereas in some least-developed countries, the value is more than 25% of their GDP (World Bank, 2023). Approximately 86.6 million people around the world (which is 26.6% of the total population) have engaged with agriculture and its allied sectors (FAOSTAT, 2021).

As the world population has already crossed the milestone of 8 billion people on November 15, 2022, and is on the way to cross 9.7 billion by 2050 (estimated) (United Nations, 2022), growth in agriculture is the only way to feed that burgeoning population and achieve Sustainable Development Goal 2: End hunger (United Nations, 2015). In the recent past, the production of agricultural and horticultural crops rose dramatically by utilizing modern inputs like fertilizers, growth promoters, pesticides, and other chemical substances, and mechanical devices for seeding, tilling, and irrigating crops (FAO, 2017). With this advancement, the incidence of different crop pests has also increased, which hugely threatens our food security. In addition to that, the advancement of technology and change in climatic conditions are joining hands in shaping this new dreaded scenario (FAO, 2021). When the threshold level is crossed, pests become harmful, destructive, and troublesome for us and our interests. Cultivated crops are forced to face challenges caused by the increasing pressure of different biotic stressors (pests), such as insects, viruses, fungi, nematodes, weeds, protozoa, MLOs, molluscs, rodents, etc. Among different stressors, insect pests contribute a

DOI: 10.1201/9781003426882-1

great proportion of crop loss, and the continuous pressure imposed by them becomes a major challenge to our food production. Aiming to combat the problem, farmers largely depend on chemical pesticides (may be in the form of insecticides, herbicides, fungicides, rodenticides, nematicides, etc.), as they are cheap, are easy to use, and offer quick visible impact in protecting the crops (Sharifzadeh et al., 2018). Likewise, for controlling insects, a huge amount of synthetic chemical insecticides has been applied every year. It has been thought that insecticides are the best control measure against insects, without taking care of the consequences associated with these synthetic chemicals on the environment and human health.

A result of the over-reliance on these toxic chemicals is coming as a boomerang and throwing a great threat to the environment, agriculture, and human health. Ultimately, it forces researchers around the world to look for economically viable and environmentally sustainable alternatives. Thus, as a result of the findings, entomopathogenic bacteria (EPB) are coming as an excellent alternative to chemical insecticides, showing great potential in controlling different insect pests. So the answer to 'Why use EPB?' for controlling insect pests mostly relies on the answer to 'Why stop using chemical insecticides?' and 'What are the benefits of using EPB?' Answers to these questions have been described in this chapter.

1.2 GLOBAL SCENARIO OF CROP LOSS DUE TO PESTS

It is estimated that world food plants have suffered from more than 1,000 species of insects, 100,000 species of diseases (including bacteria, viruses, fungi, and other microorganisms), 30,000 species of weeds, and 1,000 species of nematodes (Dhaliwal et al., 2015). An assessment of yield loss by these pests is a very difficult task, although it is estimated that without crop protection measures, 70% yield loss of major food crops may occur due to the incidence of different pests (Gordon, 2022). These losses due to pest injury may extend up to 78% for fruits, 54% for vegetables, and 32% for cereals (Cai, 2008).

Among different pests, weeds contribute the highest loss of 30%, followed by insect pests (23%), and diseases (17%) (Gordon, 2022). The FAO estimated that an average annual loss of 20–40% in crop production may take place due to the incidence of pests causing a global economic loss of around $220 billion due to plant diseases and $70 billion due to invasive insect pests (FAO, 2019). Among different pests, only herbivorous insects are reported to destroy about one-fifth of the world's total crop production annually (Dhaliwal et al., 2015). To address such losses caused by pests, demand for pesticides increased during the post–green revolution era. However, in spite of the heavy use of synthetic insecticides, losses caused by insect and mite pests are still higher in the post–green revolution era as compared to the pre–green revolution era (Benedict, 2003).

1.3 GLOBAL SCENARIO OF PESTICIDE CONSUMPTION

Pesticides are an indispensable part of agricultural production, as they are used for the production of about one-third of the world's agricultural products (Liu et al., 2002). It has shown great potential in the control of pests, and therefore, tremendous success has been achieved by using pesticides in agricultural and horticultural activities as well as in the field of public health (Aktar et al., 2009). Hence, the demand for chemical pesticides is increasing day by day. Currently, every year, about 4.2 mmt (million metric tons) of pesticides have been used worldwide, which is expected to reach 4.6 mmt by 2026, with China as the major consumer, using 1.8 mmt of pesticides in 2021 (ReportLinker, 2023). In 2020, the worldwide use of pesticides per hectare of cropland was about 1.8 kg, with the highest pesticide application rate in Saint Lucia (20.48 kg/ha), followed by Hong Kong, Maldives, and Oman (Figure 1.1) (FAO, 2022). According to recent data, the top

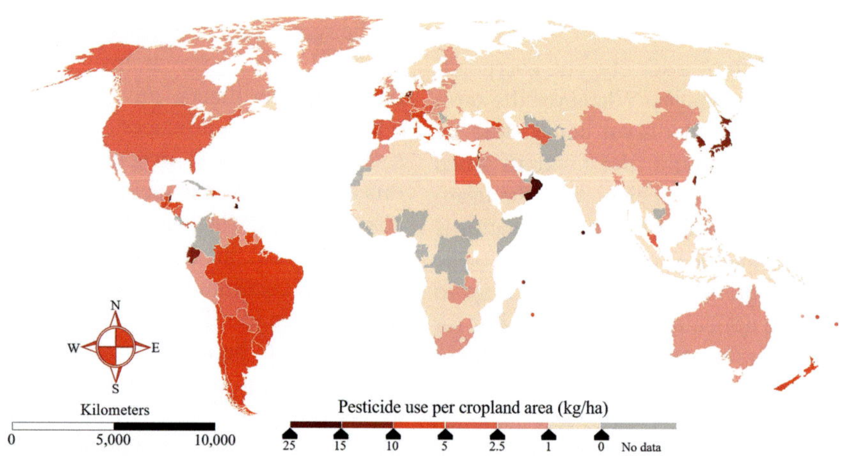

FIGURE 1.1 World scenario of pesticide use per cropland area in 2020.

Source: www.fao.org/faostat/en/#data/EP.

ten world leaders in pesticide consumption are China, USA, Brazil, Argentina, Canada, Ukraine, France, Malaysia, Australia, and Spain (Worldometer, 2023).

Globally, 2.5 mmt of pesticides have been imported annually, which is expected to reach 2.8 mmt by 2026, with an annual growth rate of 2.4% since 2016 (ReportLinker, 2023). As per the report from the World Health Organization (WHO), approximately 20% of pesticides have been used in developing countries, with increasing growth rates (Sharma et al., 2019). In African countries, where nearly 59% of the population depends on farming for living, they are still contributing only 2–4% of the global pesticide market share, also with the lowest rate of pesticide application (Abate et al., 2000; FAO, 2022), whereas the American continent accounts for the highest consumption and the highest rate of pesticide application, that is, 2.83 kg/ha. Asia has a pesticide application rate of 1.17 kg/ha, which is also below the world average. Among different groups of pesticides used globally, herbicides (52%) have the largest share, followed by fungicides (23%), then insecticides (18%), and the rest are like nematicides and rodenticides (FAO, 2022). Coming to insecticides, though their share is less compared to herbicides and fungicides, still the total insecticide market was worth $14 billion in 2021 (Straitsresearch, 2023). In Asian countries, consumption of insecticides is higher than in other parts of the world, whereas its use is lowest in European countries, which is only 12% of their total pesticide use (FAO, 2022).

1.4 POST-APPLICATION FATE OF CHEMICAL INSECTICIDES

Insecticides are applied to increase crop productivity. But like a coin has two sides, in spite of having several advantages, the non-judicious application of insecticides invites a number of problems. In due course, these chemicals get accumulated in plant parts, soil, water, air, and biota. Extensive use of insecticides contaminates these environmental resources, causing them to remain in the plants and finally enter the food chain, thereby posing a great threat to human beings (Sharma et al., 2019). Not only that, but heavy and continuous application of these chemicals also makes them non-effective against particular insects after a certain point of time. Even these insecticides show detrimental effects on different beneficial organisms, which ultimately reflect as pest resurgence and crop loss. Especially the '3Rs' (i.e. resistance, resurgence, and residue), due to the overuse of insecticides, have been considered a major concern for agricultural production. In the upcoming few points, we are going to discuss the five major fates of insecticides in agriculture, environment, and human life.

1.4.1 Development of Insecticide Resistance in Insects

One of the most important consequences of insecticide overuse is the development of resistance in insects. 'Insecticide resistance is the heritable change in susceptibility of an insect population that makes an insecticide repeatedly fail to deliver the expected level of control against the particular insect species when applied in the recommended dose' (IRAC, n.d.) (see Tabashnik et al. (2014); Whalon et al. (2008); Zalucki and Furlong (2017) for other definitions). More than 100 years ago, Melander reported the first case of field-evolved insecticide resistance in the San Jose scale, *Quadraspidiotus perniciosus* (Comstock), against sulphur-lime (Melander, 1914). From then to 1946, another 11 cases of insecticide resistance were recorded (Handley, 2019). However, the discovery of the insecticidal property of DDT gave hope to the industry that insecticide resistance will not be a problem anymore. The discovery of more new chemicals under carbamates, organophosphate, and synthetic pyrethroid groups makes this hope stronger. However, this belief quickly faded away in 1947 with the first report of DDT-resistant houseflies (Sacca, 1947). Afterwards, several new cases of insect resistance to insecticides have been documented from different parts of the world that threaten food production and human health (Coetzee & Koekemoer, 2013; Denholm et al., 1998; Enayati & Hemingway, 2010; Forgash, 1984; Heckel, 2012; Xia et al., 2018).

With the discovery of each new group of synthetic insecticides, it takes only a few years for the insects to develop resistance against them. Hitherto, a total of 629 arthropod species have been recorded to develop resistance against at least one insecticide, with ~18,236 resistant cases worldwide. Out of these 537 species are insects showing resistance against more than 400 different compounds. Diamondback moth, *Plutella xylostella* (Linnaeus), is considered as the most notorious insect in terms of resistance development, with 1,022 already reported resistant cases and against more than 100 insecticide compounds, which are higher than in other insects (Mota-Sanchez & Wise, 2023). The rate of resistance development differs for different insect species, with insecticide exposure as a key factor. Some insects appear to develop resistance at a faster rate than other species, even at the same insecticide exposure, such as the green peach aphid, *Myzus persicae* (Sulzer), which develops resistance faster against insecticides than the cotton aphid, *Aphis gossypii* Glover (IRAC, 2015).

Basically, resistance genes are present in an insect population prior to the introduction of the insecticide product. As a result of the continuous and heavy application of the same insecticide for a long period of time, the susceptible strain is killed, and only the resistant strain remains. The resistant individuals give rise to a resistant generation, which makes the insecticide completely non-effective. This situation of insecticides becoming useless is termed 'pesticide treadmill' (Figure 1.2), which is the major reason responsible for resistance development in insects against insecticides. However, this situation can be controlled by applying insecticides with different modes of action in rotation or in combination.

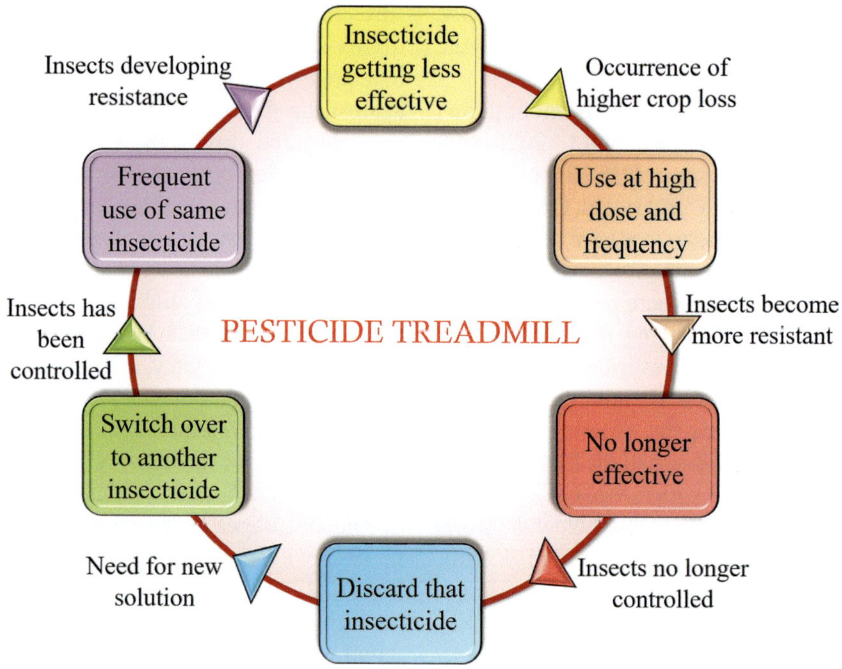

FIGURE 1.2 Schematic diagram of the pesticide treadmill.

1.4.2 Destruction of the Beneficial Arthropod Fauna

An agricultural ecosystem not only consists of harmful insects; even a number of beneficial insects (pollinators or natural enemies) are also present. The indiscriminate use of insecticides imposes a great threat to them. Natural enemies are a key component for pest suppression, and their destruction due to chemical exposure is the major factor responsible for pest resurgence and secondary pest outbreak. Insecticide exposure also reduces the rate of parasitism and predation by natural enemies. Even indirect toxicity of natural enemies may also be observed by ingesting the prey intoxicated with systemic insecticides (Chiu & Cheng, 1976). Sometimes these natural enemies are more susceptible to pesticides than the target pests due to their narrow or specific host range. *Lycosa* spider, which is an important predator of rice brown plant hopper (BPH), *Nilaparvata lugens* (Stål), is highly susceptible to insecticides that make biological control of BPH ineffective (Chu et al., 1976). For insect pests for which biological control is the only solution, the elimination of natural enemies is a great threat.

On the other hand, pollination, which is a key factor for agricultural production, relies on diverse insect pollinators like bees, butterflies, beetles, flies, etc. However, their existence is highly threatened by insecticide exposure at the time of pollination, more specifically for honeybees. Their exposure to chemical pesticides is the most debatable and controversial cause of bee decline and is considered as one of the major reasons behind the colony collapse disorder (CCD) the happened in the USA in 2006–2008 (Ellis et al., 2010), causing about 31–36% honeybee colony loss due to its nexus with other factors (VanEngelsdorp et al., 2007, 2008). Among different chemical groups, three neonicotinoids (imidacloprid, clothianidin, and thiamethoxam) and two organophosphates (chlorpyrifos and phosmate) are considered highly toxic to honeybees at the global level (Sanchez-Bayo & Goka, 2014). Exposure to these neurotoxicants attacks the central nervous system of bees, causing an acute and sublethal decline in their fitness (Stewart et al., 2014). The systemic nature of these insecticides allows them to spread through plant tissues and accumulate at a higher concentration in pollen and nectar. Even water solubility and higher persistence of these chemicals in soil and soil water make them available in the pollen and nectar of the field adjacent to wild flora at a biologically relevant concentration (Krupke et al., 2012). Intake of such intoxicated pollen and nectar is lethal to the bees. Honeybees show a short-term oral LD_{50} of 4–5 ng/bee (Lethal dose-50 (LD_{50}) indicates the amount of pesticides that causes the death of 50% of the test population) for most commonly used neonicotinoids (Suchail et al., 2000). Other pollinators are also facing the same challenges. According to a study, bumblebees that are exposed to imidacloprid showed reduced colony growth due to an 85% reduction in the queen bees' production (Whitehorn et al., 2012).

1.4.3 Insect Pest Resurgence and Secondary Pest Outbreak

Resurgence is a state where the application of an insecticide reduces the insect pest population at the initial stage, but soon after the insect pest bounces back (resurges) by increasing its level of incidence than in an untreated control or population before treatment situation (Dutcher, 2007). As mentioned earlier, the loss of natural enemies is the most important reason behind pest resurgence. However, this is not solely responsible; rather, changes in the insect's behaviour, development, dispersal, and fecundity due to insecticide treatments are also responsible for resurgence (Croft, 1990; Gerson & Cohen, 1989; Hardin et al., 1995). There were several reports of insect resurgence, among which resurgence of the BPH is an important one. Some early records of BPH resurgence were reported by the International Rice Research Institute (IRRI), Philippines, where BPH was found to be resurged after the application of decamethrin, methyl parathion, and diazinon (IRRI, 1969, 1971, 1974, 1977). A major reason behind the resurgence of BPH is the destruction of its predatory mirid bug, *Cyrtorhinus lividipennis* Reuter, due to insecticide application (Stapley et al., 1977). There were also other notable shreds of evidence, like the resurgence of *Frankliniella occidentalis* (Pergande) due to high insecticide application in onion (Dintenfass et al., 1987); the resurgence of California red scale in citrus due to DDT (Hajek, 2004); the resurgence of Cyclamen mite due to the elimination of predator *Typhlodromus reticulatus* (Oudemans) (Zalom et al., 2005); the resurgence of bollworms in cotton due to the destruction of beneficial arthropods due to pyrethroid and organophosphate insecticides (Stevensson & Matocha, 2005); the resurgence of white fly, *Bemisia tabaci* (Gennadius), due to synthetic pyrethroid application (Jeyakumar & Gupta, 2007).

Another important consequence of the indiscriminate use of pesticides and somewhat related to resurgence is the replacement of primary pests with a secondary one or secondary pest outbreak. Here, a non-target but injurious pest has increased its population when the crop has been treated with the pesticide to manage primary pests (Hajek, 2004). Generally, the secondary insect pest has occupied the niche vacated by the primary pest. This situation is driven by the heavy use of synthetic pesticides, reduction of natural enemies of non-target pests, changes in the physiology of plant or non-target species (hormoligosis), and loss of competing arthropods (Gross & Rosenheim, 2011). Cases of secondary pest outbreaks were reported from cotton fields where early-season insecticide application for controlling *Helicoverpa* resulted in the outbreak of cotton aphids, *A. gossypii*, and spider mite, *Tetranychus urticae* Koch (Wilson et al., 1998). Similarly, the application of insecticide in an apple orchard for controlling codling moth, apple maggot, and plum weevil resulted in the outbreak of white apple leafhopper, European red mite, and spotted tentiform leaf miner (Howitt, 1993). Even the application of pesticides for controlling diseases may trigger insect pest outbreaks, like the outbreak of spider mite, *Tetranychus pacificus* McGregor, due to the application of sulphur for controlling powdery mildew in grapes (Albers, 2002). Thus, controlling these secondary pests demands additional pesticide use, resulting in extra monetary loss. In cotton, a secondary pest outbreak due to early-season insecticide application for a plant bug, *Lygus hesperus* Knight, required an additional spray of late-season insecticide worth an additional $6/acre (Gross & Rosenheim, 2011).

1.4.4 Environmental Hazards

Indiscriminate use of insecticides destabilizes ecological balance, which causes environmental pollution and also threatens biodiversity. Starting from its manufacturing to its field application, and even after application, in each and every step these insecticides are hazardous to the environment and human health. While considering human health, who can forget the Bhopal gas tragedy (1984), one of the deadliest manmade disasters due to the leakage of methyl isocyanate (MIC) (a chemical compound used for producing carbamate insecticides), killing a number of people and animals? Out of the total applied pesticides, only around 0.1% is believed to reach the target organisms, and the remaining toxicants pollute the environment, creating environmental hazards (Carriger et al., 2007; Gill & Garg, 2014). Once applied, these insecticides relocate from the site of application to other non-target sites through different transfer processes, like adsorption, volatilization, spray drift, leaching, and runoff (Robinson et al., 1999). Different groups of chemicals show differences in environmental behaviour. For example, organochlorine insecticides with low acute toxicity but a higher tendency of accumulation in tissues can persist in nature for a longer time period, causing long-lasting damage, whereas organophosphate insecticides have low persistence but cause appreciable acute toxicity in mammals (Kim et al., 2017).

After field application, a large amount of insecticide reaches the soil. However, the persistence of insecticidal residues in soil depends on the properties of insecticides, like their water solubility, the octanol/water partition coefficient (Kow), soil sorption constant (Koc), and half-life in soil (DT_{50}) (Zhang et al., 2013). During sorption, insecticide binds with soil particles due to attraction between its chemical and soil particles, causing soil pollution (Qin et al., 2014). Rich organic soil, which is considered better for cultivation, shows higher adsorptive potential due to its greater surface area (Bošković et al., 2020). As the soil gets contaminated, these chemicals become harmful to soil-dwelling beneficial microbes and earthworms. After being taken up by the soil, these insecticides are leached out and pollute groundwater, for which the water solubility of the insecticide and the permeability of the soil act as key factors (Fontana et al., 2010). Generally, insecticides with low persistence have less leaching ability, whereas insecticides, like imidacloprid, with higher persistence (DT_{50}=187 days) have higher leaching potential, posing a higher environmental threat (Geng et al., 2017; Tudi et al., 2021).

These insecticides also get volatilized (converted into gas) and carried out by the wind, causing drift hazards (Singh, 2012). In the present scenario, unmanned aerial vehicles (UAVs) have been utilized in several commercial sectors for spraying insecticides at low volume in fine to very fine droplets that cause spray drift hazards and contaminate air and water (Zhang et al., 2020). Due to over-irrigation or fields with sloppy surfaces, water contaminated with insecticides moves towards water resources due to surface runoff and pollutes those, imposing hazards on aquatic life. Contamination of both surface water and groundwater raises an urgent issue regarding the shortage of drinking water (Tudi et al., 2021). According to a study conducted by the United States Geological Survey (USGS), 90% of fish and water samples collected from US streams were traced with pesticides (Rose et al., 2018). Insecticides in the soil also undergo several chemical, microbial, or photodegradation processes (Abián et al., 2002) that give rise to certain chemicals with higher toxicity. For example, chlorpyrifos, after degradation, produces 3,5,6-trichloro-2-pyridinol (TCP), which is more mobile and toxic compared to its parent insecticidal compound (Zhao et al., 2016).

1.4.5 Insecticide Residue and Health Hazards

'An apple a day keeps the doctor away'—a phrase we have been hearing since our childhood. However, this phrase will be of no use due to another consequence of the indiscriminate use of insecticides, that is, the presence of toxic chemical residues in food commodities. *Pesticide residue* is the substance detected in any matrix due to the use of pesticide, including any specified derivatives of it, like conversion and degradation products, metabolites, reaction compounds, and impurities, showing adverse effects on living organisms (Saha et al., 2022). While insecticides are applied in the field, they are taken up by different plant parts and ultimately enter into the human body with the food material. Additionally, the non-biodegradable nature of some insecticides allows their residues to be bio-accumulated and bio-magnified in the food chain, resulting in toxic effects on human health.

Food materials like cereals, pulses, spices, and some vegetables that are taken after boiling or cooking have less risk due to the conversion or degradation of toxic residual molecules into non-toxic forms. However, high risk is associated with fruits and some vegetables causing health hazards due to insecticide poisoning. As they are consumed raw, that increases the chances of consuming more toxic molecules in our daily lives (Saha et al., 2022). In a study conducted in Ghana, almost all the examined vegetable and fruit samples were found to be contaminated with pesticide residues above the maximum residue limits (MRLs) (Fosu et al., 2017). Similarly, from the Almaty region of Kazakhstan, 59% of studied vegetable samples were found to be traced with pesticide residue, and in 28% of samples, the limit is above the MRLs (Lozowicka et al., 2015). There remains a long list of signs indicating a person is affected by insecticides, though the symptoms are similar to those of other illnesses. The list starts with fever, weakness, headache, sweating, diarrhoea, restlessness, insomnia, nausea, irritation of nose and throat, fatigue, nervousness, sore joints, thirst, eye and skin irritation, loss of appetite, dizziness, vomiting, inability to breathe or increased breathing, loss of reflexes, unconsciousness, damage to nerves and organs, constriction of pupils, convulsions, reproductive problems, hormonal effects, cancer, impaired metabolism, muscle twitching, reduced ability of blood to clot, and so on. In the long term, there remains every chance for some other chronic or acute health problems with no preliminary outward symptoms (Saha et al., 2022).

The detrimental impacts of chemical insecticides have been summarized in Figure 1.3.

Considering the deleterious effects of these chemical insecticides on the environment and human life, there is the need for society to develop environmentally safe strategies for insect pest management. This forces researchers all over the world to look for alternative strategies. In this respect, concepts of integrated pest management (IPM) are getting more popular in reducing pest population below the economic injury level (EIL) in an economically viable, environmentally sustainable manner, by reducing the sole dependency on chemical pesticides and utilizing other

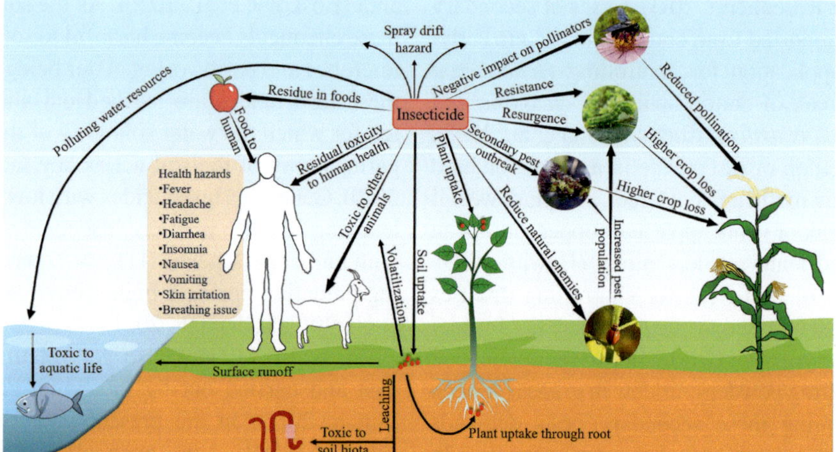

FIGURE 1.3 Schematic diagram of insecticide hazards to the environment and human health.

methods of pest management. As a component of IPM, biological control of insect pests utilizing different biopesticides is considered the most efficient alternative to chemical insecticides.

1.5 BIOPESTICIDES FOR INSECT PEST MANAGEMENT

To overcome insecticidal hazards, modern agriculture is moving towards the use of different biopesticides. These biopesticides are biological in origin, specific to their target organism, less toxic to humans and other non-target organisms, less hazardous to the environment, required in less quantity with better pest killing efficiency, found to have different modes of action, and decomposed more quickly compared to conventional pesticides (Cantrell et al., 2012). Thus, the advancement in the field of biopesticides reduces environmental pollution by replacing synthetic chemical insecticides and allows sustainable growth in agriculture as well as maintaining an eco-friendly environment. Biopesticides can be classified into three categories (Figure 1.4), viz., microbial biopesticides, biochemical pesticides, and plant-incorporated protectants (PIPs), which are also of different types.

Biochemical biopesticides include substances that are naturally occurring in the environment and kill insect pests through non-toxic mechanisms. This type includes different botanicals, semiochemicals, and growth regulators. Unlike traditional pesticides, these biochemical pesticides can kill insects indirectly by disrupting their sexual functions and hindering other metabolic activities. Under this category, different types of pheromones; some plant-derived botanicals, like alkaloids, terpenoids, phenolics, and other secondary plant metabolites; and also some vegetable oils are widely used as biopesticides (Abdollahdokht et al., 2022). Coming to PIPs, this is a newer approach to pest management, and these are the substances required for pest control that are produced by plants and genetic material needed by the plant for producing those substances (Gupta & Dikshit, 2010). PIPs like plant-incorporated Cry or Vip toxins or RNA interference (RNAi) technology, where double-stranded ribonucleic acid (dsRNA) has been incorporated into the plants to develop genetically modified (GM) crops, are expressed by the tissue of those GM crops to protect them from insect pests. A number of GM crops with PIPs have been developed and commercialized in recent days, and full-fledged research is going on in this area of pest management.

However, the group of biopesticides that have earned more popularity and are commercially utilized in different corners of the world is the microbial biopesticide. Different possible types of microbial biopesticide include entomopathogenic bacteria (EPB), entomopathogenic fungi (EPF), entomopathogenic viruses (i.e. nucleaopolyhydrosis virus, cytoplasmic polyhedrosis virus, or granulosis virus), entomopathogenic nematodes (EPN), entomopathogenic protozoa (Abdollahdokht et al., 2022), and some alveolates (like *Gregarina garnhami* Canning, a parasite of locusts). So far, more than 3,000 microorganisms have been recorded to cause diseases in insects. Among these microbial biopesticides, EPB is an important group, and our present study focuses on the role of different EPBs in the management of different insect pests.

1.6 ENTOMOPATHOGENIC BACTERIA (EPB)

Bacteria are omnipresent and have not left the most versatile class of the animal kingdom, Insecta, to be unutilized as their host. While most of them are symbiotic, some cause diseases in insects. The latter relationship has both positive and negative effects on agriculture and mankind: positive when attacked on pests and negative on beneficial insects. More than hundreds of bacteria have been identified to show pathogenicity to arthropods (Tanada & Kaya, 1993). Most of them

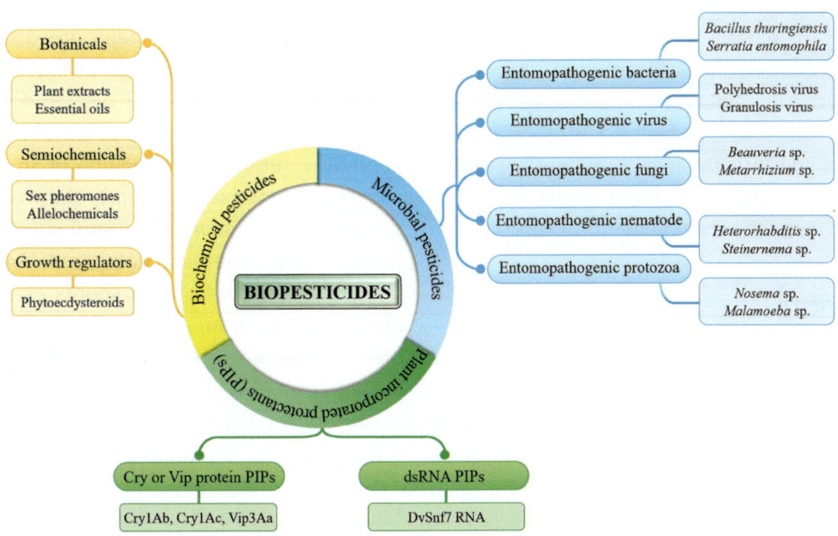

FIGURE 1.4 Classification of biopesticides, with examples, used in insect control.

are mild pathogens infecting insects under stress conditions; however, some are highly virulent against the insects (Deka et al., 2021). In this book, we are mainly concentrating on the positive use of EPBs.

EPBs are unicellular prokaryotic microorganisms, with sizes ranging from less than 1 μm to several μm in length, lacking proper nuclear structure and other defined intracellular membrane-enclosed organelles. They establish both obligate and facultative relationships with their insect host. An example of an obligate bacteria can be given as *Paenibacillus* spp., whereas *Serratia* spp. can grow in the environment without a host as facultative (detailed classification has been given in Chapter 3). However, the most advantageous ones are the toxins produced by several bacteria like *Bacillus thuringiensis* Berliner (Bt) or *Chromobacterium subtsugae* Martin et al. These toxins can be utilized against insects without living hosts, resulting in stable, non-infectious biopesticides (Glare et al., 2017). Though even after the presence of hundreds of bacteria with arthropod pathogenic potentiality, only four types of EPB under family Bacillaceae have received much attention, namely, *B. thuringiensis*, *B. cereus* Frankland and Frankland, *Lysinibacillus sphaericus* (Meyer and Neide), and *Paenibacillus popilliae* Dutky (Kalha et al., 2014).

1.7 WHY GO FOR ENTOMOPATHOGENIC BACTERIA (EPB)?

Though there are a number of potential biopesticides available, still, why have we chosen EPB as our topic for discussion, and why are researchers giving much attention to these bioagents that rely on their several advantages? Amongst all the biopesticides available till now, formulations based on EPBs dominate the biopesticides market due to several factors that made them more acceptable to consumers, that is, cost-effective mass production, specificity, persistence in the environment, safety, and most importantly, the rapid cessation of feeding by insects. These EPBs have several benefits over conventional chemical insecticides; most importantly, they are eco-friendly in nature. Another major difference between synthetic insecticides and EPBs relies on their mode of action. Generally, most synthetic insecticides kill insects by attacking their nervous system, whereas EPB shows a different mode of action and kills the insects by causing septicaemia (blood poisoning). Different advantages of EPB over synthetic chemical insecticides include the following.

1.7.1 Host Specificity

These EPBs are highly specific to their target organism, particularly the most commonly used ones. The host specificity of *B. thuringiensis* is commonly attributed either to proteinaceous parasporal crystal toxins (Cry and cytolytic toxin [Cyt]) produced by the bacteria during the stationary phase or to soluble toxins of Vip (vegetative insecticidal protein) and Sip (secreted insecticidal protein) families secreted during vegetative state (Malovichko et al., 2019). This property attributed great advantages to its use, as different serovars of Bt showed high specificity to their target group of insects, such as *Bt* var. *aizawai* and *Bt* var. *galleriae* against Lepidoptera, *Bt* var. *israelensis* against Diptera, *Bt* var. *sandiego* or *Bt* var. *tenebrionis* against Coleoptera. Not only Bt but also other major groups of bacteria show high levels of host specificity, like *L. sphaericus* against Diptera and *P. popilliae* against phytophagous scarabs.

1.7.2 Fewer Chances of Resistance Development

EPBs are considered an important tool in insecticide resistance management (IRM) strategies. While insects already developed resistance against different synthetic chemical insecticides, EPBs are still controlling insect pests efficiently. These bacteria have co-existed and co-evolved with insects for millions of years and have the ability to adapt to the insect's defence system by altering the toxin that is suitable in that situation. So there are fewer possible chances of resistance development against them. Different EPBs are also compatible synergistically with other control agents, including chemical insecticides (Musser et al., 2006) and other insect pathogens that reduce the chances of resistance development. They can be employed in rotation or in combination with chemical insecticides that will increase efficiency and environmental sustainability (Deka et al., 2021). Even their unique mode of action also reduces the chances of resistance development.

1.7.3 Higher Persistence and Long-Lasting Activity

EPBs possess long-lasting control over insect pests. Environmental persistence is one of the important factors for the field application of different bioinsecticides, and EPBs show greater persistence in the environment. After application in the natural ecosystem, these organisms are able to persist as a component of natural microflora (Kalha et al., 2014). For example, *L. sphaericus*, which is highly effective against mosquitoes, has a higher persistence in polluted water, which is the major breeding site for mosquitoes (Karch et al., 1988; Nicolas et al., 1987). This bacterium has a residual activity of 2 days to 4 weeks, depending on experimental settings (Lee et al., 1998). When Bt has been applied in protected cultivation or in greenhouses, its persistence increases significantly (Janmaat et al., 2007). *Pseudomonas fluorescens* (Trevisan) showed better persistence in open field conditions compared to their Bt counterparts (Kalha et al., 2014). Even spores of *P. popilliae* have the ability to persist in soil for protracted periods of time and only germinate, grow, and sporulate in the presence of scarab hosts (Lacey & Siegel, 2000).

1.7.4 Ability to Recycle

Some EPBs have the ability to recycle themselves in the environment, which reduces the requirement for further application. Recycling of *L. sphaericus* toxin can be carried out through the germination and multiplication of the bacteria in the dead larvae and released during the disintegration of the larval cadaver (Lee et al., 1998). It remains viable even in other non-target invertebrates without showing any toxic effect and is carried by them beyond the area of application (Yousten et al., 1991). *P. popilliae* also persists in the soil, is recycled, provides long-lasting control, and therefore does not need to be reapplied.

1.7.5 Safe for Environment and Human Health

In contrast to commercial chemical insecticides, no environmental pollution is associated with EPBs, as they are naturally occurring, already present in different habitats, and easily bio-degradable. As they are non-toxic to non-target organisms, that is why, unlike chemical insecticides, they pose less or no threat to biodiversity. When Bt formulations have been applied in crops, it does not cause any health hazard to human beings and other mammals, as their gut composition and pH differ from those of insects that cannot activate Bt protein crystals. Hence, Bt can be applied to crops up to the day of harvesting of the crop without taking care of residual toxicity and can be consumed by humans along with its Cry toxin without suffering from any health problems (Benedict & Ring, 2004). Even different GM crops are also considered safe for consumption by humans and other mammals and already accepted by different countries and cultivated worldwide (Mullaney, 2023). These EPBs can also be safely implemented by farmers in organic farming and can be used as a component in IPM practices.

1.7.6 Non-Toxic to Beneficial Arthropods

As mentioned earlier, EPBs are highly specific to their target insects, which keeps the beneficial arthropods on the safe side, unlike synthetic chemical insecticides. EPBs can be effective only when ingested and bind with proper receptors. This property limits their susceptibility to non-target insects, like pollinators and natural enemies. Different formulations of Bt apparently have little or no effect on the parasites and other natural enemies of the target insect (Heimpel, 1967). According to a study conducted to evaluate the toxicity of transgenic *Bt* var. *galleriae* toxin on pine processionary moth, *Thaumetopoea pityocampa* (Denis et Schiffermuller); wax moth, *Galleria mellonella* (Linnaeus); and their larval parasitoid, *Exorista larvarum* (Linnaeus), the toxins efficiently killed the pests without affecting the parasitoid (Marchetti et al., 2012). Moreover, feeding of *Microplitis croceipes* (Cresson) (a larval endoparasitoid of *Helicoverpa armigera* [Hübner]) with Dipel or purified Bt spores of strain HD-73 mixed with honey not only proved to be safe for the wasps, even it significantly increased the longevity of adult wasps (Blumberg et al., 1997), but, according to a number of laboratory studies, also revealed that *Bt* formulations don't show any detectable effects on adult hymenopterous (Chilcutt & Tabashnik, 1999; Ksentini et al., 2010; Ruiu et al., 2007a; Xu et al., 2004) or tachinid (Flexner et al., 1986) parasitoids, either treated through stomach or contact route. Even honeybees are also found to remain non-affected by different EPB formulations (Bailey et al., 2005; Ruiu et al., 2007b).

Having such advantages in the form of EPB is a blessing for agriculture, mankind, and ultimately, the environment. However, these criteria are not sufficient to compete with chemical pesticides in some cases due to a narrower host range and lower efficacy. Even resistance cases against different EPBs are also coming up. So our search continues to discover new bacterial insect pathogens and toxins that can serve the purpose more efficiently with a broader host range. In the initial days, the team entomopathogenic bacteria was built up mainly with the members of *Bacillus* and related genera, among which Bt proved to be the most important, then went on increasing with EPN symbionts like *Photorhabdus* and *Xenorhabdus* and other Gammaproteobacteria, like *Serratia*, *Yersinia*, and *Pseudomonas*. Newer in the team are Betaproteobacteria, like *Burkholderia* and *Chromobacterium*, which came up with a broader host range. The rising stars are Alphaproteobacteria, like *Rickettsia*, *Wolbachia*, and *Rickettsiella*, and Tenericutes, like *Spiroplasma* and *Acholeplasma* (Boucias & Pendland, 1998). Actinobacteria are also gaining more importance in the scientific and agribusiness communities for producing varieties of metabolites with the potential for insect killing. In the upcoming chapters of this book, we have provided a descriptive discussion of these major EPBs in detail.

1.8 CONCLUSION

Standing in the current scenario, where toxic chemicals inextricably linked with our day-to-day life may be present in our kitchen, our drinking water, the air that we breathe, the food that we consume, and even in the soil with which our children play, there is a need to look for solutions that will give our future generation a non-toxic, chemical-free, and safer environment. In this respect, a shift is necessary from chemical insecticides to different biopesticides to give the world residue-free foods. EPBs are emerging as an important pillar in the field of pest management. It not only provides a safer environment but also reduces health hazards and protects biodiversity. A number of EPBs are there to do this job. In the present day, people are getting aware of the toxic effects of these chemicals and demanding chemical-free and organically produced crops that have led to augmented utilization of biopesticides. As a result, the market share of biopesticides is growing much faster compared to that of synthetic chemical pesticides, with EPBs holding a dominant position in the global market. However, more research and efforts are still needed in the field of EPB. Even more focus also needs to be given to other fields of biological control. Considering this one quote from Rachel Carson's *Silent Spring*, it is truly justifiable that '[m]an is a part of nature, and his war against nature is inevitably a war against himself'.

REFERENCES

Abate, T., van Huis, A., & Ampofo, J. K. O. (2000). Pest management strategies in traditional agriculture: An African perspective. *Annual Review of Entomology*, *45*(1), 631–659. https://doi.org/10.1146/annurev.ento.45.1.631

Abdollahdokht, D., Gao, Y., Faramarz, S., Poustforoosh, A., Abbasi, M., Asadikaram, G., & Nematollahi, M. H. (2022). Conventional agrochemicals towards nano-biopesticides: An overview on recent advances. *Chemical and Biological Technologies in Agriculture*, *9*(1), 1–19. https://doi.org/10.1186/s40538-021-00281-0

Abián, J., Durand, G., & Barceló, D. (2002). Analysis of chlorotriazines and their degradation products in environmental samples by selecting various operating modes in thermospray HPLC/MS/MS. *Journal of Agricultural and Food Chemistry*, *41*(8), 1264–1273. https://doi.org/10.1021/JF00032A020

Aktar, M. W., Sengupta, D., & Chowdhury, A. (2009). Impact of pesticides use in agriculture: Their benefits and hazards. *Interdisciplinary Toxicology*, *2*(1), 1–12. http://doi.org/10.2478/v10102-009-0001-7

Albers, C. E. (2002). *Laboratory evaluation of the toxicity of four fungicides used to control Uncinula necator on the spider mite predator Metaseiulus occidentalis*. (Doctoral dissertation, California Polytechnic State University). California Polytechnic State University.

Bailey, J., Scott-Dupree, C., Harris, R., Tolman, J., & Harris, B. (2005). Contact and oral toxicity to honey bees (*Apis mellifera*) of agents registered for use for sweet corn insect control in Ontario, Canada. *Apidologie*, *36*(4), 623–633. https://doi.org/10.1051/apido:2005048

Benedict, J. H. (2003). Strategies for controlling insect, mite and nematodes pests. In M. J. Chrispeels, & D. E. Sadava (Eds.), *Plants, genes, and crop biotechnology* (pp. 414–442). Jones and Bartlett Publishers.

Benedict, J. H., & Ring, D. R. (2004). Transgenic crops expressing Bt proteins: Current status, challenges and outlook. In O. Koul, & G. S. Dhaliwal (Eds.), *Transgenic crop protection: Concepts and strategies* (pp. 15–84). Science Publishers.

Blumberg, D., Navon, A., Keren, S., Goldenberg, S., & Ferkovich, S. M. (1997). Interactions among *Helicoverpa armigera* (Lepidoptera: Noctuidae), its larval endoparasitoid *Microplitis croceipes* (Hymenoptera: Braconidae), and *Bacillus thuringiensis*. *Journal of Economic Entomology*, *90*(5), 1181–1186. https://doi.org/10.1093/jee/90.5.1181

Bošković, N., Brandstätter-Scherr, K., Sedláček, P., Bílková, Z., Bielská, L., & Hofman, J. (2020). Adsorption of epoxiconazole and tebuconazole in twenty different agricultural soils in relation to their properties. *Chemosphere*, *261*, 127637. https://doi.org/10.1016/J.CHEMOSPHERE.2020.127637

Boucias, D. G., & Pendland, J. C. (1998). *Principles of insect pathology*. Kluwer Academic Publishers.

Cai, D. W. (2008). Understand the role of chemical pesticides and prevent misuses of pesticides. *Bulletin of Agricultural Science and Technology*, *1*(6), 36–38.

Cantrell, C. L., Dayan, F. E., & Duke, S. O. (2012). Natural products as sources for new pesticides. *Journal of Natural Products*, *75*(6), 1231–1242. https://doi.org/10.1021/np300024u

Carriger, J. F., Rand, G. M., Gardinali, P. R., Perry, W. B., Tompkins, M. S., & Fernandez, A. M. (2007). Pesticides of potential ecological concern in sediment from south Florida canals: An ecological risk prioritization for aquatic arthropods. *Soil & Sediment Contamination*, *15*(1), 21–45. https://doi.org/10.1080/15320380500363095

Chilcutt, C. F., & Tabashnik, B. E. (1999). Effects of *Bacillus thuringiensis* on adults of *Cotesia plutellae* (Hymenoptera: Braconidae), a parasitoid of the diamondback moth, *Plutella xylostella* (Lepidoptera: Plutellidae). *Biocontrol Science and Technology*, *9*(3), 435–440. https://doi.org/10.1080/09583159929695

Chiu, S. C., & Cheng, C. H. (1976). Toxicity of some insecticides commonly used for rice insects control to the predators of rice hoppers. *Plant Protection Bulletin*, *18*, 256–267.

Chu, Y. I., Lin, T. H., & Mu, T. (1976). Relative toxicity of 9 insecticides against rice insect pests and their predators. *Plant Protection Bulletin*, *18*, 369–376.

Coetzee, M., & Koekemoer, L. L. (2013). Molecular systematics and insecticide resistance in the major African malaria vector *Anopheles funestus*. *Annual Review of Entomology*, *58*, 393–412. https://doi.org/10.1146/annurev-ento-120811-153628

Croft, B. A. (1990). *Arthropod biological control agents and pesticides*. John Wiley and Sons.

Deka, B., Baruah, C., & Babu, A. (2021). Entomopathogenic microorganisms: Their role in insect pest management. *Egyptian Journal of Biological Pest Control*, *31*, 121. https://doi.org/10.1186/s41938-021-00466-7

Denholm, I., Cahill, M., Dennehy, T. J., & Horowitz, A. R. (1998). Challenges with managing insecticide resistance in agricultural pests, exemplified by the whitefly *Bemisia tabaci*. *Philosophical Transactions of the Royal Society of London. Series B: Biological Sciences*, *353*(1376), 1757–1767. https://doi.org/10.1098/rstb.1998.0328

Dhaliwal, G. S., Jindal, V., & Mohindru, B. (2015). Crop losses due to insect pests: Global and Indian scenario. *Indian Journal of Entomology*, *77*(2), 165–168. https://doi.org/10.5958/0974-8172.2015.00033.4

Dintenfass, L. P., Bartell, D. P., & Scott, M. A. (1987). Predicting resurgence of western flower thrips (Thysanoptera: Thripidae) on onions after insecticide application in the Texas High Plains. *Journal of Economic Entomology*, *80*(2), 502–506. https://doi.org/10.1093/jee/80.2.502

Dutcher, J. D. (2007). A review of resurgence and replacement causing pest outbreaks in IPM. In A. Ciancio, & K. G. Mukerji (Eds.), *General concepts in integrated pest and disease management* (pp. 27–43). Springer. https://doi.org/10.1007/978-1-4020-6061-8_2

Ellis, J. D., Evans, J. D., & Pettis, J. (2010). Colony losses, managed colony population decline, and Colony Collapse Disorder in the United States. *Journal of Apicultural Research*, *49*(1), 134–136. https://doi.org/10.3896/IBRA.1.49.1.30

Enayati, A., & Hemingway, J. (2010). Malaria management: Past, present, and future. *Annual Review of Entomology*, *55*, 569–591. https://doi.org/10.1146/annurev-ento-112408-085423

FAO. (2017). *The future of food and agriculture: Trends and challenges*. Food and Agriculture Organization of the United Nations, Rome.

FAO. (2019). *New standards to curb the global spread of plant pests and diseases*. Food and Agriculture Organization of the United Nations, Rome. https://www.fao.org/news/story/en/item/1187738/icode/

FAO. (2021). *Climate change fans spread of pests and threatens plants and crops, new FAO study*. Food and Agriculture Organization of the United Nations, Rome. https://www.fao.org/news/story/en/item/1402920/icode

FAO. (2022). *Pesticides use, pesticides trade and pesticides indicators: Global, regional and country trends, 1990–2020* (FAOSTAT analytical brief 46). Food and Agriculture Organization of the United Nations, Rome.

FAOSTAT. (2021). *Population and environment*. Food and Agriculture Organization of the United Nations, Rome. Retrieved May 13, 2023, from https://www.fao.org/faostat/en/#data

Flexner, J., Lighthart, B., & Croft, B. A. (1986). The effects of microbial pesticides on non-target, beneficial arthropods. *Agriculture, Ecosystems & Environment*, *16*(3–4), 203–254. https://doi.org/10.1016/0167-8809(86)90005-8

Fontana, A. R., Lana, N. B., Martinez, L. D., & Altamirano, J. C. (2010). Ultrasound-assisted leaching-dispersive solid-phase extraction followed by liquid–liquid microextraction for the determination of polybrominated diphenyl ethers in sediment samples by gas chromatography–tandem mass spectrometry. *Talanta*, *82*(1), 359–366. https://doi.org/10.1016/J.TALANTA.2010.04.050

Forgash, A. J. (1984). History, evolution, and consequences of in secticide resistance. *Pesticide Biochemistry and Physiology*, *22*(2), 178–186. https://doi.org/10.1016/0048-3575(84)90087-7

Fosu, P. O., Donkor, A., Ziwu, C., Dubey, B., Kingsford-Adaboh, R., Asante, I., Nyarko, S., Tawiah, R., & Nazzah, N. (2017). Surveillance of pesticide residues in fruits and vegetables from Accra Metropolis markets, Ghana, 2010–2012: A case study in Sub-Saharan Africa. *Environmental Science and Pollution Research*, *24*, 17187–17205. https://doi.org/10.1007/s11356-017-9287-8

Geng, Y., Ma, J., Zhou, R., Jia, R., Li, C., & Ma, X. (2017). Assessment of insecticide risk to human health in groundwater in Northern China by using the China-PEARL model. *Pest Management Science*, *73*(10), 2063–2070. https://doi.org/10.1002/PS.4572

Gerson, U., & Cohen, E. (1989). Resurgences of spider mites (Acari: Tetranychidae) induced by synthetic pyrethroids. *Experimental and Applied Acarology*, *6*, 29–46. https://doi.org/10.1007/BF01193231

Gill, H. K., & Garg, H. (2014). Pesticides: Environmental impacts and management strategies. In M. L. Larramendy, & S. Soloneski (Eds.), *Pesticides—toxic aspects* (p. 187). IntechOpen. https://doi.org/10.5772/57399

Glare, T. R., Jurat-Fuentes, J. L., & O'callaghan, M. (2017). Basic and applied research: Entomopathogenic bacteria. In L. A. Lacey (Ed.), *Microbial control of insect and mite pests* (pp. 47–67). Academic press.

Gordon, A. (2022, September 7). *Yield losses due to pests*. AGRIVI. https://www.agrivi.com/blog/yield-losses-due-to-pests/

Gross, K., & Rosenheim, J. A. (2011). Quantifying secondary pest outbreaks in cotton and their monetary cost with causal inference statistics. *Ecological Applications*, *21*(7), 2770–2780. https://doi.org/10.1890/11-0118.1

Gupta, S., & Dikshit, A. K. (2010). Biopesticides: An eco-friendly approach for pest control. *Journal of Biopesticides*, *3*(Special Issue), 186.

Hajek, A. E. (2004). *Natural enemies: An introduction to biological control*. Cambridge University Press.

Handley, J. (2019, March 22). *Pesticides-a brief history and analysis*. Pitschcare Serving the Turfcare Industry. https://www.pitchcare.com/news-media/pesticides-a-brief-history-and-analysis.html

Hardin, M. R., Benrey, B., Coll, M., Lamp, W. O., Roderick, G. K., & Barbosa, P. (1995). Arthropod pest resurgence: An overview of potential mechanisms. *Crop Protection*, *14*(1), 3–18. https://doi.org/10.1016/0261-2194(95)91106-P

Heckel, D. G. (2012). Insecticide resistance after *Silent Spring*. *Science*, *337*(6102), 1612–1614. https://doi.org/10.1126/science.1226994

Heimpel, A. M. (1967). A critical review of *Bacillus thuringiensis* var. *thuringiensis* Berliner and other crystalliferous bacteria. *Annual Review of Entomology*, *12*(1), 287–322. https://doi.org/10.1146/annurev.en.12.010167.001443

Howitt, A. J. (1993). *Common tree fruit pests*. Michigan State University Extension Service.

IRAC. (2015). *Understanding Iinsecticide Mmode of Aaction in Ccrop Pprotection Sstrategies*. Caydee Savinelli Annual Crop Protection School.

IRAC. (n.d.). *Introduction to resistance*. Insecticide Resistance Action Committee. https://irac-online.org/training-centre/resistance/

IRRI. (1969). *The international rice research institute annual report for 1968*. p. 402.

IRRI. (1971). *The international rice research institute annual report for 1970*. p. 265.

IRRI. (1974). *The international rice research institute annual report for 1973*. p. 266.

IRRI. (1977). *The international rice research institute annual report for 1976*. p. 418.

Janmaat, A. F., Ware, J., & Myers, J. (2007). Effects of crop type on *Bacillus thuringiensis* toxicity and residual activity against *Trichoplusia ni* in greenhouses. *Journal of Applied Entomology*, *131*(5), 333–337. https://doi.org/10.1111/j.1439-0418.2007.01181.x

Jeyakumar, P., & Gupta, G. P. (2007). Insecticide induced biochemical changes in cotton and consequent resurgence of whitefly. *Pesticide Research Journal*, *19*(1), 90–92.

Kalha, C. S., Singh, P. P., Kang, S. S., Hunjan, M. S., Gupta, V., & Sharma, R. (2014). Entomopathogenic viruses and bacteria for insect-pest control. In D. P. Abrol (Ed.), *Integrated pest management* (pp. 225–244). Academic Press. https://doi.org/10.1016/B978-0-12-398529-3.00013-0

Karch, S., Monteny, N., & Coz, J. (1988). Persistence of *Bacillus sphaericus* in a mosquito breeding site 4 years after its introduction for microbial control [*Culex pipiens*]. *Comptes Rendus de l'Academie des Sciences Serie 3 Sciences de la Vie (France)*, *307*, 289–192.

Kim, K. H., Kabir, E., & Jahan, S. A. (2017). Exposure to pesticides and the associated human health effects. *Science of the Total Environment*, *575*, 525–535. https://doi.org/10.1016/J.SCITOTENV.2016.09.009

Krupke, C. H., Hunt, G. J., Eitzer, B. D., Andino, G., & Given, K. (2012). Multiple routes of pesticide exposure for honey bees living near agricultural fields. *PLoS One*, *7*(1), e29268. https://doi.org/10.1371/journal.pone.0029268

Ksentini, I., Jardak, T., & Zeghal, N. (2010). *Bacillus thuringiensis*, deltamethrin and spinosad side-effects on three *Trichogramma* species. *Bulletin of Insectology*, *63*(1), 31 37.

Lacey, L. A., & Siegel, J. P. (2000). Safety and ecotoxicology of entomopathogenic bacteria. In J.-F. Charles, A. Delecluse, & C. Nielsen-Le Roux (Eds.), *Entomopathogenic bacteria: From laboratory to field application* (pp. 253–273). Springer. https://doi.org/10.1007/978-94-017-1429-7_14

Lee, H. K., Cheong, H., & Gill, S. S. (1998). Microbial control of insects: Use of bacterial insecticides. In G. S. Dhaliwal, & E. A. Heinrichs (Eds.), *Critical issues in insect pest management* (pp. 87–117). Commonwealth Publishers.

Liu, C. J., Men, W. J., Liu, Y. J., & Zhang, H. (2002). The pollution of pesticides in soils and its bioremediation. *System Sciences and Comprehensive Studies in Agriculture*, *18*(4), 295–297.

Lozowicka, B., Abzeitova, E., Sagitov, A., Kaczynski, P., Toleubayev, K., & Li, A. (2015). Studies of pesticide residues in tomatoes and cucumbers from Kazakhstan and the associated health risks. *Environmental Monitoring and Assessment*, *187*, 1–19. https://doi.org/10.1007/s10661-015-4818-6

Malovichko, Y. V., Nizhnikov, A. A., & Antonets, K. S. (2019). Repertoire of the *Bacillus thuringiensis* virulence factors unrelated to major classes of protein toxins and its role in specificity of host-pathogen interactions. *Toxins*, *11*(6), 347. https://doi.org/10.3390/toxins11060347

Marchetti, E., Battisti, A., Squartini, A., & Dindo, M. L. (2012). Susceptibility of adult *Exorista larvarum* to conventional and transgenic *Bacillus thuringiensis galleriae* toxin. *Bulletin of Insectology*, *65*(1), 133–137.

Melander, A. L. (1914). Can insects become resistant to sprays? *Journal of Economic Entomology*, *7*(2), 167–173.

Mota-Sanchez, D., & Wise, J. C. (2023). *The arthropod pesticide resistance database*. Michigan State University. Retrieved May 14, 2023, from http://www.pesticideresistance.org

Mullaney, E. G. (2023). *Bacillus thuringiensis*. Encyclopedia Britannica. https://www.britannica.com/science/Bacillus-thuringiensis

Musser, F. R., Nyrop, J. P., & Shelton, A. M. (2006). Integrating biological and chemical controls in decision making: European corn borer (Lepidoptera: Crambidae) control in sweet corn as an example. *Journal of Economic Entomology*, *99*(5), 1538–1549. https://doi.org/10.1093/jee/99.5.1538

Nicolas, L., Dossou-Yovo, J., & Hougard, J. M. (1987). Persistence and recycling of *Bacillus sphaericus* 2362 spores in *Culex quinquefasciatus* breeding sites in West Africa. *Applied Microbiology and Biotechnology*, *25*, 341–345. https://doi.org/10.1007/BF00252544

Qin, F., Gao, Y. X., Guo, B. Y., Xu, P., Li, J. Z., & Wang, H. L. (2014). Environmental behavior of benalaxyl and furalaxyl enantiomers in agricultural soils. *Journal of Environmental Science and Health, Part B*, *49*(10), 738–746. https://doi.org/10.1080/03601234.2014.929482

ReportLinker. (2023). *Pesticide market outlook 2022–2026*. https://www.reportlinker.com/clp/global/13

Robinson, D. E., Mansingh, A., & Dasgupta, T. P. (1999). Fate and transport of ethoprophos in the Jamaican environment. *Science of the Total Environment*, *237*, 373–378. https://doi.org/10.1016/S0048–9697(99)00150–3

Rose, C. E., Coupe, R. H., Capel, P. D., & Webb, R. M. (2018). Holistic assessment of occurrence and fate of metolachlor within environmental compartments of agricultural watersheds. *Science of the Total Environment*, *612*, 708–719. https://doi.org/10.1016/j.scitotenv.2017.08.154

Ruiu, L., Floris, I., Satta, A., & Fois, X. (2007b). Susceptibility of the honeybee (*Apis mellifera* L.) to entomopathogenic bacterial toxins used for the biological control. *Redia*, *15*, 105–108.

Ruiu, L., Satta, A., & Floris, I. (2007a). Susceptibility of the house fly pupal parasitoid *Muscidifurax raptor* (Hymenoptera: Pteromalidae) to the entomopathogenic bacteria *Bacillus thuringiensis* and *Brevibacillus laterosporus*. *Biological Control*, *43*(2), 188–194. https://doi.org/10.1016/j.biocontrol.2007.08.005

Sacca, G. (1947). On the existence of house flies resistant to DDT. *Rivista di Parassitologia*, *8*(2/3), 127–128.

Saha, S., Laskar, N., & Phani, V. (2022). Pesticide residue in fruit crops and its management. In N. Laskar, & V. Phani (Eds.), *Pests of fruit crops* (pp. 445–468). NIPA.

Sanchez-Bayo, F., & Goka, K. (2014). Pesticide residues and bees—a risk assessment. *PLoS One*, *9*(4), e94482. https://doi.org/10.1371/journal.pone.0094482

Sharifzadeh, M. S., Abdollahzadeh, G., Damalas, C. A., & Rezaei, R. (2018). Farmers' criteria for pesticide selection and use in the pest control process. *Agriculture*, *8*(2), 24. https://doi.org/10.3390/agriculture8020024

Sharma, A., Kumar, V., Shahzad, B., Tanveer, M., Sidhu, G. P. S., Handa, N., Kohli, S. K., Yadav, P., Bali, A. S., Parihar, R. D., Dar, O. I., Singh, K., Jasrotia, S., Bakshi, P., Ramakrishnan, M., Kumar, S., Bharadwaj, R., & Thukral, A. K. (2019). Worldwide pesticide usage and its impacts on ecosystem. *SN Applied Sciences*, *1*, 1446. https://doi.org/10.1007/s42452-019-1485-1

Singh, D. K. (2012). Pesticides and environment. In D. K. Singh (Ed.), *Pesticide chemistry and toxicology* (pp. 114–122). Bentham Science Publisher.

Stapley, J. H., Jackson, Y. Y., & Golden, W. G. (1977). Observation in the Solomon Islands on the control of the brown planthopper, *Nilaparvata lugens* (Stål). In *The brown planthopper: Threat to rice production in Asia* (p. 14). International Rice Research Institute.

Stevensson, D. E., & Matocha, M. A. (2005). *A pest management strategic plan for cotton production in Texas*. USDA-ARS/CSREES.

Stewart, S. D., Lorenz, G. M., Catchot, A. L., Gore, J., Cook, D., Skinner, J., Mueller, T. C., Johnson, D. R., Zawislak, J., & Barber, J. (2014). Potential exposure of pollinators to neonicotinoid insecticides from the use of insecticide seed treatments in the mid-southern United States. *Environmental Science & Technology*, *48*(16), 9762–9769. https://doi.org/10.1021/es501657w

Straitsresearch. (2023). *Insecticides market*. https://straitsresearch.com/report/insecticides-market

Suchail, S., Guez, D., & Belzunces, L. P. (2000). Characteristics of imidacloprid toxicity in two *Apis mellifera* subspecies. *Environmental Toxicology and Chemistry: An International Journal*, *19*(7), 1901–1905. https://doi.org/10.1002/etc.5620190726

Tabashnik, B. E., Mota-Sanchez, D., Whalon, M. E., Hollingworth, R. M., & Carrière, Y. (2014). Defining terms for proactive management of resistance to Bt crops and pesticides. *Journal of Economic Entomology*, *107*(2), 496–507. https://doi.org/10.1603/EC13458

Tanada, Y., & Kaya, H. K. (1993). *Insect pathology*. Academic Press.

Tudi, M., Daniel Ruan, H., Wang, L., Lyu, J., Sadler, R., Connell, D., Chu, C., & Phung, D. T. (2021). Agriculture development, pesticide application and its impact on the environment. *International Journal of Environmental Research and Public Health*, *18*(3), 1112. https://doi.org/10.3390/ijerph18031112

United Nations. (2015). *Resolution adopted by the General Assembly on 25 September 2015. 70/1 transforming our world: The 2030 agenda for sustainable development* (Report No. A/RES/70/1).

United Nations. (2022). *World population prospects 2022* (UN DESA/POP/2021/TR/NO.3).

VanEngelsdorp, D., Hayes Jr., J., Underwood, R. M., & Pettis, J. (2008). A survey of honey bee colony losses in the US, fall 2007 to spring 2008. *PLoS One*, *3*(12), e4071. https://doi.org/10.1371/journal.pone.0004071

VanEngelsdorp, D., Underwood, R., Caron, D., & Hayes Jr., J. (2007). Estimate of managed colony losses in the winter of 2006–2007: A report commissioned by the Apiary Inspectors of America. *American Bee Journal*, *147*, 599–603.

Whalon, M. E., Mota-Sanchez, D., & Hollingworth, R. M. (2008). Analysis of global pesticide resistance in arthropods. In M. E. Whalon, D. Mota-Sanchez, & R. M., Hollingworth (Eds.), *Global pesticide resistance in arthropods* (pp. 5–31). CABI. https://doi.org/10.1079/9781845933531.000

Whitehorn, P. R., O'connor, S., Wackers, F. L., & Goulson, D. (2012). Neonicotinoid pesticide reduces bumble bee colony growth and queen production. *Science*, *336*(6079), 351–352. https://doi.org/10.1126/science.1215025

Wilson, L. J., Bauer, L. R., & Lally, D. A. (1998). Effect of early season insecticide use on predators and outbreaks of spider mites (Acari: Tetranychidae) in cotton. *Bulletin of Entomological Research*, *88*(4), 477–488. https://doi.org/10.1017/S000748530004222X

World Bank. (2023). *Agriculture and food.* Retrieved May 11, 2023, from https://www.worldbank.org/en/topic/agriculture/overview

Worldometer. (2023, May 11). *Pesticide use by country.* Food and Agriculture, Pesticides, Woldometer. Retrieved May 11, 2023, from https://www.worldometers.info/food-agriculture/pesticides-by-country/

Xia, X., Sun, B., Gurr, G. M., Vasseur, L., Xue, M., & You, M. (2018). Gut microbiota mediate insecticide resistance in the diamond-back moth, *Plutella xylostella* (L.). *Frontiers in Microbiology*, *9*, 25. https://doi.org/10.3389/fmicb.2018.00025

Xu, Y. Y., Liu, T. X., Leibee, G. L., & Jones, W. A. (2004). Effects of selected insecticides on *Diadegma insulare* (Hymenoptera: Ichneumonidae), a parasitoid of *Plutella xylostella* (Lepidoptera: Plutellidae). *Biocontrol Science and Technology*, *14*(7), 713–723. https://doi.org/10.1080/09583150410001682395

Yousten, A. A., Benfield, E. F., Campbell, R. P., Foss, S. S., & Genthner, F. J. (1991). Fate of *Bacillus sphaericus* 2362 spores following ingestion by nontarget invertebrates. *Journal of Invertebrate Pathology*, *58*(3), 427–435. https://doi.org/10.1016/0022-2011(91)90189-W

Zalom, F. G., Phillips, P. A., Toscano, N. C., & Boldsa, M. (2005). *Cyclamen mite.* University of California Agriculture and Natural Resources, Publication 3468.

Zalucki, M. P., & Furlong, M. J. (2017). Behavior as a mechanism of insecticide resistance: Evaluation of the evidence. *Current Opinion in Insect Science*, *21*, 19–25. https://doi.org/10.1016/j.cois.2017.05.006

Zhang, F., He, J., Yao, Y., Hou, D., Jiang, C., Zhang, X., Di, C., & Otgonbayar, K. (2013). Spatial and seasonal variations of pesticide contamination in agricultural soils and crops sample from an intensive horticulture area of Hohhot, North-West China. *Environmental Monitoring and Assessment*, *185*, 6893–6908. https://doi.org/10.1007/s10661-013-3073-y

Zhang, H., Qi, L., Wu, Y., Musiu, E. M., Cheng, Z., & Wang, P. (2020). Numerical simulation of airflow field from a six–rotor plant protection drone using lattice Boltzmann method. *Biosystems Engineering*, *197*, 336–351. https://doi.org/10.1016/j.biosystemseng.2020.07.018

Zhao, Y., Wendling, L. A., Wang, C., & Pei, Y. (2016). Behavior of chlorpyrifos and its major metabolite TCP (3,5,6-trichloro-2-pyridinol) in agricultural soils amended with drinking water treatment residuals. *Journal of Soils and Sediments*, *17*(4), 889–900. https://doi.org/10.1007/S11368-016-1586-Z

2 History of Entomopathogenic Bacteria

ABSTRACT

This chapter tries to draw a timeline for the landmark discoveries of several entomopathogenic bacteria, starting from the glorious history of the most used biopesticide, *Bacillus thuringiensis*, to the newest EPB, which is yet to be exploited commercially. The first discovered EPB, *B. thuringiensis*, is about to complete its bidecacentennial in the sphere of research. From that, the discovery of several other EPBs continued full-fledged, courtesy of the never-tired scientific community. The discovery of symbiotic bacteria of entomopathogenic nematodes and insecticides developed from the metabolites of bacteria are also taken care of in this chapter.

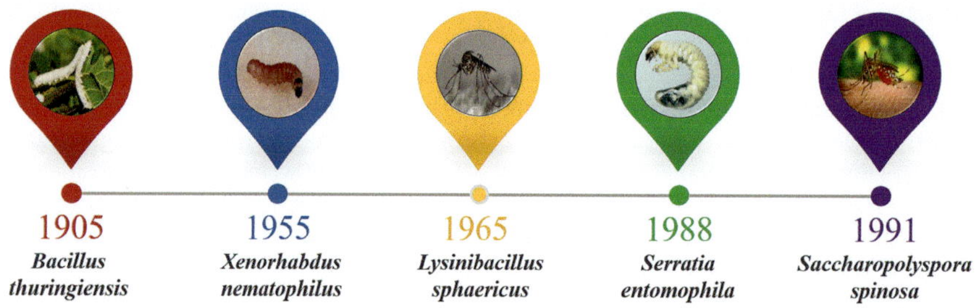

2.1 INTRODUCTION

> 'If you don't know history, then you don't know anything. You are a leaf that doesn't know it is part of a tree.'
>
> **—Michael Crichton, renowned American author**

Entomopathogenic bacteria have a long history, completing 120 years of tracing. To improve the present situation, we must know how the pioneers of this field worked throughout this extensive period and achieved success so that we can analyze those and utilize them in our own way to pave our path for a sustainable future. This chapter aims to provide a brief insight into the history of the discovery of different entomopathogenic bacteria.

2.2 *BACILLUS THURINGIENSIS*

S. Ishiwata identified a *Bacillus* sp. as the cause of flacherie, or flaccid disease of the silkworm, *Bombyx mori* (Linnaeus), in 1901 and named it *B. sotto* (the Japanese word *sotto* means 'collapse' or 'sudden death') (Ayers et al., 1977; Ishiwata, 1905; Menegatti et al., 2018). The baton of research on this bacterium was carried out by E. Berliner in Thüringen, Germany, about ten years after its discovery by Ishiwata, when another strain was found by Berliner, doing a useful job of killing Mediterranean flour moth larvae, *Ephestia* (=*Anagasta*) *kuehniella* Zeller, a pest of stored grains, by causing its death (Knowles, 1994). Based on the place of discovery, it was named *Bacillus thuringiensis* Berliner (Bt), a name now given to a large family of bacteria producing insecticidal parasporal crystals. He first noted that the bacterium can produce inclusion bodies, or 'Restkorper', alongside the endospore. But this isolate (Berliner strain of variety Bt) was lost and again isolated in the same manner by Mattes in 1927, and this was made available commercially (Reardon et al., 1994). People started to use these proteins as insecticidal sprays and dusts since the 1930s to control insect pests (Cohen, 2015). However, it is surprising to know that the insecticidal properties of Bt were not unnoticed many years before the bacterium was identified, as some accounts suggest that Bt spores may have already been in use in ancient Egypt (Sanahuja et al., 2011). The foremost attempt to use Bt, knowingly, for insect control took place in the late 1920s against the gypsy moth, *Lymantria dispar* (Linnaeus), in the north-eastern United States. Sporeine, the first commercial Bt product, was made available for the first time in 1938 in France, one year in advance of the discovery of the insecticidal properties of DDT, which still stands as the most mentionable turning point in the journey of inorganic insecticides (Reardon et al., 1994; Reiner et al., 2014). However, all these were going on without knowing the exact reason for this insecticidal property. Edward Steinhaus, who is often considered as the father of modern insect pathology, demonstrated the effectiveness of Bt in controlling insect pests in the work of his PhD dissertation, which was published as the 'Catalogue of Bacteria Associated Extracellularly with Insects and Ticks', and continued to be further explored in his future works (Steinhaus, 1942, 1959). Almost after 25 years, in 1953, Thomas Angus realized that this trait is credited to the parasporal crystals, named and defined as protein by Christopher Hannay (Hannay, 1953). It was believed for many years that the insecticidal spectrum of Bt was confined to the larvae of Lepidoptera, until screening programmes revealed new strains active against Diptera (Goldberg & Margalit,

DOI: 10.1201/9781003426882-2

1977) and Coleoptera (Krieg, 1987). In the initial period, the *kurstaki* strain of Bt (*Btk*) was considered as the most important strain and found effective against Lepidopteran pests of different fruits, vegetables, ornamental crops, and forest trees. Products of this strain have been sold since the early 1960s and account for the bulk of Bt applications. During the late 1970s, another Bt strain, called Bti or the *Bt* serovar *israelensis* strain of Bt, was identified, which was effective against Dipteran insects. However, they are now most widely utilized as transgene inserted into crop plants today. These transgenic plants (Bt crops) can express bacterial toxin genes to defend themselves against pests. These crops have effectively controlled pest populations and increased crop yield and quality. Vaeck *et al.* were the first to produce a Bt crop (tobacco) in the laboratory of a company named 'Plant Genetic Systems' in Belgium in 1985, but the crop was not commercially successful (Abbas, 2018). However, after the Environmental Protection Agency's (EPA) approval, Bt crops' flourishing development and cultivation started in several crops like brinjal, corn, cotton, poplar, potato, rice, soybean, and tomato. The list is expected to increase positively in the upcoming days. Insect-resistant crops are the second most planted commercial transgenic crops based on toxins from the bacterium Bt. Presently, the area under Bt crop cultivation is around 100 mi ha (Brookes & Barfoot, 2017). The understanding of the mode of action of Bt as an insect pathogen has been crystal clear with its genome sequencing published in 2007 (Challacombe et al., 2007).

2.3 OTHER ENTOMOPATHOGENIC BACTERIA FROM PHYLUM FIRMICUTES

Here, firstly, we would like to draw the readers' attention to the times of the 1910s in the USA, where an accidentally introduced pest, the Japanese beetle, *Popillia japonica* Newman, was creating severe damage to a wide range of crops and ornamentals. People were in an emergency need of a solution. During the 1930s, the extensive search resulted in the discovery of some diseased larvae with milky white haemolymph. The milky spores collected from the larvae affected with the milky disease led to the establishment of the causal agent, *B. popilliae* Dutky, as the first insect pathogen to be registered in the United States as a microbial control agent in 1948, which is commercially available in the name of Doom or Japademic (Boucias & Pendland, 1998). The pathogen was described by Dutky in 1940 along with *B. lentimorbus* with a detailed morphological description (Dutky, 1940; Khachatourians, 2009). Both the species were later reclassified as separate genus, *Paenibacillus* (Latin word *Paene* means 'almost'), in 1993 (Ash et al., 1993; Grady et al., 2016). Another important discovery amongst entomopathogenic bacteria was exploring *Lysinibacillus sphaericus* (Meyer and Neide), formerly known as *B. sphaericus*, as lethal against mosquitoes. The team of W. R. Kellen, T. B. Clark, J. E. Lindegren, and B. C. Ho collected some dead fourth instar larvae of *Culiseta incidens* (Thomson) on March 16, 1963, from water standing in rock holes near Big Creek, Fresno County, California, and found *B. sphaericus* as the cause of the death (Kellen et al., 1965). Although this discovery prompted further investigation of the bacterium, it was not until 1973 that strains with the potential for use in insect control programmes were discovered based on the research on infected larval mosquitoes from Delhi, India (Singer, 1973). *Brevibacillus laterosporus* (Laubach) was first discovered with entomopathogenic properties in one of the most beneficial insects, honeybees. White, in 1912, isolated this bacterium as a secondary infection associated with European foulbrood and named it *B. orpheus* (White, 1912). However, the name *B. laterosporus* was given and described by Laubach, who isolated a different strain from fresh water (Laubach, 1916). In 1946, Steinhaus mentioned this bacterium as a spore-forming entomopathogenic bacteria effective against different insect orders, *viz.* Coleoptera, Lepidoptera, Diptera, and against nematodes and molluscs (Ruiu, 2013). *B. subtilis* (Ehrenberg) is a well-known bacterium for nutrient cycling by producing several enzymes (Abd El-Salam et al., 2011). But the entomopathogenic nature of this bacteria was unknown till the discovery of pesticidal activity against the larvae of root-knot nematode *Meloidogyne incognita* (Kofold and White) and cyst nematodes *Heterodera cajani* Koshy; *H. zeae* Koshy, Swarup, and Sethi; and *H. avenae* Wollenweber (Gotke & Swarup, 1988). Against insects, it was first reported from *Drosophila melanogaster* Meigen (Lazare et al., 1998). Later, it was further reported against corn rootworms, mosquitoes, other insects, and nematodes (Abd El-Salam et al., 2011). In an experiment of isolation of bacteria associated with the death of the fourth instar larvae of *Culiseta longiareolata* (Macquart) carried by Khyami-Horani et al., along with *B. sphaericus*, *B. cereus*, and *B. brevis*, another *Bacillus* sp. was discovered for the first time to have an entomopathogenic property. It is *B. megaterium* de Bary (Khyami-Horani et al., 1999). However, two years before this, the mutants of *B. megaterium* were proved to be capable of synthesizing the mosquitocidal binary toxin of *B. sphaericus* (England et al., 1997). From the recent discoveries, another worth mentioning is the work of Ming Yan, Michael H. Roehrl, and Julia Y. Wang in 2007, where they found that crystalline inclusions in *B. licheniformis* (Weigmann) resemble parasporal crystals of Bt (Yan et al., 2007). They foresaw rightly that it could be a powerful biopesticide in the future. This was proved not even five years before that it performs very efficiently against *Aedes aegypti* (Linnaeus) which has started developing resistance against *B. sphaericus* and *Bt* serovar *israelensis* (Rajagopal et al., 2020). Another very recent discovery is the coming up of *B. aryabhattai* Shivaji et al. as an entomopathogen of *Cassida vittata* Villers in sugar beet (Bazazo et al., 2019), a bacterium first isolated from cryotubes used to collect air samples at an altitude of 27 to 41 km. (Shivaji et al., 2009). Let us carry forward our discussion from *Bacillus* sp. to other Firmicutes. In 1990, de Barjac et al. isolated a strain of *Clostridium bifermentans* (Weinberg and Séguin) from a mixed bacterial sample originating from Malaysia and showed that it exhibits high toxicity against mosquito and blackfly larvae. They named it *C. bifermentans* serovar *malaysia (C.b.m.)* (de Barjac et al., 1990; Nicolas et al., 1990).

2.4 PHYLUM ACTINOBACTERIA

This section demands the focus to be on the history of the emergence of Spinosad, the first biological insecticide. A bacterial genus, *Saccharopolyspora spinosa* Mertz and Yao, was discovered in a sugar mill in crushed sugarcane, the type of strain being strain A83543 (Mertz & Yao, 1990). After some days, it was found that upon fermentation, this bacterium produces some unique-structured tetracyclic macrolides, which they named the A83543 factor (Kirst et al., 1991). Insecticidal activity was recorded when the A83543 culture broth was screened through mosquito larvicide assay on *Aedes aegypti* (Kirst et al., 1992). Later, the name *spinosyns* was assigned to the products (Thompson et al., 1995). Spinosyn A and spinosyn D are active ingredients effective against different insect orders (Waldron et al., 2000), especially Lepidoptera and Diptera and other members, like planthoppers, leafhoppers, spider mites, and cockroaches (Kirst, 2010). These two components constituting the major and minor portions were released commercially as Spinosad, first time launched in 1997 (Salgado & Sparks, 2005). The broad-spectrum activity of a substance isolated from *Streptomyces tanashiensis* Hata et al., named flavensomycin, is effective against *Musca domestica* Linnaeus and *Locusta migratoria* (Linnaeus), along with fungi like *Saccharomyces* and *Penicillium* and vertebrate pest mouse (Craveri & Giolitti, 1957).

2.5 PHYLUM PROTEOBACTERIA

According to the Bible, King Herod had the blood of thousands of male children on his hands. Here we are going to discuss one such bacterium which has been contributing towards reducing the male population of insects in nature. *Wolbachia* is responsible for male killing, feminization, parthenogenesis, and cytoplasmic incompatibility of several insects and disrupts the sex ratio, thus known as Herod bug, a bug living parasitically inside bugs (Knight, 2001). The genus was first identified in 1924 by Marshall Hertig and Simeon Burt Wolbach in the common house mosquito, *Culex pipiens* Linnaeus (Hertig & Wolbach, 1924). Hertig formally described the species in 1936 and proposed both the generic and specific names: *Wolbachia pipientis* Hertig (Hertig, 1936).

After seven years of the coming up of *B. subtilis* as entomopathogenic to *D. melanogaster*, one non-*Bacillus* bacteria was isolated from the same insect species of strain L48T, that is, *Pseudomonas entomophila* Mulet et al. It was first isolated from a female adult and has been reported to be equally effective against larvae and adults (Vodovar et al., 2005). In 2007, a broad-spectrum entomopathogen was discovered by Martin et al. In a single year, they reported the bacteria *Chromobacterium subtsugae* Martin et al. from the Colorado potato beetle (*Leptinotarsa decemlineata* [Say]), diamondback moth (*Plutella xylostella* [Linnaeus]), gypsy moth (*L. dispar*), small hive beetle (*Aethina tumida* Murray), Southern corn rootworm (*Diabrotica undecimpunctata* Mannerheim), Southern green stink bug (*Nezara viridula* [Linnaeus]), sweet potato whitefly (*Bemisia tabaci* [Gennadius]), tobacco hornworm (*Manduca sexta* [Linnaeus]), and Western corn rootworm (*Diabrotica virgifera* LeConte) having lethal or sublethal effects (Martin et al., 2007). A somewhat newer discovery is *Yersinia entomophaga* sp. nov. Hurst et al., which was isolated for the first time from the New Zealand grass-grub, *Costelytra zealandica* (White) (Hurst et al., 2011). Another important member of this group is *Serratia entomophila* Grimont et al., which is also used as a biocontrol agent against grass-grub (*C. zealandica*) and was first reported by Grimont et al. in 1988 (Chattopadhyay & Sen, 2013; Grimont et al., 1988).

Other than bacteria, one among the other groups of entomopathogens is famous for its efficiency in killing insects, and those are entomopathogenic nematodes. However, here also, bacteria play a major role behind the scenes. This association was first described for the bacterium *Achromobacter nematophilus* Poinar and Thomas associated with *Neoaplectana carpocapsae* Weiser, strain DD136 (Poinar & Thomas, 1965). This nematode was first found as the cause of death of codling moth larvae in Eastern North America. The bacterium was later transferred to a new genus, *Xenorhabdus*, with the type species *X. nematophilus* (Poinar and Thomas). They also reported *X. luminescens* sp. nov (Thomas and Poinar) to be associated with another entomopathogenic nematode, *Heterorhabditis bacteriophora* Poinar, which was first isolated from dead *Capnodis tenebricosa* (Olivier) (Thomas & Poinar, 1979). It has a unique character in that the infected insect cadavers glow in the dark (Poinar & Grewal, 2012). In 1993, N. E. Boemare, R. J. Akhurst, and R. G. Mourant conducted a DNA relatedness experiment among the *Xenorhabdus* strains and, based on the results, transferred *X. luminiscens* to a new genus, *Photorhabdus*. Since then, many species and subspecies of this genus have been discovered to be associated with diverse entomopathogenic nematodes (Boemare et al., 1993).

2.6 PHYLUM TENERICUTES

Another broad-spectrum entomopathogenic bacteria is *Acholeplasma entomophilum* Tully et al., which was, in its first report, found from the gut contents of representatives of

TABLE 2.1
Entomopathogenic Bacteria: List of Insects in the First Report

Phylum	Entomopathogenic Bacteria	Name of Insect Where It Was First Reported	Reference
Firmicutes	*Bacillus thuringiensis* Berliner	*Bombyx mori* (Linnaeus)	(Ishiwata, 1905)
	Paenibacillus popilliae Dutky	*Popillia japonica* Newman	(Dutky, 1940)
	Lysinibacillus sphaericus (Meyer and Neide)	*Culiseta incidens* (Thomson)	(Kellen et al., 1965)
	Brevibacillus laterosporus (Laubach)	*Apis mellifera* Linnaeus	(White, 1912)
	Bacillus subtilis (Ehrenberg)	*Drosophila melanogaster* Meigen	(Lazare et al., 1998)
	Bacillus megaterium de Bary	*Culiseta longiareolata* Macquart	(Khyami-Horani et al., 1999)
	Bacillus licheniformis (Weigmann)	*Aedis aegypti* (Linnaeus)	(Rajagopal et al., 2020)
	Bacillus aryabhattai Shivaji et al.	*Cassida vittata* Villers	(Bazazo et al., 2019)
	Clostridium bifermentans (Weinberg and Séguin)	*Anopheles stephensi* Liston	(de Barjac et al., 1990)
Actinobacteria	*Streptomyces tanaschiensis* Hata et al. (associated insecticidal substance: flavensomycin)	*Musca domestica* Linnaeus and *Locusta migratoria* (Linnaeus)	(Craveri & Giolitti, 1957)
	Saccharopolyspora spinosa Mertz and Yao (associated insecticidal substance-Spinosyn)	*Aedes aegypti* (Linnaeus)	(Kirst et al., 1991)
Proteobacteria	*Xenorhabdus nematophilus* (Poinar and Thomas)	*Cydia pomonella* (Linnaeus) (associated nematode: *Neoaplectana carpocapsae* Weiser)	(Weiser, 1955; Poinar & Thomas, 1965)
	Photorhabdus luminiscens (Thomas and Poinar)	*Capnodis tenebricosa* (Olivier) (associated nematode: *Heterorhabditis bacteriophora* Poinar)	(Thomas & Poinar, 1979)
	Yersinia entomophaga Hurst et al.	*Costelytra zealandica* (White)	(Hurst et al., 2011)
	Pseudomonas entomophila Mulet et al.	*Drosophila melanogaster* Meigen	(Vodovar et al., 2005)
	Chromobacter subtsugae Martin et al.	*Leptinotarsa decemlineata* Say	(Martin et al., 2007)
	Serratia entomophila Grimont et al.	*Costelytra zealandica* (White)	(Grimont et al., 1988)
	Wolbachia pipientis Hertig	*Culex pipiens* Linnaeus	(Hertig & Wolbach, 1924; Hertig, 1936)
Tenericutes	*Acholeplasma entomophilum* Tully et al.	*Tabanus catenatus* Walker	(Tully et al., 1988)

nine insect genera, among which tabanid fly (*Tabanus catenatus* Walker), beetle, honeybee, and some lepidopteran hosts are most significant (Boucias & Pendland, 1998; Tully et al., 1988).

So insect pathologists are not at rest, and the list of discoveries as new bacteria, newly identified entomopathogenic traits of known bacteria, and their standardization for application is increasing and expected to increase more. Recorded entomopathogenic bacteria and their host are listed in Table 2.1.

2.7 CONCLUSION

Every discovery, research, innovation, and invention ask for lots of patience, courage, struggle, and most importantly, failure, maybe multiple times as well. Within this brief span of discussion, we could hardly bring some mentionable success stories, many stories of struggles behind these successes and many more stories of failure we could not discuss here, but we duly acknowledge their contribution for paving the path to a better future for humanity. Our deep homage to all the researchers working day and night in laboratories across the world to provide us quality food. There are so many things to learn and inculcate through these significant achievements. Hope this chapter inspires researchers to work more passionately in this field.

REFERENCES

Abbas, M. S. T. (2018). Genetically engineered (modified) crops (*Bacillus thuringiensis* crops) and the world controversy on their safety. *Egyptian Journal of Biological Pest Control*, *28*(1), 52. https://doi.org/10.1186/s41938-018-0051-2

Abd El-Salam, A. M. E., Nemat, A. M., & Magdy, A. (2011). Potency of *Bacillus thuringiensis* and *Bacillus subtilis* against the cotton leafworm, *Spodoptera littoralis* (Bosid.) larvae. *Archives of Phytopathology and Plant Protection*, *44*(3), 204–215. https://doi.org/10.1080/03235400902952129

Ash, C., Priest, F. G., & Collins, M. D. (1993). Molecular identification of rRNA group 3 bacilli (Ash, Farrow, Wallbanks and Collins) using a PCR probe test—Proposal for the creation of a new genus *Paenibacillus*. *Antonie van Leeuwenhoek*, *64*(3–4), 253–260. https://doi.org/10.1007/BF00873085

Ayers, J. H., Blue, T. A., Bramhall, R. R., Brownstain, T. J., Davis, E. E., Degraw, J. I., Elward, T. E., Inman, R. E., Johnson, O. H., Leaf, E. B., Offensend, F. L., & Stent, P. D. (1977). *New innovative pesticides—an evaluation of incentives and disincentives for commercial development by industry*. Environmental Protection Agency, Office of Pesticide Programs Criteria and Evaluation Division, Economic Analysis Branch. https://www.google.co.in/books/edition/_/ZR3uH7O713QC?hl=en&gbpv=0

Bazazo, K., Ekram, A., & El-Sheikh, M. (2019). New record of entomopathogenic bacteria, *Bacillus aryabhattai* strain b8w 22, isolated from *Cassida vittata* vill. and its pathogenicity against this insect in Egyptian sugar beet fields. *Zagazig Journal of Agricultural Research*, *46*(6), 2247–2254. https://doi.org/10.21608/zjar.2019.65077

Boemare, N. E., Akhurst, R. J., & Mourant, R. G. (1993). DNA relatedness between *Xenorhabdus* spp. (Enterobacteriaceae), symbiotic bacteria of entomopathogenic nematodes, and a proposal to transfer *Xenorhabdus luminescens* to a New Genus, *Photorhabdus* gen. *nov*. *International Journal of Systematic and Evolutionary Microbiology*, *43*(2), 249–255. https://doi.org/10.1099/00207713-43-2-249

Boucias, D. G., & Pendland, J. C. (1998). *Principles of insect pathology* (1st ed.). Springer. https://doi.org/10.1007/978-1-4615-4915-4

Brookes, G., & Barfoot, P. (2017). *GM crops: Global socio-economic and environmental impacts 1996–2015*. PG Economics Ltd.

Challacombe, J., Altherr, M., Xie, G., Bhotika, S., Brown, N., Bruce, D., Campbell, C., Campbell, M., Chen, J., Chertkov, O., Cleland, C., Dimitrijevic, M., Doggett, N., Fawcett, J., Glavina, T., Goodwin, L., Green, L., Han, C., Hill, K., . . . Brettin, T. (2007). The complete genome sequence of *Bacillus thuringiensis* Al Hakam. *Journal of Bacteriology*, *189*(9), 3680–3681. https://doi.org/10.1128/JB.00241-07

Chattopadhyay, P., & Sen, S. K. (2013). Systemic infestation of *Serratia entomophila* AB2 through plant tissue inferred protection against insect pest and fungal pathogens. *African Journal of Microbiology Research*, *7*(21), 2651–2655. https://doi.org/10.5897/AJMR2013.5743

Cohen, Z. P. (2015). *Bacillus thuringiensis*. Cornell University. https://biocontrol.entomology.cornell.edu/pathogens/bacillus.php

Craveri, R., & Giolitti, G. (1957). An antibiotic with fungicidal and insecticidal activity produced by *Streptomyces*. *Nature*, *179*(4573), 1307–1307. https://doi.org/10.1038/1791307a0

de Barjac, H., Sebald, M., Charles, J. F., & Cheong, W. H. (1990). *Clostridium bifermentans* serovar *malaysia*, a new anaerobic bacterium toxic to mosquito and blackfly larvae. *Comptes Rendus de l'Academie Des Sciences Serie 3 Sciences de La Vie*, *310*(9), 383–387.

Dutky, S. R. (1940). Two new spore-forming bacteria causing milky diseases of Japanese Beetle larvae. *Journal of Agricultural Research*, *61*(1), 57–68.

England, D. F., Penfold, R. J., Delaney, S. F., & Rogers, P. L. (1997). Isolation of *Bacillus megaterium* mutants that produce high levels of heterologous protein, and their use to construct a highly mosquitocidal strain. *Current Microbiology*, *35*(2), 71–76. https://doi.org/10.1007/s002849900215

Goldberg, L. J., & Margalit, J. (1977). A bacterial spore demonstrating rapid larvicidal activity against *Anopheles sergentii, Uranotaenia unguiculata, Culex univitattus, Aedes aegypti* and *Culex pipiens*. *Mosquito News*, *37*(3), 355–358.

Gotke, N., & Swarup, G. (1988). On the potential of some bacterial biocides against root-knot and cyst nematodes. *Indian Journal of Nematology*, *18*(1), 152–153.

Grady, E. N., MacDonald, J., Liu, L., Richman, A., & Yuan, Z. C. (2016). Current knowledge and perspectives of *Paenibacillus*: A review. *Microbial Cell Factories*, *15*, 203. https://doi.org/10.1186/S12934-016-0603-7/TABLES/1

Grimont, P. A. D., Jackson, T. A., & Ageron, E. N. M. (1988). *Serratia entomophila* sp. *nov*. associated with amber disease in the New Zealand grass grub *Costelytra zealandica*. *International Journal of Systematic Bacteriology*, *38*, 1–6. https://doi.org/10.1099/00207713-38-1-1

Hannay, C. L. (1953). Crystalline inclusions in aerobic spore-forming bacteria. *Nature*, *172*(4387), 1004–1004. https://doi.org/10.1038/1721004a0

Hertig, M. (1936). The Rickettsia, *Wolbachia pipientis* (gen. et sp.n.) and Associated inclusions of the mosquito, *Culex pipiens*. *Parasitology*, *28*(4), 453–486. https://doi.org/10.1017/S0031182000022666

Hertig, M., & Wolbach, S. B. (1924). Studies on Rickettsia-like micro-organisms in insects. *The Journal of Medical Research*, *44*(3), 329–374.7.

Hurst, M. R. H., Becher, S. A., Young, S. D., Nelson, T. L., & Glare, T. R. (2011). *Yersinia entomophaga* sp. *nov*., isolated from the New Zealand grass grub *Costelytra zealandica*. *International Journal of Systematic and Evolutionary Microbiology*, *61*(4), 844–849. https://doi.org/10.1099/ijs.0.024406-0

Ishiwata, S. (1905). Concerning "Sotto Kin" a bacillus of a disease of the silkworm. *Report to Association of Sericulture Japan*, *160*, 1–8.

Kellen, W. R., Clark, T. B., Lindegren, J. E., Ho, B. C., Rogoff, M. H., & Singer, S. (1965). *Bacillus sphaericus* Neide as a pathogen of mosquitoes. *Journal of Invertebrate Pathology*, *7*(4), 442–448. https://doi.org/10.1016/0022–2011(65)90120–5

Khachatourians, G. G. (2009). Insecticides, microbial. In M. Schaechter (Ed.), *Encyclopedia of microbiology* (3rd ed., pp. 95–109). Academic Press. https://doi.org/10.1016/B978-0-12-809633-8.13066-3

Khyami-Horani, H., Katbeh-Bader, A., & Mohsen, Z. H. (1999). Isolation of endospore-forming bacilli toxic to *Culiseta longiareolata* (Diptera: Culicidae) in Jordan. *Letters in Applied Microbiology*, *28*(1), 57–60. https://doi.org/10.1046/j.1365–2672.1999.00469.x

Kirst, H. A. (2010). The spinosyn family of insecticides: Realizing the potential of natural products research. *Journal of Antibiotics*, *63*(3), 101–111. https://doi.org/10.1038/ja.2010.5

Kirst, H. A., Michel, K. H., Martin, J. W., Creemer, L. C., Chio, E. H., Yao, R. C., Nakatsukasa, W. M., Boeck, L. V. D., Occolowitz, J. L., Paschal, J. W., Deeter, J. B., Jones, N. D., & Thompson, G. D. (1991). A83543A-D, unique fermentation-derived tetracyclic macrolides. *Tetrahedron Letters*, *32*(37), 4839–4842. https://doi.org/10.1016/S0040-4039(00)93474-9

Kirst, H. A., Michel, K. H., Mynderase, J. S., Chio, E. H., Yao, R. C., Nakasukasa, W. M., Boeck, L. D., Occlowitz, J. L., Paschal, J. W., Deeter, J. B., & Thompson, G. D. (1992). Discovery, isolation, and structure elucidation of a family of structurally unique, fermentation-derived tetracyclic macrolides. In D. Baker et al. (Ed.), *Synthesis and chemistry of agrochemicals III* (pp. 214–225). ACS Symposium Series. https://doi.org/10.1021/bk-1992-0504.ch020

Knight, J. (2001). Meet the Herod bug. *Nature*, *412*(6842), 12–14. https://doi.org/10.1038/35083744

Knowles, B. H. (1994). Mechanism of action of *Bacillus thuringiensis* insecticidal δ-endotoxins. In P. D. Evans (Ed.), *Advances in insect physiology* (Vol. 24, pp. 275–308). Academic Press. https://doi.org/10.1016/S0065-2806(08)60085-5

Krieg, A. (1987). *Bacillus thuringiensis* subsp. *tenebrionis*, strain BI 256–82: A third pathotype within the H-serotype 8a8b. *Systematic and Applied Microbiology*, *9*(1–2), 138–141. https://doi.org/10.1016/S0723-2020(87)80068-1

Laubach, C. A. (1916). Spore-bearing bacteria in water. *Journal of Bacteriology*, *1*(5), 505–512. https://pubmed.ncbi.nlm.nih.gov/16558712/

Lazare, K., Haubruge, E., Lienard, V., & Gaspar, C. (1998). Use of *Bacillus subtilis* (Cohn) as insecticide against *Drosophila melanogaster* (Meigen). *Mededelingen—Faculteit Landbouwkundige En Toegepaste Biologische*

Wetenschappen Universiteit Gent (Belgium), *61*(3a), 887–893.

Martin, P. A. W., Gundersen-Rindal, D., Blackburn, M., & Buyer, J. (2007). *Chromobacterium subtsugae* sp. *nov*., a betaproteobacterium toxic to Colorado potato beetle and other insect pests. *International Journal of Systematic and Evolutionary Microbiology*, *57*, 993–999. https://doi.org/10.1099/ijs.0.64611-0

Menegatti, C., Da Paixão Melo, W. G., Carrão, D. B., De Oliveira, A. R. M., Do Nascimento, F. S., Lopes, N. P., & Pupo, M. T. (2018). *Paenibacillus polymyxa* associated with the stingless bee *Melipona scutellaris* produces antimicrobial compounds against entomopathogens. *Journal of Chemical Ecology*, *44*(12), 1158–1169. https://doi.org/10.1007/s10886-018-1028-z

Mertz, F. R., & Yao, R. C. (1990). *Saccharopolyspora spinosa* sp. *nov*. isolated from soil collected in a sugar mill rum still. *International Journal of Systematic Bacteriology*, *40*(1), 34–39. https://doi.org/10.1099/00207713-40-1-34/CITE/REFWORKS

Nicolas, L., Hamon, S., Frachon, E., Sebald, M., & de Barjac, H. (1990). Partial inactivation of the mosquitocidal activity of *Clostridium bifermentans* serovar *malaysia* by extracellular proteinases. *Applied Microbiology and Biotechnology*, *34*(1), 36–41. https://doi.org/10.1007/BF00170920

Poinar, G. O., Jr., & Grewal, P. S. (2012). History of entomopathogenic nematology. *Journal of Nematology*, *44*(2), 153–161.

Poinar, G. O., Jr., & Thomas, G. M. (1965). A new bacterium, *Achromobacter nematophilus* sp. nov. (Achromobacteriaceae: Eubacteriales) associated with a nematode. *International Bulletin of Bacteriological Nomenclature and Taxonomy*, *15*(4), 249–252.

Rajagopal, G., Jeyavani, J., & Ilango, S. (2020). Larvicidal and histopathological efficacy of inhabitant pathogenic bacterial strains to reduce the dengue vector competence. *Pest Management Science*, *76*(11), 3587–3595. https://doi.org/10.1002/ps.5892

Reardon, R. C., Dubois, N. R., & McLane, W. (1994). *Bacillus thuringiensis for managing gypsy moth: A review*. USDA Forest Service, National Center of Forest Health Management. https://archive.org/details/CAT10861975/page/n7/mode/2up

Reiner, E. J., Jobst, K. J., Megson, D., Dorman, F. L., & Focant, J. F. (2014). Analytical methodology of POPs. In G. O'Sullivan, & C. Sandau (Eds.), *Environmental forensics for persistent organic pollutants* (pp. 59–139). Elsevier. https://doi.org/10.1016/B978-0-444-59424-2.00003-7

Ruiu, L. (2013). *Brevibacillus laterosporus*, a pathogen of invertebrates and a broad-spectrum antimicrobial species. *Insects*, *4*(3), 476–492. https://doi.org/10.3390/insects4030476

Salgado, V. L., & Sparks, T. C. (2005). The Spinosyns: Chemistry, biochemistry, mode of action, and resistance. In L. I. Gilbert (Ed.), *Comprehensive molecular insect science* (Vol. 6, pp. 137–173). Elsevier. https://doi.org/10.1016/B0-44-451924-6/00078-8

Sanahuja, G., Banakar, R., Twyman, R. M., Capell, T., & Christou, P. (2011). *Bacillus thuringiensis*: A century of research, development and commercial applications. *Plant Biotechnology Journal*, *9*(3), 283–300. https://doi.org/10.1111/J.1467-7652.2011.00595.X

Shivaji, S., Chaturvedi, P., Begum, Z., Pindi, P. K., Manorama, R., Padmanaban, D. A., Shouche, Y. S., Pawar, S., Vaishampayan, P., Dutt, C. B. S., Datta, G. N., Manchanda, R. K., Rao, U. R., Bhargava, P. M., & Narlikar, J. V. (2009). *Janibacter hoylei* sp. *nov*., *Bacillus isronensis* sp. *nov*. and *Bacillus aryabhattai* sp. *nov*., isolated from cryotubes used for collecting air from the upper atmosphere. *International Journal of Systematic and Evolutionary Microbiology*, *59*(12), 2977–2986. https://doi.org/10.1099/ijs.0.002527-0

Singer, S. (1973). Insecticidal activity of recent bacterial isolates and their toxins against mosquito larvae. *Nature*, *244*(5411), 110–111. https://doi.org/10.1038/244110a0

Steinhaus, E. A. (1942). *Catalogue of bacteria associated extracellularly with insects and ticks*. Burgess Publishing Co. https://doi.org/10.5962/bhl.title.7268

Steinhaus, E. A. (1959). On the improbability of *Bacillus thuringiensis* Berliner mutating to forms pathogenic for vertebrates. *Journal of Economic Entomology*, *52*(3), 506–508. https://doi.org/10.1093/JEE/52.3.506

Thomas, G. M., & Poinar, G. O. (1979). *Xenorhabdus* gen. nov., a genus of entomopathogenic, nematophilic bacteria of the family Enterobacteriacease. *International Journal of Systematic Bacteriology*, *29*(4), 352–360. https://doi.org/10.1099/00207713-29-4-352

Thompson, G. D., Busacca, J. D., Jantz, O. K., Kirst, H. A., Larson, L., & Sparks, T. C. (1995). Spinosyns: An overview of new natural insect management systems. *Proceedings of Beltwide Cotton Production Conference* (pp. 1039–1043), San Antonio, TX.

Tully, J. G., Rose, D. L., Carle, P., Bové, J. M., Hackett, K. J., & Whitcomb, R. F. (1988). *Acholeplasma entomophilum* sp. *nov*. from gut contents of a wide range of host insects. *International Journal of Systematic and Evolutionary Microbiology*, *38*(2), 164–167. https://doi.org/10.1099/00207713-38-2-164

Vodovar, N., Vinals, M., Liehl, P., Basset, A., Degrouard, J., Spellman, P., Boccard, F., & Lemaitre, B. (2005). *Drosophila* host defense after oral infection by an entomopathogenic *Pseudomonas* species. *Proceedings of the National Academy of Sciences*, *102*(32), 11414–11419. https://doi.org/10.1073/PNAS.0502240102

Waldron, C., Madduri, K., Crawford, K., Merlo, D. J., Treadway, P., Broughton, M. C., & Baltz, R. H. (2000). A cluster of genes for the biosynthesis of spinosyns, novel macrolide insect control agents produced by *Saccharopolyspora spinosa*. *Antonie van Leeuwenhoek*, *78*, 385–390.

Weiser, J. (1955). *Neoaplectana carpocapsae* n. sp. (Anguillata, Steinernematidae) novy cizopasnik housenek obalece jablecneho, *Carpocapsae pomonella* L. *Vestnik Ceskoslovenske Spolecnosti Zoologicke*, *19*, 44–52.

White, G. F. (1912). *The cause of European foul brood*. U.S. Dept. of Agriculture, Bureau of Entomology, Washington, DC. https://www.worldcat.org/title/cause-of-european-foul-brood/oclc/811139145

Yan, M., Roehrl, M. H., & Wang, J. Y. (2007). Discovery of crystalline inclusions in *Bacillus licheniformis* that resemble parasporal crystals of *Bacillus thuringiensis*. *Canadian Journal of Microbiology*, *53*(9), 1111–1115. https://doi.org/10.1139/W07-076

3 Classification of Entomopathogenic Bacteria

ABSTRACT

Classification is necessary for the simple identification of organisms and the description of their distinctive traits. This chapter deals with the classification of entomopathogenic bacteria (EPB) based on their properties and taxonomic positions. The EPB can be classified based on their response to Gram reactions. They can also be classified based on their host dependency, spore, and crystal production. Four phyla, viz., Firmicutes, Actinobacteria, Proteobacteria, and Tenericutes, possess such types of bacteria.

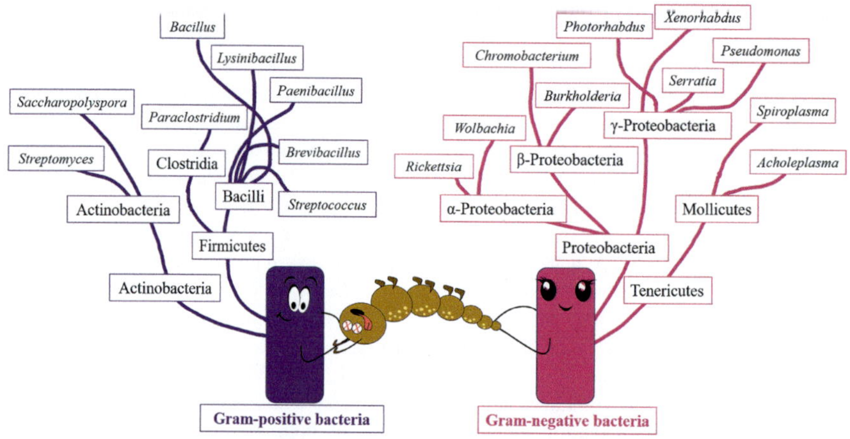

3.1 INTRODUCTION

Classification is the scientific procedure of arranging organisms into a hierarchical series of groups and subgroups based on their similarities and dissimilarities. It is essential for entomopathogenic bacteria and other organisms, for their easy identification, the description of their characteristic features, and the formulation of strategies for their utilization for human benefit.

Entomopathogenic bacteria are broadly classified based on the production of endospores by the bacteria. They are divided into spore-formers and non-spore-formers, further subdivided into obligate and facultative. In addition, the spore-forming bacteria may be crystalliferous (crystal-forming) or non-crystalliferous. They can also be classified based on Gram reactions, whether they are Gram-positive or Gram-negative.

3.2 BACTERIAL SPORES

Bacterial spores are small oval or spherical structures resistant to high temperatures, radiation, desiccation, and chemical agents. They are the most dormant form of bacteria due to minimal metabolism, minimal respiration, and reduced enzyme production (Basta & Annamaraju, 2023; Erkmen, 2021). Several bacteria produce different types of spores: endospores, exospores, conidia, akinetes, myxospores, cysts, and bdellocysts (Mondal, 2011). In the case of entomopathogenic bacteria, endospores are the vital ones among all types of spores. Still, there are some entomopathogenic Actinobacteria that can produce conidia as spores. Gram-positive bacteria generally produce endospores (Basta & Annamaraju, 2023). In addition to endospores, these spore-forming bacteria may also produce proteinaceous parasporal structures under adverse circumstances. On the other hand, non-spore-forming bacteria cannot produce any spores and are unstable in the storage (Soumia et al., 2020). The spores are formed at the end of the active exponential phase of the bacterial growth (Mondal, 2011).

3.3 OBLIGATE AND FACULTATIVE BACTERIA

Entomopathogenic bacteria can be further divided into obligate and facultative based on their nature of in vivo and/or in vitro multiplication (Figure 3.1). For example, *Rickettsia* are obligate pathogens; hence, this group of microorganisms is not cultivated on artificial nutrient media, which is a strong limitation for their production and use in the biocontrol industry (Gouli et al., 2021). *Paenibacillus popilliae* Dutky and *P. lentimorbus* (Dutky) are obligate pathogens only found as resistant spores outside hosts; thus, they are obligate spore-forming bacteria (Glare et al., 2017). Facultative

DOI: 10.1201/9781003426882-3

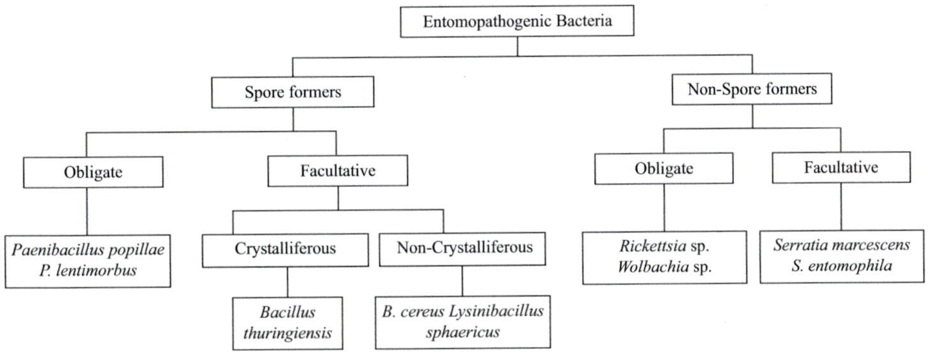

FIGURE 3.1 Phenotypic classification of entomopathogenic bacteria.

bacteria are easy to culture in vitro, so we have so many commercial products from this type of bacteria (Figure 3.1).

3.4 BACTERIAL CRYSTAL PROTEIN

Crystal (Cry) proteins are pore-forming toxins that specifically target invertebrates (insects and nematodes) (Schachter, 2007). Cry protein expression might be sporulation-dependent or independent, but 80% of the crystal proteins are synthesized simultaneously with the sporulation (Keshavareddy & Kumar, 2016). There are several types of crystal proteins effective against different hosts, and almost all of them are produced by *Bacillus thuringiensis* Berliner. A detailed discussion regarding this protein is provided in the next chapter.

Based on the presence of crystals, facultative spore-forming bacteria can be classified into two major groups: crystalliferous and non-crystalliferous (Figure 3.1).

3.5 PHYLA COMPRISING ENTOMOPATHOGENIC BACTERIA

Entomopathogenic bacteria generally belong to four different phyla: Firmicutes, Actinobacteria, Proteobacteria, and Tenericutes (Table 3.1). While the first two phyla possess thick layers of peptidoglycan, thus proving themselves to be Gram-positive, Proteobacteria are rich in lipid content, which gets dissolved by alcohol during Gram staining; thus, they become Gram-negative. The last phylum, Tenericutes, does not possess a cell wall, so they cannot retain the purple colour of crystal violet during Gram staining and also comes under Gram-negative bacteria. Though the detailed description of the aforesaid phyla is beyond the scope and objective of the book, the general characteristics of all these four phyla are described briefly in what follows for the sake of a better understanding of the classification of entomopathogenic bacteria.

3.5.1 Firmicutes

The term *Firmicutes* found its origin from Latin root *firmus*, 'strong'. and *cutis*, 'skin', indicating the presence of a thick peptidoglycan layer (Margulis & Chapman, 2009). Phylum Firmicutes possess Gram-positive bacteria with low guanine + cytosine (G + C) content. Among them, all anaerobic bacteria belong to the class Clostridia, whereas class Bacilli possess bacteria which are obligate or facultative aerobe (Dube, 2021). The organisms in this phylum have round cells, called cocci (singular coccus), or rod-shaped forms (Margulis & Chapman, 2009). Firmicutes produce endospores, which are resistant to desiccation and can survive extreme conditions. The motility is determined mostly by the flagella (Mukherjee & Kearns, 2014).

3.5.2 Actinobacteria

Actinomycete is a Greek word that translates to 'ray fungus' (Sowani et al., 2017). Actinobacteria are widespread Gram-positive bacteria with high G + C concentrations and a distinctive filamentous appearance. They have long been acknowledged as major earthy odour producers (Ortenberg & Telsch, 2003). The morphologies of Actinobacteria are quite diverse, including coccoid and rod-coccoid forms; fragmenting hyphal forms; forms with persistent, highly distinct branching mycelia; and more (Barka et al., 2016). These filamentous bacteria are demonstrated to have an unparallel ability to synthesize a variety of beneficial secondary metabolites due to the abundance of biosynthetic gene clusters. It is thought that each actinobacterial strain has the genetic capacity to create 15–25 secondary metabolites. Actinobacteria are the source of almost 10,000 antibiotics, which count for 45% of all known bioactive microbial compounds. The genus *Streptomyces* only describes 80% of the richest drug-prolific family among all the kingdoms (Puttaswamygowda et al., 2019).

3.5.3 Proteobacteria

The phylum Proteobacteria now represents the biggest and most phenotypically diversified evolutionary branch within the domain of bacteria. To represent the tremendous range of morphologies and physiologies found within this bacterial phylum, Stackebrandt et al. named the Proteobacteria after the Greek God of the sea, Proteus, who could take on many various shapes, in 1988 (Kersters et al., 2006). The members exhibit significant variation in structure and metabolism and are grouped into five classes: α-, ß-, γ-, δ-, and ε-Proteobacteria (Dube, 2021). Proteobacteria are all Gram-negative bacteria. The majority of organisms move by utilizing the flagella; however, others are not mobile or rely on bacterial gliding. Most members are either facultatively or obligatorily anaerobic,

TABLE 3.1
Classification of Bacterial Entomopathogens

Gram Reaction	Phylum	Class	Order	Family	Genus	Species
Gram-positive	Firmicutes	Bacilli	Bacillales	Bacilliaceae	*Bacillus*	*B. thurigiensis* Berliner
						B. cereus Frankland and Frankland
						B. subtilis (Ehrenberg)
						B. amyloliquefaciens (Fukumoto)
						B. megaterium de Bary
						B. licheniformis (Weigmann)
						B. aryabhattai Shivaji et al.
					Lysinibacillus	*L. sphaericus* (Meyer and Neide)
				Paenibacilliaceae	*Paenibacillus*	*P. polymyxa* (Prazmowski)
						P. popillae Dutky
					Brevibacillus	*B. laterosporus* (Laubach)
			Lactobacillales	Enterococcaceae	*Melissococcus*	*M. plutonius* (White)
					Enterococcus	*E. faecalis* (Andrewes and Horder)
		Clostridia	Eubacteriales	Clostridiaceae	*Paraclostridium*	*P. bifermentans* (Weinberg and Seguin)
	Actinobacteria	Actinobacteria	Kitasatosporales	Streptomycetaceae	*Streptomyces*	*S. avermitilis* (Burg et al.)
			Pseudonocardiales	Pseudonocardiaceae	*Sachharopolyspora*	*S. spinosa* Mertz and Yao
			Actinobacteriales	Micrococcaceae	*Kocuria*	*K. kristinae* (Kloos et al.)
					Arthrobacter	*A. gandavensis* Storms et al.
Gram negative	Proteobacteria	Gammaproteobacteria	Enterobacteriales	Enterobacteriaceae	*Xenorhabdus*	*X. nematophilus* (Poinar and Thomas)
					Photorhabdus	*P. luminiscens* (Thomas and Poinar)
					Serratia	*S. entomophila* Grimont et al.
						S. marcescens Bizio
						S. liquefaciens (Grimes and Hennerty)
					Yersinia	*Y. entomophaga* Hurst et al.

(Continued)

TABLE 3.1 (*Continued*)
Classification of Bacterial Entomopathogens

Gram Reaction	Phylum	Class	Order	Family	Genus	Species
					Enterobacter	*E. cloacae* (Jordan)
			Pseudomonadales	Pseudomonadaceae	*Pseudomonas*	*P. entomophila* Mulet et al.
						P. fluorescens Migula
						P. aeruginosa (Schroeter)
						P. taiwanensis Wang et al.
						P. cedrina Dabboussi et al.
						P. putida (Trevisan)
			Legionellales	Coxiellaceae	*Rickettsiella*	*R. popilliae* (Dutky and Gooden)
		Betaproteobacteria	Neisseriales	Chromobacteriaceae	*Chromobacterium*	*C. subtsugae* Martin et al.
			Burkholderiales	Burkholderiaceae	*Burkholderia*	*B. rinojensis* Cordova-Kreylos et al.
		Alphaproteobacteria	Rickettsiales	Rickettsiaceae	*Rickettsia*	*R. prowazekii* da Rocha-Lima
				Anaplasmataceae	*Wolbachia*	*W. pipientis* Hertig
	Tenericutes	Mollicutes	Entomoplasmatales	Spiroplasmataceae	*Spiroplasma*	*S. leptinotarsae* Hackett et al.
			Acholeplasmatales	Acholeplasmataceae	*Acholeplasma*	*A. entomophilum* Tully et al.

chemolithotrophs, chemoorganotrophs, heterotrophs, and some are also autotrophs (Kersters et al., 2006).

3.5.4 Tenericutes

The phylum Tenericutes found its name from the Latin word *tener*, which means 'soft' or 'tender', indicating the absence of a cell wall in such bacteria (Margulis & Chapman, 2009). The genera *Spiroplasma* and *Acholeplasma* belong to the phylum Tenericutes, composed of the class Mollicutes, which evolved from the phylum Firmicutes, represented by *Bacillus* and *Clostridium*. Mollicutes have some of the smallest genome sizes among all culturable organisms and lack a peptidoglycan (PG) layer, unlike other bacteria. These unique characteristics were established during the evolutionary process of Mollicutes from Firmicutes. Many bacterial species, including *Escherichia coli* (Migula) and *Bacillus subtilis* (Ehrenberg), can grow in the L-form, which does not synthesize the PG layer under stresses, inhibiting peptidoglycan maintenance. Mollicutes may have survived in the L-form due to their ability to escape the innate immune system of their hosts by halting the synthesis of PG, a major target of the natural immune system of hosts. During the evolution from a Firmicutes-like ancestor to extant Mollicutes, they established stable parasitism with the acquisition of adhesion ability, modulation of antigenic properties, and reduction in metabolic pathways. In the absence of the PG, the flagella-based motility common in Firmicutes was lost because the machinery is anchored to the PG layer. Hence, Mollicutes may have evolved new motility systems because motility is beneficial for parasitic life. Interestingly, in addition to swimming, Mollicutes have two types of unique gliding motilities (Sasajima & Miyata, 2021). The placing of Mollicutes in this phylum is a debatable issue, as some scientists keep the view to place Mollicutes as a class of phylum Firmicutes. But we must consider several points from both sides of the coin, viz., some members possess pseudo-outer membranes but unanimously all lack the second membrane, which is the characteristic of Gram-negative bacteria, whereas they have low G + C content, proving their origin from Firmicutes (Dube, 2021). Another important characteristic of the Tenericutes is obligate parasitism in eukaryotic organisms, like animals, plants, or insects. They are never found as free-living organisms in nature (Stülke et al., 2009). Based on 16S rDNA sequences, Mollicutes are divided into four main clades (Davis et al., 2013). The detailed classification of entomopathogenic bacteria starting from Gram reaction to species level has been displayed in Table 3.1.

In order of importance, several of these bacteria have been discussed in the coming chapters to represent the diversity of bacteria responsible for causing pathogenicity in insects.

3.6 CONCLUSION

Numerous entomopathogenic bacteria exist on Earth, and so analyzing the diversity within such a sizable population at the individual level is nearly impossible. Therefore, it is crucial to organize the different types of entomopathogenic bacteria for study purposes by classifying them in a logical manner. The similarities and differences help us comprehend the relationships among diverse entomopathogenic bacteria, as well as their evolutionary processes and geographical distribution.

REFERENCES

Barka, E. A., Vatsa, P., Sanchez, L., Gaveau-Vaillant, N., Jacquard, C., Klenk, H.-P., Clément, C., Ouhdouch, Y., & van Wezel, G. P. (2016). Taxonomy, physiology, and natural products of Actinobacteria. *Microbiology and Molecular Biology Reviews*, *80*(1), 1–43. https://doi.org/10.1128/mmbr.00019-15

Basta, M., & Annamaraju, P. (2023, January). *Bacterial spores*. StatPearls Publishing. https://www.ncbi.nlm.nih.gov/books/NBK556071/

Davis, J. J., Xia, F., Overbeek, R. A., & Olsen, G. J. (2013). Genomes of the class Erysipelotrichia clarify the firmicute origin of the class Mollicutes. *International Journal of Systematic and Evolutionary Microbiology*, *63*(Pt 7), 2727–2741. https://doi.org/10.1099/IJS.0.048983-0

Dube, H. C. (2021). Bacteria as plant pathogens. In *Modern plant pathology* (Student ed., pp. 46–57). Agrobios.

Erkmen, O. (2021). Bacterial endospore staining techniques. In *Laboratory practices in microbiology* (pp. 107–113). AcademicPress.https://doi.org/10.1016/B978-0-323-91017-0.00026-3

Glare, T. R., Jurat-Fuentes, J.-L., & O'Callaghan, M. (2017). Basic and applied research. In L. A. Lacey (Ed.), *Microbial control of insect and mite pests* (pp. 47–67). Elsevier. https://doi.org/10.1016/B978-0-12 803527 6.00004-4

Gouli, V. V., Marcelino, J. A. P., & Gouli, S. Y. (2021). The basic biological resources for the production of microbial pesticides. In *Microbial pesticides* (pp. 1–134). Academic Press. https://doi.org/10.1016/B978-0-12-824451-7.00001-2

Kersters, K., De Vos, P., Gillis, M., Swings, J., Vandamme, P., & Stackebrandt, E. (2006). Introduction to the Proteobacteria. In M. Dworkin, S. Falkow, E. Rosenberg, K. H. Schleifer, & E. Stackebrandt (Eds.), *The prokaryotes* (pp. 3–37). Springer. https://doi.org/10.1007/0-387-30745-1_1

Keshavareddy, G., & Kumar, A. R. V. (2016). Bacillus thuringiensis. In Omkar (Ed.), *Ecofriendly pest management for food security* (pp. 443–473). Academic Press. https://doi.org/10.1016/B978-0-12-803265-7.00014-2

Margulis, L., & Chapman, M. J. (2009). Kingdom Prokaryotae (Bacteria, Monera, Prokarya). In *Kingdoms and domains an illustrated guide to the Phyla of life on earth* (4th ed., pp. 35–107). Academic Press. https://doi.org/10.1016/B978-0-12-373621-5.00001-5

Mondal, K. K. (2011). Bacterial spores: Types, structure and formation of endospore. In *Plant bacteriology* (pp. 109–113). Kalyani Publishers.

Mukherjee, S., & Kearns, D. B. (2014). The structure and regulation of flagella in *Bacillus subtilis*. *Annual Review of Genetics*, *48*,319–340.https://doi.org/10.1146/annurev-genet-120213-092406

Ortenberg, E., & Telsch, B. J. (2003). Taste and odour problems in potable water. In D. Mara, & N. Horan (Eds.), *Handbook of water and wastewater microbiology* (pp. 777–793). Academic Press. https://doi.org/10.1016/B978-012470100-7/50043-1

Puttaswamygowda, G. H., Olakkaran, S., Antony, A., & Purayil, A. K. (2019). Present status and future perspectives of marine actinobacterial metabolites. In V. Buddolla (Ed.), *Recent developments in applied microbiology and biochemistry* (pp. 307–319). Academic Press. https://doi.org/10.1016/B978-0-12-816328-3.00022-2

Sasajima, Y., & Miyata, M. (2021). Prospects for the mechanism of *Spiroplasma* swimming. *Frontiers in Microbiology*, *12*, 706426. https://doi.org/10.3389/fmicb.2021.706426

Schachter, H. (2007). Glycobiology of *Caenorhabditis elegans*. *Comprehensive Glycoscience: From Chemistry to Systems Biology*, *4*, 81–100. https://doi.org/10.1016/B978-044451 967-2/00083-0

Soumia, P. S., Krishna, R., Jaiswal, D. K., Verma, J. P., Yadav, J., & Singh, M. (2020). Entomopathogenic microbes for sustainable crop protection: Future perspectives. In A. N. Yadav, J. Singh, C. Singh, & N. Yadav (Eds.), *Current trends in microbial biotechnology for sustainable agriculture* (pp. 469–497). Springer. https://doi.org/10.1007/978-981-15-6949-4_19

Sowani, H., Kulkarni, M., Zinjarde, S., & Javdekar, V. (2017). Gordonia and related genera as opportunistic human pathogens causing infections of skin, soft tissues, and bones. In K. Kon, & M. Rai (Eds.), *The microbiology of skin, soft tissue, bone and joint infections* (pp. 105–121). Academic Press. https://doi.org/10.1016/B978-0-12-811079-9.00007-0

Stülke, J., Eilers, H., & Schmidl, S. R. (2009). *Mycoplasma* and *Spiroplasma*. In M. Schaechter (Ed.), *Encyclopedia of microbiology* (3rd ed., pp. 208–219). Academic Press. https://doi.org/10.1016/B978-012373944-5.00229-7

4 Phylum Firmicutes
Bacillus thuringiensis, an Overview

ABSTRACT

Starting from discovery, *Bacillus thuringiensis* has gained the most importance as entomopathogenic bacteria. This soil bacterium can survive in diverse environments and has the ability to infect a large number of insects from various insect orders, like Lepidoptera, Coleoptera, Diptera, and Hemiptera, as well as nematodes, and causes bacteraemia or septicaemia due to the action of its bacterial toxins. Particularly, it is the crystal (Cry) toxin produced by the bacterium having much typical role in pest management; however, several other toxins are also produced by this bacterium. It possesses little or no threat to non-target species due to the specificity of different strains against selective insects and its unique mode of action, which makes it non-toxic to humans and other mammals. Such advantages of this bacterium result in its worldwide acceptance for commercial use. However, certain drawbacks are also there. Still, this bacterium is considered as one of the most eco-friendly tools for insect pest management.

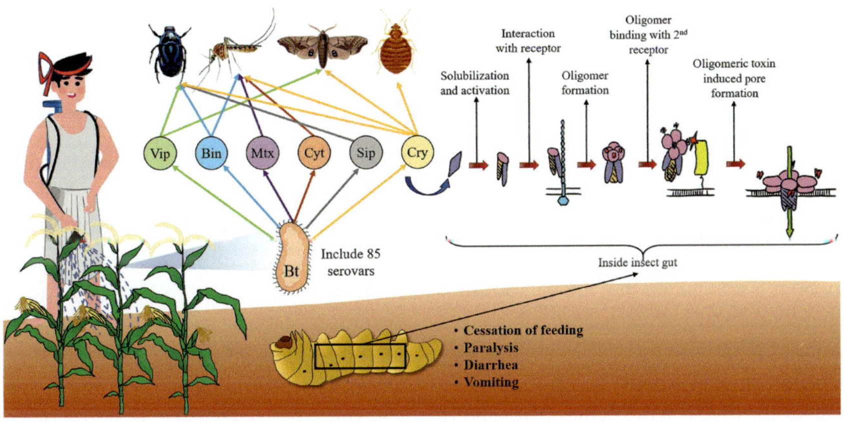

4.1 INTRODUCTION

The phylum Firmicutes includes all low-G + C-content Gram-positive bacteria. The phylum has two major classes: Bacilli and Clostridia. *Bacillus thuringiensis* Berliner (Bt), the oldest known and most used entomopathogenic bacteria to date, belongs to the class Bacilli.

Bt is a facultative anaerobic, Gram-positive, rod-shaped bacterium having dimensions of 1.0–1.2 × 3.0–5.0 µm in its vegetative stage and occurs in long to short tangled chains (Boucias & Pendland, 1998). During sporulation, the ellipsoidal spore, measuring 0.5–1.2 × 1.3–2.5 µm, resides in the unswollen sporangium with parasporal crystalline toxins (Cohen, 2015). Its colonies are often uneven in shape, hazy in colour, dry in texture, and flat with a rough surface. Bt strains typically have genomes that range in size from 2.4 to 5.7 million base pairs. There are 85 recognized subspecies of Bt, categorized into 71 serotypes based on flagellar serotyping. Some of the most noticeable are *Bt* subsp. *thuringiensis*, *kurstaki*, *aizawai*, *israelensis*, *morrisoni*, *thompsoni*, *tenebrionis*, *japonensis*, etc. (Anil et al., 2017). A wide range of insecticidal proteins have been produced by Bt strains which are effective against larvae of different insect orders and, even in certain cases, species. This makes Bt-based products the best-selling biological insecticides sold in the modern era (Roh et al., 2007; Schnepf et al., 1998). The genes expressing the insecticidal proteins have been successfully utilized for formulating novel insecticidal formulations and incorporating them in transgenic crops (Sanchis, 2011). At the beginning of sporulation, and also during the stationary growth phase as parasporal crystalline inclusions, Bt strains produce crystal (Cry) toxins and cytolytic (Cyt) toxins, which are also referred to as δ-endotoxins (Figure 4.1). After ingestion by the insects, crystals are dissolved in the insect's alkaline midgut and start to express their toxic effects. There are 16 classes of pesticidal proteins available in Bt. Crickmore et al. (2021) advised the use of the term *pesticidal protein* in place of *toxins* as Bt is being vigorously used in agriculture and the *toxin* term gives a negative impression concerning human health.

DOI: 10.1201/9781003426882-4

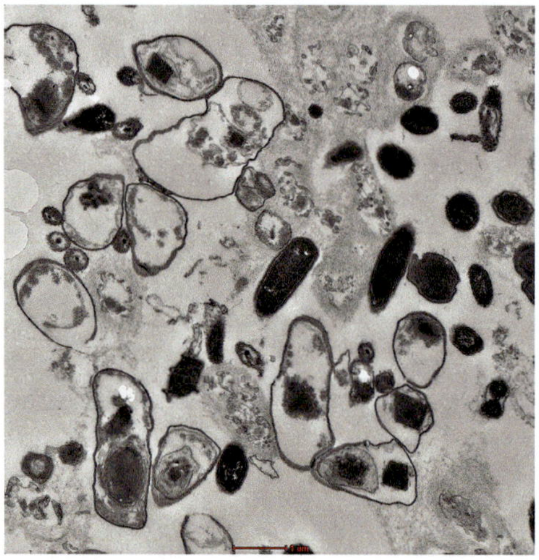

FIGURE 4.1 Bipyramidal protein crystals of Bt mixed with spores.

4.2 HABITAT AND HOST RANGE

Bt is a ubiquitous soil microorganism that has been frequently isolated from diverse environments, like soil, insect cadavers, foliar plant surface, aquatic environment, animal faeces, flour mills, and grain elevators in different parts of the world, including marine sediments and even from soils of Antarctica (DeLucca et al., 2011; Forsyth & Logan, 2000; Ishiwata, 1905a; Knowles, 1994; Maeda et al., 2000; Maheswaran et al., 2010; Martin & Travers, 1989; Smith & Couche, 1991; Venter & Bøhn, 2016). So it is concluded that the bacterium is almost omnipresent, but its natural habitat is soil, which acts as a reservoir, as it cannot multiply in the soil as such in significant numbers and needs an insect body for multiplication (Anil et al., 2017; Yara et al., 1997). After being discovered, for many years, Bt was used only against Lepidoptera, using a highly potent strain (*Bt* var. *kurstaki*). This strain continues to serve as the basis for numerous Bt formulations. However, further screening of many other Bt strains revealed that some of them are active against larvae of Coleoptera (beetles) (Krieg et al., 1983) or Diptera (small flies, mosquitoes) (de Barjac, 1978a; Goldberg & Margalit, 1977). Although their range of insect hosts varies, the majority of them have the same fundamental toxin structure. Different levels of binding affinity to the receptors in the insect's stomach may be the cause. For instance, *Bt* var. *aizawai* produces toxins that are very specific to Lepidoptera and have no impact on other insects, but they differ somewhat from those of *Bt* var. *kurstaki*. Furthermore, *Bt* var. *israelensis* strains generate toxins that are extremely effective against Diptera. In this instance, we can use the Simulium blackfly, which is a vector of several tropical illnesses, fungus gnat larvae, and some mosquito species (particularly *Aedes* spp.); however, *Culex* spp. and *Anopheles* spp. require greater dosages of the toxin. Strains of *Bt* var. *San Diego* or *Bt* var. *tenebrionis* are marketed to control some Coleoptera, especially for controlling Colorado potato beetles (Deacon, n.d.). The effectiveness of all the known δ-endotoxins originating from Bt strains is shown in Figure 4.2.

4.3 CLASSIFICATION OF *BACILLUS THURINGIENSIS* STRAINS

The discovery of Bt's entomopathogenic potential against a wide range of insect groups has increased the search for new isolates of Bt with toxicity against specific groups of economically important insect pests (Bernhard et al., 1997; Martin & Travers, 1989). Different Bt strains have shown variation in their pathogenicity, which demands a reliable method of classification. Keeping this in view, Bt has been classified into many strains, subspecies, serotypes, varieties, serovars, pathovars, biovars, or crystovars by different workers worldwide. Various methods are used to differentiate the subspecies, such as flagella serology, biochemical

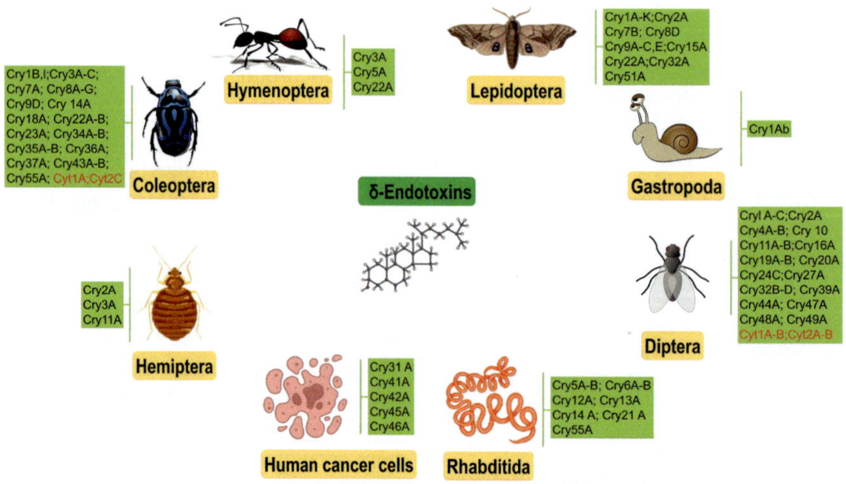

FIGURE 4.2 Summarized view displaying the known host spectrum of Bt δ-endotoxins (Cry and Cyt). **Note:** Cry1A-C (separated by a hyphen) denotes a group of Cry1A, Cry1B, and Cry1C toxins. Cry1B, I (separated by a colon) denotes different Cry1B and Cry1I toxins. Groups or individual poisons are separated by semicolons.

tests, parasporal body (crystal) antigen, molecular techniques, and plasmid profiling (Adams et al., 1996).

4.3.1 Flagella Serology

Flagellar agglutination is considered as the most popular technique to differentiate Bt strains. de Barjac and Bonnefoi (1962) first performed a detailed study comparing proteolytic abilities, carbohydrate utilization, pigmentation, and other microbiological characteristics of different strains of Bt. In this method, based on a particularly consistent and trustworthy characteristic, the H antigen, serological typing has brought some order to the Bt group. For serological typing, antisera against different flagellar antigens (also known as H antigens) of each Bt strain has also been raised by them. They reported that flagellar serology is strongly correlated with the more tedious biochemical and microbiological examination of Bt strains. Importantly, the 24 isolates were reduced to 6 different serotypes in their original study. Before the introduction of this method, much confusion occurred in classification, as various isolates were often given the same name, and sometimes the same strain was given several names. As a result of these misidentifications, failure occurred in some of the initial Bt field applications. However, the introduction of serological typing put an end to this confusion in Bt strains classification (de Barjac & Frachon, 1990).

The catalogue from the International Entomopathogenic Bacillus Centre (IEBC), Pasteur Institute, located in Paris, has assigned thousands of Bt isolates from every part of the world to 55 antigenic groups and 68 serovars (IEBC, 1996). After that, by the end of 1999, the most prevalent and revised list of the H-antigen classification was published by Lecadet et al. (1999), including several new serotypes with a total of 69 H-antigen groups and 13 subgroups. Later on, three more serovars were isolated and described (Khyami-Horani et al., 2003; Li et al., 2000; Roh et al., 2009), resulting in the current list consisting of 85 Bt serovars (Table 4.1).

TABLE 4.1
Classification of *Bacillus thuringiensis* Strains According to the H Serotype

Sl. No.	H Antigen	Serovar	Abbreviation	First Isolated From	First Mention and/or First Validly Described
1	1	*thuringiensis*	THU	*Ephestia kuehniella* Zeller	(Berliner, 1915*; Heimpel & Angus, 1958)
2	2	*finitimus*	FIN	*Malacosoma disstria* Hübner	(Heimpel & Angus, 1958)
3	3a, 3c	*alesti*	ALE	*Bombyx mori* Linnaeus	(Heimpel & Angus, 1958; Toumanoff & Vago, 1951)
4	3a, 3b, 3c	*kurstaki*	KUR	*Ephestia kuehniella* Zeller	(de Barjac & Lemille, 1970)
5	3a, 3b, 3d	*mogi*	MOG	Fallen leaves (Korea)	(Roh et al., 2009)
6	3a, 3d	*sumiyoshiensis*	SUM	Soil (Sumiyoshi, Fukuoka, Japan)	(Ohba & Aizawa, 1989)
7	3a, 3d, 3e	*fukuokaensis*	FUK	Soil (Nishi Koen Park, Fukuoka, Japan)	(Ohba & Aizawa, 1989)
8	4a, 4b	*sotto*	SOT	*Bombyx mori* Linnaeus	(Heimpel & Angus, 1958; Ishiwata, 1905b)
9	4a, 4c	*kenyae*	KEN	*Trichophaga tapetzella* (Linnaeus)	(Bonnefoi & de Barjac, 1963; Norris & Burges, 1963)
10	5a, 5b	*galleriae*	GAL	*Galleria mellonella* (Linnaeus)	(de Barjac & Bonnefoi, 1962; Shvetsova, 1959)*
11	5a, 5c	*canadensis*	CAN	n.d. (Canada)	(de Barjac & Bonnefoi, 1972)
12	6	*entomocidus*	ENT	*Aphomia gularis* (Zeller)	(Heimpel & Angus, 1958)
13	7	*aizawai*	AIZ	*Heliothis assulta* Hübner (Japan)	(Bonnefoi & de Barjac, 1963)
14	8a, 8b	*morrisoni*	MOR	*Aglossa caprealis* (Hübner)	(Bonnefoi & de Barjac, 1963)
15	8a, 8c	*ostriniae*	OST	n.d. (China)	(Ren et al., 1975)
16	8b, 8d	*nigeriensis*	NIG	Soil from mosquito habitats (Nigeria)	(Weiser & Prasertphon, 1984)
17	9	*tolworthi*	TOL	*Plodia interpunctella* (Hübner)	(de Barjac & Bonnefoi, 1968; Norris, 1964)
18	10a, 10b	*darmstadiensis*	DAR	*Galleria mellonella* (Linnaeus)	(Krieg et al., 1968)
19	10a, 10c	*londrina*	LON	n.d. (Brazil)	Arantes et al. (unpublished)
20	11a, 11b	*toumanoffi*	TOU	*Galleria mellonella* (Linnaeus)	(Krieg, 1969)
21	11a, 11c	*kyushuensis*	KYU	Silkworm litter (Fukushima Prefecture)	(Ohba & Aizawa, 1979)
22	12	*thompsoni*	THO	*Galleria mellonella* (Linnaeus)	(de Barjac & Thompson, 1970)
23	13	*pakistani*	PAK	*Cydia pomonella* (Linnaeus) (Pakistan)	(de Barjac et al., 1977)
24	14	*israelensis*	ISR	Culicidae larvae (Israel)	(de Barjac, 1978b)
25	15	*dakota*	DAK	Soil of Oat, barley and wheat field (North Dakota, USA)	(DeLucca et al., 1979)

(Continued)

TABLE 4.1 *(Continued)*

Classification of *Bacillus thuringiensis* Strains According to the H Serotype

Sl. No.	H Antigen	Serovar	Abbreviation	First Isolated From	First Mention and/or First Validly Described
26	16	*indiana*	IND	Soil of soybean field (Indiana, USA)	(DeLucca et al., 1979)
27	17	*tohokuensis*	TOH	Silkworm litter (Fukushima Prefecture, Japan)	(Ohba et al., 1981a)
28	18a, 18b	*kumamotoensis*	KUM	Silkworm litter (Kumamoto Prefecture, Japan)	(Ohba et al., 1981b)
29	18a, 18c	*yosoo*	YOS	Soil (Korea)	(Lee et al., 1995)
30	19	*tochigiensis*	TOC	Soil from mulberry field (Tochigi Prefecture, Japan)	(Ohba et al., 1981b)
31	20a, 20b	*yunnanensis*	YUN	Soil (China)	(Wang, 1979*; Yu et al., 1984)
32	20a, 20c	*pondicheriensis*	PON	Soil (India)	Rajagopalan *et al.* (unpublished)
33	21	*colmeri*	COL	Grain dust (Louisiana, USA)	(DeLucca et al., 1984)
34	22	*shandongiensis*	SHA	Soil (China)	(Ying et al., 1986)
35	23	*japonensis*	JAP	Soil (Japan)	(Ohba & Aizawa, 1986)
36	24a, 24b	*neoleonensis*	NEO	Soil (Mexico)	(Rodriguez-Padilla et al., 1990)
37	24a, 24c	*novosibirsk*	NOV	Soil (Western Siberia)	(Burtseva et al., 1995)
38	25	*coreanensis*	COR	Soil (Korea)	(Lee et al., 1994)
39	26	*silo*	SIL	Grain silo (France)	de Barjac & Lecadet (unpublished)
40	27	*mexicanensis*	MEX	Soil (Mexico)	(Rodriguez-Padilla & Galan-Wong, 1989)
41	28a, 28b	*monterrey*	MON	n.d. (Mexico)	Rodriguez-Padilla et al. (unpublished)
42	28a, 28c	*jegathesan*	JEG	Soil (Malaysia)	(Seleena et al., 1995)
43	29	*amagiensis*	AMA	Soil (Tokyo, Japan)	(Ishii & Ohba, 1993)
44	30	*medellin*	MED	Soil from the mosquito breeding sites (Colombia)	(Orduz et al., 1992)
45	31	*toguchini*	TOG	Soil (Novosibirsk, Russia)	(Khodyrev, 1990)
46	32	*cameroun*	CAM	Soil (Cameroon)	Jacquemard (unpublished)*; Juarez-Perez et al., 1994)
47	33	*leesis*	LEE	Soil (Korea)	(Lee et al., 1994)
48	34	*konkukian*	KON	Soil (Korea)	(Lee et al., 1994)
49	35	*seoulensis*	SEO	Soil (Korea)	(Lee et al., 1995)
50	36	*malaysiensis*	MAL	n.d. (Malaysia)	Ho (unpublished)
51	37	*andaluciensis*	AND	Soil (Spain)	(Quesada-Moraga et al., 2004)
52	38	*oswaldocruzi*	OSW	Powdered *Cuminum cyminum* L. and *Piper nigrum* L. (Brazil)	(Rabinovitch et al., 1995)
53	39	*brasiliensis*	BRA	Powdered *Cuminum cyminum* L. and *Piper nigrum* L. (Brazil)	(Rabinovitch et al., 1995)
54	40	*huazhongensis*	HUA	Soil from radish field and cornfield (China)	(Jingyuan et al., 1996)
55	41	*sooncheon*	SOO	Soil (Korea)	(Lee et al., 1995)
56	42	*jinghongiensis*	JIN	Soil (China)	(Li et al., 1999)
57	43	*guiyangiensis*	GUI	Soil (China)	(Li et al., 1999)
58	44	*higo*	HIG	Silkworm litter (Kumamoto, Japan)	(Ohba et al., 1995)
59	45	*roskildiensis*	ROS	Ivy leaves (Denmark)	Hinrinschen et al. (unpublished)
60	46	*chanpaisis*	CHA	Soil of rice-growing areas, rice stemborer, rice dust, and rice bran from rice mill (Thailand)	(Chanpaisaeng, 1995)
61	47	*wratislaviensis*	WRA	Soil (Wroclaw, Poland)	(Lonc et al., 1997)
62	48	*balearica*	BAL	Soil (Spain)	(Iriarte et al., 2000)
63	49	*muju*	MUJ	Leaves (Korea)	(Park et al., 1997)
64	50	*navarrensis*	NAV	Soil (Spain)	(Iriarte et al., 2000)
65	51	*xiaguangiensis*	XIA	Bean (China)	Jian et al. (unpublished)
66	52	*kim*	KIM	Soil (Korea)	(Kim et al., 1999)
67	53	*asturiensis*	AST	Soil (Spain)	(Quesada-Moraga et al., 2004)
68	54	*poloniensis*	POL	*Lymantria monacha* (Linnaeus) (Denmark)	Damgaard et al. (unpublished)

TABLE 4.1 *(Continued)*
Classification of *Bacillus thuringiensis* Strains According to the H Serotype

Sl. No.	H Antigen	Serovar	Abbreviation	First Isolated From	First Mention and/or First Validly Described
69	55	*palmanyolensis*	PAL	Soil (Spain)	(Quesada-Moraga et al., 2004)
70	56	*rongseni*	RON	Soil (China)	(Li et al., 1999)
71	57	*pirenaica*	PIR	Dust sample in a corn silo (Elorz, Spain)	(Porcar et al., 1999)
72	58	*argentinensis*	ARG	n.d. (Argentina)	Campos-Dias et al. (unpublished)
73	59	*iberica*	IBE	Horticultural soil (Viella, Spain)	(Porcar et al., 1999)
74	60	*pingluonsis*	PIN	Soil (China)	(Li et al., 1999)
75	61	*sylvestriensis*	SYL	Scotch pine (Denmark)	Damgaard (unpublished)
76	62	*zhaodongensis*	ZHA	Soil (China)	(Li et al., 1999)
77	63	*bolivia*	BOL	*Rhigopsidius tucumanus* Heller in infested potato tuber (Bolivia)	(Ferrandis et al., 1999)
78	64	*azorensis*	AZO	Soil (Portugal)	Santiago-Alvarez et al. (unpublished)
79	65	*pulsiensis*	PUL	Grain field (Pakistan)	Khalique & Khalique (unpublished)
80	66	*graciosensis*	GRA	Soil (Portugal)	Santiago-Alvarez et al. (unpublished)
81	67	*vazensis*	VAZ	Soil (Portugal)	Santiago-Alvarez et al. (unpublished)
82	68	*thailandensis*	THA	n.d. (Thailand)	Chanpaisaeng et al. (unpublished)
83	69	*pahangi*	PAH	n.d. (Malaysia)	Seleena & Lee (unpublished)
84	70	*sinensis*	SIN	Soil (Yunnan Province, China)	(Li et al., 2000)
85	71	*jordanica*	JOR	Soil (Jordon)	(Khyami-Horani et al., 2003)

Note: *First mention or designation of the serovar.
Source: Modified from Lecadet et al. (1999).

4.3.2 Other Classification Approaches

Though flagellar serology is the most reliable and popular method for classifying Bt strains, there are several other approaches of Bt classification. The first approach of Bt classification was based on morphological and biochemical characterization of Bt by using conventional microbiological techniques (Heimpel & Angus, 1958). In 1962, in parallel with serological typing, this method was applied to 24 isolates and used to differentiate them into six subgroups that were in accordance with the previously mentioned six described serotypes (de Barjac & Bonnefoi, 1962; de Barjac & Frachon, 1990). Another method of classification is based on susceptibility to different bacteriophages by using 'phage types' (de Barjac & Frachon, 1990). Various other methods for the classification of Bt used by several other researchers include crystal serology (Krywienczyk & Fast, 1980; Lynch & Baumann, 1985; Pendleton & Morrison, 1967; Smith, 1987), esterase pattern (Norris, 1964), plasmid pattern (Lereclus et al., 1982), multilocus enzyme electrophoresis (Selander et al., 1986), etc.

4.4 CRY GENE CLASSIFICATION SCHEME

The insecticidal toxins that each Bt serotype produces determine its insecticidal action rather than its serology. The genes that produce the proteinaceous crystal are found on large plasmids. Approximately 6,600 Cry protein sequences of bacteria Bt were successfully retrieved from UniProtKB database (UniProt, 2023). The proteins coded from these genes are composed of 123–1,169 amino acids in case of the order Coleoptera, 643–1,180 amino acids in Diptera, and mostly more than 1,100 amino acids in Lepidoptera (Das et al., 2021). Hofte and Whiteley (1989) gave the classification scheme of *Cry* genes on the basis of both the host specificity of the toxins as well as their structural similarities. A nomenclature scheme was proposed by these authors, keeping the *Cry* epithet (for designating crystal) proteins, that is, Cry I (Lepidoptera-specific toxin), Cry II (Lepidoptera- and Diptera-specific toxin), Cry III (Coleoptera-specific toxin), Cry IV (Diptera-specific toxin), and Cyt A, which is a cytolytic and haemolytic toxin. Most Bt researchers immediately accepted this nomenclature. However, this system of classification became onerous as numerous new toxins were discovered. A revised system of nomenclature for the Bt Cry genes was suggested by Crickmore et al. (1998) that is solely based on the identity of amino acid, and in this system, Roman numerals are replaced by Arabic numerals, as well as having the parentheses removed. For example, the old style of representation Cry IA(a) was modified as Cry1Aa. In recent days, 818 genes belonging to 81 different families have been listed. Among them, the *Cry* genes take up 78 families, whereas *Cyt* genes take up 3 families (Mendoza-Almanza et al., 2020). For the naming of new pesticidal proteins, the new sequence should be matched with the holotype via pairwise alignment using the Needleman algorithm (Crickmore et al., 2021).

4.5 SYMPTOMS OF *BT* INFECTION

It is very important to recognize and analyze the symptoms produced by entomopathogens to assign them as the responsible ones and how efficiently they are meeting their responsibility of killing insect pests. Depending on the host, bacterium, or stage of pathogenesis, the infected organism will exhibit a variety of outward signs following the entry and establishment of illness by entomopathogenic bacteria. Vomiting, diarrhoea, paralysis, and feeding stoppage are the usual symptoms. Following their passage past the gut epithelial barrier, pathogens multiply in the haemocoel, causing bacteraemia or septicaemia as a result of the action of pathogenic agents and bacterial toxins; this usually causes the host to change in colour and tissue. Although the integument initially stays intact, insects killed by bacteria are often black in colour, soft, and flaccid. In addition, the cadaver shrivels, dries, and hardens.

Husz first tested the pathogenicity of Bt against corn borer in 1927, where he found infection through the mouth causes cent percent mortality within 1–1.5 dpi under experimental conditions (Steinhaus, 1942). In the case of aphids, different Bt strains start showing effects within 48–60 hours of infection. During eating, they first move slowly and involuntarily, and their abdominal segments begin to constrict. The aphids stop moving after 60 to 72 hours, become brown instead of green, grow larger and develop black patches on the first abdominal segments, and lose moisture. After 80 hours, the black discolouration spreads toward the metathorax, mesothorax, prothorax, and pleural region, and the abdomen is completely compressed and dehydrated (Torres-Quintero et al., 2015). Insects that have been infected with Bt exhibit paralysis of the mouth and stomach, feeding stoppage, midgut microvillus enlargement and destruction, and disturbance of the ion, pH, and glucose balance. Midgut structure and function collapse as midgut cells separate from basement membranes, and these separated cells burst in the midgut lumen and cause septicaemia. This ultimately results in total paralysis and death (Boucias & Pendland, 1998).

4.6 MODE OF ACTION

As a general characteristic of most entomopathogens, the insecticidal proteins generated by Bt strains are highly selective for certain insect species. This has contributed to the great advantage of its use, as some only affect caterpillars while others affect beetles. They are regarded as environmentally safe for birds, mammals, aquatic species, and natural insect enemies that aid in controlling pests, although none of them have an impact on mammals. Thus, it has led to the huge acceptance of Bt formulations and transgenic Bt crops in the farming community and, along with it, resistance against Bt toxins in some pest species (Tabashnik et al., 2013). The entomopathogenic nature of Bt is all contributed by different types of toxins which are produced at the beginning of sporulation and throughout the stationary growth phase. In this phase, Bt produces cytoplasmic inclusions called parasporal crystal bodies, which are made up of one or more proteins (inactive protoxins). The presence of a parasporal crystal is the phenotypic trait of Bt and is used to separate this bacterium from other *Bacillus* species (Sansinenea, 2012). Crickmore (2017) described the schematic overview of the current nomenclature system used by the Bt toxin nomenclature committee for δ-endotoxins, also known as insecticidal crystal proteins (ICP) (Cry and Cytl) and secretable toxins (Vip and Sip). To date, more than 700 Cry gene sequences coding for Cry proteins are reported (Palma et al., 2014), and large plasmids are the usual location for these genes (Sansinenea, 2012).

When the crystals are ingested by the susceptible larvae while feeding, the crystal toxin undergoes a different course of action, like (1) solubilization in higher alkaline midgut pH, (2) activation of the linearized protoxins by the gut protease, (3) binding of the activated toxins to the receptors found in the midgut columnar epithelial cells, and (4) creation of ion channels or pores by insertion of the activated toxin into the apical membrane of the columnar epithelial cell (Schnepf et al., 1998).

4.6.1 Solubilization of the Toxin

When the Bt crystal is dissolved in alkali in vitro through the reduction of the disulfide linkages, the dimer (230 kDa) is converted to the 130–140 kDa protoxin (in case of lepidopteran protoxin), and then the protoxins are solubilized in the alkaline pH (pH 9 to 12) of the midgut environment (Knowles, 1994). A pH above 9.5 is required for the proteins specific to Lepidoptera (Knowles & Dow, 1993). Cryl, Cry3A, Cry4A, and Cry5B proteins are soluble at or above pH 9.5, whereas Cry2 and Cry4D proteins are soluble at pH 12 (Jain et al., 2016).

4.6.2 Activation of the Protoxin

A significant factor influencing the particular activation of the Cry toxins is the pH of the midgut. Cleavage of the Cry toxins is a crucial step for the activation of toxins, and also for the specificity in different insects (Schünemann et al., 2014). After solubilization, the midgut proteases cleave the inactive protoxin into 60–70 kDa protease-resistant proteins (Bravo et al., 2005). The toxins are proteolytically activated when the N-terminal peptide (25–30 amino acids for Cryl toxins, 49 residues for Cry2Aa, and 58 residues for Cry3A) is removed. About half of the residual protein from the C-terminus has been eliminated in the case of lengthy Cry protoxins and the N-terminus. The cleavage of Cry1A at residue 51 is the cause of pre–pore formation and the loss of helix α-1. Cry4B cleaves into two pieces, each weighing 18 and 46 kDa. Cleavage of Cry11A forms two fragments of 32 and 34 kDa. The cleavage points of different Cry protoxins are displayed in Figure 4.3. The crystal structure of activated Cry protein contains three distinct domains. Domain I comprises seven α-helices, Domain II is composed of 3 antiparallel β-plates, and Domain III forms a β-sandwich (Bravo et al., 2007).

4.6.3 Receptor Binding in Lepidopteran Larvae

Four distinct binding proteins for Cry1A toxins have been identified across multiple lepidopteran insects. These are a cadherin-like protein (CADR), a glycosylphosphatidylinositol (GPI)–anchored aminopeptidase-N (APN), a GPI-anchored alkaline phosphatase (ALP), and a 270 kDa glycoconjugate (GCR) (Jurat-Fuentes & Adang, 2004; Knight et al., 1994; Vadlamudi et al., 1995; Valaitis et al., 2001). The function of the toxin–receptor interaction in the insect *Manduca sexta* (Linnaeus) has been thoroughly documented. In this insect, there are at least two Cry1A-binding proteins, that is, a CADR protein (Bt-R1) and a GPI-anchored APN, which act as receptors (Knight et al., 1994; Vadlamudi et al., 1995). Surface plasmon resonance experiments have shown that the binding affinity of monomeric Cry1A toxins to the *M. sexta* Bt-R1 is in the range of 1 nM (Vadlamudi et al., 1995), while in the case of APN, it is in the range of 100 nM (Jenkins et al., 2000). The interaction between Cry1A toxins and Bt-R1 receptor is extremely complicated because it involves at least three binding epitopes in two molecules. A change in receptor binding results in the change in resistance to Cry toxin (Ferré & Van Rie, 2002). In the case of the laboratory-selected *Heliothis virescens* (Fabricius) Cry1Ac-resistant line YHD2, it showed that a single mutation was responsible for the 40–80% of Cry1Ac resistance levels and linked with a retrotransposon insertion in the CADR gene (Gahan et al., 2001). Also, the characterization of CADR alleles of field-derived and laboratory-selected strains of *Pectinophora gossypiella* (Saunders) (pink bollworm) and *Helicoverpa armigera* (Hübner) revealed different mutated CADR alleles associated with Cry1Ac resistance (Morin et al., 2003; Xu et al., 2005). In the case of APN, the domain III of Cry1Ac toxin ties with the APN receptor, which specifically recognizes N-acetylgalactosamine (GalNAc) moieties in contrast to the Cry1Aa and Cry1Ab toxins, which show no binding capacities to GalNAc (Masson et al., 1995). In the lepidopteran insect *Lymantria dispar* Linnaeus, a sequential binding mechanism was postulated for the interaction between Cry1Ac and APN. First, domain III interacts with APN GalNAcc sugar moieties, and then it facilitates the subsequent interaction of domain II loop regions with another region in the APN receptor (Jenkins et al., 2000). Another type of receptor is ALP, where it was shown that the *H. virescens*–resistant line YHD2 has a part of the Cry1Ac-resistant phenotype that brings down the creation of GPI-anchored ALP receptor due to mutation (Jurat-Fuentes & Adang, 2004). In *M. sexta*, proteomic examination of BBMV (brush border membrane vesicles) of Cry1Ac-binding proteins uncovered that ALP is a putative receptor particle (McNall & Adang, 2003).

4.6.4 Pre–Pore Formation or Oligomerization

Before membrane insertion, several PFTs (pore-forming toxins) form a soluble oligomeric structure. When Cry1Ab toxin binds to *M. sexta* Bt-R1, it undergoes an extra proteolytic cleavage in its N-terminal end (helix α-1), which facilitates pre-pore oligomeric structure formation. It is important for membrane insertion and toxicity (Gómez et al., 2002b; Rausell et al., 2003). Cry1Ab protoxin was

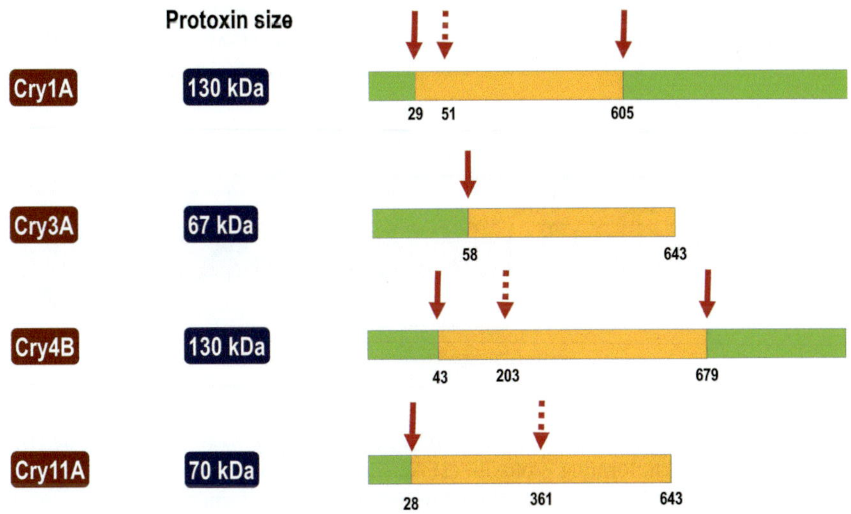

FIGURE 4.3 Relative length of Cry protoxin and the site of protease digestion. **Note:** Solid arrows indicate the amino- and carboxy-terminal cleavage sites of activated toxins; dotted arrows indicate intramolecular cleavages; green boxes indicate prototoxins; yellow boxes indicate active toxins.

incubated with the single-chain scFv73 antibody, which mimics the CADR receptor or the toxin-binding peptides of Bt-R_1 (CADR repeats 7 and 11), and upon treatment with *M. sexta* midgut juice, it resulted in the preparations of toxin which form a 250 kDa oligomer lacking in the helix α-1 of domain I (Gómez et al., 2002a, 2003). It was reported that the binding affinity of the oligomeric structures of Cry1Ab and Cry1Ac to the APN receptor increased 100- to 200-fold, showing an apparent dissociation constant of 0.75–1 nM (Gómez et al., 2003; Pardo-López et al., 2006). In comparison to the 60 kDa monomer, the oligomer was appropriate for membrane insertion, as determined by measurement of the toxin membrane insertion. Analysis of the pore activity of the Cry1Ab oligomeric structure in synthetic planar lipid bilayers showed distinct kinetic properties from the monomeric Cry1Ab toxin. Firstly, compared to monomeric toxins, pure oligomer preparations generate holes at lower toxin concentrations. Second, the oligomeric Cry1Ab toxin exhibited persistent channels with a strong opening propensity, which differed from the kinetics of a monomeric toxin. Therefore, it was demonstrated that the oligomeric complexes containing Cry toxin had substantial pore activity, while the monomeric samples had negligible pore-formation activity. It proves that oligomer formation is very much needed for the action of Cry toxins.

4.6.5 Membrane Insertion and Formation of Active Channel

A crucial step in oligomerization and insertion of bacterial PFT into the membrane is that receptors located in lipid rafts interact with different bacterial PFTs (Cabiaux et al., 1997). The APN receptor has been involved in the insertion of Cry1A toxins. GalNac is a binding determinant in the Cry1Ac-APN interaction. GalNAc interacts with oligomeric Cry1Ac *in vitro* due to a conformational shift in the toxin, which also improves its insertion into lipid membranes. This indicates that to facilitate membrane insertion, the interaction between the pre-pore oligomeric structure of Cry1A toxins and the APN is important (Pardo-López et al., 2006). After a conformational change, the alkaline pH of the midgut lumen of lepidopteran and dipteran insects induces the molten globule state of the pre-pore complex. The monomeric toxins do not have such flexible conformation as that of the membrane-inserted oligomer (Rausell et al., 2004). The molten globule state may be connected to the conformational shift that was seen following the contact between the pre-pore oligomer and APN because previously it was mentioned that this interaction facilitated membrane insertion. Although it was not proven, it may be possible (Pardo-López et al., 2006). Additionally, only domain I of the membrane-inserted pore could resist heat denaturation, indicating that, unlike domains II and III, only domain I may be inserted into the membrane. Finally, a comparatively looser conformation of membrane-inserted domain I is induced at alkaline pH, which is important for forming an active channel (Rausell et al., 2004). This ultimately leads to ionic imbalance, excessive loss of cellular ATPs, and septicaemia, as the disruption of the gut may permit the invasion of bacteria into the haemolymph (Bravo et al., 2011). About 15 minutes after the formation of the channels, cellular metabolism ceases, the insect stops feeding, gets dehydrated, and eventually dies. These steps in the mode of action of Bt Cry toxins are illustrated in Figure 4.4.

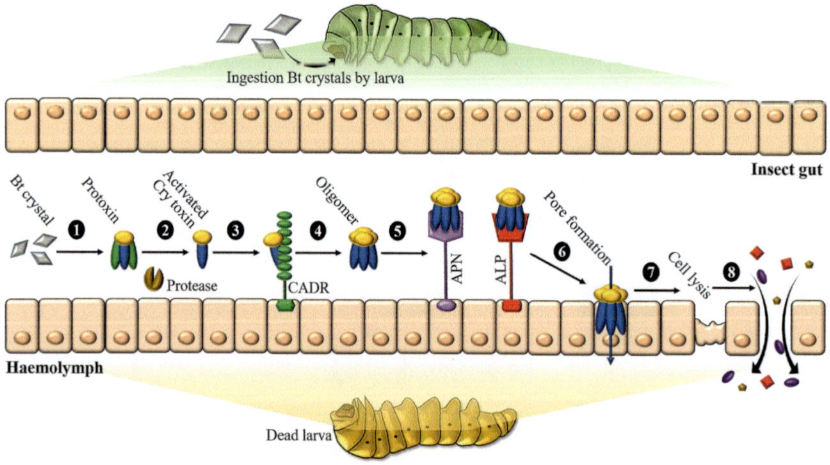

FIGURE 4.4 Schematic representation of the mode of action of Bt Cry toxin. **Note:** First Bt crystal ingested by the larvae. (1) Crystals dissolve in the alkaline pH of the stomach, forming protoxin. (2) Activation of protoxin by protease. (3) The monomeric Cry toxin interacts with the first receptor (CADR or GCR), causing the toxin to undergo a conformational change. (4) Formation of oligomer. (5) Oligomeric toxin binds to the second receptor (GPI-APN or GPI-ALP), resulting in the occurrence of a conformational change which induced the molten globule state of the toxin. (6) Oligomeric toxins inserted into the membrane and induce pore formation. (7) Disruption of epithelial cell layer. (8) Causes ionic imbalance, loss of ATP, and septicaemia. It ultimately results in the breakdown of cells, which causes larval death.

4.7 OTHER GROUPS OF BT TOXIN

Though Cry toxin is the major responsible one in conferring Bt its well-known applicability in insect management, there are several other toxins secreted by the bacterium to act on different insects. Their importance cannot be neglected. Some of the important other groups of toxins are discussed in what follows.

4.7.1 Cyt Toxins (Cytolytic Toxins)

This class of ICP has widespread cytolytic (haemolytic) action *in vitro*. Cyts are often mildly poisonous. However, when combined with some Cry toxins, the toxicity of Cyt may get enhanced, as it is supposed that the Cyt protein, binding to the target cell membrane, can function as a receptor for that Cry toxin (Argôlo-Filho & Loguercio, 2014). Even Cyt proteins act as a synergist that enhance the insecticidal effects of Cry or Vip3 toxins. Thus, Cyt toxins reduce the chances of resistance development to some Cry proteins in some insect species (Soberon et al., 2013; Yu et al., 2012). For instance, Cyt1Aa toxin, which is active against the cottonwood leaf beetle, *Chrysomela scripta* Fabricius, restricts the resistance development in insects to Cry3Aa (Federici & Bauer, 1998). Similarly, Cyt1Aa proteins have inhibited the development of resistance to Cry4 and Cry11Aa toxins in larvae of laboratory-selected populations of *Culex quinquefasciatus* Say. The oligomerization of Cry11Aa toxin and pore formation both have been facilitated due to the binding of Cry11Aa toxin to Cyt1Aa toxin, and this is considered as the mechanism that leads to synergism (Soberon et al., 2013). For the Cyt proteins, two different mechanisms of action have been proposed, with the first one suggesting a pore-formation model, whereas the second one supports a less-specific detergent mechanism of action (Butko, 2003; Soberon et al., 2013). Cyt toxins are mostly found in Bt strains active against Diptera (Bravo et al., 2007). To date, Cyt proteins have been classified by the Bt Toxin Nomenclature Committee into three different families (Cyt1, Cyt2, and Cyt3), showing toxic effect mostly against some black flies and mosquitoes (Soberon et al., 2013). However, strains of some diverse Bt subspecies, such as *Bt* subsp. *morrisoni*, bear *Cyt* genes having toxicity against a wider range of insect pests belonging to the order Diptera, Lepidoptera, and Coleoptera (Guerchicoff et al., 2001). Van Frankenhuyzen (2009) reported that Cyt2C may also be active against certain nematodes, like Rhabditida, and cancer cells.

4.7.2 VIP (Vegetative Insecticidal Proteins)

These represent another class of toxins that are expressed and released when Bt is in its vegetative condition. The Bt toxin Nomenclature Committee has currently discovered and categorized Vip proteins into four families. They are Vip1, Vip2, Vip3, and a recently identified novel family of Vip proteins that are classified as Vip4 by the Committee (Crickmore, 2017). The groups of Vip1 and Vip2 each separately are weakly toxic but can enhance each other's toxicity strongly, and the group of Vip3 proteins binds to other cell membrane receptors than those of the Cry toxins (Argôlo-Filho & Loguercio, 2014). The homology of Vip1 and Vip2 to other binary toxins of bacteria suggests that a typical A + B type of binary toxins has been formed by Vip1 and Vip2 toxins, where Vip2 toxin is the cytotoxic A-domain, whereas Vip1 toxin is the receptor-binding domain, which is responsible for the translocation of the cytotoxic Vip2 toxin into the cell of the host (Barth et al., 2004; De Maagd et al., 2003). The proposed mode of action of Vip toxins consists of the proteolytic activation of the cell-binding B precursor (Vip1 toxin), along with its monomeric interaction with cell surface receptor(s) by the toxin, which is followed by homoheptamer formation, subsequently translocating the A (Vip2) component of the toxin into the cytoplasm through acid endosomes (Barth et al., 2004). Vip1 and Vip2 heterodimer toxins are effective against Coleopteran and Hemipteran insect orders, and Vip3, which is a single-chain non-binary toxin, is effective against many species of Lepidoptera, and crops like cotton and maize have successfully been transformed with various Vip3 toxins (Syed et al., 2020).

4.7.3 SIP (Secreted Insecticidal Proteins)

These are toxins produced by Bt that are around 41 kDa in size. Like Vip proteins, Sip is also synthesized with a signal peptide sequence of 30 amino acids, which is processed by proteases and releases an active protein (Mendoza Almanza et al., 2020). Sip protein was first identified as Sip1Aa1 when obtained from EG2158 Bt strain culture supernatants. The homology of this toxin strongly suggests that the toxicity of Sip1Aa1 may be caused by the formation of pores, but still, its mechanism of action remains unknown (Donovan et al., 2006). These proteins have insecticidal activity against Coleopterans, such as *Leptinotarsa decemlineata* (Say), *Diabrotica undecimpunctata howardi* Barber, and *Diabrotica virgifera virgifera* LeConte (Mendoza-Almanza et al., 2020). However, till present day, Sip1Aa1 is the only protein which is reported with published insecticidal activity against this aforementioned insect pest (Palma et al., 2014).

4.7.4 Parasporins

Some Cry proteins are non-haemolytic but capable of preferentially killing cancer cells (Ohba et al., 2009). These Cry proteins are known as parasporins and are made by Bt or similar bacteria. The parasporin proteins are all Cry proteins. Among these parasporins, PS1, PS3, and PS6 are 3D-Cry proteins, and PS4 probably belongs to the Mtx group (Argôlo-Filho & Loguercio, 2014).

4.7.5 Bin (Binary) Toxins

These proteins are binary. To become synergistically toxic, these proteins need to be paired with a complementary protein. The proteins in this category are designated as BinA and BinB, and they are members of the Cry protein types Cry35, Cry36, and Cry49. Cry toxins are highly similar to *B. sphaericus* BinA/BinB toxins. The toxicity of these toxins in combination with their conjugative complement is considerably larger than the toxicity of each of these toxins separately (Palma et al., 2014).

4.7.6 Mtx (Mosquitocidal) Toxins

It is a group of Cry proteins that resemble β pore-forming toxins, similar to the ε-toxin of *Clostridium perfringens* (Veillon and Zuber). In most cases, it is toxic to mosquitoes (Mtx). However, for some Mtx proteins, toxicity is still not proven, and some of these are apparently non-toxic.

4.7.7 α-Exotoxin

The proteinaceous and thermolabile α-exotoxin exhibits increased toxicity to some insects when administered orally and intra-haemocoelically. It also shows toxicity to mice and some other vertebrates (Tanada & Kaya, 1993).

4.7.8 β-Exotoxin

Some Bt strains have the ability of producing β-exotoxin, which is a thermostable, non-proteinaceous, and secretable secondary metabolite expressing non-specific toxic effects, as the toxin is not only harmful to a wide range of insects but also controls different mammalian pests (Liu et al., 2010, 2014). β-exotoxins, which are also known as Thuringiensin (Thu), have a low molecular weight (700 Da) and are analogues of the adenine nucleotide. It is proved to be lethal against a wide range of insect pests, including insects that come under the orders Lepidoptera, Diptera, Coleoptera, Hymenoptera, Orthoptera, and Isoptera, even when it shows toxicity to several nematode species also. Thu has the chemical formula $C_{22}H_{32}O_{19}N_5P$ and is composed of adenosine, phosphoric acid, glucose, and gluconic diacid (Liu et al., 2014). Some strains of Bt producing a related β-exotoxin type II have also been reported, and this toxin may be the uracil analogue of Thuringiensin (Levinson et al., 1990). To date, the mode of insecticidal toxicity of Thu is not fully understood. However, it has been found that β-exotoxins express their toxicity by inhibiting the DNA-dependent RNA polymerase (Gohar & Perchat, 2001), and this may be the reason for its toxicity against mammals also (McClintock et al., 1995). It acts as an ATP analogue (Farkas et al., 1969), interfering with the RNA polymerase (Beebee et al., 1972; Sebesta & Horska, 1970). Synthesis of RNA has been inhibited by Thu as it competes with ATP on the binding sites, which ultimately affect moulting and pupation of insects and cause teratological effects at sublethal doses (Burgerjon et al., 1969; Espinasse et al., 2002; Ignoffo & Gregory, 1972). Production of β-exotoxin is more prevalent in some serovars compared to others (Hernandez et al., 2003).

4.7.9 Enhancin-Like Protein

A pairwise similarity of 20–30% has been shown by Enhancin-like proteins to viral enhancin proteins (metalloproteases). This promotes viral infections by degrading the mucin of the protective peritrophic matrix. One of the commonly known enhancin-like proteins is Bel enhancin, which can effectively enhance Cry1Ac toxicity against American bollworm, *H. armigera* (Fang et al., 2009). But an identical enhancin-like protein that shares 100% of pairwise identity with Bel enhancin does not show any significant increase in level of toxicity of the Cry9Ea toxin (when used in combination) against *Spodoptera exigua* (Hübner) and *Trichoplusia ni* (Hübner) (Zhao et al., 2011).

4.7.10 Thuringiolysin and Haemolysin II

Both of these toxins are cytolytic peptides (cholesterol-dependent cytolysin [CDC] family). Thuringiolysin is produced by Bt having a molecular weight of 47 kDa and very similar to cerolysin of *B. cereus*. Its activity is inhibited by a low concentration of cholesterol. On the other hand, haemolysin II is secreted by both Bt and *B. cereus*, having a molecular weight of 29 kDa. Unlike thuringiolysin, its activity is not inhibited by low concentrations of cholesterol (Pendleton et al., 1973). These cytolytic toxins showed haemolytic activity against erythrocytes and other eukaryotic cells by forming pores in the planar lipid bilayers and thus are found to be lethal to humans and other organisms (Bernheimer & Rudy, 1986).

4.7.11 Unrelated Cry

This diverse group has little sequence homology with other Cry proteins. Some can act as a conjugative toxin to an Mtx toxin (Cry37 with Mtx toxin Cry23) or a Bin toxin (Cry34 with Bin toxin Cry35) (van der Hoeven, 2014).

4.8 ADVANTAGES

It is always required to consider the advantages and disadvantages before implementing or adopting any strategy. Bt is considered as an important tool in insecticide resistance management (IRM) strategies. Bt toxin can be effective only when ingested and gets its proper receptors. This property limits the susceptibility of non-target insects, like pollinators, natural enemies, and other animals, and human beings to this insecticide to a great extent (ESA, 2018). Moreover, environmental pollution can be efficiently controlled by Bt formulations as compared to commercial chemical insecticides. Also, this can be safely implemented in IPM practices (Mullaney, n.d.). Bt can be applied topically to crops while they are still being harvested and sold to consumers. Organic farming practitioners highly prefer Bt insecticides as they are considered natural and safe (Hellmich & Hellmich, 2012). Humans can consume this bacterium along with its Cry proteins and DNA repeatedly by feeding on uncooked vegetables, especially salads, for years without suffering from any harmful effects (Benedict & Ring, 2004). Bt transgenic crops aid in increasing crop yield and, as a result, farmer income. This leads to increased farm output and better livelihood of farmers' community (Kathage & Qaim, 2012). Through a decrease in the usage of synthetic pesticides, Bt crops also aid in pollution control. They also aid in the protection of beneficial insects (Betz et al., 2000).

4.9 DRAWBACKS

The characteristic of Bt that confers its advantage fetch its disadvantage too. Other pesticides are frequently needed to protect the crops from damage by other pests, as controlling major pests with Bt can lead to the coming up of minor ones as major. However, the usefulness of Cry toxins for commercial applications has some limitations due to their relatively higher cost, limited crop coverage, and chances of rapid environmental inactivation. In comparison to other low-cost conventional chemical insecticides, the pest control efficiency of *Bt* is relatively less. When applied in spray or liquid form, Bt is prone to degradation by sunlight or is washed away by rain (Behle et al., 1997). Most formulations are applicable for foliage and remain effective for less than a week following application (GRDC, 2023). Some of the newer strains, developed for controlling leaf beetles, lose their activity in about 24 hours. The effect of Bt proteins on the environment is rather unclear. According to the EPA, more study is required to investigate the effects of Bt protein accumulation in soils, the risks towards non-targets, and the chances of gene transfer from Bt crops to wild relatives. In 2003, a study was done at Ohio State University, where they experimentally cross-pollinated wild sunflowers with genetically modified Bt sunflowers, and the results suggested that modified genes in cultivated crops may drift into closely related populations and increase the hardiness of these plants, and we cannot avoid the risk of potential weed species (Mullaney, n.d.). Bt toxin is often sprayed on the surface of a host plant, making it less probable for insects attacking plant roots to consume it. In addition, Bt toxin can only kill a susceptible insect during a particular stage of development, making it less effective than chemical pesticides. Moreover, it costs 1.5–3 times more than the latter one. Crop productivity may decrease as a result of pests developing a resistance to the toxins produced by Bt crops. (Chapter 13 dedicatedly discusses this vital issue.) So there are many challenges to be met by insect pathologists to make Bt replace chemical control.

4.10 CONCLUSION

Bacillus thuringiensis is the brightest star in the field of entomopathogenic bacteria, both in research and commercial utilization. This organism brought revolution in the management of insect pests via non-chemical methods. Bt can exist in a variety of environments in both vegetative and sporulated forms. It can, however, interact with and colonize the body of hosts from a variety of taxa in any habitat. As a result, Bt has a great capacity for proliferation and dispersion. This is mostly because it can grow and generate spores practically everywhere, as well as passively reach many kinds of habitats. Rapid co-evolutionary processes, along with a considerable amount of genomic flexibility, transformed some of these isolates into host-interacting pathogens at different levels. We must acknowledge the increasing number of problems as a result of the widespread use of Bt, but the scientific community is coming up with innovative solutions as well and trying hard to defeat smart insects, who are also continuing their evolution against our management tactics. Further ecologically focused studies on Bt will help in establishing the bacterium as a more successful biocontrol agent.

REFERENCES

Adams, L. A., Liu, C. L., McIntosh, S. C., & Starnes, R. L. (1996). Diversity and biological activity of *Bacillus thuringiensis*. In L. G. Copping (Ed.), *Crop protection agents from nature: Natural products and analogues* (pp. 360–388). The Royal Society of Chemistry.

Anil, Kashyap, L., Goswami, T. N., Patel, V. K., & Sharma, R. K. (2017). *Bacillus thuringiensis* and insect pest management. In Md. A. Anwer (Ed.), *Biopesticides and bioagents* (pp. 331–369). Apple Academic Press. https://doi.org/10.1201/9781315365558-13

Argôlo Filho, Costa, R., & Loguercio, L. L. (2014). *Bacillus thuringiensis* is an environmental pathogen and host-specificity has developed as an adaptation to human-generated ecological niches. *Insects*, *5*(1), 62. https://doi.org/10.3390/INSECTS5010062.

Barth, H., Aktories, K., Popoff, M. R., & Stiles, B. G. (2004). Binary bacterial toxins: Biochemistry, biology, and applications of common *Clostridium* and *Bacillus* proteins. *Microbiology and Molecular Biology Review*, *68*, 373–402. https://doi.org/10.1128/mmbr.68.3.373-402.20

Beebee, T., Korner, A., & Bond, R. P. (1972). Differential inhibition of mammalian ribonucleic acid polymerases by an exotoxin from *Bacillus thuringiensis*. The direct observation of nucleoplasmic ribonucleic acid polymerase activity in intact nuclei. *Biochemistry Journal*, *127*, 619–634.

Behle, R. W., McGuire, M. R., & Shasha, B. S. (1997). Effects of sunlight and simulated rain on residual activity of *Bacillus thuringiensis* formulations. *Journal of Economic Entomology*, *90*(6), 1560–1566. https://doi.org/10.1093/JEE/90.6.1560

Benedict, J. H., & Ring, D. R. (2004). Transgenic crops expressing Bt proteins: Current status, challenges and outlook. In O. Koul, & G. S. Dhaliwal (Eds.), *Transgenic crop protection: Concepts and strategies* (pp. 15–84). Science Publishers, Inc.

Berliner, E. (1915). Ueber die Schlaffsucht der Mehlmottenraupe (*Ephestia kühniella* Zell.) und ihren Erreger *Bacillus thuringiensis* n. sp. *Zeitschrift für Angewandte Entomologie*, *2*, 29–56.

Bernhard, K., Jarrett, P., Meadows, M., Butt, J., Ellis, D. J., Roberts, G. M., Pauli, S., Rodgers, P., & Burges, H. D. (1997). Natural isolates of *Bacillus thuringiensis*: Worldwide distribution, characterization, and activity against insect pests. *Journal of Invertebrate Pathology*, *70*(1), 59–68. https://doi.org/10.1006/jipa.1997.4669

Bernheimer, A. W., & Rudy, B. (1986). Interactions between membranes and cytolytic peptides. *Biochimica et Biophysica Acta (BBA)-Reviews on Biomembranes*, *864*(1), 123–141. https://doi.org/10.1016/0304-4157(86)90018-3

Betz, F. S., Hammond, B. G., & Fuchs, R. L. (2000). Safety and advantages of *Bacillus thuringiensis*-protected plants to control insect pests. *Regulatory Toxicology and Pharmacology*, *32*(2), 156–173. https://doi.org/10.1006/RTPH.2000.1426

Bonnefoi, A., & de Barjac, H. (1963). Classification des souches du groupe *Bacillus thuringiensis* par la de´termination de l'antige'ne *flagellaire*. *Entomophaga*, *8*, 223–229.

Boucias, D. G., & Pendland, J. C. (1998). *Principles of insect pathology* (1st ed., Vol. 9). Springer. https://doi.org/10.1007/978-1-4615-4915-4

Bravo, A., Gill, S. S., & Soberón, M. (2005). *Bacillus thuringiensis* mechanisms and use. In L. I. Gilbert (Ed.), *Comprehensive molecular insect science* (Vol. 6, pp. 175–205). Elsevier. https://doi.org/10.1016/B0-44-451924-6/00081-8

Bravo, A., Gill, S. S., & Soberón, M. (2007). Mode of action of *Bacillus thuringiensis* Cry and Cyt toxins and their potential for insect control. *Toxicon*, *49*(4), 423–435. https://doi.org/10.1016/J.TOXICON.2006.11.022.

Bravo, A., Likitvivatanavong, S., Gill, S., & Soberón, M. (2011). *Bacillus thuringiensis*: A story of a successful bioinsecticide. *Insect Biochemistry and Molecular Biology*, *41*(7), 423–431. https://doi.org/10.1016/j.ibmb.2011.02.006

Burgerjon, A., Biache, G., & Cals, P. (1969). Teratology of the Colorado potato beetle, *Leptinotarsa decemlineata*, as provoked by larval administration of the thermostable toxin of *Bacillus thuringiensis*. *Journal of Invertebrate Pathology*, *14*, 274–278. https://doi.org/10.1016/0022-2011(69)90117-7

Burtseva, L. I., Burlak, V. A., Kalmikova, G. V., De Barjac, H., & Lecadet, M. M. (1995). *Bacillus thuringiensis novosibirsk* (serovar H24a24c) a new subspecies from the West Siberian plain. *Journal of Invertebrate Pathology*, *66*(1), 92–93. https://doi.org/10.1006/jipa.1995.1067

Butko, P. (2003). Cytolytic toxin Cyt1A and its mechanism of membrane damage: Data and hypotheses. *Applied Environmental Microbiology*, *69*, 2415–2422. https://doi.org/10.1128/AEM.69.5.2415–2422.2003

Cabiaux, V., Wolff, C., & Ruysschaert, J. M. (1997). Interaction with a lipid membrane: A key step in bacterial toxins virulence. *International Journal of Biological Macromolecules*, *21*(4), 285–298. https://doi.org/10.1016/S0141-8130(97)00078 0

Chanpaisaeng, J. (1995). Strain diversity of *Bacillus thuringiensis* in Thailand. *Workshop on Application of Microbial in Plant Pest Management*. Bangkok (Thailand).

Cohen, Z. P. (2015, February 15). *Bacillus thuringiensis*. Cornell University. Retrieved June 21, 2021, from https://biocontrol.entomology.cornell.edu/pathogens/bacillus.php

Crickmore, N. (2017). *Bacillus thuringiensis* toxin classification. In *Bacillus thuringiensis and Lysinibacillus sphaericus* (pp. 41–52). Springer International Publishing. https://doi.org/10.1007/978-3-319-56678-8_3

Crickmore, N., Berry, C., Panneerselvam, S., Mishra, R., Connor, T. R., & Bonning, B. C. (2021). A structure-based nomenclature for *Bacillus thuringiensis* and other bacteria-derived pesticidal proteins. *Journal of Invertebrate Pathology*, *186*, 107438. https://doi.org/10.1016/J.JIP.2020.107438

Crickmore, N., Zeigler, D. R., Feitelson, J., Schnepf, E., Van Rie, J., Lereclus, D., Baum, J., & Dean, D. H. (1998). Revision of the nomenclature for the *Bacillus thuringiensis* pesticidal crystal proteins. *Microbiology & Molecular Biology Review*, *62*(3), 807–813. https://doi.org/10.1128/mmbr.62.3.807-813.1998

Das, S. K., Pradhan, S. K., Samal, K. C., & Singh, N. R. (2021). Structural, functional, and evolutionary analysis of Cry toxins of *Bacillus thuringiensis*: An in-silico study. *Egyptian Journal of Biological Pest Control*, *31*, 44. https://doi.org/10.1186/s41938-021-00394-6

Deacon, J. (n.d.). *The microbial world: Bacillus thuringiensis*. Institute of Cell and Molecular Biology, The University of Edinburgh. Retrieved July 23, 2021, from http://archive.bio.ed.ac.uk/jdeacon/microbes/bt.htm

de Barjac, H. (1978a). Un nouveau candidat a la lutte biologique contre les moustiques: *Bacillus thuringiensis* var. *israelensis*. *Entomophaga*, *23*, 309–319. https://doi.org/10.1007/BF02373047

de Barjac, H. (1978b). A new variety of *Bacillus thuringiensis* very toxic to mosquitoes: *B. thuringiensis* var. *israelensis* serotype 14. *Compte Rendus de l'Acade´mie des Sciences de Paris*, *286D*, 797–800.

de Barjac, H., & Bonnefoi, A. (1962). Essai de classification biochimique et se´rologique de 24 souches de *Bacillus* du type *B. thuringiensis*. *Entomophaga*, *7*, 5–31.

de Barjac, H., & Bonnefoi, A. (1968). A classification of strains of *Bacillus thuringiensis* Berliner with a key to their differentiation. *Journal of Invertebrate Pathology*, *11*(3), 335–347. https://doi.org/10.1016/0022-2011(68)90182-1

de Barjac, H., & Bonnefoi, A. (1972). Presence of H-antigenic subfactors in serotype 5 of B*acillus thuringiensis* var. *canadensis*. *Journal of Invertebrate Pathology*, *20*, 212–213.

de Barjac, H., Cosmao Dumanoir, V., Shaik, R., & Viviani, G. (1977). *Bacillus thuringiensis* var. *pakistani*: une nouvelle sousespe' ce correspondant au serotype 13. *Comptes-Rendus de l'Acade ´mie Des Sciences de Paris, 284D*, 2051–2053.

de Barjac, H., & Frachon, E. (1990). Classification of *Bacillus thuringiensis* strains. *Entomophaga*, *35*, 233–240. https://doi.org/10.1007/BF02374798

de Barjac, H., & Lemille, F. (1970). Presence of antigenic subfactors in serotype 3 of *Bacillus thuringiensis*. *Journal of Invertebrate Pathology*, *15*(1), 139–140. https://doi.org/10.1016/0022-2011(70)90113-8

de Barjac, H., & Thompson, J. V. (1970). A new serotype of *Bacillus thuringiensis: B. thuringiensis* var. *thompsoni* (serotype 11). *Journal of Invertebrate Pathology*, *15*(1), 141–144. https://doi.org/10.1016/0022–2011(70)90114-X

DeLucca, A. J., II, Palmgren, M. S., & Ciegler, A. (2011). Bacillus thuringiensis in grain elevator dusts. *Canadian Journal of Microbiology*, *28*(4), 452–456. https://doi.org/10.1139/M82-068

DeLucca, A. J., II, Palmgren, M. S., & de Barjac, H. (1984). A new serovar of *Bacillus thuringiensis* from grain dusts: *Bacillus thuringiensis* var. *colmeri* (serovar H-21). *Journal of Invertebrate Pathology*, *43*(3), 437–438. https://doi.org/10.1016/0022-2011(84)90095-8

DeLucca, A. J., II, Simonson, J. L., & Larson, A. D. (1979). Two new serovars of *Bacillus thuringiensis*: Serovars *dakota* and *indiana* (serovars 15 and 16). *Journal of Invertebrate Pathology*, *34*(3), 323–324. https://doi.org/10.1016/0022-2011(79)90085-5

De Maagd, R. A., Bravo, A., Berry, C., Crickmore, N., & Schnepf, H. E. (2003). Structure, diversity, and evolution of protein toxins from spore-forming entomopathogenic bacteria. *Annual Review of Genetics*, *37*, 409–433. https://doi.org/10.1146/annurev.genet.37.110801.143042

Donovan, W. P., Engleman, J. T., Donovan, J. C., Baum, J. A., Bunkers, G. J., Chi, D. J., Clinton, W. P., English, L., Heck, G. R., Ilagan, O. M., Krasomil-Osterfeld, K. C., Pitkin, J.

W., Roberts, J. K., & Walters, M. R. (2006). Discovery and characterization of Sip1A: A novel secreted protein from *Bacillus thuringiensis* with activity against coleopteran larvae. *Applied Microbiological Biotechnology*, *72*, 713–719. https://doi.org/10.1007/s00253-006-0332-7

ESA. (2018, May). *Is Bt safe for humans to eat?* Entomological Society of America. https://www.entsoc.org/sites/default/files/files/Science-Policy/2018/ESA-Factsheet-Bt.pdf

Espinasse, S., Gohar, M., Chaufaux, J., Buisson, C., Perchat, S., & Sanchis, V. (2002). Correspondence of high levels of beta-exotoxin I and the presence of *cry1B* in *Bacillus thuringiensis*. *Applied Environmental Microbiology*, *68*, 4182–4186. https://doi.org/10.1128/AEM.68.9.4182-4186.2002

Fang, S. L., Wang, L., Guo, W., Zhang, X., Peng, D. H., Luo, C. P., Yu, Z. I., & Sun, M. (2009). *Bacillus thuringiensis* Bel protein enhances the toxicity of Cry1Ac protein to *Helicoverpa armigera* larvae by degrading insect intestinal mucin. *Applied Environmental Microbiology*, *75*(16), 5237–5243. https://doi.org/10.1128/AEM.00532-09

Farkas, J., Sebesta, K., Horska, K., Samek, Z., Dollijs, J., & Storm, F. (1969). The structure of exotoxin of *Bacillus thuringiensis* var. *gelechiae*. *Collection of Czechoslovak Chemical Communications*, *34*, 1118–1120. https://doi.org/10.1135/cccc19691118

Federici, B. A., & Bauer, L. S. (1998). Cyt1Aa protein of *Bacillus thuringiensis* is toxic to the cottonwood leaf beetle, *Chrysomela scripta*, and suppresses high levels of resistance to Cry3Aa. *Applied and Environmental Microbiology*, *64*(11), 4368–4371. https://doi.org/10.1128/AEM.64.11.4368-4371.1998

Ferrandis, M. D., Andrew, R., Porcar, M., Iriarte, J., Cosmao-Dumanoir, V., Lecadet, M. M., Caballero, P., & Ferré, J. (1999). Characterization of *Bacillus thuringiensis* serovar bolivia (serotype H63), a novel serovar isolated from the Bolivian high valleys. *Letters in Applied Microbiology*, *28*(6), 440–444. https://doi.org/10.1046/j.1365-2672.1999.00558.x

Ferré, J., & Van Rie, J. (2002). Biochemistry and genetics of insect resistance to *Bacillus thuringiensis*. *Annual Review of Entomology*, *47*, 501–533. https://doi.org/10.1146/ANNUREV.ENTO.47.091201.145234

Forsyth, G., & Logan, N. A. (2000). Isolation of *Bacillus thuringiensis* from Northern Victoria Land, Antarctica. *Letters in Applied Microbiology*, *30*(3), 263–266. https://doi.org/10.1046/J.1472-765X.2000.00706.X

Gahan, L. J., Gould, F., & Heckel, D. G. (2001). Identification of a gene associated with Bt resistance in *Heliothis virescens*. *Science*, *293*(5531), 857–860. https://doi.org/10.1126/SCIENCE.1060949

Gohar, M., & Perchat, S. (2001). Sample preparation for beta-exotoxin determination in *Bacillus thuringiensis* cultures by reversed-phase high-performance liquid chromatography. *Annals of Biochemistry*, *298*(1), 112–117. https://doi.org/10.1006/abio.2001.5373

Goldberg, L. J., & Margalit, J. (1977). A bacterial spore demonstrating rapid larvicidal activity against *Anopheles sergentii*, *Uranotaenia unguiculata*, *Culex univitattus*, *Aedes aegypti* and *Culex pipiens*. *Mosquito News*, *37*(3), 355–358.

Gómez, I., Dean, D. H., Bravo, A., & Soberón, M. (2003). Molecular basis for *Bacillus thuringiensis* Cry1Ab toxin specificity: Two structural determinants in the *Manduca sexta* Bt-R1 receptor interact with loops alpha-8 and 2 in domain II of Cy1Ab toxin. *Biochemistry*, *42*(35), 10482–10489. https://doi.org/10.1021/BI034440P.

Gómez, I., Miranda-Rios, J., Rudiño-Piñera, E., Oltean, D. I., Gill, S. S., Bravo, A., & Soberón, M. (2002a). Hydropathic complementarity determines interaction of epitope (869) HITDTNNK (876) in *Manduca sexta* Bt-R (1) receptor with loop 2 of domain II of *Bacillus thuringiensis* Cry1A toxins. *The Journal of Biological Chemistry*, *277*(33), 30137–30143. https://doi.org/10.1074/JBC.M203121200.

Gómez, I., Sánchez, J., Miranda, R., Bravo, A., & Soberón, M. (2002b). Cadherin-like receptor binding facilitates proteolytic cleavage of helix alpha-1 in domain I and oligomer pre-pore formation of *Bacillus thuringiensis* Cry1Ab toxin. *FEBS Letters*, *513*(2–3), 242–246. https://doi.org/10.1016/S0014-5793(02)02321-9

GRDC. (2023). *Bt—IPM Guidelines for Grains*. Grands Research and Development Corporation; Queensland Government. https://ipmguidelinesforgrains.com.au/ipm-information/biological-control/pathogens/bt/

Guerchicoff, A., Delecluse, A., & Rubinstein, C. P. (2001). The *Bacillus thuringiensis cyt* genes for hemolytic endotoxins constitute a gene family. *Applied Environmental Microbiology*, *67*, 1090–1096. https://doi.org/10.1128/AEM.67.3.1090–1096.2001

Heimpel, A. M., & Angus, T. A. (1958). The taxonomy of insect pathogens related to *Bacillus cereus* Frankland and Frankland. *Canadian Journal of Microbiology*, *4*(5), 531–541. https://doi.org/10.1139/m58-058

Hellmich, R. L., & Hellmich, K. A. (2012). Use and impact of Bt Maize. *Nature Education Knowledge*, *3*(10), 4.

Hernandez, C. S., Martinez, C., Porcar, M., Caballero, P., & Ferre, J. (2003). Correlation between serovars of *Bacillus thuringiensis* and type I beta-exotoxin production. *Journal of Invertebrate Pathology*, *82*(1), 57–62. https://doi.org/10.1016/S0022-2011(02)00199-4

Hofte, H., & Whiteley, H. R. (1989). Insecticidal crystal proteins of *Bacillus thuringiensis*. *Microbiological Review*, *53*(2), 242–255. https://doi.org/10.1128/mr.53.2.242–255.1989

IEBC. (1996). *Collection of Bacillus turingiensis and Bacillus sphericus (Classified by H Serotypes)* (Catalogue No.1). Intrernational Entomopathogenic Bacillus Centre, Institute Pasteur, Paris.

Ignoffo, C. M., & Gregory, B. G. (1972). Effects of *Bacillus thuringiensis* β-exotoxin on larval maturation, adult longevity, fecundity, and egg viability in several species of lepidoptera. *Environmental Entomology*, *1*(3), 269–272. https://doi.org/10.1093/ee/1.3.269

Iriarte, J., Dumanoir, V. C., Bel, Y., Porcar, M., Ferrandis, M. D., Lecadet, M. M., Ferré, J., & Caballero, P. (2000). Characterization of *Bacillus thuringiensis* ser. *balearica* (Serotype H48) and ser. *navarrensis* (Serotype H50): Two novel serovars isolated in Spain. *Current Microbiology*, *40*(1), 17–22. https://doi.org/10.1007/s002849910004

Ishii, T., & Ohba, M. (1993). Characterization of mosquito-specific *Bacillus thuringiensis* strains coisolated from a soil population. *Systematic and Applied Microbiology*, *16*(3), 494–499. https://doi.org/10.1016/S0723-2020(11)80285-7

Ishiwata, S. (1905a). Sotto bacillus study. *Dainihon Sunshi Kaiho*, *14*, 24–28.

Ishiwata, S. (1905b). Concerning 'Sotto-Kin' a Bacillus of a disease of the silkworm. *Report to Association of Sericulture. Japan*, 160–161.

Jain, D., Saharan, V., & Pareek, S. (2016). Current status of *Bacillus thuringiensis*: Insecticidal crystal proteins and transgenic crops. In J. M. Al-Khayri, S. M. Jain, & D. V. Johnson (Eds.),

Advances in plant breeding strategies: Agronomic, abiotic and biotic stress traits (Vol. 2, pp. 657–698). Springer. https://doi.org/10.1007/978-3-319-22518-0_18

Jenkins, J. L., Lee, M. K., Valaitis, A. P., Curtiss, A., & Dean, D. H. (2000). Bivalent sequential binding model of a *Bacillus thuringiensis* toxin to gypsy moth aminopeptidase N receptor. *Journal of Biological Chemistry*, *275*(19), 14423–14431. https://doi.org/10.1074/JBC.275.19.14423.

Jingyuan, D., Ling, Y., Bo, W., Xixia, L., Ziniu, Y., & Lecadet, M. M. (1996). *Bacillus thuringiensis* subspecies *huazhongensis*, serotype H40, isolated from soils in the People's Republic of China. *Letters in Applied Microbiology*, *22*(1), 42–45. https://doi.org/10.1111/j.1472-765X.1996.tb01105.x

Juarez-Perez, V. M., Jacquemard, P., & Frutos, R. (1994). Characterization of the type strain of *Bacillus thuringiensis* subsp. *cameroun* serotype H-32. *FEMS Microbiology Letters*, *122*(1–2), 43–48. https://doi.org/10.1111/j.1574-6968.1994.tb07141.x

Jurat-Fuentes, J. L., & Adang, M. J. (2004). Characterization of a Cry1Ac-Receptor alkaline phosphatase in susceptible and resistant *Heliothis virescens* larvae. *European Journal of Biochemistry*, *271*(15), 3127–3135. https://doi.org/10.1111/J.1432-1033.2004.04238.X

Kathage, J., & Qaim, M. (2012). Economic impacts and impact dynamics of Bt (*Bacillus thuringiensis*) cotton in India. *Proceedings of the National Academy of Sciences*, *109*(29), 11652–11656. https://doi.org/10.1073/pnas.1203647109

Khodyrev, V. P. (1990). *Bacillus thuringiensis* subsp. *toguchini*-A new subspecies of crystal-forming bacteria. *Izvestiia Akademii Nauk SSSR. Seriia Biologicheskaia*, *5*, 789–791.

Khyami-Horani, H., Hajaij, M., & Charles, J. F. (2003). Characterization of *Bacillus thuringiensis* ser. *jordanica* (Serotype H71), a novel serovariety isolated in Jordan. *Current Microbiology*, *47*, 26–31. https://doi.org/10.1007/s00284-002-3940-1

Kim, S. Y., Kang, M. H., Choi, H. B., Lee, J. U., Charles, J. F., Dumanoir, V. C., Lecadet, M. M., & Lee, H. H. (1999). Characteristics of Thirty-Six *Bacillus thuringiensis* Isolates and a New Serovar of *B. thuringiensis* subsp. *kim* (Serotype H52). *Journal of Microbiology and Biotechnology*, *9*(5), 534–540.

Knight, P. J. K., Crickmore, N., & Ellar, D. J. (1994). The receptor for *Bacillus thuringiensis* CrylA(c) delta-endotoxin in the brush border membrane of the lepidopteran *Manduca sexta* is aminopeptidase N. *Molecular Microbiology*, *11*(3), 429–436. https://doi.org/10.1111/J.1365-2958.1994.TB00324.X

Knowles, B. H. (1994). Mechanism of Action of *Bacillus thuringiensis* Insecticidal δ-Endotoxins. In P. D. Evans (Ed.), *Advances in insect physiology* (Vol. 24, pp. 275–308). Elsevier. https://doi.org/10.1016/S0065-2806(08)60085-5

Knowles, B. H., & Dow, J. A. T. (1993). The crystal δ-endotoxins of *Bacillus thuringiensis*: Models for their mechanism of action on the insect gut. *BioEssays*, *15*(7), 469–476. https://doi.org/10.1002/BIES.950150706

Krieg, A. (1969). Transformations in the *Bacillus cereus—Bacillus thuringiensis* group. Description of a new subspecies: *Bacillus thuringiensis* var. *toumanoffii*. *Journal of Invertebrate Pathology*, *14*(2), 279–281. https://doi.org/10.1016/0022-2011(69)90118-9

Krieg, A., de Barjac, H., & Bonnefoi, A. (1968). A new serotype of *Bacillus thuringiensis* isolated in Germany: *Bacillus thuringiensis* var. *darmstadiensis*. *Journal of Invertebrate Pathology*, *10*(2), 428–430. https://doi.org/10.1016/0022-2011(68)90104-3

Krieg, A., Huger, A. M., Langenbruch, G. A., & Schnetter, W. (1983). *Bacillus thuringiensis* var. *tenebrionis*, a new pathotype effective against larvae of Coleoptera. *Zeitschrift für angewandte Entomologie*, *96*, 500–508.

Krywienczyk, J., & Fast, P. G. (1980). Serological relationships of the crystals of *Bacillus thuringiensis* var. *israelensis*. *Journal of Invertebrate Pathology*, *36*(1), 139–140. https://doi.org/10.1016/0022-2011(80)90150-0

Lecadet, M. M., Frachon, E., Cosmao Dumanoir, V., Ripouteau, H., Hamon, S., Laurent, P., & Thiery, I. (1999). Updating the H-antigen classification of *Bacillus thuringiensis*. *Journal of Applied Microbiology*, *86*(4), 660–672. https://doi.org/10.1046/j.1365-2672.1999.00710.x

Lee, H. H., Jung, J. D., Yoon, M. S., Lee, K. K., Lecadet, M. M., Charles, J. F., Cosmao Dumanoir, V., Franchon, E., & Shim, J. C. (1995). Distribution of *Bacillus thuringiensis* in Korea. In T. Y. Feng (Ed.), *Bacillus thuringiensis biotechnology and environmental benefits* (pp. 201–215). Hua Shiang Yuan Publishing Co.

Lee, H. H., Lee, J. A., Lee, K. Y., Chung, J. D., De Barjac, H., & Charles, J. F. (1994). New serovars of *B. thuringiensis*: *B. thuringiensis* serovar *coreanensis* (serotype H-25). *B. thuringiensis* serovar *leesis* (serotype H-33) and *B. thuringiensis* serovar *konkukian* (serotype H-34). *Journal of Invertebrate Pathology*, *63*(2), 217–219. https://doi.org/10.1006/jipa.1994.1041

Lereclus, D., Lecadet, M. M., Ribier, J., & Dedonder, R. (1982). Molecular relationships among plasmids of *Bacillus thuringiensis*: Conserved sequences through 11 crystalliferous strains. *Molecular and General Genetics MGG*, *186*, 391–398. https://doi.org/10.1007/BF00729459.

Levinson, B. L., Kasyan, K. J., Chiu, S. S., Currier, T. C., & Gonzalez, J. M. (1990). Identification of beta-exotoxin production, plasmids encoding beta-exotoxin, and a new exotoxin in *Bacillus thuringiensis* by using high-performance liquid chromatography. *Journal of Bacteriology*, *172*(6), 3172–3179. https://doi.org/10.1128/jb.172.6.3172-3179.1990

Li, R., Dai, S., & Gao, M. (2000). A new serovar of *Bacillus thuringiensis*: *Bacillus thuringiensis* subsp. *sinensis*. *Virology Singapore*, *15*, 224–225.

Li, R., Gao, M., Dai, S., & Li, X. (1999). The identification of 5 new serotypes of *Bacillus thuringiensis* from soil in China. *Acta Microbiologica Sinica*, *39*, 154–159.

Liu, X. Y., Ruan, L. F., Hu, Z. F., Peng, D. H., Cao, S. Y., Yu, Z. N., Liu, Y., Zheng, J. S., & Sun, M. (2010). Genome-wide screening reveals the genetic determinants of an antibiotic insecticide in *Bacillus thuringiensis*. *Journal of Biological Chemistry*, *285*, 39191–39200. https://doi.org/10.1074/jbc.M110.148387

Liu, X. Y., Ruan, L., Peng, D., Li, L., Sun, M., & Yu, Z. (2014). Thuringiensin: A thermostable secondary metabolite from *Bacillus thuringiensis* with insecticidal activity against a wide range of insects. *Toxins*, *6*(8), 2229–2238. https://doi.org/10.3390/toxins6082229

Lonc, E., Lecadet, M. M., Lachowicz, T. M., & Panek, E. (1997). Description of *Bacillus thuringiensis wratislaviensis* (H-47), a new serotype originating from Wroclaw (Poland) and other *Bt* soil isolates from the same area. *Letters in Applied Microbiology*, *24*(6), 467–473. https://doi.org/10.1046/j.1472-765X.1997.00039.x

Lynch, M. J., & Baumann, P. (1985). Immunological comparisons of the crystal protein from strains of *Bacillus thuringiensis*. *Journal of Invertebrate Pathology*, *46*(1), 47–57. https://doi.org/10.1016/0022-2011(85)90128-4

Maeda, M., Mizuki, E., Nakamura, Y., Hatano, T., & Ohba, M. (2000). Recovery of *Bacillus thuringiensis* from marine sediments of Japan. *Current Microbiology*, *40*(6), 418–422. https://doi.org/10.1007/S002840010080

Maheswaran, S., Sreeramanan, S., Reena Josephine, C. M., Marimuthu, K., & Xavier, R. (2010). Occurrence of *Bacillus thuringiensis* in faeces of herbivorous farm animals. *African Journal of Biotechnology*, *9*(47), 8013–8019. https://doi.org/10.5897/AJB10.1253

Martin, P. A., & Travers, R. S. (1989). Worldwide abundance and distribution of *Bacillus thuringiensis* isolates. *Applied and Environmental Microbiology*, *55*(10), 2437–2442. https://doi.org/10.1128/aem.55.10.2437-2442.1989

Masson, L., Lu, Y. J., Mazza, A., Brousseau, R., & Adang, M. J. (1995). The CryIA(c) receptor purified from *Manduca sexta* displays multiple specificities. *The Journal of Biological Chemistry*, *270*(35), 20309–20315. https://doi.org/10.1074/JBC.270.35.20309.

McClintock, J., Stone, T. B., & Sjoblad, R. D. (1995). A comparative review of the mammalian toxicity of *Bacillus thuringiensis*-based pesticides. *Pesticide Science*, *45*(2), 95–105. https://doi.org/10.1002/ps.2780450202

McNall, R. J., & Adang, M. J. (2003). Identification of novel *Bacillus thuringiensis* Cry1Ac binding proteins in *Manduca sexta* midgut through proteomic analysis. *Insect Biochemistry and Molecular Biology*, *33*(10), 999–1010. https://doi.org/10.1016/S0965-1748(03)00114-0

Mendoza-Almanza, G., Esparza-Ibarra, E. L., Ayala-Luján, J. L., Mercado-Reyes, M., Godina-González, S., Hernández-Barrales, M., & Olmos-Soto, J. (2020). The cytocidal spectrum of *Bacillus thuringiensis* toxins: From insects to human cancer cells. *Toxins*, *12*(5), 1–22. https://doi.org/10.3390/toxins12050301.

Morin, S., Biggs, R. W., Sisterson, M. S., Shriver, L., Ellers-Kirk, C., Higginson, D., Holley, D., Gahan, L. J., Heckel, D. G., Carrière, Y., Dennehy, T. J., Brown, J. K., & Tabashnik, B. E. (2003). Three cadherin alleles associated with resistance to *Bacillus thuringiensis* in pink bollworm. *Proceedings of the National Academy of Sciences*, *100*(9), 5004–5009. https://doi.org/10.1073/PNAS.0831036100

Mullaney, E. G. (n.d.). *Bacillus thuringiensis*. Encyclopedia Britannica. https://www.britannica.com/science/Bacillus-thuringiensis

Norris, J. R. (1964). The classification of *Bacillus thuringiensis*. *Journal of Applied Microbiology*, *27*(3), 439–447. https://doi.org/10.1111/j.1365 2672.1964.tb05053.x

Norris, J. R., & Burges, H. D. (1963). Esterases of crystalliferous bacteria pathogenic for insects-epizootiological applications. *Journal of Insect Pathology*, *5*(4), 460.

Ohba, M., & Aizawa, K. (1979). A new subspecies of *Bacillus thuringiensis* possessing 11a, 11c flagellar antigenic structure: *B. thuringiensis* subsp. *kyushuensis*. *Journal of Invertebrate Pathology*, *33*(3), 387–388.

Ohba, M., & Aizawa, K. (1986). Frequency of Cry—spore forming *B. cereus* possessing flagellar antigens of *Bt*. *Journal of Basic Microbiology*, *26*, 185–188.

Ohba, M., & Aizawa, K. (1989). New flagellar H antigenic subfactors in *Bacillus thuringiensis* H-serotype 3 with description of two new subspecies: *B. thuringienis sumiyoshiensis* (H-serotype 3a3d) and *B. thuringiensis* subspecies *fukuokaensis* (H-serotype 3a3d3c). *Journal of Invertebrate Pathology*, *54*(2), 208–212. https://doi.org/10.1016/0022-2011(89)90030-X

Ohba, M., Aizawa, K., & Shimizu, S. (1981a). A new subspecies of *Bacillus thuringiensis* isolated in Japan: *Bacillus thuringiensis tohokuensis* (serotype H-17). *Journal of Invertebrate Pathology*, *38*, 307–309.

Ohba, M., Mizuki, E., & Uemori, A. (2009). Parasporin, a new anticancer protein group from *Bacillus thuringiensis*. *Anticancer Research*, *29*(1), 427–433.

Ohba, M., Ono, K., Aizawa, K., & Iwana, M. I. (1981b). Two new subspecies of *Bacillus thuringiensis* isolated in Japan; *B. thuringiensis* subsp. *kumamotoensis* (serotype H-18) and *B. thuringiensis* subsp. *tochigiensis* (serotype H-19). *Journal of Invertebrate Pathology*, *38*(2), 184–190. https://doi.org/10.1016/0022-2011(81)90121-X

Ohba, M., Saitoh, H., Miyamoto, K., Higuchi, K., & Mizuki, E. (1995). *Bacillus thuringiensis* serovar *higo* (flagellar serotype 44), a new serogroup with a larvicidal activity preferential for the anopheline mosquito. *Letters in Applied Microbiology*, *21*(5), 316–318. https://doi.org/10.1111/j.1472-765X.1995.tb01068.x

Orduz, S., Rojas, W., Correa, M. M., Montoya, A. E., & de Barjac, H. (1992). A new serotype of *Bacillus thuringiensis* from Colombia toxic to mosquito larvae. *Journal of Invertebrate Pathology*, *59*(1), 99–103. https://doi.org/10.1016/0022-2011(92)90118-N

Palma, L., Muñoz, D., Berry, C., Murillo, J., & Caballero, P. (2014). *Bacillus thuringiensis* toxins: An overview of their biocidal activity. *Toxins*, *6*(12), 3296. https://doi.org/10.3390/TOXINS6123296.

Pardo-López, L., Gómez, I., Rausell, C., Sanchez, J., Soberón, M., & Bravo, A. (2006). Structural changes of the Cry1Ac oligomeric pre-pore from *Bacillus thuringiensis* induced by n-acetylgalactosamine facilitates toxin membrane insertion. *Biochemistry*, *45*(34), 10329–10336. https://doi.org/10.1021/BI060297Z

Park, S. H., Koo, B. T., Shin, B. S., Choi, S. K., Jeong, Y. M., Pan, J. G., & Kim, J. I. (1997). 국내 식물시료에서 분리한 *Bacillus thuringiensis* 균주의 다양성. *Microbiology and Biotechnology Letters*, *25*(2), 159–165.

Pendleton, I. R., Bernheimer, A. W., & Grushoff, P. (1973). Purification and partial characterization of hemolysins from *Bacillus thuringiensis*. *Journal of Invertebrate Pathology*, *21*(2), 131–135. https://doi.org/10.1016/0022-2011(73)90192-4

Pendleton, I. R., & Morrison, R. B. (1967). Antigenic analysis of the digests of the crystal toxins of *Bacillus thuringiensis*. *Journal of Applied Bacteriology*, *30*(2), 402–405. https://doi.org/10.1111/j.1365-2672.1967.tb00315.x

Porcar, M., Iriarte, J., Cosmao Dumanoir, V., Ferrandis, M. D., Lecadet, M. M., Ferre, J., & Caballero, P. (1999). Identification and characterization of the new *Bacillus thuringiensis* serovars *pirenaica* (serotype H57) and *iberica* (serotype H59). *Journal of Applied Microbiology*, *87*(5), 640–648. https://doi.org/10.1046/j.1365-2672.1999.00863.x

Quesada-Moraga, E., García-Tóvar, E., Valverde-García, P., & Santiago-Alvarez, C. (2004). Isolation, geographical diversity and insecticidal activity of *Bacillus thuringiensis* from soils in Spain. *Microbiological Research*, *159*(1), 59–71. https://doi.org/10.1016/j.micres.2004.01.011

Rabinovitch, L., Fuchs de Jesus, F., Cavados, C. F., Zahner, V., Momen, H., Da Silva, M. H. L., Cosmao Dumanoir, V., Frachon, E., & Lecadet, M. M. (1995). *Bacillus thuringiensis* subsp. *oswaldocruzi* and *Bacillus thuringiensis brasiliensis*, two novel Brazilian strains which determine new serotypes H-38 and H-39 respectively. *Memorias Instutut Oswaldo-Cruz*, *90*, 41–42.

Rausell, C., Muñoz-Garay, C., Miranda-CassoLuengo, R., Gómez, I., Rudiño-Piñera, E., Soberón, M., & Bravo, A. (2003).

Tryptophan spectroscopy studies and black lipid bilayer analysis indicate that the oligomeric structure of Cry1Ab toxin from *Bacillus thuringiensis* is the membrane-insertion intermediate. *Biochemistry*, *43*(1), 166–174. https://doi.org/10.1021/BI035527D

Rausell, C., Pardo-López, L., Sánchez, J., Muñoz-Garay, C., Morera, C., Soberón, M., & Bravo, A. (2004). Unfolding events in the water-soluble monomeric Cry1Ab toxin during transition to oligomeric pre-pore and membrane-inserted pore channel. *Journal of Biological Chemistry*, *279* (53), 55168–55175. https://doi.org/10.1074/JBC.M406279200

Ren, G. X., Li, K. T., Ying, M. H., & Yi, X. M. (1975). The classification of the strains of *Bacillus thuringiensis* group. *Acta Microbiologica Sinica*, *15*, 291–305.

Rodriguez-Padilla, C., & Galan-Wong, L. J. (1989). Serotypes of *Bacillus thuringiensis* in Mexico. In *International symposium on biological control implementation* (p. 185). North American Plant Protection Organization (NAPPO).

Rodriguez-Padilla, C., Galan-Wong, L., Barjac, H. D., Roman-Calderon, E., Tamez-Guerra, R., & Dulmage, H. (1990). *Bacillus thuringiensis* subspecies *neoleonensis* serotype H-24, a new subspecies which produces a triangular crystal. *Journal of Invertebrate Pathology*, *56*(2), 280–282. https://doi.org/10.1016/0022-2011(90)90111-I

Roh, J. Y., Choi, J. Y., Li, M. S., Jin, B. R., & Je, Y. H. (2007). *Bacillus thuringiensis* as a specific, safe, and effective tool for insect pest control. *Journal of Molecular Biology*, *17*(4), 547–559.

Roh, J. Y., Liu, Q., Lee, D. W., Tao, X., Wang, Y., Shim, H. J., Choi, J. Y., Seo, J. B., Ohba, M., Mizuki, E., & Je, Y. H. (2009). *Bacillus thuringiensis* serovar *mogi* (Flagellar Serotype 3a3b3d), a novel serogroup with a mosquitocidal activity. *Journal of Invertebrate Pathology*, *102*(3), 266–268. https://doi.org/10.1016/j.jip.2009.08.013

Sanchis, V. (2011). From microbial sprays to insect-resistant transgenic plants: History of the biospesticide *Bacillus thuringiensis*: A review. *Agronomy of Sustainable Development*, *31*, 217–231. https://doi.org/10.1051/agro/2010027

Sansinenea, E. (2012). Discovery and description of *Bacillus thuringiensis*. In E. Sansinenea (Ed.), *Bacillus thuringiensis Biotechnology* (pp. 3–18). Springer.

Schnepf, E., Crickmore, N., Van Rie, J., Lereclus, D., Baum, J., Feitelson, J., Zeigler, D. R., & Dean, D. H. (1998). *Bacillus thuringiensis* and its pesticidal crystal proteins. *Microbiology and Molecular Biology Reviews*, *62*(3), 775. https://doi.org/10.1128/mmbr.62.3.775-806.1998

Schünemann, R., Knaak, N., & Fiuza, L. M. (2014). Mode of action and specificity of *Bacillus thuringiensis* toxins in the control of caterpillars and stink bugs in soybean culture. *International Scholarly Research Notices*, *2014*(1), 135675. https://doi.org/10.1155/2014/135675.

Sebesta, K., & Horska, K. (1970). Mechanism of inhibition of DNA-dependent RNA polymerase by exotoxin of *Bacillus thuringiensis*. *Biochimica et Biophysica Acta*, *209*, 357–376. https://doi.org/10.1016/0005-2787(70)90734-3

Selander, R. K., Caugant, D. A., Ochman, H., Musser, J. M., Gilmour, M. N., & Whittam, T. S. (1986). Methods of multilocus enzyme electrophoresis for bacterial population genetics and systematics. *Applied and Environmental Microbiology*, *51*(5), 873–884. https://doi.org/10.1128/aem.51.5.873-884.1986

Seleena, P., Lee, H. L., & Lecadet, M. M. (1995). A new serovar of *Bacillus thuringiensis* possessing 28a28c flagellar antigenic structure: *Bacillus thuringiensis* serovar *jegathesan*, selectively toxic against mosquito larvae. *Journal of American Mosquito Control Association*, *11*, 471–473.

Shvetsova, O. I. (1959). Méthodes biologiques de lutte contre les insectes nuisibles aux plantes. In *Monografías* (pp. 192–201). U.R.S.S.

Smith, R. A. (1987). Use of crystal serology to differentiate among varieties of *Bacillus thuringiensis*. *Journal of Invertebrate Pathology*, *50*(1), 1–8. https://doi.org/10.1016/0022-2011(87)90138-8

Smith, R. A., & Couche, G. A. (1991). The phylloplane as a source of *Bacillus thuringiensis* variants. *Applied and Environmental Microbiology*, *57*(1), 311–315. https://doi.org/10.1128/AEM.57.1.311-315.1991

Soberon, M., Lopez-Diaz, J. A., & Bravo, A. (2013). Cyt toxins produced by *Bacillus thuringiensis*: A protein fold conserved in several pathogenic microorganisms. *Peptides*, *41*, 87–93. https://doi.org/10.1016/j.peptides.2012.05.023

Steinhaus, E. A. (1942). *Catalogue of bacteria associated extracellularly with insects and ticks*. Burgess Publishing Co. https://doi.org/10.5962/bhl.title.7268

Syed, T., Askari, M., Meng, Z., Li, Y., Abid, M. A., Wei, Y., Guo, S., Liang, C., & Zhang, R. (2020). Current insights on vegetative insecticidal proteins (Vip) as Next generation pest killers. *Toxins*, *12*(8), 522. https://doi.org/10.3390/TOXINS12080522

Tabashnik, B. E., Brévault, T., & Carrière, Y. (2013). Insect resistance to Bt crops: Lessons from the first billion acres. *Nature Biotechnology*, *31*(6), 510–521. https://doi.org/10.1038/NBT.2597.

Tanada, Y., & Kaya, H. K. (1993). *Insect pathology*. Academic Press.

Torres-Quintero, M. C., Peña-Chora, G., Hernández-Velázquez, V. M., & Arenas-Sosa, I. (2015). Signs of *Bacillus thuringiensis* (Bacillales: Bacillaceae) infection in *Myzus persicae* (Hemiptera: Aphididae): Koch's ostulates. *Florida Entomologist*, *98*(2), 799–802. https://doi.org/10.1653/024.098.0264

Toumanoff, C., & Vago, C. (1951). L'agent pathoge'ne de la flacherie du Ver a' soie ende´mique dans la re´gion des Ce´vennes: *Bacillus cereus* var. *Alesti*. *Comptes Rendus de l'Acade´mie des Sciences*, *233*, 1504–1506.

UniProt. (2023). *Cry Bacillus thuringiensis in UniProtKB search (6608)*. UniProt. Retrieved September 9, 2023, from https://www.uniprot.org/uniprotkb?query=cry+bacillus+thuringiensis

Vadlamudi, R. K., Weber, E., Ji, I., Ji, T. H., & Bulla, L. A. (1995). Cloning and expression of a receptor for an insecticidal toxin of *Bacillus thuringiensis*. *Journal of Biological Chemistry*, *270*(10), 5490–5494. https://doi.org/10.1074/JBC.270.10.5490

Valaitis, A. P., Jenkins, J. L., Lee, M. K., Dean, D. H., & Garner, K. J. (2001). Isolation and partial characterization of gypsy moth BTR-270, an anionic brush border membrane glycoconjugate that binds *Bacillus thuringiensis* Cry1A toxins with high affinity. *Archives of Insect Biochemistry and Physiology*, *46*(4), 186–200. https://doi.org/10.1002/ARCH.1028.

van der Hoeven, N. (2014). *Bacillus thuringiensis toxins: Their mode of action and the potential for interaction between them*. Ecostat-Statistical Consultancy in Ecology, Ecotoxicology and Agricultural Research.

Van Frankenhuyzen, K. (2009). Insecticidal activity of *Bacillus thuringiensis* crystal proteins. *Journal of Invertebrate Pathology*, *101*, 1–16. https://doi.org/10.1016/j.jip.2009.02.009

Venter, H. J., & Bøhn, T. (2016). Interactions between Bt crops and aquatic ecosystems: A review. *Environmental Toxicology and Chemistry*, *35*(12), 2891–2902. https://doi.org/10.1002/ETC.3583

Wang, W. (1979). A new variety of *Bacillus thuringiensis*: *B. thuringiensis* var. *yunnanensis*. *Acta Microbiologica Sinica*, *19*, 117–121.

Weiser, J., & Prasertphon, S. (1984). Entomopathogenic sporeformers from soil samples of mosquito habitats in Northern Nigeria. *Zentralblatt fu¨r Microbiologie*, *139*, 49–55.

Xu, X., Yu, L., & Wu, Y. (2005). Disruption of a cadherin gene associated with resistance to Cry1Ac delta-endotoxin of *Bacillus thuringiensis* in *Helicoverpa armigera*. *Applied and Environmental Microbiology*, *71*(2), 948–954. https://doi.org/10.1128/AEM.71.2.948-954.2005

Yara, K., Kunimi, Y., & Iwahana, H. (1997). Comparative studies of growth characteristic and competitive ability in *Bacillus thuringiensis* and *Bacillus cereus* in soil. *Applied Entomology and Zoology*, *32*(4), 625–634. https://doi.org/10.1303/AEZ.32.625

Ying, W., Jie, W., & Xichang, F. (1986). A new serovar of *Bacillus thuringiensis*. *Acta Microbiologica Sinica*, *26*(1), 1–6.

Yu, X., Liu, T., Sun, Z., Guan, P., Zhu, J., Wang, S., Li, S., Deng, Q., Wang, L., Zheng, A., & Li, P. (2012). Co-expression and synergism analysis of Vip3Aa29 and Cyt2Aa3 insecticidal proteins from *Bacillus thuringiensis*. *Current Microbiology*, *64*, 326–331. https://doi.org/10.1007/s00284-011-0070-7

Yu, Z. N., Dai, J. Y., Zhou, H. B., & Dong, Z. R. (1984). A new serotype of *Bacillus thuringiensis*. *Acta Microbiologica Sinica*, *24*(2), 117–121.

Zhao, D., Guo, W., Sun, W., Xu, D., & Liu, D. (2011). Identification of a novel enhancin-like gene from *Bacillus thuringiensis*. *Frontiers of Agriculture in China*, *5*, 423–429. https://doi.org/10.1007/s11703-011-1117-z

5 Phylum Firmicutes
Lysinibacillus sphaericus, an Overview

ABSTRACT

The entomopathogenic bacterium that is gaining attention due to its growing significance is *Lysinibacillus sphaericus* (Ls). Ls has a relatively narrow host range, showing toxicity mainly against mosquitoes. Its effectiveness significantly varies for different groups of mosquitoes. Highly toxic strains of Ls, like 1593, 2362, and C3-41, show a high level of toxicity against *Culex* mosquitoes, with little or no effectiveness against *Aedes* spp. This bacterium produces a variety of toxins, such as binary, Mtx, sphaericolysin, Cry48/Cry49, and S-layer protein, with the binary toxin being primarily responsible for its mosquitocidal activity. This binary toxin shows its activity through a sequence of steps that include solubilization and activation, receptor binding, and pore formation in midgut epithelium and, thus, finally shows physiological changes in intoxicated mosquito larvae. Though this bacterium has several advantages, there are still certain limitations, such as the development of resistance in mosquitoes, safety issues against non-target organisms, etc.; seek more future studies to develop it as a novel, effective, and environmentally safe biopesticide against a range of insect pests and disease vectors.

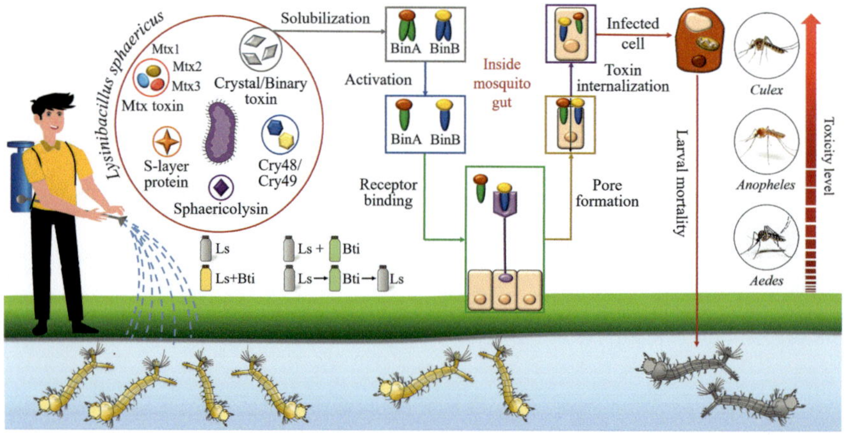

5.1 INTRODUCTION

After *Bacillus thuringiensis* Berliner (Bt), another entomopathogenic bacterium that is frequently reported is *Lysinibacillus sphaericus* (Ls). This is a spore-forming, Gram-positive bacteria that was originally described by Meyer and Neide and named *Bacillus sphaericus* Meyer and Neide (Neide, 1904) but later reassigned to the new genus *Lysinibacillus* on the basis of phenotypic characteristics, like production of peptidoglycans containing lysine and aspartic acid (Ahmed et al., 2007). Ls has a worldwide distribution and can be readily isolated from soils (Boucias & Pendland, 1998) and is also waterborne (Becker & Lüthy, 2017).

Ls has potential applicability in a variety of fields (Bustos et al., 2018; Farrugia et al., 2019; Go et al., 2014; Gou et al., 2015; Marei et al., 2018; Naureen et al., 2017; Shabanamol et al., 2017, 2018; Vargas & Dussán, 2016; Vega-Páez et al., 2019) (Figure 5.1). However, its contribution as a bioagent for controlling mosquitoes is enormous and is also the focus of this chapter. Kellen et al. (1965) first initiated the characterization of this species as a mosquito control agent after isolating this bacterium from the fourth instar larval cadaver of the fishpond mosquito, *Culiseta incidens* (Thomson), near Fresno, California. Ls is a heterogeneous group of bacteria consisting of hundreds of both non-pathogenic and pathogenic strains. However, irrespective of pathogenicity, all the strains possess some common features, like strictly aerobic, mesophilic, rod-shaped spore formers, producing spherical terminal endospores, and the ability to metabolize various amino and organic acids as carbon sources, but do not readily metabolize sugars (Berry, 2012). Certain strains of this bacteria are able to produce highly toxic mosquitocidal proteins, due to which different commercial products based on Ls gained a huge popularity and are widely utilized as successful biological control agents against the larvae of mosquitoes and blackflies since the 1980s (Silva-Filha et al., 2021).

DOI: 10.1201/9781003426882-5

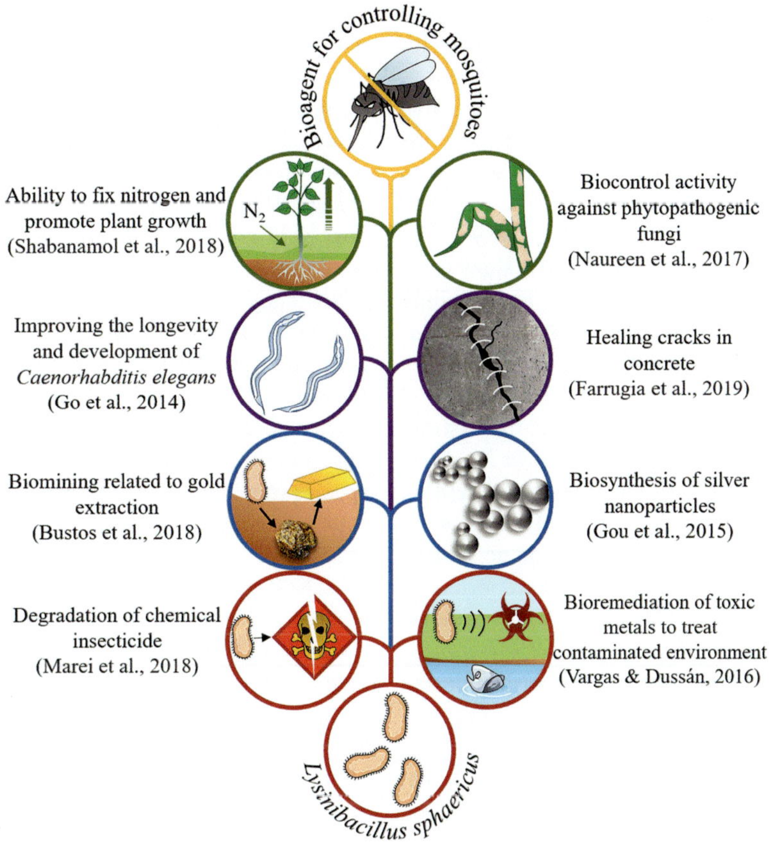

FIGURE 5.1 Potential fields of application of *Lysinibacillus sphaericus* for human welfare.

5.2 HOST RANGE

The host range of Ls is relatively narrow and mainly shows toxicity against mosquitoes. Strains with high toxicity are utilized in mosquito control programmes to reduce the population of disease transmitting mosquito vectors and thus minimize the occurrence of different mosquito-borne tropical diseases, like malaria, dengue, filariasis, yellow fever, West Nile viral disease, etc. (Hu et al., 2008a). In the earliest work, Kellen et al. (1965) tested the susceptibility of ten species of mosquito larvae belonging to *Aedes*, *Culex*, *Culiseta*, and *Orthropodomyia* genera under laboratory conditions, where all the species showed susceptibility to that bacterial strain, with *Culex* larvae as the most susceptible one. However, in the field test, mortality has occurred only for *Culex* larvae. Under laboratory and field conditions, toxicity of Ls has been evaluated for about 40 mosquito species under 8 genera (*Aedes* [*Ae.*], *Anopheles* [*An.*], *Armigeres*, *Culex* [Cx.], *Mansonia*, *Psorophora*, *Toxorhynchites*, and *Uranotaenia*) (Yuan, 2002). The strains with higher toxicity show strong activity against *Culex*, *Psorophora*, and *Mansonia* spp. (Cheong & Yap, 1985; Ramoska et al., 1977; Yuan, 2002), whereas their activity is medium against *Anopheles* spp., and it shows little or no activity against *Aedes* spp. and *Toxorhynchites* spp. (Lacey, 1983; Lacey et al., 1988; Yuan, 2002). *Toxorhynchites* are important predators of larvae of various mosquito species, which can significantly alter the composition of the mosquito community (McMillan et al., 2022) and be mass-reared and released as a successful biocontrol agent in different mosquito control programmes (Vinogradov et al., 2022). Such low toxicity of Ls to *Toxorhynchites* makes it amenable for integrated vector control. In addition, Ls was also reported to reduce the oviposition rate and kill the adult of *Cx. quinquefasciatus* Say (Zahiri & Mulla, 2005), which ultimately reduced the chances of disease transmission.

Other than *Culex*, different medically important *Anopheles* species, such as *An. gambiae* Giles, *An. stephensi* Liston, *An. albimanus* Weide., *An. quadrimaculatus* Say, *An. darling* Root, and *An. nuneztovari* Gabaldón, were reported to show susceptibility to Ls toxins (Silva-Filha et al., 2014). Ls was reported to reduce the potentiality of *An. quadrimaculatus*, which is a potent vector of rodent malaria parasite *Plasmodium berghei* Vincke and Lips (Young et al., 1990). As mentioned earlier, Ls shows little or no effectiveness against the genus *Aedes*. However, *Ae. vexans* (Meigen) was reported to be controlled by Ls. Moreover, mosquito species belonging to the subgenus *Ochlerotatus* (under the genus *Aedes*) (Savage & Strickman, 2004), such as *O. atropalpus* (Coquillett), *O. fitchii* (Felt and Young), *O. intrudens* (Dyar), *O. nigromaculis* (Ludlow), and *O. stimulans* (Walker), were also reported to be controlled by Ls (Mulligan III et al., 1978; Wraight et al., 1987). So the generalist concept that Ls is not a good component for controlling *Aedes* mosquitoes is not always true. Moreover, susceptibility depends on the mosquito species, and that is why there is variation in susceptibility within the same genus (Berry et al., 1993). Toxicity levels also vary with the mosquitocidal serotype, and even there can be variation within the same serotype (Thiery & de Barjac, 1989).

Highly toxic strains of Ls were also reported to show varying levels of toxicity and reduce the fecundity and hatching ability of different sandfly species, like *Phlebotomus papatasi* (Scopoli) (strain 1593, Pener & Wilamowski, 1996; and 2362, Wahba, 2000), *P. duboscqi* Neveu-Lemaire and *Sergentomyia schwetzi* (Adler, Theodor & Parrot) (strain 2362) (Robert et al., 1997, 1998), and *Lutzomyia longipalpis* (Lutz & Neiva) (strain 2362) (Wermelinger et al., 2000). Different commercial formulations based on Ls strains are also effective against insects like blackflies and non-biting midges (Ruiu, 2015). Lethal effects were also reported on eggs of nematode *Trichostrongylus colubriformis* (Giles) (Bone & Tinelli, 1987) and grass shrimp, *Palaemonetes pugio* Holthuis (Key & Scott, 1992). Ls is also reported to have sub-toxic effects against the water scorpion, *Laccotrephes griseus* (Guer.), a predator of mosquito larvae (Mathavan et al., 1987). Ls also produces certain toxins that enable it to kill the larvae of *Chironomus riparius* Meigen (Partridge & Berry, 2002); German cockroach, *Blattella germanica* (Linnaeus); and cutworm, *Spodoptera litura* (Fabricius) (Nishiwaki et al., 2007).

5.3 STRAINS AND SEROTYPES

This bacterium is a heterogeneous species, and for its classification, numerous methods have been employed. Krych et al. (1980) identified five homology groups (I to V) by examining the DNA homology of Ls strains. The homology group I consists of ATCC 14577, a type of strain of Ls, along with three other strains. This group has less relation with other homology groups. Homology group II is again subdivided into two subgroups, namely, IIA and IIB. Among them, subgroup IIA (having over 79% homology relation) contains all the strains with pathogenicity towards mosquito larvae (Krych et al., 1980), except strain A3-2, which is non-toxic to mosquitoes but toxic to German cockroach and the common cutworm (Nishiwaki et al., 2007). The strains of this subgroup can be easily distinguished from the strains of all other homology groups due to its mosquito pathogenicity. Studies revealed that the largest number of strains comes under homology group III, while homology group V contains the minimum number of strains. There is a close relation between homology groups III and IV, which creates difficulties while distinguishing the strain from these groups (Krych et al., 1980). Ls has also been classified based on two other systems, namely, bacteriophage typing (Yousten, 1984; Yousten et al., 1980) and serotyping using flagellar antigens (de Barjac et al., 1985; de Barjac, 1990). From both these methods, a similar grouping yielded for the strains of the IIA subgroup.

The relatively low levels of homology between the groups might represent a distinct species, but due to a lack of convenient tests to distinguish them, all remained designated as Ls (Alexander & Priest, 1990). That is why some other techniques, like numerical classification on the basis of taxonomy of phenotypic features (Alexander & Priest, 1990; Guerineau et al., 1991), analysis of cellular fatty acid composition (Frachon et al., 1991), ribotyping (de Muro et al., 1992), and random amplified polymorphic DNA analysis (Woodburn et al., 1995), have been developed, indicating most of the pathogenic strains have recovered from few groups. Phylogenetic analyses based on 16S ribosomal DNA and lipid profile comparison support the existence of seven subgroups within *Ls sensu lato*, including *Ls* (*sensu stricto*), *L. fusiformis*, and four possibly new species (Nakamura, 2000). Among these seven subgroups, four are similar to those of previous descriptions by Krych et al. (1980). Subgroup IIB is reclassified under *L. fusiformis* (Nakamura, 2000). Later, group V and group VI are again reclassified as different species, namely, *B. pycnus* sp. *nov*. and *B. neidei* sp. *nov*., respectively (Nakamura et al., 2002).

Some common entomopathogenic strains of Ls include Kellen (K), SSII-1, 1593, 2362, 2297, etc. (Porter et al., 1993). The first mosquitocidal Ls strain discovered by Kellen et al. (1965) was named strain K. Though this strain possesses a very low level of toxicity and has gained very little interest in its development as a potential biocontrol agent (Silva-Filha et al., 2014), this discovery still opens up a new area of investigation on the pathogenic potentiality of this bacterium. As a consequence, the first strain of Ls, that is, strain SSII-1, with 10,000 times more potential toxicity to mosquitoes than strain K was isolated in India in 1973 (Singer, 1973). This initiated the search for more active mosquitocidal strains, and as a result, several new strains have been discovered later on. Since then, various mosquito larvicidal strains with significantly higher toxicity have been discovered all over the world. Among them, strain 1593, originating from Indonesia (Charles et al., 1996); 2297, from Sri Lanka (Wickremesinghe & Mendis, 1980); and 2362, from Nigeria (Weiser, 1984), showed the highest degree of toxicity against mosquitoes. Among them, strain 2362 earned more attention than others as it is capable of producing binary larvicidal toxin even at higher temperatures and became an important factor for its production in most tropical areas (Lee et al., 1998). In 1987, from a mosquito breeding site in China, another strain, C3-41, was isolated, showing a high level of toxicity to different species of mosquitoes, specifically the *Culex* spp., as compared to the commercialized strain 2362 (Zhang et al., 1987). Even in many cities of China, the formulation of this strain is considered the sole larvicidal agent for treating mosquito breeding sites under integrated mosquito control programmes (Hu et al., 2008a). A comparative genome analysis of various Ls strains having different levels of toxicity showed that the strains with high toxicity possess strong syntenic relationships and share a 'chromosome backbone' from a common ancestor, with ~4,470–4,701 number of predicted genes (Silva-Filha et al., 2021).

Based on flagella agglutination, all the strains of Ls have been differentiated into 49 serotypes. The relationship between sporulation, formation of crystal, and mosquito larvicidal activity was clearly established for strains belonging to the serotypes H5a5b (Broadwell & Baumann, 1986; Myers et al., 1979), H25 (de Barjac & Charles, 1983; Kalfon et al., 1984; Yousten & Davidson, 1982), and H6 (de Barjac et al., 1988). Toxicity and serotype were found to be correlated, where highly toxic strains were classified under these three serotypes, and strains under serotypes 1a, 2a2b, and 9a9c were reported to have a low level of toxicity (Priest et al., 1997). However, after the discovery of mosquitocidal strains under serotypes H3, H6, and H48 from Ghana (Thiery et al., 1992), the generalization of the correlation between serotype and toxicity became irrelevant and may not be that straightforward. Among the 49 serotypes, most are reported to be non-entomopathogenic, whereas only 9 serotypes (H1, H2, H3, H5, H6, H9, H25, H26, and H48) express insecticidal activity against mosquito larvae (Yuan, 2002). Information on these serotypes is presented in Table 5.1. By hybridizing the specific gene probes to chromosomal DNA, it was found that all the examined strains of serotype 5a5b contained binary toxin (*btx*) genes (Priest et al., 1997). The *btx* genes were also present in some strains of the serotype 1a, 3, 6, 25, and 48. The mosquitocidal toxin (*mtx*) genes were found in the strains of serotypes 1a, 2a2b, 5a5b, 6, 9a9c, 25, and 48. But the strains of serotypes 3, 6, and 26a26b were lacking in both toxin genes (Priest et al., 1997).

5.4 TOXINS

Generally, two kinds of larvicidal toxins are produced by the toxic strains of Ls. The first one is binary toxin (Btx), produced by some highly active strains and related to the crystal toxins of Bt. The second one is Mtx toxin, which is present in weakly active strains. Both of these toxins have differences in their composition and the time of their synthesis (Charles et al., 1996). Apart from that, some strains of Ls are capable of producing certain other toxins, like sphaericolysin, Cry48/Cry49 toxin, and S-layer protein.

5.4.1 Binary Toxin (Btx/Bin) or Crystal Toxin

This is the binary toxin (Bin), which is responsible for the mosquitocidal activity of Ls and is the primary active ingredient of different commercial products based on the highly toxic Ls strains, like 1593, 2362, C3-41, etc. (Lacey, 2007; Regis et al., 2001). At the time of stage III sporulation, binary toxin is synthesized in an equimolar proportion and assembled in crystal forms (Baumann et al., 1985). This toxin is a heteromorphic protein consisting of 41.9~42 kDa (kilodalton) (BinA; renamed Tppl) and 51.4~51 kDa (BinB; renamedTpp2) active peptides containing 370 and 448 amino acid residues and encoded by *binA* (1,113 bp) and *binB* (1,347 bp) genes, respectively (Baumann et al., 1988; Charles et al., 1996; Hindley & Berry, 1987). Based on the partial DNA sequence of *bin* genes, Priest et al. (1997) classified Bin toxins into four types, namely, Bin1, Bin2, Bin3, and Bin4. Here they classified the *Bin* genes from strains of serotypes H1, H3, H6, and H48 as type 1 (e.g. strain IAB59); serotypes H5a5b (e.g. strain 1593, 2362) and H25 (e.g. strain 2297) as type 2 and type 3, respectively; and from strain LP1-G of serotype 3 as type 4 (Priest et al., 1997). Among them, Bin1 and Bin2 are highly toxic due to their high binding affinity in the larval midgut (Silva-Filha et al., 2004) and are analyzed in most of the studies to evaluate the mode of action (MoA) of the Bin toxin. Berry and Crickmore (2017) recently classified insecticidal proteins based on their structure, classifying BinA and BinB toxins as 'Beta sheet toxin' and placing them in the 'Toxin_10 family'.

For binary toxin, the toxic action requires both BinA and BinB components (Charles et al., 1996), which originated from the same ancestor and share identical areas (Baumann et al., 1988). Both of these toxin components are co-transcribed from a single operon, prior to the end of exponential growth and sporulation toxin, in the form of crystals (Ahmed et al., 1995; Baumann & Baumann, 1989). Strains that are inhibited in the early stage of sporulation cannot accumulate toxic crystals (Charles et al., 1988; El-Bendary et al., 2005). Genes responsible for expressing these toxins are chromosomal, at least in the Ls strains 1593, 2297, and 2362, although plasmids are present in these strains (de Muro et al., 1992). Assessment through DNA hybridization showed that all the highly toxic strains had a similar sequence of toxins (de Muro & Priest, 1994). Some other members of the Bin toxin family (Toxin_10 family) include Cry49 toxin from Ls (mentioned later), Cry35 and Cry36 toxins from Bt, and a putative protein from *Chlorobium phaeobacteroides* Pfennig (Berry & Crickmore, 2017; Jones et al., 2007).

Subcloning experiments revealed that the BinA protein can show toxicity by itself. However, its activity is weaker compared to crystals containing both BinA and BinB proteins (Nicolas et al., 1993). Conversely, the BinB protein does not show toxicity alone; however, its presence may augment the mosquitocidal activity of the BinA protein, suggesting the synergy between these two subunits (Broadwell et al., 1990a; Nicolas et al., 1993). In vivo, slow conversion of BinA occurs, leading to the formation of a ~39 kDa stable protein. In contrast, the BinB converts rapidly to form a stable ~43 kDa fragment (Baumann et al., 1985; Broadwell & Baumann, 1987). However, in vitro assay suggests no synergy between these protein components (Baumann & Baumann, 1991). In vitro, minimal active fragments from both the proteins have been delineated, which actually correspond to activated fragments (Broadwell et al., 1990b; Clark & Baumann, 1990; Šebo et al., 1990). When larvae ingest the Bin toxin crystal, it becomes solubilized in the alkaline gut of the larvae and gets activated by proteolytic cleavage into derivatives having lower molecular weight (Aly et al., 1989; Broadwell & Baumann, 1987; Davidson et al., 1990). This proteolytic activation increases the toxicity (Broadwell & Baumann, 1987)

TABLE 5.1
Details of Some Toxic and Some Non-Toxic Strains of *Lysinibacillus sphaericus*

				Toxic Genes[c]							
Strain	**Origin**	**Source[a]**	**Serotype[b]**	***BinA***	***BinB***	***Mtx1***	***Mtx2***	***Mtx3***	***Cry48Aa***	***Cry49Aa***	**Toxicity[d]**
Kellen K	USA	*Culiseta incidens*	1	–	–	+	–	–	–	–	L
Bs–197	India	Nd	1	+	+	+	+	+	–	–	M
Kellen Q	USA	*Culiseta incidens*	1	–	–	+	+	+	–	–	L
NRS1693	USA	Nd	2	–	–	–	+	–	–	–	N
SSII–1	India	*Culex fatigans*	2	–	–	+	+	+	–	–	L
IAB881	Ghana	*Culex* breeding site	3	+	+	–	–	+	–	–	H
LP1–G	Singapore	Breeding site	3	+	+	–	–	+	+	+	H
Dak614	France	Nd	4	–	–	–	–	–	–	–	N
C3–41	China	Breeding site	5	+	+	+	+	+	–	–	H
2362	Nigeria	*Simulium damnosum*	5	+	+	+	+	+	–	–	H
1593	Indonesia	*Culex fatigans*	5	+	+	+	+	+	–	–	H
2317–2	USA	Nd	5	+	+	+	+	+	–	–	H
Bs44–4	China	Nd	5	+	+	+	+	+	–	–	H
Bs77–5	China	Nd	5	+	+	+	+	+	–	–	H
Bs90–2	China	Nd	5	+	+	+	+	+	–	–	M
Bs91–5	China	Nd	5	+	+	+	+	+	–	–	H
Bs150–3	China	Nd	5	+	+	+	+	+	–	–	M
Bs356–11	China	Nd	5	+	+	+	+	+	–	–	H
Bs436–8	China	Nd	5	+	+	+	+	+	–	–	H
S–35	China	Nd	5	+	+	+	+	+	–	–	H
S–70	China	Nd	5	+	+	+	+	+	–	–	H
S–128	China	Nd	5	+	+	+	+	+	–	–	H
Bs117–3	China	Nd	5	+	+	–	–	+	–	–	M
2317–3	Thailand	Nd	5	+	+	+	+	+	–	–	H
1691	El Salvador	*Anopheles albimanus*	5	+	+	+	+	+	–	–	H
47–6b	China	Nd	6	+	+	+	+	+	+	+	H
IAB59	Ghana	*Anopheles gambiae*	6	+	+	+	+	+	+	+	H
IAB763	Ghana	Nd	6	+	+	+	+	+	+	+	H
IAB769	Ghana	Nd	6	+	+	+	+	+	+	+	H
2314–2	Thailand	Nd	9	–	–	–	–	–	–	–	N
Cok31	Turkey	Breeding site	9	–	–	+	+	+	–	–	L
2297 (MR4)	Sri Lanka	Coconut husk pit	25	+	+	+	+	+	–	–	H
2173 (ISPC5)	India	*Culex fatigans*	26	–	–	–	–	+	+	+	M
IAB872	Ghana	*Culex* breeding site	48	+	+	+	+	+	–	–	H
ATCC 14577d	USA	Nd	48	–	–	–	–	–	–	–	N
Bs208–6	China	Nd	ND	–	–	–	–	–	–	–	N
Bs225–2	China	Nd	ND	–	–	–	–	–	–	–	N
SO7001	France	Nd	ND	–	–	–	+	–	–	–	N
Bs227–3	China	Nd	ND	–	–	–	–	–	–	–	N

Note:

[a] Nd = no data available.

[b] Serotype is determined based on de Barjac et al. (1985); ND = not determined.

[c] +, present; –, absent.

[d] Mosquitocidal activity based on lethal concentration (LC_{50}) on fourth instar *Cx. pipiens* Linnaeus larvae: L ($LC_{50} \geq 1 \times 10^{-3}$ sporulation culture dilution), M ($LC_{50} \sim 1 \times 10^{-4}$), H ($LC_{50} \leq 1 \times 10^{-6}$), and N (non-toxic).

Source: Taken and modified from Charles et al. (1996); Ge et al. (2011).

and occurs in both susceptible and non-susceptible insects (Nicolas et al., 1990). The detailed mechanism is described in later part. BinA components can withstand a deletion of 17 amino acids from both C- and N-terminus without any loss in toxicity (Broadwell et al., 1990b; Oei et al., 1990), whereas for BinB, 53 and 32 amino acids from C- and N-terminus of BinB, respectively, are not essential for toxicity (Clark & Baumann, 1990).

5.4.2 Mosquitocidal (Mtx) Toxin

The second toxin produced by Ls during the vegetative growth period is a mosquitocidal toxin (Mtx) (Charles et al., 1996). Generally, Btx toxin is synthesized by the highly toxic strains of Ls, which may also contain one or more Mtx toxins. In contrast, less toxic strains are able to synthesize only Mtx toxins (El-Bendary, 2006). Basically, the Mtx toxin is of three types, including Mtx1 (also referred to as Mtx) with ~100 kDa (100.557 kDa) molecular mass (Thanabalu et al., 1991), Mtx2 with 31.8 kDa molecular mass (Thanabalu & Porter, 1996), and Mtx3 having a molecular mass of 35.8 kDa (Liu et al., 1996). Recently, another toxin has been added to this group, that is, Mtx4, but its activity has yet not been assessed so far (Berry, 2012). Both Mtx and Mtx2 proteins possess short N-terminal leader sequences, a characteristic feature of Gram-positive bacterial signal peptides (Thanabalu et al., 1991; Thanabalu & Porter, 1995), whereas Mtx3 lacks a putative leader sequence (Liu et al., 1996). The Mtx toxin is produced as a single chain containing 870 amino acid residues (Thanabalu et al., 1991), whereas Mtx2 contains 292 amino acid sequences (Thanabalu & Porter, 1996) and Mtx3 contains 326 amino acid sequences (Liu et al., 1996). Among these three types, Mtx is well characterized.

Out of the Mtx protein's 870 amino acid residues, 29 at the N-terminus function as a signal peptide (Thanabalu et al., 1991). Deletion of the putative signal sequence results in a 97 kDa form of Mtx toxin (Mtx21) in strain SSII-1, having a high degree of toxicity against the larvae of *Cx. quinquefasciatus* (LC_{50} ~15 ng/mL), comparable to that of binary toxins of highly toxic strains (Thanabalu et al., 1992). Due to the function of the crude gut extract of mosquito larvae, and by trypsin or chymotrypsin digestion, the 97 kDa protein is further processed into two fragments, an N-terminal, ~27 kDa fragment, and a C-terminal, ~70 kDa fragment (Schirmer et al., 2002; Thanabalu et al., 1992). The ~70 kDa fragment contains three repeated regions having 90 amino acids each. The ~27 kDa fragment contains a short region that is analogous to a transmembrane domain (Charles et al., 1996). Both of the fragments are required for mosquitocidal activity (Thanabalu et al., 1993). This ~27 kDa fragment displays distinct similarity in sequence with the bacterial ADP-ribosyltransferases enzyme domains, including the A subunits of cholera toxin, the S1 subunit of pertussis toxin, and the heat-labile toxin of *Escherichia coli* (Migula) (Thanabalu et al., 1991). Fusion proteins produced in *E. coli* are used to evaluate the toxicity of the Mtx toxin (Thanabalu et al., 1992).

The Mtx2 and Mtx3 toxins share no similarities with Mtx toxin, crystal toxins, or any other insecticidal toxins (Liu et al., 1996; Thanabalu et al., 1991, 1992). However, similarity is reported between Mtx2 and Mtx3 proteins, where Mtx3 exhibits 38% similarity to the Mtx2 over a stretch of 135 amino acids (Liu et al., 1996). The Mtx2 displays similarity to two toxins, which show activity against mammalian cells. This includes 34% similarity to the ε-toxin from *Clostridium perfringens* (Veillon and Zuber) and 29% similarity to the cytotoxin from *Pseudomonas aeruginosa* (Schröter) (Thanabalu & Porter, 1995). Similar to Mtx2, Mtx3 also shows 30% identity to the ε-toxin from *C. perfringens* and 27% identity to the cytotoxin from *P. aeruginosa* over a stretch of 144 and 63 amino acids, respectively (Liu et al., 1996).

5.4.3 Sphaericolysin

Nishiwaki et al. (2007) isolated the Ls strain A3-2 from the crop of antlion, *Myrmeleon bore* (Tjeder). Despite belonging to DNA homology subgroup IIA, this strain is unable to kill mosquitoes because it lacks mosquitocidal (*bin*, *mtx*) genes. Instead, it possesses an unknown insecticidal protein containing the undecapeptide motif ECTGLAWEWWR related to the C-terminal sequence seen in cholesterol-dependent cytolysins (CDCs) (Nishiwaki et al., 2007). CDCs are a family of pore-forming proteins (exotoxins) produced by Gram-positive bacteria, like *Arcanobacterium*, *Brevibacillus*, *Clostridium*, *Paenibacillus*, *Listeria*, and *Streptococcus* (Tweten, 2005). Several CDCs have also been described under the genus *Bacillus*, like thuringiolysin from Bt, cereolysin from *B. cereus* Frankland and Frankland, anthrolysin from *B. anthracis* Cohn, etc. (From et al., 2008). Likewise, the insecticidal toxin expressed by the A3-2 strain is an addition to the list of CDCs with similar activity (Nishiwaki et al., 2007). As the name suggests, CDCs require the presence of cholesterol in target membrane for the formation of pore (exception intermedilysin from *Streptococcus intermedius* Prévot and pyolysin from *Arcanobacterium pyogenes* [Glage]) (From et al., 2008). This toxin from the A3-2 strain is a ~53 kDa (53.03717 kDa) protein called sphaericolysin, consisting of 478 amino acid residues and formed by the cleavage of the N-terminal sequence of a ~56 kDa (56.11899 kDa) precursor protein. It exhibits rapid toxicity to *B. germanica* at a dose of 100 ng/insect within 5 minutes and slow toxicity to *S. litura* at a dose of 500 ng/insect after 16 hours (Nishiwaki et al., 2007).

Another Ls strain, B354, was isolated from water sources in Norway (From et al., 2005), showing cytotoxicity to Vero cell epithelia, and was identified under DNA homology group V (From et al., 2008). This strain also lacks mosquitocidal genes but is able to produce a 51 kDa protein with high amino acid sequence similarity (91.4%) to sphaericolysin of strain A3-2, followed by other CDCs, like cereolysin O, anthrolysin, and thuringiolysin (From et al., 2008). The genome sequence of the C3-41 strain also has sphaericolysin (Hu et al., 2008a). Even the type strain 1593 also possesses the sphaericolysin encoding gene (Nishiwaki et al., 2007), indicating the wide presence of the sphaericolysin encoding gene in different Ls strains.

Sphaericolysin has a pore-forming action, and tyrosine 159 of sphaericolysin plays a vital role in this aspect by inducing haemolysis. This toxin shows haemolytic action against the erythrocytes, resulting in the formation of many pores on the erythrocyte membrane. Thus, the toxin strongly damages the neurons, which become lysed and get dissociated from the thoracic ganglia of *B. germanica*. Sphaericolysin-induced damage in the ganglia of *B. germanica* is much higher than that of *S. litura*, indicating higher sensitivity of the cockroaches towards the sphaericolysin compared to the cutworms. Though the nervous system of insects is the possible target site of the sphaericolysin, it may also attack other parts due to the wide distribution of cholesterol in different insect tissues. However, the insecticidal activity of sphaericolysin abolishes when co-injected with cholesterol. Additionally, the Tryl59Ala (or Y159A) mutation, in which tyrosine 159 is substituted with alanine, also significantly reduces sphaericolysin's insecticidal activity (minimum paralysis dose for cockroach >860 ng/insect) (Nishiwaki et al., 2007).

According to many reports, Ls is unique to insects, especially mosquitoes, and has no negative effects on non-target organisms (Brown et al., 2004; Lacey, 2007; Tietze et al., 1993). However, while considering sphaericolysin-producing Ls strains, the major concern is that it is not only toxic to insect neurons but can also show toxicity to non-target organisms. The toxins under the CDC family were found to have haemolytic activity against erythrocytes of humans, cows, horses, sheep, and many other organisms with higher toxicity against human cells, forming pores in planar lipid bilayers. This makes the sphaericolysin lethal to humans and other vertebrates or invertebrates (Nishiwaki et al., 2007; From et al., 2008). So further studies need to be carried out before implementing these toxins for commercial utilization.

5.4.4 Cry48/Cry49 Toxin

During sporulation, some Ls strains produce a set of two-component Cry toxin, that is, Cry48 and Cry49 (renamed as Tpp49). These two toxins were characterized in the strain IAB59. Cry48 and Cry49 proteins have a molecular mass of 135 kDa and 53 kDa, respectively (Jones et al., 2007). Cry48Aa is the first reported three-domain Cry protein family of Ls, having ~33% similarity with the mosquitocidal Cry4Aa1 and Cry4Ba1 toxins from *Bt* var. *israelensis* (Bti). Despite this similarity, Cry48Aa is unable to produce toxic effects alone. Conversely, Cry49Aa is a Bin-like protein, having 30% pairwise similarity with each Bin protein (Jones et al., 2007) and related to its relatives from the Toxin_10 family, that is, Cry36 (Rupar et al., 2003) (34% pairwise identity; Jones et al., 2007) and Cry35 (Baum et al., 2004; Ellis et al., 2002) (24% pairwise identity; Jones et al., 2007). Cry36 alone is weakly toxic to coleopteran larvae (Rupar et al., 2003), whereas Cry35 requires partner protein to express toxicity and is encoded in a single operon along with its ~14 kDa Cry34 partner protein (Ellis et al., 2002). Likewise, Cry49Aa is also dependent on a partner protein to express its toxicity (Jones et al., 2007).

As stated earlier, neither Cry48Aa nor Cry49Aa can cause toxicity in mosquitoes when fed individually. An equimolar concentration of both units is required to express toxicity. The LC_{50} for the third instar *Cx. quinquefasciatus* larvae equates to 6.3 ng/mL for Cry49Aa and 15.9 ng/mL for Cry48Aa, which is comparable to that of Bin toxin from Ls (Jones et al., 2007) (LC_{50} value ~30 ng/mL for each BinA and BinB; Oei et al., 1992). Therefore, these two toxins represent a new combination having insecticidal properties by exploiting an interaction between two toxin families that remained unassociated before (Jones et al., 2008). Such two-component dependency between a three-domain Cry and a Bin-like protein is novel and is also noticed between Cry toxins of Bti and Bin toxins of Ls (Wirth et al., 2004). However, unlike Bin toxin, which is naturally formed in equimolar concentration, the synthesis of Cry48Aa in native Ls strains is low and likely unstable, resulting in low toxicity (Jones et al., 2007).

Though the MoA of Cry48Aa/Cry49Aa is unclear, the toxic effects produced by these toxins cause morphological changes in target cells of the *Culex* larvae. This effect is similar to the synergistic interaction between Bin and Cry11 toxins, in which some changes are similar to Bin intoxication, such as endoplasmic reticulum fragmentation or cytoplasmic vacuolization, while others are similar to three-domain Cry toxin activity, such as mitochondrial swelling (de Melo et al., 2009). But it is not clear which of these two components possess putative pore-forming and receptor-binding properties (Berry, 2012; Jones et al., 2007). However, these two toxins possess a remarkable feature as they are able to combat resistance developed against the Bin toxins by producing extra crystal proteins on sporulation and have a strategic role in resistance management (Jones et al., 2007; Nielsen-Leroux et al., 2002; Pei et al., 2002). Consequently, toxins having a distinct action mechanism, like Cry48Aa/Cry49Aa and Bti toxins (Cry4Aa, Cry4Ba, Cry11Aa, and Cyt1Aa), do not face any cross-resistance in larvae resistance to Bin toxin (Georghiou & Wirth, 1997; Pei et al., 2002; Yuan et al., 2003). However, the activity of Cry48Aa/Cry49Aa toxins is highly specific to *Cx. quinquefasciatus*, and it does not express any toxic effect towards lepidopteran, coleopteran, and other dipteran insects, including the closely related *Anopheles* and *Aedes* mosquitoes (Jones et al., 2008).

5.4.5 S-Layer Protein

Surface-layer (S-layer) protein is a paracrystalline structure associated with the surface of bacteria and archaea cells. S-layer forms up to 15% of a cell's protein. However, its purpose is not clear in all bacteria. It might function in molecular exchange between environment and cell, surface recognition, cellular adhesion, heavy metal binding, virulence factors, and larvicidal activity (Lozano & Dussán, 2017). Some strains of Ls, such as OT4b25, OT4b26, III(3)7, C7, and 2362, produce this S-layer protein (SlpC) (120–130 kDa) during the vegetative growth stage. This protein from Ls shows mosquitocidal activity against *Cx. quinquefasciatus* (Allievi et al., 2014; Lozano et al., 2011; Lozano &

Dussán, 2017). Allievi et al. (2014) and Lozano and Dussán (2017) reported synergistic effects between purified S-layer protein and spore-crystal preparations (SCP) from Ls against *Culex* sp. and *Ae. aegypti* (Linnaeus). However, in an earlier study, Bowditch et al. (1989) found S-layer protein from Ls strain 2362 to show no toxic effect on the larvae of *Cx. pipiens*. The S-layer protein from the strain C3-41 also does not have any direct toxicity against mosquito larvae (Hu et al., 2008b; Li et al., 2013). This S-layer protein is also responsible for the pathogenicity of other bacteria under the *Bacillus* group, like *B. cereus* (Wang et al., 2013), *B. anthracis* (Kern & Schneewind, 2010), Bt (Pena et al., 2006; Zhou et al., 2011), *Paenibacillus alvei (Cheshire and Cheyne)* (Janesch et al., 2013), etc.

5.4.6 Other Toxins

A recent study revealed that some highly toxic strains of Ls, such as strain OT4b.25, III(3)7, and 2362, are capable of expressing haemolysin D (*hlyD*) and chitin-binding protein genes. Both of these toxins were evaluated against *Ae. aegypti* larvae and found to be effective. This indicates that haemolysin D and the chitin-binding protein could be two novel mosquitocidal toxins from Ls and can be utilized alongside other Ls toxins to make this bacterium a suitable biocontrol agent against *Ae. aegypti* (Rojas-Pinzón & Dussán, 2017).

5.5 MODE OF ACTION

Ls and Bti have a more or less similar MoA. As mentioned earlier, binary toxin (Bin) is the major component responsible for the toxicity of Ls. Numerous studies conducted globally reveal the following processes involved in the MoA of binary toxins against susceptible insect targets (Figure 5.2).

5.5.1 Solubilization and Activation of Toxin

The initial step in the MoA of Ls crystals containing Bin protoxin has similarity to that of Bti. First, mosquito larvae ingest the spore crystals from water. After being ingested, these crystals rapidly solubilize in the alkaline lumen of the anterior stomach and release the protoxin (BinA and BinB) (Charles, 1987). The protoxins get activated by the proteolytic cleavage of the 42 kDa (BinA) and 51 kDa (BinB) peptides and produce 39 kDa and 43 kDa active proteins, respectively (Baumann et al., 1985). This proteolytic activation increases toxicity (Broadwell & Baumann, 1987) and occurs in both susceptible and non-susceptible insects (Nicolas et al., 1990). Though BinA and BinB both are monomeric, after activation, both may combine to form a heterodimer (Surya et al., 2016). Unlike Cry toxins (Bravo et al., 2007), there is no evidence of oligomer formation in the MoA of Bin toxins (Smith et al., 2005; Surya et al., 2016). Rather, they possess a trefoil domain and another domain for pore formation (Colletier et al., 2016).

5.5.2 Receptor Binding

In Ls, this step is completely different from that of Bti. In Bti, there are multiple receptors in the midgut epithelium for toxin binding (Fernandez-Luna et al., 2010; Lee et al., 2015; Likitvivatanavong et al., 2011). In contrast, the interaction between activated Bin toxins of Ls and midgut

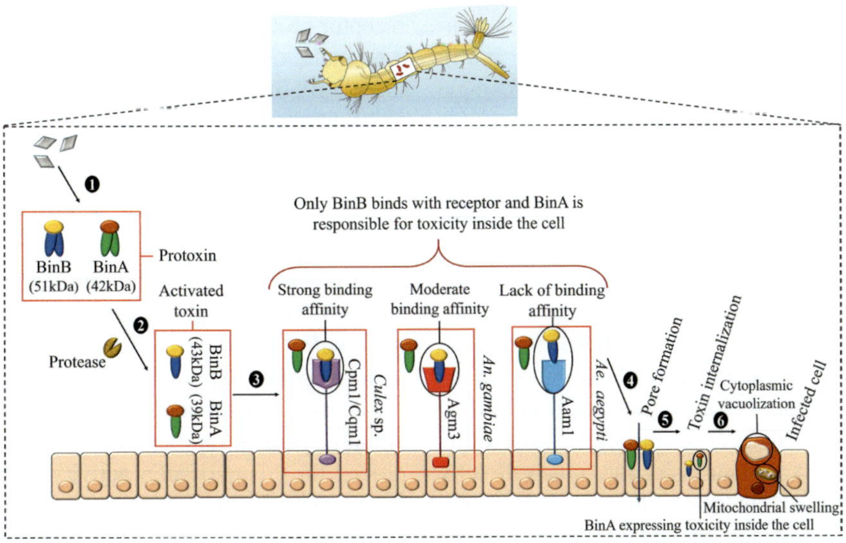

FIGURE 5.2 Schematic representation of the mode of action of Bin toxin from *Lysinibacillus sphaericus* in mosquito larvae: (1) Solubilization of the ingested crystals in the alkaline lumen of the anterior stomach to release protoxin. (2) Proteolytic activation of the protoxin. (3) Binding of BinB subunit with specific receptor in the midgut epithelium with species-specific variation in binding affinity. (4) Formation of pores in the midgut epithelium. (5) Internalization of both the subunits, where BinA subunit causes toxicity inside the cell. (6) Toxicity causes physiological changes in the cell.

epithelial cells is entirely dependent on the toxins' binding affinity to a single class of receptors. This reliance on a single receptor is a disadvantage of using Ls in mosquito control programmes since mosquitoes develop resistance easily (Darboux et al., 2001; Silva-Filha et al., 1999). However, the variation in the MoAs of Ls and Bti toxins allows us to utilize them jointly to overcome resistance development (Silva-Filha et al., 2021), as described in the later part.

Studies on the functional domains of Bin subunits revealed the ability of only the BinB component in receptor binding, whereas BinA is responsible for toxicity inside the cell, and an equimolar ratio of both of these components is required for optimal toxicity (Nicolas et al., 1993; Oei et al., 1992). Here, both the N- and C-terminal regions of the BinB subunit play a crucial role in toxicity. The N-terminal region is essential for receptor binding (Sanitt et al., 2008) and contains some residues, crucial for this interaction (Silva-Filha et al., 2021). This N-terminal domain contains two cysteine amino acid residues, required for its toxicity (Boonyos et al., 2010). Conversely, the C-terminal region is essential for the interaction of the BinB subunit with the BinA subunit, followed by toxin complex internalization into the target cells (Sanitt et al., 2008). This region contains an aromatic residue cluster, which is critical for the proper toxin conformation and membrane insertion (Chooduang et al., 2018). However, this receptor binding is species-specific and correlated with the in vivo species susceptibility to the toxin (Nielsen-Leroux & Charles, 1992; Silva-Filha et al., 2004). Different Bin toxin receptors have been identified, specific to different mosquito species. These Bin toxin receptors are ortholog midgut-bound α-glucosidases (Silva-Filha et al., 2021). In the most susceptible *Culex* mosquito species, BinB toxin binds specifically to the gastric caeca and posterior part of the midgut, whereas the BinA component binds non-specifically throughout the midgut (Charles et al., 1996). This BinB binding is specifically associated with a single receptor (Charles et al., 1997), denominated as Cpm1 (*Cx. pipiens* maltase 1) for *Cx. pipiens pipiens*, corresponding to maltase 1 (Darboux et al., 2001; Silva-Filha et al., 1999), and Cqm1 (*Cx. quinquefasciatus* maltase 1) for *Cx. pipiens quinquefasciatus*, corresponding to maltase 1 (Romão et al., 2006). According to the localization studies of both Bin subunits in the treated larvae of *Cx. quinquefasciatus*, binding of the BinB subunit to the Cqm1 receptor is the necessary step for the internalization of both the subunits, which may happen by endocytosis (Lekakarn et al., 2015; Silva-Filha & Peixoto, 2003). These studies showed that the presence of the BinA subunit inside the cells is directly responsible for the toxicity, which further depends on the binding of the BinB with the specific receptor (Charles et al., 1997; Lekakarn et al., 2015).

In *Anopheles* larvae, the Bin toxins bind to a specific midgut receptor (Silva-Filha et al., 1997), that is, Agm3 (*An. gambiae* maltase 3) for *An. gambiae*, corresponding to maltase 3 (Opota et al., 2008). Here, the toxins have a higher binding affinity to the brush border membrane fractions (BBMF) of *An. gambiae* than *An. stephensi*. However, in comparison to *Culex* mosquitoes, this binding affinity is much lower in *An. gambiae*, causing its less sensitivity to the Bin toxins (Opota et al., 2008; Silva-Filha et al., 1997). In *An. gambiae*, the BinA subunit is able to bind to the receptor independent of BinB; still, maximum binding is observed at an equimolar ratio of BinA and BinB (Charles et al., 1997).

Among these receptors, the first characterized one was Cpm1, which showed 97% identity with Cqm1 (Romão et al., 2006) and 66.4% with Agm3 (Opota et al., 2008). Aam1 (*Ae. aegypti* maltase 1) is another ortholog from *Ae. aegypti* that shares 74% similarity with Cqm1 and is produced as a membrane protein without the ability to bind with Bin toxin, causing larvae refractoriness (Ferreira et al., 2010). This ortholog shows high homology with Cpm1 and Agm1 (*An. gambiae* maltase 1), mainly in the NH2 terminus, which is believed to enclose the Bin toxin binding site. However, there is no specific binding in *Aedes* BBMFs. The initial steps in Bin toxin MoA, like ingestion, solubilization, and activation, are effective in *Ae. aegypti*. This clearly indicates that the Bin binding to the receptor is the most critical event in the Bin toxin mechanism and is responsible for the narrow host range of Ls. So it can be suggested that the Bin receptor is a Cpm1-like protein. Slight species-to-species variation in the amino acid sequence in this receptor protein leads to differences in the Bin binding affinity and consequently results in differences in susceptibility (Opota et al., 2008). The *cpm1/cqm1* and *agm3* genes encode proteins of 580 and 588 amino acid residues, respectively, displaying four conserved motifs of α-glucosidase, predicted glycosylation sites, and a signalling sequence for a glycosylphosphatidylinositol (GPI) anchor at the C-terminal end (Silva-Filha et al., 2014). The expression of these genes as receptor proteins is a must for the binding of the Bin toxin. Any mutation in these genes prevents them from expressing as GPI-anchored proteins, which is the most important resistance mechanism in mosquitoes to the Bin toxin (Silva-Filha et al., 2021).

5.5.3 Pore Formation

After receptor binding, the series of events that cause cell death are not clear enough. In a patch clamping test on *Cx. quinquefasciatus* cells, the toxin components were found to form pores (Cokmus et al., 1997). Research employing artificial membranes has yielded comparable findings, although studies vary about which component may be most crucial for this particular activity (Boonserm et al., 2006; Schwartz et al., 2001). So it is possible that the pore-forming ability of the Bin toxin is responsible for its MoA. Here, the C-terminal region of the BinA subunit governs cell toxicity (Hire et al., 2009) and also might be involved in the formation of pores in the epithelium of intestine, causing toxin internalization (Elangovan et al., 2000; Schwartz et al., 2001). This subunit contains some specific amino acid residues responsible for its toxicity (Silva-Filha et al., 2021). Moreover, studies showed that both Bin subunits have this capacity of forming pores in culture cells or artificial membranes (Silva-Filha et al., 2021). In cultured *Cx. quinquefasciatus* cells, Ls toxins produce lesions with a pore radius of 0.5–0.6 nm (Davidson, 1989), which is within the range of pore radius found for the Bt toxins (Knowles & Ellar, 1987). Transforming the mammalian epithelial cell line MDCK (Madin and Darby canine kidney cells) to express the Cpm1 receptor gene leads to pore formation and vacuole appearance after adding Bin protein, although cell lysis does not occur (Pauchet et al., 2005). In a detailed study, Bin toxin was found to induce cationic pores, to produce large vacuoles, and for both toxin components to internalize along with their receptors (Opota et al., 2011).

5.5.4 Cytopathology and Physiological Changes

Singer (1973) first reported that Ls kills the target organism by toxaemia rather than septicaemia. Similar to Bt, Ls also targets the microvillar epithelial cells in the midgut. However, larval midgut epithelium completely breaks down by the Bti toxins, whereas it does not happen in the case of Ls toxins (Charles et al., 1996). When mosquito larvae ingest high concentrations of Ls spores, the midgut cells of the infected larvae get swollen and lie against the body wall, throwing the peritrophic membrane along with the food contents into zigzag folds. The infected cells become rounded up, get swollen, phase dark, and ultimately get detached from their support (Yuan, 2002). Usually, within 15 minutes of spore ingestion, midgut alterations start. Highly toxic strains like 1593 and 2297 both cause similar damage in the midgut of *Cx. pipiens* (Charles et al., 1996; Davidson, 1981), whereas there are differences in the toxic symptoms produced by these two strains in other mosquito species. Large vacuoles (and/or cytolysosomes) form in the midgut cells of *Cx. pipiens*, while in the midgut cells of *An. stephensi*, large areas with low electron density have appeared. Mitochondrial swelling is a general symptom in *An. stephensi*, *Cx. pipiens*, and *Ae. aegypti* when intoxicated with a high dose of spore crystals (Charles, 1987). The intoxicated cells are characterized by dilation of the mitochondrial cristae and endoplasmic reticulum, followed by vacuole enlargement and mitochondrial matrix condensation within a few minutes of treatment (Davidson & Titus, 1987). In addition, the intracellular ATP is markedly reduced, and the cell membranes gradually lose their integrity (Berry, 2012). After severe disruption of midgut cells of the posterior stomach and gastric caeca, neural and skeletal tissue cells also get damaged, probably inducing the onset of paralysis in the infected larvae (Singh & Gill, 1988). Oxygen uptake by mitochondria is inhibited in larvae of *Cx. quinquefasciatus* treated with Ls due to the inactivity of larval choline acetyltransferase in the presence of Ls toxin (Narasu & Gopinathan, 1988).

5.6 APPLICATION OF LS FOR MOSQUITO CONTROL

Since the 1980s, commercial formulations based on insecticidal toxins from Ls have been in use to control mosquitoes. Isolation of highly toxic mosquitocidal Ls strains that produce crystals having Bin toxin (such as 2362, 1593, C3-41, etc.) has uplifted its commercial use in mosquito control programmes in several countries. This bacterium was first introduced in southern France during 1987 to control *Cx. pipiens* (Thiery et al., 1996). Thereafter, the World Health Organization (WHO) supported field trials in some filariasis-endemic countries to evaluate the effectiveness of Ls in controlling the main vector of filariasis, that is, *Cx. quinquefasciatus*, in urban areas characterized by poor sanitation and high mosquito proliferation (Silva-Filha et al., 2021). Thus, it becomes a promising agent for controlling *Culex* mosquitoes in both rural and urban areas of developing countries, where formulations are commercially produced to use in vector control programmes and applied in spore formulations (Berry, 2012). Some tropical countries, such as Brazil, Cameroon, Côte d'Ivoire, India, Sri Lanka, Tanzania, and Thailand, have been using Ls to control this filariasis vector for many years. Every year, about 150 tons of Ls formulation have been used in China as a larvicidal agent throughout the past two decades, with nearly 8,000 ha of different mosquito breeding sites having been treated with these formulations each year (Yuan, 2002).

Basically, Ls-based larvicides are mostly used against mosquitoes due to their certain advantages, like persistency in polluted water and ability to be recycled. Moreover, these biolarvicides can efficiently control mosquitoes in their natural habitats. However, the occurrence of resistance in field-treated mosquito populations to Ls demands urgent solutions. Conversely, Bti crystals that target different receptors in the midgut still remain effective against the Bin-resistant larvae. That's why, nowadays, Ls is used either in rotation, in combination, or as a mixture with Bti (Silva-Filha et al., 2021). These strategies avoid or delay the chances of resistance development in mosquitoes against Bin toxin and thus increase its efficacy in mosquito control programmes. Since the discovery of Ls's mosquitocidal activity, several small- to medium-scale field trials have been conducted to assess the efficacy of Ls-based products against mosquitoes and their toxicity against non-target aquatic organisms under various environmental conditions. These field trials gathered a significant amount of information on appropriate formulations, doses, methods for product application, target species of mosquitoes, and the effectiveness of control measures on mosquito populations. Table 5.2 provides examples of field experiments undertaken around the world to assess the performance of Ls, Bti, and Ls-Bti combined commercial products in mosquito control programmes.

TABLE 5.2
Field Trials Using *Lysinibacillus sphaericus*–Based Larvicides Either in Solo Application or in Rotation or in Combination or as Mixture with Bti and Other Insecticides for Mosquito Control

Scheme	Country	Product	Strain Used/Active Ingredient	Target Species	Scenario	Outcome	Reference
Ls solo application	Guatemala	GRISELESF®	2362 (strain ABTS 1743)	*An. albimanus*	Urban	Achieved 94.57% larval reduction in maturity stage	(Castro et al., 2000)
	Ghana	VectoLex® (Ls)	Do	*An. gambiae*	Urban	100% larval mortality within 24 hours achieved after 50-fold concentration of LC_{95} treatment	(Baffour-Awuah et al., 2014)
	USA	Do	Do	*Ae. triseriatus* *An. punctipennis* *Cx. restuans* *Cx. pipiens*	Urban waste tiers and catch basins	Complete larval reduction achieved	(Siegel & Novak, 1997)
	Australia	Do	Do	*Cx. annulirostris* *Cx. quinquefasciatus*	Urban	100% mortality till second week without affecting non-target aquatic animals and water	(Brown et al., 2004)
	Burkina Faso	—	Do	*An. gambiae*	Urban ponds	Full control of third and fourth instar larvae with treatment persistency for a day	(Nicolas et al., 1987a)
	Colombia	—	Strain 2362 and III(3)7	*Ae. aegypti*	Simulated field	Over 90% larval mortality achieved at 48 hours	(Santana-Martínez et al., 2020)
	India	—	Strain B-101	*Cx. quinquefasciatus* *Cx. tritaeniorhynchus* *Cx. vishnui*	Urban	Significant reduction in larval–pupal count and percentage larval habitats than routine antilarval measures	(Yadav et al., 1997)
Ls and Bti in rotation	Kenya	VectoLex® and VectoBac® (Bti)	2362 (strain ABTS 1743) (Ls) and AM65-52 (Bti)	*An. gambiae* *An. funestus*	Rural	Overall reduction of larval density by 95% and adults biting to humans by 92%	(Fillinger & Lindsay, 2006)
	Côte d'Ivoire	Do	Do	*An. gambiae* *An. funestus*	Urban	A decline in the biting rate of the species recorded	(Tchicaya et al., 2009)
	Gambia	Do	Do	*An. gambiae*	Rural	100% larval mortality and 94% less pupal development	(Majambere et al., 2007)
	Tanzania	Do	Do	*An. gambiae* *Cx. quinquefasciatus*	Urban	Reduced larval population and malaria transmission	(Fillinger et al., 2008)
Mixture of Ls and Bti	Turkey	Do	Do	*Cx. pipiens*	Urban	At higher dosage, >90% control achieved for 28 days	(Cetin et al., 2007)

Ls-Bti conjugated product	Brazil	VectoMax® (2.7% Ls + 4.5% Bti)	2362 (strain ABTS 1743) (Ls) and AM65-52 (Bti)	*Ae. aegypti* *Cx. quinquefasciatus*	Urban	Significant reduction in adult density achieved	(de Mendonça Santos et al., 2018)
		Do	Do	*An. darlingi* *An. albitarsis*	Peri-urban	Tenfold less average larval density and malaria reduction	(Fontoura et al., 2021)
	USA	Do	Do	*Ae. melanimon* *Cx. tarsalis*	Sylvatic	Control *Cx. tarsalis* for 28 days post-treatment and also reduce *Ae. melanimon* density	(Dritz et al., 2011)
	Spain	Do	Do	*Ae. albopictus*	Urban/indoor catch basin	Close to 100% efficacy achieved for up to 345 days post-treatment	(Eritja, 2013)
	Switzerland	Do	Do	*Ae. albopictus*	Urban	>98% larval reduction achieved	(Guidi et al., 2013)
		Do	Do	*Ae. albopictus* *Cx. pipiens*	Urban	100 and 60% less adult of *Cx. pipiens* and *Ae. albopictus*	(Ravasi et al., 2023)
	Italy	VectoMax® + other measure	Do	*Ae. albopictus*	Urban	Significant reduction in egg density	(Baldacchino et al., 2017)
	Netherlands	VectoMax® + Source reduction	Do	*Ae. japonicus*	Peri-urban/ allotment garden	Larval abundance reduced significantly	(Ibañez-Justicia et al., 2018)
	Kenya	FourStar® and LL3 (6% Ls + 1% Bti)	2362 (strain AML 614) (Ls) and BMP 144(Bti)	*An. gambiae* *An. funestus*	Rural	Both products significantly reducing larval density with similar efficacy	(Kahindi et al., 2018)
		FourStar®	Do	*An. gambiae*	Rural	No pupa for first 2–3 months, and 60% less pupa for 5 months	(Afrane et al., 2016)
		CulinexCombi® (1 × 10^6 ITUs of Bti + 2.5 × 10^4 ITUs of Ls)	—	*An. gambiae* *Cx. quinquefasciatus*	Urban/peri-urban	Significant reduction in larval density achieved	(Mwangangi et al., 2011)
Mixture of Ls and a pyrethroid through thermal fogging	Malaysia	VectoLex® and Resigen® (pyrethroid)	S-bioallethrin+ Permethrin+ Piperonyl butoxide w/w (pyrethroid)	*Cx. sitiens* *Cx. quinquefasciatus*	Simulated field	Effective against both larvae and adults without any antagonistic effect; simultaneously thermal fogging of the mixture improving efficiency	(Lee et al., 2016)
Ls–Bti combined, Bti solo	Kenya	CulinexCombi and VectoBac®	Do	*Cx. sitiens* *Cx. quinquefasciatus*	Unused swimming pool	VectoBac® effective against early *Culex* and CulinexCombi against all stages of both	(Kahindi et al., 2008)
Ls-Bti combined, Ls solo	Brazil	VectoMax® and VectoLex®	Do	*An. darlingi*	Rural/fish ponds	>95% reduction in late instar larval density recorded up to 7 days post-application	(Fontoura et al., 2020)

(Continued)

TABLE 5.2 *(Continued)*
Field Trials Using *Lysinibacillus sphaericus*–Based Larvicides Either in Solo Application or in Rotation or in Combination or as Mixture with Bti and Other Insecticides for Mosquito Control

Scheme	Country	Product	Strain Used/Active Ingredient	Target Species	Scenario	Outcome	Reference
Ls-Bti combined, Bti solo, Methoprene	Senegal	VectoMax®, VectoBac®, and MetaLarv®	*4.25% (S)-methoprene* (MetaLarv®)	*An. arabiensis*	Urban	Complete control of larvae obtained 48 hours after treatment	(Diédhiou et al., 2017)
Ls-Bti combined, Triflumuron	Colombia	VectoMax® and Starycide® (Triflumuron)	Triflumuron 39.7% w/w (Starycide®)	*Ae. aegypti* *Cx. quinquefasciatus*	Urban	Vectomax significantly reducing both species; Triflumuron significantly reducing only *Ae. aegypti*	(Ocampo et al., 2008)
Ls-Bti combined, Bti solo, Ls solo, Spinosad	USA	VectoMax®, VectoBac®, VectoLex®, Natular™ XRT (Spinosad)	6.25% spinosad (Natular™ XRT)	*Cx. pipiens* *Cx. restuans* *Ae. japonicus*	Urban	Significant reduction in larval population achieved in all cases	(Anderson et al., 2011)
Ls-Bti combined, Spinosad	USA	FourStar® and Natular™ XRT tablet	Do	*Cx. pipiens* *Cx. restuans*	Urban	Both products reducing the immature numbers for 8 consecutive weeks	(Harbison et al., 2014)
Ls-Bti combined, Ls solo, Spinosad	USA	FourStar®, VectoLex® and Natular™ XRT, Natular™ T30, Natular™ G30 (Spinosad)	6.25%, 8.33%, and 2.5% spinosad in Natular™ XRT, Natular™ T30, and Natular™ G30	*Cx. pipiens*	Urban	Pupal production reduced by about 48% in FourStar-, Natular XRT–, and Natular T30–treated basins, and by 87% and 99% in VectoLex- and Natular G30–treated basins, respectively	(Nasci et al., 2017)

5.7 ADVANTAGES

One of the most important features of using Ls in mosquito control programme is its host specificity and safety towards non-target organisms. These biolarvicides efficiently control mosquitoes in their natural habitats without causing any adverse effects to the non-target aquatic organisms (Brown et al., 2004; Lacey, 2007; Merritt et al., 2005; Mulla et al., 1984; Thierry et al., 2023; Yousten et al., 1991). It is also safe for beneficial insects like honeybees (Davidson et al., 1977; Vandenbergi, 1990) and different eukaryotic microorganisms (Tietze et al., 1993). Investigations on the fate of Ls spores in non-target invertebrates showed that these spores never accumulate and show no active growth in these organisms (Yousten et al., 1991, 1992). Environmental persistency is another key advantage of Ls. This factor is important in the field application of different bioinsecticides. Ls has a higher persistency in the polluted water, which is the major breeding site for mosquitoes, particularly for the *Culex* species (Karch et al., 1988; Nicolas et al., 1987b). Ls exhibits a residual activity of 2 days to 4 weeks, depending on the experimental settings (Lee et al., 1998). Overall persistency of Ls is much longer compared to Bti (Silapanuntakul et al., 1983). There may be a relation of persistency with ultraviolet (UV) resistance and sensitivity to UV-B for both spore and toxin (Hadapad et al., 2008). By adding UV protectants, stability of both spore and crystal proteins can be enhanced to some extent (Hadapad et al., 2009). Its insecticidal activity remains even after the reduction of spore viability (Burke et al., 1983). Spores of Ls also have a lower tendency to stick to and sediment with other particulates in the water column than Bti spores. This property increases their persistency in the feeding zones of mosquito larvae (Yousten et al., 1992).

Ls is also capable of recycling in the environment. Recycling of the toxin can be carried out through the germination and multiplication of the Ls bacteria in the dead larvae intoxicated with the toxic spores. Sporulation of these vegetative cells occurs upon exhaustion of nutrients and is released during the disintegration of larval cadaver (Lee et al., 1998). When any invertebrate other than mosquitoes ingests the spores, they remain viable within that organism without reproduction. It was found that midges emerging out of Ls-treated water can carry the spores to a new location (Yousten et al., 1991). This ability of recycling seems to be the characteristic feature of highly toxic strains that have the ability of rapid germination in insects. Thus, the ultimate number of Ls cells in the cadavers can increase by approximately 1–1,000 times to the number of originally ingested spores (Charles & Nicolas, 1986; Correa & Yousten, 1995). In addition to the mosquitocidal activity, Bin toxin is also found to induce strong cytocidal activity and apoptotic events in HK1 cancer cells. So as an anti-cancer agent, this toxin may possess high therapeutic potential (Kanwal & Boonserm, 2023).

5.8 DRAWBACKS

Resistance development in insects against bacterial toxins is a major problem of its use. Like Bt, many cases of mosquito resistance to Ls have already been documented from different corners of the world (Rao et al., 1995; Silva-Filha et al., 1995; Sinègre et al., 1994; Su et al., 2018, 2019; Yuan et al., 2000). As mentioned earlier, it becomes relatively easy for the mosquitoes to develop a high level of resistance against Ls strains that produce only one spore-associated toxin (Bin) and also due to the binding of the toxin to a specific group of receptors. Such mechanisms of resistance development in mosquito larvae against Ls and the way to get rid of them are elaborately described in Chapter 12. Another key disadvantage of employing Ls for mosquito control programmes is that the spore formulations utilized in field trials have a tendency to settle quickly from the water surface, that is, mosquito breeding grounds, to the mud bottom. For that, a slow-release formulation that continuously releases Ls and Bti at high concentrations within the mosquito larval feeding zone could enhance the activity of both bacilli. Such a floating bead formulation with controlled release of toxins was formulated from the spore–toxin complex of both bacilli, providing over 90% control against the third/fourth instar larvae of *An. stephensi* at the Central Drug Research Institute, Lucknow (Saxena et al., 1994). Another drawback is related to its effect on non-target organisms. Even after being safe for beneficial arthropods, a recent study showed that Ls can be pathogenic to the Australian stingless bee, *Tetragonula carbonaria* (Smith). Ls shows symptoms of brood disease within 22 days from infection. It causes queenless colonies with reduced adult bee populations and little foraging intensity. Dead larvae become fluid-like or dry, and larval remains may create a ropy mass similar to that of dead honeybee larvae infested with *Paenibacillus larvae* (White) (Shanks et al., 2017).

5.9 CONCLUSION

With rising environmental and health concerns over the indiscriminate use of broad-spectrum synthetic chemical insecticides, as well as the development of insect resistance to these chemicals, pest control industries are prioritizing the development of entomopathogenic bacteria as an alternative to toxic chemicals for controlling various insect pests. In this respect, the role of Ls in controlling mosquitoes is enormous, as it shows high efficiency in different mosquito control programmes. In recent days, with increasing understanding about this organism and its entomopathogenic potentiality, its importance in mosquito control is also increasing. However, still a number of issues remain to be clarified. Development of resistance in mosquitoes against the Bin toxin of Ls is one of them, which can be solved by using Ls in combination or rotation or as a mixture with other mosquitocidal agents, such as Bti. Another issue lies in its safety against non-target organisms. Although it is safe for other organisms, still new cases of its pathogenicity against non-target organisms are coming out. Such situations demand further studies on the potential doses, host specificity, MoA, and method of application of Ls. Further, by using recombinant technology, the development of highly toxic Ls strains or other organisms carrying its toxins can delay resistance development and also increases its field persistency. Moreover, in addition to the Bin toxin, Ls also produces certain other toxins, which still remain underutilized. Further studies on these toxins can explore their efficiency in pest management. Thus, it will be possible to make this bacterium a novel, effective, and environmentally safe biopesticide against a range of insect pests and disease vectors.

REFERENCES

Afrane, Y. A., Mweresa, N. G., Wanjala, C. L., Gilbreath III, T. M., Zhou, G., Lee, M. C., Githeko, A. K., & Yan, G. (2016). Evaluation of long-lasting microbial larvicide for malaria vector control in Kenya. *Malaria Journal*, *15*, 577. https://doi.org/10.1186/s12936-016-1626-6

Ahmed, H. K., Mitchell, W. J., & Priest, F. G. (1995). Regulation of mosquitocidal toxin synthesis in *Bacillus sphaericus*. *Applied Microbiology and Biotechnology*, *43*, 310–314. https://doi.org/10.1007/BF00172830

Ahmed, I., Yokota, A., Yamazoe, A., & Fujiwara, T. (2007). Proposal of *Lysinibacillus boronitolerans* gen. nov. sp. nov., and transfer of *Bacillus fusiformis* to *Lysinibacillus fusiformis* comb. nov. and *Bacillus sphaericus* to *Lysinibacillus sphaericus* comb. nov. *International Journal of Systematic and Evolutionary Microbiology*, *57*(5), 1117–1125. https://doi.org/10.1099/ijs.0.63867-0

Alexander, B., & Priest, F. G. (1990). Numerical classification and identification of *Bacillus sphaericus* including some strains pathogenic for mosquito larvae. *Journal of General Microbiology*, *136*(2), 367–376. https://doi.org/10.1099/00221287-136-2-367

Allievi, M. C., Palomino, M. M., Prado Acosta, M., Lanati, L., Ruzal, S. M., & Sánchez-Rivas, C. (2014). Contribution of S-layer proteins to the mosquitocidal activity of *Lysinibacillus sphaericus*. *PLoS One*, *9*(10), e111114. https://doi.org/10.1371/journal.pone.0111114

Aly, C., Mulla, M. S., & Federici, B. A. (1989). Ingestion, dissolution, and proteolysis of the *Bacillus sphaericus* toxin by mosquito larvae. *Journal of Invertebrate Pathology*, *53*(1), 12–20. https://doi.org/10.1016/0022-2011(89)90068-2

Anderson, J. F., Ferrandino, F. J., Dingman, D. W., Main, A. J., Andreadis, T. G., & Becnel, J. J. (2011). Control of mosquitoes in catch basins in Connecticut with *Bacillus thuringiensis israelensis*, *Bacillus sphaericus*, and Spinosad. *Journal of the American Mosquito Control Association*, *27*(1), 45–55. https://doi.org/10.2987/10-6079.1

Baffour-Awuah, S., Owusu-Dabo, E., Kruppa, T., Annan, A., Nartey, R., Dogbe, J., Pandam, S., Becker, N., Oppong, S., & Obiri-Danso, K. (2014). *Lysinibacillus sphaericus* biolarvicide, an efficacious tool in the control of *Anopheles gambiae* in Kumasi, Ghana. *East African Journal of Public Health*, *11*(4), 841–861.

Baldacchino, F., Bussola, F., Arnoldi, D., Marcantonio, M., Montarsi, F., Capelli, G., Rosa, R., & Rizzoli, A. (2017). An integrated pest control strategy against the Asian tiger mosquito in Northern Italy: A case study. *Pest Management Science*, *73*(1), 87–93. https://doi.org/10.1002/ps.4417

Baum, J. A., Chu, C. R., Rupar, M., Brown, G. R., Donovan, W. P., Huesing, J. E., Ilagan, O., Malvar, T. M., Pleau, M., Walters, M., & Vaughn, T. (2004). Binary toxins from *Bacillus thuringiensis* active against the western corn rootworm, *Diabrotica virgifera virgifera* LeConte. *Applied and Environmental Microbiology*, *70*(8), 4889–4898. https://doi.org/10.1128/AEM.70.8.4889-4898.2004

Baumann, L., & Baumann, P. (1989). Expression in *Bacillus subtilis* of the 51-and 42-kilodalton mosquitocidal toxin genes of *Bacillus sphaericus*. *Applied and Environmental Microbiology*, *55*(1), 252–253. https://doi.org/10.1128/aem.55.1.252-253.1989

Baumann, L., & Baumann, P. (1991). Effects of components of the *Bacillus sphaericus* toxin on mosquito larvae and mosquito-derived tissue culture-grown cells. *Current Microbiology*, *23*(1), 51–57. https://doi.org/10.1007/BF02092309

Baumann, L., Broadwell, A. H., & Baumann, P. (1988). Sequence analysis of the mosquitocidal toxin genes encoding 51.4-and 41.9-kilodalton proteins from *Bacillus sphaericus* 2362 and 2297. *Journal of Bacteriology*, *170*(5), 2045–2050. https://doi.org/10.1128/jb.170.5.2045-2050.1988

Baumann, P., Unterman, B. M., Baumann, L., Broadwell, A. H., Abbene, S. J., & Bowditch, R. D. (1985). Purification of the larvicidal toxin of *Bacillus sphaericus* and evidence for high-molecular-weight precursors. *Journal of Bacteriology*, *163*(2), 738–747. https://doi.org/10.1128/jb.163.2.738-747.1985

Becker, N., & Lüthy, P. (2017). Mosquito control with entomopathogenic bacteria in Europe. In L. A. Lacey (Ed.), *Microbial control of insect and mite pests: From theory to practice* (pp. 379–392). Academic Press. https://doi.org/10.1016/B978-0-12-803527-6.00026-3

Berry, C. (2012). The bacterium, *Lysinibacillus sphaericus*, as an insect pathogen. *Journal of Invertebrate Pathology*, *109*(1), 1–10. https://doi.org/10.1016/j.jip.2011.11.008

Berry, C., & Crickmore, N. (2017). Structural classification of insecticidal proteins—towards an in silico characterisation of novel toxins. *Journal of Invertebrate Pathology*, *142*, 16–22. https://doi.org/10.1016/j.jip.2016.07.015

Berry, C., Hindley, J., Ehrhardt, A. F., Grounds, T., De Souza, I., & Davidson, E. W. (1993). Genetic determinants of host ranges of *Bacillus sphaericus* mosquito larvicidal toxins. *Journal of Bacteriology*, *175*(2), 510–518. https://doi.org/10.1128/jb.175.2.510-518.1993

Bone, L. W., & Tinelli, R. (1987). *Trichostrongylus colubriformis*: Larvicidal activity of toxic extracts from *Bacillus sphaericus* (strain 1593) spores. *Experimental Parasitology*, *64*(3), 514–516. https://doi.org/10.1016/0014-4894(87)90066-X

Boonserm, P., Moonsom, S., Boonchoy, C., Promdonkoy, B., Parthasarathy, K., & Torres, J. (2006). Association of the components of the binary toxin from *Bacillus sphaericus* in solution and with model lipid bilayers. *Biochemical and Biophysical Research Communications*, *342*(4), 1273–1278. https://doi.org/10.1016/j.bbrc.2006.02.086

Boonyos, P., Soonsanga, S., Boonserm, P., & Promdonkoy, B. (2010). Role of cysteine at positions 67, 161 and 241 of a *Bacillus sphaericus* binary toxin BinB. *Bmb Reports*, *43*(1), 23–28. https://doi.org/10.5483/BMBRep.2010.43.1.023

Boucias, D. G., & Pendland, J. C. (1998). *Principles of insect pathology*. Springer US. https://doi.org/10.1016/0022-2011(67)90149-8

Bowditch, R. D., Baumann, P., & Yousten, A. A. (1989). Cloning and sequencing of the gene encoding a 125-kilodalton surface-layer protein from *Bacillus sphaericus* 2362 and of a related cryptic gene. *Journal of Bacteriology*, *171*(8), 4178–4188. https://doi.org/10.1128/jb.171.8.4178-4188.1989

Bravo, Alejandra, Gill, S. S., & Soberón, M. (2007). Mode of action of *Bacillus thuringiensis* Cry and Cyt toxins and their potential for insect control. *Toxicon*, *49*(4), 423–435. https://doi.org/10.1016/J.TOXICON.2006.11.022

Broadwell, A. H., Baumann, L., & Baumann, P. (1990a). Larvicidal properties of the 42 and 51 kilodalton *Bacillus sphaericus* proteins expressed in different bacterial hosts: Evidence for a binary toxin. *Current Microbiology*, *21*(6), 361–366. https://doi.org/10.1007/BF02199438

Broadwell, A. H., & Baumann, P. (1986). Sporulation-associated activation of *Bacillus sphaericus* larvicide. *Applied and Environmental Microbiology*, *52*(4), 758–764. https://doi.org/10.1128/aem.52.4.758-764.1986

Broadwell, A. H., & Baumann, P. (1987). Proteolysis in the gut of mosquito larvae results in further activation of the *Bacillus sphaericus* toxin. *Applied and Environmental Microbiology*, *53*(6), 1333–1337. https://doi.org/10.1128/aem.53.6.1333-1337.1987

Broadwell, A. H., Clark, M. A., Baumann, L., & Baumann, P. (1990b). Construction by site-directed mutagenesis of a 39-kilodalton mosquitocidal protein similar to the larva-processed toxin of *Bacillus sphaericus* 2362. *Journal of Bacteriology*, *172*(7), 4032–4036. https://doi.org/10.1128/jb.172.7.4032-4036.1990

Brown, M. D., Watson, T. M., Carter, J., Purdie, D. M., & Kay, B. H. (2004). Toxicity of VectoLex (*Bacillus sphaericus*) products to selected Australian mosquito and nontarget species. *Journal of Economic Entomology*, *97*(1), 51–58. https://doi.org/10.1093/jee/97.1.51

Burke Jr., W. F., McDonald, K. O., & Davidson, E. W. (1983). Effect of UV light on spore viability and mosquito larvicidal activity of *Bacillus sphaericus* 1593. *Applied and Environmental Microbiology*, *46*(4), 954–956. https://doi.org/10.1128/aem.46.4.954-956.1983

Bustos, M. C., Ibarra, H., & Dussán, J. (2018). The golden activity of *Lysinibacillus sphaericus*: New insights on gold accumulation and possible nanoparticles biosynthesis. *Materials*, *11*(9), 1587. https://doi.org/10.3390/ma11091587

Castro, S. B., Arias, A. M., Velásquez, O. C., Granados, R. T., & Mendoza, I. (2000). Introduction of *Bacillus sphaericus* strain-2362 (GRISELESF) for biological control of malaria vectors in Guatemala. *Revista Cubana de Medicina Tropical*, *52*(1), 37–43.

Cetin, H., Dechant, P., & Yanikoglu, A. (2007). Field trials with tank mixtures of *Bacillus thuringiensis* subsp. *israelensis* and *Bacillus sphaericus* formulations against *Culex pipiens* larvae in septic tanks in Antalya, Turkey. *Journal of the American Mosquito Control Association*, *23*(2), 161–165. https://doi.org/10.2987/8756-971X(2007)23[161:FTWTMO]2.0.CO;2

Charles, J. F. (1987). Ultrastructural midgut events in Culicidae larvae fed with *Bacillus sphaericus* 2297 spore/crystal complex. *Annales de l'Institut Pasteur/Microbiologie*, *138*(4), 471–484. https://doi.org/10.1016/0769-2609(87)90064-0

Charles, J. F., Kalfon, A., Bourgouin, C., & de Barjac, H. (1988). *Bacillus sphaericus* asporogenous mutants: Morphology, protein pattern and larvicidal activity. *Annales de l'Institut Pasteur/Microbiologie*, *139*(2), 243–259. https://doi.org/10.1016/0769-2609(88)90010-5

Charles, J. F., & Nicolas, L. (1986). Recycling of *Bacillus sphaericus* 2362 in mosquito larvae: A laboratory study. *Annales de l'Institut Pasteur/Microbiologie*, *137*(1), 101–111. https://doi.org/10.1016/S0769-2609(86)80097-7

Charles, J. F., Nielson-LeRoux, C., & Delecluse, A. (1996). *Bacillus sphaericus* toxins: Molecular biology and mode of action. *Annual Review of Entomology*, *41*(1), 451–472. https://doi.org/10.1146/annurev.en.41.010196.002315

Charles, J. F., Silva-Filha, M. H., Nielsen-LeRoux, C., Humphreys, M. J., & Berry, C. (1997). Binding of the 51-and 42-kDa individual components from the *Bacillus sphaericus* crystal toxin to mosquito larval midgut membranes from *Culex* and *Anopheles* sp. (Diptera: Culicidae). *FEMS Microbiology Letters*, *156*(1), 153–159. https://doi.org/10.1111/j.1574-6968.1997.tb12721.x

Cheong, W. C., & Yap, H. H. (1985). Bioassays of *Bacillus sphaericus* (strain 1593) against mosquitoes of public health importance in Malaysia. *The Southeast Asian Journal of Tropical Medicine and Public Health*, *16*(1), 54–58.

Chooduang, S., Surya, W., Torres, J., & Boonserm, P. (2018). An aromatic cluster in *Lysinibacillus sphaericus* BinB involved in toxicity and proper in-membrane folding. *Archives of Biochemistry and Biophysics*, *660*, 29–35. https://doi.org/10.1016/j.abb.2018.10.006

Clark, M. A., & Baumann, P. (1990). Deletion analysis of the 51-kilodalton protein of the *Bacillus sphaericus* 2362 binary mosquitocidal toxin: Construction of derivatives equivalent to the larva-processed toxin. *Journal of Bacteriology*, *172*(12), 6759–6763. https://doi.org/10.1128/jb.172.12.6759-6763.1990

Cokmus, C., Davidson, E. W., & Cooper, K. (1997). Electrophysiological effects of *Bacillus sphaericus* binary toxin on cultured mosquito cells. *Journal of Invertebrate Pathology*, *69*(3), 197–204. https://doi.org/10.1006/jipa.1997.4660

Colletier, J. P., Sawaya, M. R., Gingery, M., Rodriguez, J. A., Cascio, D., Brewster, A. S., Michels-Clark, T., Hice, R. H., Coquelle, N., Boutet, S., Williams, G. J., Messerschmidt, M., DePonte, D. P., Sierra, R. G., Laksmono, H., Koglin, J. E., Hunter, M. S., Park, H. W., Uervirojnangkoorn, M., Bideshi, D. K., . . . Eisenberg, D. S. (2016). De novo phasing with X-ray laser reveals mosquito larvicide BinAB structure, *Nature*, *539*(7627), 43–47. https://doi.org/10.1038/nature19825

Correa, M., & Yousten, A. A. (1995). *Bacillus sphaericus* spore germination and recycling in mosquito larval cadavers. *Journal of Invertebrate Pathology*, *66*(1), 76–81. https://doi.org/10.1006/jipa.1995.1064

Darboux, I., Nielsen-LeRoux, C., Charles, J. F., & Pauron, D. (2001). The receptor of *Bacillus sphaericus* binary toxin in *Culex pipiens* (Diptera: Culicidae) midgut: Molecular cloning and expression. *Insect Biochemistry and Molecular Biology*, *31*(10), 981–990. https://doi.org/10.1016/S0965-1748(01)00046-7

Davidson, E. W. (1981). A review of the pathology of bacilli infecting mosquitoes, including an ultrastructural study of larvae fed *Bacillus sphaericus* 1593 spores. *Developments in Industrial Microbiology*, *22*, 69–81.

Davidson, E. W. (1989). Variation in binding of *Bacillus sphaericus* toxin and wheat germ agglutinin to larval midgut cells of six species of mosquitoes. *Journal of Invertebrate Pathology*, *53*(2), 251–259. https://doi.org/10.1016/0022-2011(89)90015-3

Davidson, E. W., Morton, H. L., Moffett, J. O., & Singer, S. (1977). Effect of *Bacillus sphaericus* strain SSII-1 on honey bees, *Apis mellifera*. *Journal of Invertebrate Pathology*, *29*(3), 344–346. https://doi.org/10.1016/S0022-2011(77)80041-4

Davidson, E. W., Oei, C., Meyer, M., Bieber, A. L., Hindley, J., & Berry, C. (1990). Interaction of the *Bacillus sphaericus* mosquito larvicidal proteins. *Canadian Journal of Microbiology*, *36*(12), 870–878. https://doi.org/10.1099/00221287-138-7-1515

Davidson, E. W., & Titus, M. (1987). Ultrastructural effects of the *Bacillus sphaericus* mosquito larvicidal toxin on cultured mosquito cells. *Journal of Invertebrate Pathology*, *50*(3), 213–220. https://doi.org/10.1016/0022-2011(87)90085-1

de Barjac, H. (1990). Classification of *Bacillus sphaericus* strains and comparative toxicity to mosquito larvae. In H. de Barjac, & D. J. Sutherland (Eds.), *Bacterial control of mosquitoes & black flies* (pp. 228–236). Springer Dordrecht. https://doi.org/10.1007/978-94-011-5967-8_14

de Barjac, H., & Charles, J. F. (1983). Une nouvelle toxine active sur les moustiques, présente dans des inclusions cristallines produites par *Bacillus sphaericus*. *Comptes Rendus de l'Académie des Sciences—Serie3*, *296*(19), 905–910.

de Barjac, H., Larget-Thiery, I., Dumanoir, V. C., & Ripouteau, H. (1985). Serological classification of *Bacillus sphaericus* strains on the basis of toxicity to mosquito larvae. *Applied Microbiology and Biotechnology*, *21*(1), 85–90. https://doi.org/10.1007/BF00252367

de Barjac, H., Thiery, I., Cosmao-Dumanoir, V., Frachon, E., Laurent, P. H., Charles, J. F., Hamon, S., & Ofori, J. (1988). Another *Bacillus sphaericus* serotype harbouring strains very toxic to mosquito larvae: Serotype H6. *Annales de l'Institut Pasteur/Microbiologie*, *139*(3), 363–377. https://doi.org/10.1016/0769-2609(88)90028-2

de Melo, J. V., Jones, G. W., Berry, C., Vasconcelos, R. H. T., De Oliveria, C. M. F., Furtado, A. F., Peixoto, C. A., & Silva-Filha, M. H. N. L. (2009). Cry48Aa/Cry49Aa binary toxin from *Bacillus sphaericus* displays cytopathological effects on susceptible and binary toxin-resistant *Culex quinquefasciatus* larvae. *Applied and Environmental Microbiology*, *75*(14), 4782–4789. https://doi.org/10.1128/AEM.00811-09

de Mendonça Santos, E. M., Regis, L. N., Silva-Filha, M. H. N. L., Barbosa, R. M. R., de Melo-Santos, M. A. V., Gomes, T. C. S., & de Oliveira, C. M. F. (2018). The effectiveness of a combined bacterial larvicide for mosquito control in an endemic urban area in Brazil. *Biological Control*, *121*, 190–198. https://doi.org/10.1016/j.biocontrol.2018.03.006

de Muro, M. A., Mitchell, W. J., & Priest, F. G. (1992). Differentiation of mosquito-pathogenic strains of *Bacillus sphaericus* from non-toxic varieties by ribosomal RNA gene restriction patterns. *Microbiology*, *138*(6), 1159–1166. https://doi.org/10.1099/00221287-138-6-1159

de Muro, M. A., & Priest, F. G. (1994). A colony hybridization procedure for the identification of mosquitocidal strains of *Bacillus sphaericus* on isolation plates. *Journal of Invertebrate Pathology*, *63*(3), 310–313. https://doi.org/10.1006/jipa.1994.1058

Diédhiou, S. M., Konaté, L., Doucouré, S., Samb, B., Niang, E. A., Sy, O., Thiaw, O., Konate, A., Wotodjo, A. N., Diallo, M., Gadiaga, L., Sokhna, C., & Faye, O. (2017). Effectiveness of three biological larvicides and of an insect growth regulator against *Anopheles arabiensis* in Senegal. *Bulletin de la Société de Pathologie Exotique*, *110*, 102–115. https://doi.org/10.1007/s13149-016-0531-4

Dritz, D. A., Lawler, S. P., Evkhanian, C., Graham, P., Baracosa, V., & Dula, G. (2011). Control of mosquito larvae in seasonal wetlands on a wildlife refuge using Vectomax™ CG. *Journal of the American Mosquito Control Association*, *27*(4), 398–403. https://doi.org/10.2987/11-6161.1

Elangovan, G., Shanmugavelu, M., Rajamohan, F., Dean, D. H., & Jayaraman, K. (2000). Identification of the functional site in the mosquito larvicidal binary toxin of *Bacillus sphaericus* 1593M by site-directed mutagenesis. *Biochemical and Biophysical Research Communications*, *276*(3), 1048–1055. https://doi.org/10.1006/bbrc.2000.3575

El-Bendary, M. A. (2006). *Bacillus thuringiensis* and *Bacillus sphaericus* biopesticides production. *Journal of Basic Microbiology*, *46*(2), 158–170. https://doi.org/10.1002/jobm.200510585

El-Bendary, M. A., Priest, F. G., Charles, J. F., & Mitchell, W. J. (2005). Crystal protein synthesis is dependent on early sporulation gene expression in *Bacillus sphaericus*. *FEMS Microbiology Letters*, *252*(1), 51–56. https://doi.org/10.1016/j.femsle.2005.08.027

Ellis, R. T., Stockhoff, B. A., Stamp, L., Schnepf, H. E., Schwab, G. E., Knuth, M., Russell, J., Cardineau, G. A., & Narva, K. E. (2002). Novel *Bacillus thuringiensis* binary insecticidal crystal proteins active on western corn rootworm, *Diabrotica virgifera virgifera* LeConte. *Applied and Environmental Microbiology*, *68*(3), 1137–1145. https://doi.org/10.1128/AEM.68.3.1137-1145.2002

Eritja, R. (2013). Laboratory tests on the efficacy of VBC60035, a combined larvicidal formulation of *Bacillus thuringiensis israelensis* (strain AM65–52) and *Bacillus sphaericus* (strain 2362) against *Aedes albopictus* in simulated catch basins. *Journal of the American Mosquito Control Association*, *29*(3), 280–283. https://doi.org/10.2987/12-6308R.1

Farrugia, C., Borg, R. P., Ferrara, L., & Buhagiar, J. (2019). The application of *Lysinibacillus sphaericus* for surface treatment and crack healing in mortar. *Frontiers in Built Environment*, *5*, 62. https://doi.org/10.3389/fbuil.2019.00062

Fernandez-Luna, M. T., Lanz-Mendoza, H., Gill, S. S., Bravo, A., Soberon, M., & Miranda-Rios, J. (2010). An α-amylase is a novel receptor for *Bacillus thuringiensis* ssp. *israelensis* Cry4Ba and Cry11Aa toxins in the malaria vector mosquito *Anopheles albimanus* (Diptera: Culicidae). *Environmental Microbiology*, *12*(3), 746–757. https://doi.org/10.1111/j.1462-2920.2009.02117.x

Ferreira, L. M., Romão, T. P., de-Melo-Neto, O. P., & Silva-Filha, M. H. N. L. (2010). The orthologue to the Cpm1/Cqm1 receptor in *Aedes aegypti* is expressed as a midgut GPI-anchored α-glucosidase, which does not bind to the insecticidal binary toxin. *Insect Biochemistry and Molecular Biology*, *40*(8), 604–610. https://doi.org/10.1016/j.ibmb.2010.05.007

Fillinger, U., Kannady, K., William, G., Vanek, M. J., Dongus, S., Nyika, D., Geissbühler, Y., Chaki, P. P., Govella, N. J., Mathenga, E. M., Singer, B. H., Mshinda, H., Lindsay, S. W., Tanner, M., Mtasiwa, D., de Castro, M. C., & Killeen, G. F. (2008). A tool box for operational mosquito larval control: Preliminary results and early lessons from the Urban Malaria Control Programme in Dar es Salaam, Tanzania. *Malaria Journal*, *7*, 20. https://doi.org/10.1186/1475-2875-7-20

Fillinger, U., & Lindsay, S. W. (2006). Suppression of exposure to malaria vectors by an order of magnitude using microbial larvicides in rural Kenya. *Tropical Medicine & International Health*, *11*(11), 1629–1642. https://doi.org/10.1111/j.1365-3156.2006.01733.x

Fontoura, P. S., da Costa, A. S., Ribeiro, F. S., Ferreira, M. S., Castro, M. C., & Ferreira, M. U. (2020). Field efficacy of VectoMax FG and VectoLex CG biological larvicides for malaria vector control in northwestern Brazil. *Journal of Medical Entomology*, *57*(3), 942–946. https://doi.org/10.1093/jme/tjz220

Fontoura, P. S., Silva, M. F., da Costa, A. S., Ribeiro, F. S., Ferreira, M. S., Ladeia-Andrade, S., Tonini, J., Rodrigues, P. T., Cstro, M. C., & Ferreira, M. U. (2021). Monthly biological larviciding associated with a tenfold decrease in larval density in fish farming ponds and reduced community-wide malaria incidence in northwestern Brazil. *Parasites & Vectors*, *14*, 445. https://doi.org/10.1186/s13071-021-04964-3

Frachon, E., Hamon, S., Nicolas, L., & de Barjac, H. (1991). Cellular fatty acid analysis as a potential tool for predicting mosquitocidal activity of *Bacillus sphaericus* strains. *Applied and Environmental Microbiology*, *57*(11), 3394–3398. https://doi.org/10.1128/aem.57.11.3394–3398.1991

From, C., Granum, P. E., & Hardy, S. P. (2008). Demonstration of a cholesterol-dependent cytolysin in a noninsecticidal *Bacillus sphaericus* strain and evidence for widespread distribution of the toxin within the species. *FEMS Microbiology Letters*, *286*(1), 85–92. https://doi.org/10.1111/j.1574-6968.2008.01256.x

From, C., Pukall, R., Schumann, P., Hormazábal, V., & Granum, P. E. (2005). Toxin-producing ability among *Bacillus* spp. outside the *Bacillus cereus* group. *Applied and Environmental Microbiology*, *71*(3), 1178–1183. https://doi.org/10.1128/AEM.71.3.1178-1183.2005

Ge, Y., Hu, X., Zheng, D., Wu, Y., & Yuan, Z. (2011). Allelic diversity and population structure of *Bacillus sphaericus* as revealed by multilocus sequence typing. *Applied and Environmental Microbiology*, *77*(15), 5553–5556. https://doi.org/10.1128/AEM.00207-11

Georghiou, G. P., & Wirth, M. C. (1997). Influence of exposure to single versus multiple toxins of *Bacillus thuringiensis* subsp. *israelensis* on development of resistance in the mosquito *Culex quinquefasciatus* (Diptera: Culicidae). *Applied and Environmental Microbiology*, *63*(3), 1095–1101. https://doi.org/10.1128/aem.63.3.1095-1101.1997

Go, J., Lee, K. M., Park, Y., & Yoon, S. S. (2014). Extended longevity and robust early-stage development of *Caenorhabditis elegans* by a soil microbe, *Lysinibacillus sphaericus*. *Environmental Microbiology Reports*, *6*(6), 730–737. https://doi.org/10.1111/1758-2229.12196

Gou, Y., Zhou, R., Ye, X., Gao, S., & Li, X. (2015). Highly efficient in vitro biosynthesis of silver nanoparticles using *Lysinibacillus sphaericus* MR-1 and their characterization. *Science and Technology of Advanced Materials*. https://doi.org/10.1088/1468-6996/16/1/015004

Guerineau, M., Alexander, B., & Priest, F. G. (1991). Isolation and identification of *Bacillus sphaericus* strains pathogenic for mosquito larvae. *Journal of Invertebrate Pathology*, *57*(3), 325–333. https://doi.org/10.1016/0022-2011(91)90136-E

Guidi, V., Lüthy, P., & Tonolla, M. (2013). Comparison between diflubenzuron and a *Bacillus thuringiensis israelensis*—and *Lysinibacillus sphaericus*—based formulation for the control of mosquito larvae in urban catch basins in Switzerland. *Journal of the American Mosquito Control Association*, *29*(2), 138–145. https://doi.org/10.2987/12-6301R.1

Hadapad, A. B., Hire, R. S., Vijayalakshmi, N., & Dongre, T. K. (2009). UV protectants for the biopesticide based on *Bacillus sphaericus* Neide and their role in protecting the binary toxins from UV radiation. *Journal of Invertebrate Pathology*, *100*(3), 147–152. https://doi.org/10.1016/j.jip.2008.12.003

Hadapad, A. B., Vijayalakshmi, N., Hire, R. S., & Dongre, T. K. (2008). Effect of ultraviolet radiation on spore viability and mosquitocidal activity of an indigenous ISPC-8 *Bacillus sphaericus* Neide strain. *Acta Tropica*, *107*(2), 113–116. https://doi.org/10.1016/j.actatropica.2008.04.024

Harbison, J. E., Henry, M., Xamplas, C., Berry, R., Bhattacharya, D., & Dugas, L. R. (2014). A comparison of FourStar™ briquets and Natular™ XRT tablets in a North Shore suburb of Chicago, IL. *Journal of the American Mosquito Control Association*, *30*(1), 68–70. https://doi.org/10.2987/13-6355.1

Hindley, J., & Berry, C. (1987). Identification, cloning and sequence analysis of the *Bacillus sphaericus* 1593 41.9 kD larvicidal toxin gene. *Molecular Microbiology*, *1*(2), 187–194. https://doi.org/10.1111/j.1365-2958.1987.tb00511.x

Hire, R. S., Hadapad, A. B., Dongre, T. K., & Kumar, V. (2009). Purification and characterization of mosquitocidal *Bacillus sphaericus* BinA protein. *Journal of Invertebrate Pathology*, *101*(2), 106–111. https://doi.org/10.1016/j.jip.2009.03.005

Hu, X., Fan, W., Han, B., Liu, H., Zheng, D., Li, Q., Dong, W., Yan, J., Gao, M., Berry, C., & Yuan, Z. (2008a). Complete genome sequence of the mosquitocidal bacterium *Bacillus sphaericus* C3–41 and comparison with those of closely related *Bacillus* species. *Journal of Bacteriology*, *190*(8), 2892–2902. https://doi.org/10.1128/JB.01652-07.

Hu, X., Li, J., Hansen, B. M., & Yuan, Z. (2008b). Phylogenetic analysis and heterologous expression of surface layer protein SlpC of *Bacillus sphaericus* C3–41. *Bioscience, Biotechnology, and Biochemistry*, *72*(5), 1257–1263. https://doi.org/10.1271/bbb.70747

Ibañez-Justicia, A., Teekema, S., Den Hartog, W., Jacobs, F., Dik, M., & Stroo, A. (2018). The effectiveness of Asian bush mosquito (*Aedes japonicus japonicus*) control actions in colonised peri-urban areas in the Netherlands. *Journal of Medical Entomology*, *55*(3), 673–680. https://doi.org/10.1093/jme/tjy002

Janesch, B., Messner, P., & Schäffer, C. (2013). Are the surface layer homology domains essential for cell surface display and glycosylation of the S-layer protein from *Paenibacillus alvei* CCM 2051T? *Journal of Bacteriology*, *195*(3), 565–575. http://dx.doi.org/10.1128/JB.01487-12

Jones, G. W., Nielsen-Leroux, C., Yang, Y., Yuan, Z., Fiúza Dumas, V., Monnerat, R. G., & Berry, C. (2007). A new Cry toxin with a unique two-component dependency from *Bacillus sphaericus*. *ASEB Journal*, *21*(14), 4112–4120. https://doi.org/10.1096/fj.07-8913com

Jones, G. W., Wirth, M. C., Monnerat, R. G., & Berry, C. (2008). The Cry48Aa-Cry49Aa binary toxin from *Bacillus sphaericus* exhibits highly restricted target specificity. *Environmental Microbiology*, *10*(9), 2418–2424. https://doi.org/10.1111/j.1462-2920.2008.01667.x

Kahindi, S. C., Midega, J. T., Mwangangi, J. M., Kibe, L. W., Nzovu, J., Luethy, P., Githure, J., & Mbogo, C. M. (2008). Efficacy of Vectobac DT and Culinexcombi against mosquito larvae in unused swimming pools in Malindi, Kenya. *Journal of the American Mosquito Control Association*, *24*(4), 538–542. https://doi.org/10.2987/5734.1

Kahindi, S. C., Muriu, S., Derua, Y. A., Wang, X., Zhou, G., Lee, M. C., Mwangangi, J., Atieli, H., Githeko, A. K., & Yan, G. (2018). Efficacy and persistence of long-lasting microbial larvicides against malaria vectors in western Kenya highlands. *Parasites & Vectors*, *11*, 438. https://doi.org/10.1186/s13071-018-3009-z

Kalfon, A., Charles, J. F., Bourgouin, C., & de Barjac, H. (1984). Sporulation of *Bacillus sphaericus* 2297: An electron microscope study of crystal-like inclusion biogenesis and toxicity to mosquito larvae. *Microbiology*, *130*(4), 893–900. https://doi.org/10.1099/00221287-130-4-893

Kanwal, S., & Boonserm, P. (2023). Study on cellular localization of bin toxin and its apoptosis-inducing effect on human nasopharyngeal carcinoma cells. *Current Cancer Drug Targets*, *23*(5), 388–399. https://doi.org/10.2174/1568009623666221124102524

Karch, S., Monteny, N., & Coz, J. (1988). Persistence of *Bacillus sphaericus* in a mosquito breeding site 4 years after its introduction for microbial control. *Comptes Rendus de l'Académie des Sciences, III (Sciences de la Vie)*, *307*(5), 289–292.

Kellen, W. R., Clark, T. B., Lindegren, J. E., Ho, B. C., Rogoff, M. H., & Singer, S. (1965). *Bacillus sphaericus* Neide as a pathogen of mosquitoes. *Journal of Invertebrate Pathology*, *7*(4), 442–448. https://doi.org/10.1016/0022-2011(65)90120-5

Kern, J., & Schneewind, O. (2010). BslA, the S-layer adhesin of *B. anthracis*, is a virulence factor for anthrax pathogenesis. *Molecular Microbiology*, *75*(2), 324–332. https://doi.org/10.1111/j.1365-2958.2009.06958.x

Key, P. B., & Scott, G. I. (1992). Acute toxicity of the mosquito larvicide, *Bacillus sphaericus*, to the grass shrimp, *Palaemonetes pugio*, and mummichog, *Fundulus heteroclitus*. *Bulletin of Environmental Contamination and Toxicology*, *49*(3), 425–430. https://doi.org/10.1007/BF01239647

Knowles, B. H., & Ellar, D. J. (1987). Colloid-osmotic lysis is a general feature of the mechanism of action of *Bacillus thuringiensis* δ-endotoxins with different insect specificity. *Biochimica et Biophysica Acta (BBA)-General Subjects*, *924*(3), 509–518. https://doi.org/10.1016/0304-4165(87)90167-X

Krych, V. K., Johnson, J. L., & Yousten, A. A. (1980). Deoxyribonucleic acid homologies among strains of *Bacillus sphaericus*. *International Journal of Systematic and Evolutionary Microbiology*, *30*, 476–484. https://doi.org/10.1099/00207713-30-2-476

Lacey, L. A. (1983). Larvicidal activity of *Bacillus* pathogens against *Toxorhynchites* mosquitoes (Diptera: Culicidae). *Journal of Medical Entomology*, *20*(6), 620–624. https://doi.org/10.1093/jmedent/20.6.620

Lacey, L. A. (2007). *Bacillus thuringiensis* serovariety israelensis and *Bacillus sphaericus* for mosquito control. *Journal of the American Mosquito Control Association*, *23*(sp2), 133–163. https://doi.org/10.2987/8756-971X(2007)23[133:BTSIAB]2.0.CO;2

Lacey, L. A., Lacey, C. M., Peacock, B., & Thiery, I. (1988). Mosquito host range and field activity of *Bacillus sphaericus* isolate 2297 (serotype 25). *Journal of the American Mosquito Control Association*, *4*(1), 51–56.

Lee, H. K., Cheong, H., & Gill, S. S. (1998). Microbial control of insects: Use of bacterial insecticides. In G. S. Dhaliwal, & E. A. Heinrichs (Eds.), *Critical issues in insect pest management* (pp. 87–117). Commonwealth Publishers.

Lee, H. L., David, L., Nazni, W. A., Rozilawati, H., Nurulhusna, H., Afizah, A. N., Rosilawati, R., Roziah, A., The, C. H., & Seleena, B. (2016). Thermally applied *Lysinibacillus sphaericus* and pyrethroids against *Culex sitiens* Wiedemann and *Culex quinquefasciatus* Say in Malaysia. *Southeast Asian Journal of Tropical Medicine and Public Health*, *47*(4), 747–758.

Lee, S. B., Chen, J., Aimanova, K. G., & Gill, S. S. (2015). *Aedes* cadherin mediates the in vivo toxicity of the Cry11Aa toxin to *Aedes aegypti*. *Peptides*, *68*, 140–147. https://doi.org/10.1016/j.peptides.2014.07.015

Lekakarn, H., Promdonkoy, B., & Boonserm, P. (2015). Interaction of *Lysinibacillus sphaericus* binary toxin with mosquito larval gut cells: Binding and internalization. *Journal of Invertebrate Pathology*, *132*, 125–131. https://doi.org/10.1016/j.jip.2015.09.010

Li, J., Yang, L., Hu, X., Zheng, D., Yan, J., & Yuan, Z. (2013). Nanoscale mono-and multi-layer cylinder structures formed by recombinant S-layer proteins of mosquitocidal *Bacillus sphaericus* C3–41. *Applied Microbiology and Biotechnology*, *97*, 7275–7283. https://doi.org/10.1007/s00253-012-4664-1

Likitvivatanavong, S., Chen, J., Evans, A. M., Bravo, A., Soberon, M., & Gill, S. S. (2011). Multiple receptors as targets of Cry toxins in mosquitoes. *Journal of Agricultural and Food Chemistry*, *59*(7), 2829–2838. https://doi.org/10.1021/jf1036189

Liu, J. W., Porter, A. G., Wee, B. Y., & Thanabalu, T. (1996). New gene from nine *Bacillus sphaericus* strains encoding highly conserved 35.8-kilodalton mosquitocidal toxins. *Applied and Environmental Microbiology*, *62*(6), 2174–2176. https://doi.org/10.1128/aem.62.6.2174-2176.1996

Lozano, L. C., Ayala, J. A., & Dussán, J. (2011). *Lysinibacillus sphaericus* S-layer protein toxicity against *Culex quinquefasciatus*. *Biotechnology Letters*, *33*, 2037–2041. https://doi.org/10.1007/s10529-011-0666-9

Lozano, L. C., & Dussán, J. (2017). Synergistic activity between S-layer protein and spore–crystal preparations from *Lysinibacillus sphaericus* against *Culex quinquefasciatus* larvae. *Current Microbiology*, *74*, 371–376. https://doi.org/10.1007/s00284-016-1185-7

Majambere, S., Lindsay, S. W., Green, C., Kandeh, B., & Fillinger, U. (2007). Microbial larvicides for malaria control in The Gambia. *Malaria Journal*, *6*, 76. https://doi.org/10.1186/1475-2875-6-76

Marei, G. I. K., Mohammed, Y. M., Rabea, E. I., & Badawy, M. E. (2018). Isolation, characterisation and efficacy of the bacterial strain *Lysinibacillus sphaericus* YMM in biodegradation of malathion insecticide in liquid media. *International Journal of Environmental Studies*, *76*(4), 616–633. https://doi.org/10.1080/00207233.2018.1522922

Mathavan, S., Velpandi, A., & Johnson, J. C. (1987). Sub-toxic effects of *Bacillus sphaericus* 1593 M on feeding growth and reproduction of *Laccotrephes griseus* (Hemiptera: Nepidae). *Experimental Biology*, *46*(3), 149–153.

McMillan, J. R., Olson, M. M., Petruff, T., Shepard, J. J., & Armstrong, P. M. (2022). Impacts of *Lysinibacillus sphaericus* on mosquito larval community composition and larval competition between *Culex pipiens* and *Aedes albopictus*. *Scientific Reports*, *12*(1), 18013. https://doi.org/10.1038/s41598-022-21939-1

Merritt, R. W., Lessard, J. L., Wessell, K. J., Hernandez, O., Berg, M. B., Wallace, J. R., Novak, J. A., Ryan, J., & Merritt, B. W. (2005). Lack of effects of *Bacillus sphaericus* (Vectolex®) on nontarget organisms in a mosquito-control program in southeastern Wisconsin: A 3-year study. *Journal of the American Mosquito Control Association*, *21*(2), 201–212. https://doi.org/10.2987/8756-971X(2005)21[201:LOEOBS]2.0.CO;2

Mulla, M. S., Darwazeh, H. A., Davidson, E. W., Dulmage, H. T., & Singer, S. (1984). Larvicidal activity and field efficacy of *Bacillus sphaericus* strains against mosquito larvae and their safety to nontarget organisms. *Mosquito News*, *44*(3), 336–342.

Mulligan III, F. S., Schaefer, C. H., & Miura, T. (1978). Laboratory and field evaluation of *Bacillus sphaericus* as a mosquito control agent. *Journal of Economic Entomology*, *71*(5), 774–777. https://doi.org/10.1093/jee/71.5.774

Mwangangi, J. M., Kahindi, S. C., Kibe, L. W., Nzovu, J. G., Luethy, P., Githure, J. I., & Mbogo, C. M. (2011). Wide-scale application of Bti/Bs biolarvicide in different aquatic habitat types in urban and peri-urban Malindi, Kenya. *Parasitology Research*, *108*, 1355–1363. https://doi.org/10.1007/s00436-010-2029-1

Myers, P., Yousten, A. A., & Davidson, E. W. (1979). Comparative studies of the mosquito-larval toxin of *Bacillus sphaericus* SSII-1 and 1593. *Canadian Journal of Microbiology*, *25*(11), 1227–1231. https://doi.org/10.1139/m79-193

Nakamura, L. K. (2000). Phylogeny of *Bacillus sphaericus*-like organisms. *International Journal of Systematic and Evolutionary Microbiology*, *5*(5), 1715–1722. https://doi.org/10.1099/00207713-50-5-1715

Nakamura, L. K., Shida, O., Takagi, H., & Komagata, K. (2002). *Bacillus pycnus* sp. nov. and *Bacillus neidei* sp. nov., round-spored bacteria from soil. *International Journal of Systematic and Evolutionary Microbiology*, *52*(2), 501–505. https://doi.org/10.1099/00207713-52-2-501

Narasu, M. L., & Gopinathan, K. P. (1988). Effect of *Bacillus sphaericus* 1593 toxin on choline acetyl transferase and mitochondrial oxidative activities of the mosquito larvae. *Indian Journal of Biochemistry & Biophysics*, *25*(3), 253–256.

Nasci, R. S., Runde, A. B., Henry, M., & Harbison, J. E. (2017). Effectiveness of five products to control *Culex pipiens* larvae in urban stormwater catch basins. *Journal of the American Mosquito Control Association*, *33*(4), 309–317. https://doi.org/10.2987/17-6686.1

Naureen, Z., Rehman, N. U., Hussain, H., Hussain, J., Gilani, S. A., Al Housni, S. K., Mabood, F., Khan, A. L., Farooq, S., Abbas, G., & Harrasi, A. A. (2017). Exploring the potentials of *Lysinibacillus sphaericus* ZA9 for plant growth promotion and biocontrol activities against phytopathogenic fungi. *Frontiers in Microbiology*, *8*, 1477. https://doi.org/10.3389/fmicb.2017.01477

Neide, E. (1904). Botanische beschreibund einiger sporenbildenden bacterien. *Zentralblatt für Bakteriologie, Parasitenkunde, Infektionskrankheiten und Hygiene Abteilung II*, *12*, 337–352.

Nicolas, L., Darriet, F., & Hougard, J. M. (1987a). Efficacy of *Bacillus sphaericus* 2362 against larvae of *Anopheles gambiae* under laboratory and field conditions in West Africa. *Medical and Veterinary Entomology*, *1*(2), 157–162. https://doi.org/10.1111/j.1365-2915.1987.tb00337.x

Nicolas, L., Dossou-Yovo, J., & Hougard, J. M. (1987b). Persistence and recycling of *Bacillus sphaericus* 2362 spores in *Culex quinquefasciatus* breeding sites in West Africa. *Applied Microbiology and Biotechnology*, *25*, 341–345. https://doi.org/10.1007/BF00252544

Nicolas, L., Lecroisey, A., & Charles, J. F. (1990). Role of the gut proteinases from mosquito larvae in the mechanism of action and the specificity of the *Bacillus sphaericus* toxin. *Canadian Journal of Microbiology*, *36*(11), 804–807. https://doi.org/10.1139/m90-138

Nicolas, L., Nielsen-Leroux, C., Charles, J. F., & Delécluse, A. (1993). Respective role of the 42-and 51-kDa components of the *Bacillus sphaericus* toxin overexpressed in *Bacillus thuringiensis*. *FEMS Microbiology Letters*, *106*(3), 275–279. https://doi.org/10.1111/j.1574-6968.1993.tb05976.x

Nielsen-Leroux, C., & Charles, J. F. (1992). Binding of *Bacillus sphaericus* binary toxin to a specific receptor on midgut brush-border membranes from mosquito larvae. *European Journal of Biochemistry*, *210*(2), 585–590. https://doi.org/10.1111/j.1432-1033.1992.tb17458.x

Nielsen-Leroux, C., Pasteur, N., Prètre, J., Charles, J. F., Ben Sheikh, H., & Chevillon, C. (2002). High resistance to *Bacillus sphaericus* binary toxin in *Culex pipiens* (Diptera: Culicidae): The complex situation of West Mediterranean countries. *Journal of Medical Entomology*, *39*(5), 729–735. https://doi.org/10.1603/0022-2585-39.5.729

Nishiwaki, H., Nakashima, K., Ishida, C., Kawamura, T., & Matsuda, K. (2007). Cloning, functional characterization, and mode of action of a novel insecticidal pore-forming toxin, sphaericolysin, produced by *Bacillus sphaericus*. *Applied and Environmental Microbiology*, *73*(10), 3404–3411. https://doi.org/10.1128/AEM.00021-07

Ocampo, C. B., Calderon, G. I. G., Perez, M., & Morales, C. A. (2008). Evaluation of the triflumuron and the mixture of *Bacillus thuringiensis* plus *Bacillus sphaericus* for control of the immature stages of *Aedes aegypti* and *Culex quinquefasciatus* (Diptera: Culicidae) in catch basins. *Biomedica*, *28*(2), 224–233. https://doi.org/10.7705/biomedica.v28i2.93

Oei, C., Hindley, J., & Berry, C. (1990). An analysis of the genes encoding the 51.4-and 41.9-kDa toxins of *Bacillus sphaericus* 2297 by deletion mutagenesis: The construction of fusion proteins. *FEMS Microbiology Letters*, *72*(3), 265–273. https://doi.org/10.1111/j.1574-6968.1990.tb03900.x

Oei, C., Hindley, J., & Berry, C. (1992). Binding of purified *Bacillus sphaericus* binary toxin and its deletion derivatives to *Culex quinquefasciatus* gut: Elucidation of functional binding domains. *Microbiology*, *138*(7), 1515–1526. https://doi.org/10.1099/00221287-138-7-1515

Opota, O., Charles, J. F., Warot, S., Pauron, D., & Darboux, I. (2008). Identification and characterization of the receptor for the *Bacillus sphaericus* binary toxin in the malaria vector mosquito, *Anopheles gambiae*. *Comparative Biochemistry and Physiology Part B: Biochemistry and Molecular Biology*, *149*(3), 419–427. https://doi.org/10.1016/j.cbpb.2007.11.002

Opota, O., Gauthier, N. C., Doye, A., Berry, C., Gounon, P., Lemichez, E., & Pauron, D. (2011). *Bacillus sphaericus* binary toxin elicits host cell autophagy as a response to intoxication. *PLoS One*, *6*(2), e14682. https://doi.org/10.1371/journal.pone.0014682

Partridge, M. R., & Berry, C. (2002). Insecticidal activity of the *Bacillus sphaericus* Mtx1 toxin against *Chironomus riparus*. *Journal of Invertebrate Pathology*, *79*(2), 135–136. https://doi.org/10.1016/S0022-2011(02)00025-3

Pauchet, Y., Luton, F., Castella, C., Charles, J. F., Romey, G., & Pauron, D. (2005). Effects of a mosquitocidal toxin on a mammalian epithelial cell line expressing its target receptor. *Cellular Microbiology*, *7*(9), 1335–1344. https://doi.org/10.1111/j.1462-5822.2005.00560.x

Pei, G., Oliveira, C. M., Yuan, Z., Nielsen-LeRoux, C., Silva-Filha, M. H., Yan, J., & Regis, L. (2002). A strain of *Bacillus sphaericus* causes slower development of resistance in *Culex quinquefasciatus*. *Applied and Environmental Microbiology*, *68*(6), 3003–3009. https://doi.org/10.1128/AEM.68.6.3003-3009.2002

Pena, G., Miranda-Rios, J., de la Riva, G., Pardo-López, L., Soberón, M., & Bravo, A. (2006). A *Bacillus thuringiensis* S-layer protein involved in toxicity against *Epilachna varivestis* (Coleoptera: Coccinellidae). *Applied and Environmental Microbiology*, *72*(1), 353–360. https://doi.org/10.1128/AEM.72.1.353-360.2006

Pener, H., & Wilamowski, A. (1996). Susceptibility of larvae of the sandfly *Phlebotomus papatasi* (Diptera: Psychodidae) to *Bacillus sphaericus*. *Bulletin of Entomological Research*, *86*(2), 173–175. https://doi.org/10.1017/S000748530005241X

Porter, A. G., Davidson, E. W., & Liu, J. W. (1993). Mosquitocidal toxins of bacilli and their genetic manipulation for effective biological control of mosquitoes. *Microbiological Reviews*, *57*(4), 838–861. https://doi.org/10.1128/mr.57.4.838-861.1993

Priest, F. G., Ebdrup, L., Zahner, V., & Carter, P. (1997). Distribution and characterization of mosquitocidal toxin genes in some strains of *Bacillus sphaericus*. *Applied and Environmental Microbiology*, *63*(4), 1195–1198. https://doi.org/10.1128/aem.63.4.1195-1198.1997

Ramoska, W. A., Singer, S., & Levy, R. (1977). Bioassay of three strains of *Bacillus sphaericus* on field-collected mosquito larvae. *Journal of Invertebrate Pathology*, *30*(2), 151–154. https://doi.org/10.1016/0022-2011(77)90212-9

Rao, D. R., Mani, T. R., Rajendran, R., Joseph, A. S., Gajanana, A., & Reuben, R. (1995). Development of a high level of resistance to *Bacillus sphaericus* in a field population of *Culex quinquefasciatus* from Kochi, India. *Journal of the American Mosquito Control Association*, *11*(1), 1–5.

Ravasi, D., Parrondo Monton, D., Tanadini, M., Campana, V., & Flacio, E. (2023). Efficacy of biological larvicide VectoMax® FG against *Aedes albopictus* and *Culex pipiens* under field conditions in urban catch basins. *Journal of the European Mosquito Control Association*, *42*(1), 51–60. https://doi.org/10.52004/JEMCA2023.0005

Regis, L., Silva-Filha, M. H., Nielsen-LeRoux, C., & Charles, J. F. (2001). Bacteriological larvicides of dipteran disease vectors. *TRENDS in Parasitology*, *17*(8), 377–380. https://doi.org/10.1016/S1471-4922(01)01953-5

Robert, L. L., Perich, M. J., Schlein, Y., & Jacobson, J. L. (1998). *Bacillus sphaericus* inhibits hatching of phlebotomine sand fly eggs. *Journal of the American Mosquito Control Association*, *14*(3), 351–352.

Robert, L. L., Perich, M. J., Schlein, Y., Jacobson, R. L., Wirtz, R. A., Lawyer, P. G., & Githure, J. I. (1997). Phlebotomine sand fly control using bait-fed adults to carry the larvicide *Bacillus sphaericus* to the larval habitat. *Journal of the American Mosquito Control Association*, *13*(2), 140–144.

Rojas-Pinzón, P. A., & Dussán, J. (2017). Contribution of *Lysinibacillus sphaericus* hemolysin and chitin-binding protein in entomopathogenic activity against insecticide resistant *Aedes aegypti*. *World Journal of Microbiology and Biotechnology*, *33*, 181. https://doi.org/10.1007/s11274-017-2348-9

Romão, T. P., de Melo Chalegre, K. D., Key, S., Ayres, C. F. J., de Oliveira, C. M. F., de-Melo-Neto, O. P., & Silva-Filha, M. H. N. L. (2006). A second independent resistance mechanism to *Bacillus sphaericus* binary toxin targets its α-glucosidase receptor in *Culex quinquefasciatus*. *The FEBS Journal*, *273*(7), 1556–1568. https://doi.org/10.1111/j.1742-4658.2006.05177.x

Ruiu, L. (2015). Insect pathogenic bacteria in integrated pest management. *Insects*, *6*(2), 352–367. https://doi.org/10.3390/insects6020352

Rupar, M. J., Donovan, W. P., Chu, C. R., Pease, E., Tan, Y., Slaney, A. C., Malvar, T. M., & Baum, J. A. (2003). *Nucleic acids encoding coleopteran-toxic polypeptides and insect-resistant transgenic plants comprising them* (U.S. Patent No. 7,078,592). U.S. Patent and Trademark Office. https://patents.google.com/patent/US7078592B2/en

Sanitt, P., Promdonkoy, B., & Boonserm, P. (2008). Targeted mutagenesis at charged residues in *Bacillus sphaericus* BinA toxin affects mosquito-larvicidal activity. *Current Microbiology*, *57*, 230–234. https://doi.org/10.1007/s00284-008-9180-2

Santana-Martínez, J. C., Rojas-Pinzón, P. A., & Dussán, J. (2020). Susceptibility of three Colombian *Aedes aegypti* Linnaeus (Diptera: Culicidae) populations to a *Lysinibacillus sphaericus* formulation. *Journal of Applied Entomology*, *144*(1–2), 54–63. https://doi.org/10.1111/jen.12702

Savage, H. M., & Strickman, D. (2004). The genus and subgenus categories within Culicidae and placement of *Ochlerotatus* as a subgenus of *Aedes*. *Journal of the American Mosquito Control Association*, *20*(2), 208–214.

Saxena, N., Bihari, V., & Basu, S. K. (1994). Sustained release floating bead formulation of *Bacilus sphaericus* 1593 and *Bacillus thuringiensis* H-14 for controlling mosquitoes including disease vectors. *Pesticide Research Journal*, *6*, 37–43.

Schirmer, J., Just, I., & Aktories, K. (2002). The ADP-ribosylating mosquitocidal toxin from *Bacillus sphaericus*: Proteolytic activation, enzyme activity, and cytotoxic effects. *Journal of Biological Chemistry*, *277*(14), 11941–11948. https://doi.org/10.1074/jbc.M108463200

Schwartz, J. L., Potvin, L., Coux, F., Charles, J. F., Berry, C., Humphreys, M. J., Jones, A. F., Bernhart, I., Dalla Serra, M., & Menestrina, G. (2001). Permeabilization of model lipid membranes by *Bacillus sphaericus* mosquitocidal binary toxin and its individual components. *The Journal of Membrane Biology*, *184*, 171–183. https://doi.org/10.1007/s00232-001-0086-1

Šebo, P., Bennardo, T., De La Torre, F., & Szulmajster, J. (1990). Delineation of the minimal portion of the *Bacillus sphaericus* 1593M toxin required for the expression of larvicidal activity. *European Journal of Biochemistry*, *194*(1), 161–165. https://doi.org/10.1111/j.1432-1033.1990.tb19440.x

Shabanamol, S., Divya, K., George, T. K., Rishad, K. S., Sreekumar, T. S., & Jisha, M. S. (2018). Characterization and in planta nitrogen fixation of plant growth promoting endophytic diazotrophic *Lysinibacillus sphaericus* isolated from rice (*Oryza sativa*). *Physiological and Molecular Plant Pathology*, *102*, 46–54. https://doi.org/10.1016/j.pmpp.2017.11.003

Shabanamol, S., Sreekumar, J., & Jisha, M. S. (2017). Bioprospecting endophytic diazotrophic *Lysinibacillus sphaericus* as biocontrol agents of rice sheath blight disease. *3 Biotech*, *7*(5), 337. https://doi.org/10.1007/s13205-017-0956-6

Shanks, J. L., Haigh, A. M., Riegler, M., & Spooner-Hart, R. N. (2017). First confirmed report of a bacterial brood disease in stingless bees. *Journal of Invertebrate Pathology*, *144*, 7–10. https://doi.org/10.1016/j.jip.2017.01.004

Siegel, J. P., & Novak, R. J. (1997). Field trials of Vectolex CG, a *Bacillus sphaericus* larvicide, in Illinois waste tires and catch basins. *Journal of the American Mosquito Control Association*, *13*(4), 305–310.

Silapanuntakul, S., Pantuwatana, S., Bhumiratana, A., & Charoensiri, K. (1983). The comparative persistence of toxicity of *Bacillus sphaericus* strain 1593 and *Bacillus thuringiensis* serotype H-14 against mosquito larvae in different kinds of environments. *Journal of Invertebrate Pathology*, *42*(3), 387–392. https://doi.org/10.1016/0022-2011(83)90179-9

Silva-Filha, M. H. N. L., Berry, C., & Regis, L. (2014). *Lysinibacillus sphaericus*: Toxins and mode of action, applications for mosquito control and resistance management. In T. Dhadialla, & S. Gill (Eds.), *Advances in insect physiology: Vol. 47. Insects midgut and insecticidal proteins* (pp. 89–176). Elsevier. https://doi.org/10.1016/B978-0-12-800197-4.00003-8

Silva-Filha, M. H. N. L., Fontes de Oliveira, C. M., Regis, L., Yuan, Z., Rico, C. M., & Nielsen-LeRoux, C. (2004). Two *Bacillus sphaericus* binary toxins share the midgut receptor binding site: Implications for resistance of *Culex pipiens* complex (Diptera: Culicidae) larvae. *FEMS Microbiology Letters*, *241*(2), 185–191. https://doi.org/10.1016/j.femsle.2004.10.018

Silva-Filha, M. H. N. L., Nielsen-LeRoux, C., & Charles, J. F. (1997). Binding kinetics of *Bacillus sphaericus* binary toxin to midgut brush-border membranes of

Anopheles and *Culex* sp. mosquito larvae. *European Journal of Biochemistry*, *247*(3), 754–761. https://doi.org/10.1111/j.1432-1033.1997.00754.x

Silva-Filha, M. H. N. L., Nielsen-LeRoux, C., & Charles, J. F. (1999). Identification of the receptor for *Bacillus sphaericus* crystal toxin in the brush border membrane of the mosquito *Culex pipiens* (Diptera: Culicidae). *Insect Biochemistry and Molecular Biology*, *29*(8), 711–721. https://doi.org/10.1016/S0965-1748(99)00047-8

Silva-Filha, M. H. N. L., & Peixoto, C. A. (2003). Immunocytochemical localization of the *Bacillus sphaericus* binary toxin components in *Culex quinquefasciatus* (Diptera: Culicidae) larvae midgut. *Pesticide Biochemistry and Physiology*, *77*(3), 138–146. https://doi.org/10.1016/j.pestbp.2003.07.002

Silva-Filha, M. H. N. L., Regis, L., Nielsen-Leroux, C., & Charles, J. F. (1995). Low-level resistance to *Bacillus sphaericus* in a field-treated population of *Culex quinquefasciatus* (Diptera: Culicidae). *Journal of Economic Entomology*, *88*(3), 525–530. https://doi.org/10.1093/jee/88.3.525

Silva-Filha, M. H. N. L., Romão, T. P., Rezende, T. M. T., Carvalho, K. D. S., Gouveia de Menezes, H. S., Alexandre do Nascimento, N., Soberón, M., & Bravo, A. (2021). Bacterial toxins active against mosquitoes: Mode of action and resistance. *Toxins*, *13*(8), 523. https://doi.org/10.3390/toxins13080523

Sinègre, G., Babinot, M., Quermel, J. M., & Gaven, B. (1994). First field occurrence of *Culex pipiens* resistance to *Bacillus sphaericus* in Southern France. *Proceedings of the 8th European Meeting of Society for Vector Ecology* (pp. 5–8). Society for Vector Ecology.

Singer, S. (1973). Insecticidal activity of recent bacterial isolates and their toxins against mosquito larvae. *Nature*, *244*(5411), 110–111. https://doi.org/10.1038/244110a0

Singh, G. J. P., & Gill, S. S. (1988). An electron microscope study of the toxic action of *Bacillus sphaericus* in *Culex quinquefasciatus* larvae. *Journal of Invertebrate Pathology*, *52*(2), 237–247. https://doi.org/10.1016/0022-2011(88)90131-0

Smith, A. W., Camara-Artigas, A., Brune, D. C., & Allen, J. P. (2005). Implications of high-molecular-weight oligomers of the binary toxin from *Bacillus sphaericus*. *Journal of Invertebrate Pathology*, *88*(1), 27–33. https://doi.org/10.1016/j.jip.2004.10.005

Su, T., Thieme, J., Ocegueda, C., Ball, M., & Cheng, M. L. (2018). Resistance to *Lysinibacillus sphaericus* and other commonly used pesticides in *Culex pipiens* (Diptera: Culicidae) from Chico, California. *Journal of Medical Entomology*, *55*(2), 423–428. https://doi.org/10.1093/jme/tjx235

Su, T., Thieme, J., White, G. S., Lura, T., Mayerle, N., Faraji, A., Cheng, M. L., & Brown, M. Q. (2019). High resistance to *Bacillus sphaericus* and susceptibility to other common pesticides in *Culex pipiens* (Diptera: Culicidae) from Salt Lake City, UT. *Journal of Medical Entomology*, *56*(2), 506–513. https://doi.org/10.1093/jme/tjy193

Surya, W., Chooduang, S., Choong, Y. K., Torres, J., & Boonserm, P. (2016). Binary toxin subunits of *Lysinibacillus sphaericus* are monomeric and form heterodimers after in vitro activation. *PLoS One*, *11*(6), e0158356. https://doi.org/10.1371/journal.pone.0158356

Tchicaya, E. S., Koudou, B. G., Keiser, J., Adja, A. M., Cissé, G., Tanner, M., Tano, Y., & Utzinger, J. (2009). Effect of repeated application of microbial larvicides on malaria transmission in central Côte d'Ivoire. *Journal of the American Mosquito Control Association*, *25*(3), 382–385. https://doi.org/10.2987/08-5809.1

Thanabalu, T., Berry, C., & Hindley, J. (1993). Cytotoxicity and ADP-ribosylating activity of the mosquitocidal toxin from *Bacillus sphaericus* SSII-1: Possible roles of the 27-and 70-kilodalton peptides. *Journal of Bacteriology*, *175*(8), 2314–2320. https://doi.org/10.1128/jb.175.8.2314-2320.1993

Thanabalu, T., Hindley, J., & Berry, C. (1992). Proteolytic processing of the mosquitocidal toxin from *Bacillus sphaericus* SSII-1. *Journal of Bacteriology*, *174*(15), 5051–5056. https://doi.org/10.1128/jb.174.15.5051-5056.1992

Thanabalu, T., Hindley, J., Jackson-Yap, J., & Berry, C. (1991). Cloning, sequencing, and expression of a gene encoding a 100-kilodalton mosquitocidal toxin from *Bacillus sphaericus* SSII-1. *Journal of Bacteriology*, *173*(9), 2776–2785. https://doi.org/10.1128/jb.173.9.2776-2785.1991

Thanabalu, T., & Porter, A. G. (1995). Efficient expression of a 100-kilodalton mosquitocidal toxin in protease-deficient recombinant *Bacillus sphaericus*. *Applied and Environmental Microbiology*, *61*(11), 4031–4036. https://doi.org/10.1128/aem.61.11.4031-4036.1995

Thanabalu, T., & Porter, A. G. (1996). A *Bacillus sphaericus* gene encoding a novel type of mosquitocidal toxin of 31.8 kDa. *Gene*, *170*(1), 85–89. https://doi.org/10.1016/0378-1119(95)00836-5

Thierry, D. N., Djamouko-Djonkam, L., Gisèle, F. D., Audrey, M. M. P., Timoléon, T., Hubert, Z. T. S., Sinclair, W. C., & Christophe, A. N. (2023). Assessment of the impact of the biological larvicide VectoMax G: Combination of *Bacillus thuringiensis* and *Lysinibacillus sphaericus* on non-target aquatic organisms in Yaoundé-Cameroon. *Heliyon*, *9*(8), e17723. https://doi.org/10.1016/j.heliyon.2023.e17723

Thiery, I., Back, C., Barbazan, P., & Sinègre, G. (1996). Applications de *Bacillus thuringiensis* et de *B sphaericus* dans la démoustication et la lutte contre les vecteurs de maladies tropicales. *Annales de l'Institut Pasteur/Actualités*, *7*(4), 247–260. https://doi.org/10.1016/S0924-4204(97)86393-7

Thiery, I., & de Barjac, H. (1989). Selection of the most potent *Bacillus sphaericus* strains based on activity ratios determined on three mosquito species. *Applied Microbiology and Biotechnology*, *31*(5), 577–581. https://doi.org/10.1007/BF00270798

Thiery, I., Ofori, J., Cosmao Dumanoir, V., Hamon, S., & de Barjac, H. (1992). New mosquitocidal strains from Ghana belonging to serotypes H3, H6 and H48 of *Bacillus sphaericus*. *Applied Microbiology and Biotechnology*, *37*, 718–722. https://doi.org/10.1007/BF00174834

Tietze, N. S., Olson, M. A., Hester, P. G., & Moore, J. J. (1993). Tolerance of sewage treatment plant microorganisms to mosquitocides. *Journal of the American Mosquito Control Association*, *9*, 477–479.

Tweten, R. K. (2005). Cholesterol-dependent cytolysins, a family of versatile pore-forming toxins. *Infection and Immunity*, *73*(10), 6199–6209. https://doi.org/10.1128/iai.73.10.6199-6209.2005

Vandenbergi, J. D. (1990). Safety of four entomopathogens for caged adult honey bees (Hymenoptera: Apidae). *Journal of Economic Entomology*, *83*(3), 755–759. https://doi.org/10.1093/jee/83.3.755

Vargas, J. E., & Dussán, J. (2016). Adsorption of toxic metals and control of mosquitos-borne disease by *Lysinibacillus sphaericus*: Dual benefits for health and environment. *Biomedical and Environmental Sciences*, *29*(3), 187–196. https://doi.org/10.3967/bes2016.023

Vega-Páez, J. D., Rivas, R. E., & Dussán-Garzón, J. (2019). High efficiency mercury sorption by dead biomass of *Lysinibacillus sphaericus*—new insights into the treatment of contaminated water. *Materials*, *12*(8), 1296. https://doi.org/10.3390/ma12081296

Vinogradov, D. D., Sinev, A. Y., & Tiunov, A. V. (2022). Predators as control agents of mosquito larvae in micro-reservoirs (review). *Inland Water Biology*, *15*, 39–53. https://doi.org/10.1134/S1995082922010138

Wahba, M. M. (2000). The influence of *Bacillus sphaericus* on the biology and histology of *Phlebotomus papatasi*. *Journal of the Egyptian Society of Parasitology*, *30*(1), 315–323.

Wang, Y. T., Oh, S. Y., Hendrickx, A. P., Lunderberg, J. M., & Schneewind, O. (2013). *Bacillus cereus* G9241 S-layer assembly contributes to the pathogenesis of anthrax-like disease in mice. *Journal of Bacteriology*, *195*(3), 596–605. http://dx.doi.org/10.1128/JB.02005-12

Weiser, J. (1984). A mosquito-virulent *Bacillus sphaericus* in adult *Simulium damnosum* from northern Nigeria. *Zentralblatt für Mikrobiologie*, *139*(1), 57–60. https://doi.org/10.1016/S0232-4393(84)80033-5

Wermelinger, E. D., Zanuncio, J. C., Rangel, E. F., Cecon, P. R., & Rabinovitch, L. (2000). Toxicity of *Bacillus* species to larvae of *Lutzomyia longipalpis* (L. & N.) (Diptera: Psychodidae: Phlebotominae). *Anais da Sociedade Entomológica do Brasil*, *29*, 609–614. https://doi.org/10.1590/S0301-80592000000300025

Wickremesinghe, R. S. B., & Mendis, C. L. (1980). *Bacillus sphaericus* spore from Sri Lanka demonstrating rapid larvicidal activity on *Culex quinquefasciatus*. *Mosquito News*, *40*(3), 387–389.

Wirth, M. C., Jiannino, J. A., Federici, B. A., & Walton, W. E. (2004). Synergy between toxins of *Bacillus thuringiensis* subsp. *israelensis* and *Bacillus sphaericus*. *Journal of Medical Entomology*, *41*(5), 935–941. https://doi.org/10.1603/0022-2585-41.5.935

Woodburn, M. A., Yousten, A. A., & Hilu, K. H. (1995). Random amplified polymorphic DNA fingerprinting of mosquito-pathogenic and nonpathogenic strains of *Bacillus sphaericus*. *International Journal of Systematic and Evolutionary Microbiology*, *45*(2), 212–217. https://doi.org/10.1099/00207713-45-2-212

Wraight, S. P., Molloy, D. P., & Singer, S. (1987). Studies on the culicine mosquito host range of *Bacillus sphaericus* and *Bacillus thuringiensis* var *israelensis* with notes on the effects of temperature and instar on bacterial efficacy. *Journal of Invertebrate Pathology*, *49*(3), 291–302. https://doi.org/10.1016/0022-2011(87)90061-9

Yadav, R. S., Sharma, V. P., & Upadhyay, A. K. (1997). Field trial of *Bacillus sphaericus* strain B-101 (serotype H5a, 5b) against filariasis and Japanese encephalitis vectors in India. *Journal of the American Mosquito Control Association*, *13*(2), 158–163.

Young, M. D., Undeen, A. H., Dame, D. A., & Wing, S. R. (1990). The effect of *Bacillus sphaericus* upon the susceptibility of *Anopheles quadrimaculatus* to *Plasmodium berghei*. *Journal of the American Mosquito Control Association*, *6*(1), 139–140.

Yousten, A. A. (1984). Bacteriophage typing of mosquito pathogenic strains of *Bacillus sphaericus*. *Journal of Invertebrate Pathology*, *43*(1), 124–125.

Yousten, A. A., Benfield, E. F., Campbell, R. P., Foss, S. S., & Genthner, F. J. (1991). Fate of *Bacillus sphaericus* 2362 spores following ingestion by nontarget invertebrates. *Journal of Invertebrate Pathology*, *58*(3), 427–435. https://doi.org/10.1016/0022-2011(91)90189-W

Yousten, A. A., & Davidson, E. W. (1982). Ultrastructural analysis of spores and parasporal crystals formed by *Bacillus sphaericus* 2297. *Applied and Environmental Microbiology*, *44*(6), 1449–1455. https://doi.org/10.1128/aem.44.6.1449-1455.1982

Yousten, A. A., de Barjac, H., Hedrick, J., & Myers, P. (1980). Comparison between bacteriophage typing and serotyping for the differentiation of *Bacillus sphaericus* strains. *Annales de Microbiologie*, *131B*(3), 297–308.

Yousten, A. A., Genthner, F. J., & Benfield, E. F. (1992). Fate of *Bacillus sphaericus* and *Bacillus thuringiensis* serovar *israelensis* in the aquatic environment. *Journal of the American Mosquito Control Association*, *8*(2), 143–148.

Yuan, Z. (2002). *Bacillus sphaericus*: Mechanism and application as a mosquito larvicide. In R. K. Upadhyay (Ed.), *Advances in microbial control of insect pests*. Springer. https://doi.org/10.1007/978-1-4757-4437-8_3

Yuan, Z. M., Pei, G. F., Regis, L., Nielsen-Leroux, C., & Cai, Q. X. (2003). Cross-resistance between strains of *Bacillus sphaericus* but not *B. thuringiensis israelensis* in colonies of the mosquito *Culex quinquefasciatus*. *Medical and Veterinary Entomology*, *17*(3), 251–256. https://doi.org/10.1046/j.1365-2915.2003.00429.x

Yuan, Z. M., Zhang, Y., Cai, Q., & Liu, E. Y. (2000). High-level field resistance to *Bacillus sphaericus* C3–41 in *Culex quinquefasciatus* from Southern China. *Biocontrol Science and Technology*, *10*(1), 41–49. https://doi.org/10.1080/09583150029378

Zahiri, N. S., & Mulla, M. S. (2005). Non-larvicidal effects of *Bacillus thuringiensis israelensis* and *Bacillus sphaericus* on oviposition and adult mortality of *Culex quinquefasciatus* Say (Diptera: Culicidae). *Journal of Vector Ecology*, *30*(1), 155–162.

Zhang, Y., Liu, E., Cai, C., & Chen, Z. (1987). Isolation of two highly toxic *Bacillus sphaericus* strains. *Insecticidal Microorg*, *1*, 98–99.

Zhou, Z., Peng, D., Zheng, J., Guo, G., Tian, L., Yu, Z., & Sun, M. (2011). Two groups of S-layer proteins, SLP1s and SLP2s, in *Bacillus thuringiensis* co-exist in the S-layer and in parasporal inclusions. *BMB Reports*, *44*(5), 323–328. https://doi.org/10.5483/BMBRep.2011.44.5.323

6 Phylum Firmicutes
Paenibacillus popilliae, an Overview

ABSTRACT

Paenibacillus popilliae is a spore-forming bacterium very effective in controlling Japanese beetles. The spores from soil get inserted in insect bodies and slowly kill them via septicaemia and produce characteristic milky disease. They are very specific in host selection and difficult to grow on artificial media, never producing spores in vitro. This limited the commercial production and spread of insect pathogen product Doom® and Japademic®, once acting as saviour. They also need a large area under treatment for effective control. But we cannot deny its importance in the management of Japanese beetles when all other measures failed, and hopefully, soon scientists will come up with solutions to overcome the disadvantages.

6.1 INTRODUCTION

Paenibacillus popilliae Dutky is a Gram-variable, spore-forming rod of 1.3–5.2 µm × 0.5–0.8 µm belonging to class Bacilli, catalase-negative, facultative anaerobes whose spores can survive for several years in the soil (Koppenhöfer & Wu, 2017). The vegetative cells are Gram-negative, but prespore cells and sporangia show Gram-positive reactions (Boucias & Pendland, 1998). The bacterium has a genome of a smaller size (3.83 Mbp) (Grady et al., 2016). It is extremely fastidious. This was proved by experimentation on J-medium that variations in the medium led to limited and transient growth of the vegetative stage (Bulla et al., 1978). This bacterium shows resistance to antibiotic vancomycin and can be used to screen and isolate the same (Patel et al., 2000). It is noteworthy that though *P. lentimorbus* (Dutky) was considered a variety of *P. popilliae* (Boucias & Pendland, 1998), it is now confirmed as a distinct species at the molecular level (Rippere et al., 1998). Morphological differences between the species include the spindle-shaped sporangium of *P. popilliae* versus the lemon-shaped sporangium of *P. lentimorbus* and, more importantly, the presence of a parasporal body in *P. popilliae* that is absent from sporulated *P. lentimorbus* (Dutky, 1940).

6.2 HABITAT AND HOST RANGE

This bacterium grows only in rich media composed of yeast extract, casein hydrolysate, or an equivalent source of amino acid, and sugars. This amino acid source is considered vital for the growth of this bacterium, along with some vitamins, like barbituric acid and thiamine. Trehalose, a sugar found in insect haemolymph, is a favourite carbon source of *P. popilliae*, although they can grow on glucose (Deacon, 1998). The bacterium apparently lacks a complete tricarboxylic acid cycle and catalase, which may contribute to their inability to sporulate in vitro on nutrient broth or agar, distinguishing the genus from *Bacillus* (Glare et al., 2017; Khachatourians, 2009). This is an obligate pathogen and is found to be associated only with phytophagous scarab hosts of about 70 species of insects within the subfamilies Melolonthinae, Rutelinae, and Dynastinae (Khachatourians, 2009). There are four different subgroups, namely, A_1, A_2, B_1, and B_2, under which there are several different varieties. Strains of this bacterium have been detected in at least 29 species of Scarabaeidae (Klein & Jackson, 1992), which include the Japanese beetle and the chafers—important pasture pests—but also the beneficial dung beetles (Deacon, 1998). Strains show very less to no cross-infectivity against

DOI: 10.1201/9781003426882-6

insect species apart from their source of isolation. It is important to note that the specificity of infection seems to be dependent on the success of disrupting the gut epithelial layer from invading the haemocoel, because infection of spores is proved to be independent of strain specificity (Franken et al., 1996).

6.3 MODE OF ACTION

Grubs use to ingest the *P. popilliae* spores residing in the soil. In their gut, spores germinate within two days, followed by the proliferation of the bacterial vegetative cells, which attain the highest numbers within three to five days by initial replication at the alimentary tract. By this time, some motile vegetative cells penetrate the gut cell wall and appear to undergo cell division within the intracellular vacuoles (Splittstoesser et al., 1978). Next, progeny vegetative cells are released, begin to grow, and by day 5 to 10, the haemolymph contain large numbers of cells. However, the problem is that symptoms do not start showing even at this point, and the host scarab continuously feeds for an extended period as septicaemia is not elicited immediately even after the presence of large numbers of bacterial cells in scarabs. Few spores are also formed in this stage, but the main sporulation phase takes place later on and ends in 14–21 days. By this time, the larva develops the typical milky appearance. In laboratory conditions, the larva remains alive until this stage and usually contains about 5×10^{10} spores (Boucias & Pendland, 1998). However, according to different reports, in field conditions, sometimes larvae die earlier prior to the completion of the main sporulation phase. This is a matter of concern because sporulation stops with the death of the host and the larva ultimately releases fewer spores to maintain the level of infestation of a site (Deacon, 1998). This will lead to a low spread of epizootics and also might cause resistance development. In addition, when the insect dies, it releases endospores in the soil, which serves as the reservoir of milky disease for the next susceptible host. The bacterium produces chitinase enzyme hydrolyzing chitin, a structural polysaccharide of insect exoskeleton and gut lining, which also may be responsible for insect death (Grady et al., 2016). Knowledge of the mechanism of death in *P. popilliae*–infected insects is still unclear. The growth of bacterial cells in the haemolymph causes physiological starvation, which seems to be the most likely explanation, and the fat reserves of diseased larvae are much reduced compared to those of healthy larvae (Boucias & Pendland, 1998). However, there may be the involvement of the toxins, as they have been found in the bacterial culture filtrates and shown lethal effects on injection. CryBP1 is a crystal protein from *P. popilliae* sporulating cells having ~40% similarities to Cry2 toxins of *Bacillus thuringiensis*. This toxin contributes to pathogenic invasion through the gut wall. It is activated in the midgut of the host and interacts with the midgut, making it more susceptible to invasion by the vegetative cells (Zhang et al., 1997). However, there is a lack of clear difference in pathogenesis between *P. popilliae* and paraspore-free *P. lentimorbus*, which supports a limited role of parasporal toxins in infection (Glare et al., 2017). *P. popillae*'s mode of action has been described pictorially in Figure 6.1 for more clarification.

6.4 RELATIVE EFFECTIVENESS

The use of *P. popilliae* is limited due to its inability to produce infective spores in vitro. Along with in vivo products like Doom®, there are some putative in vitro products, like Grub Attack®. Though results are comparable, in soil confinement assays, infectivity by feeding (per os) was greater for the in vivo ones made by maceration of diseased grubs than for the putative in vitro formulation (Redmond & Potter, 1995).

This bacterial treatment is most effective when applied on a wide region or state basis to reduce overall levels of scarab infestation (Deacon, 2013). Therefore, for small landowners, it is of no use. Also, the obligate dependency of *P. popilliae* on hosts for its sporulation, and no ingestion or ingestion of fewer spores (not enough to cause disease) by some larvae create chances of periodic resurgence to take place.

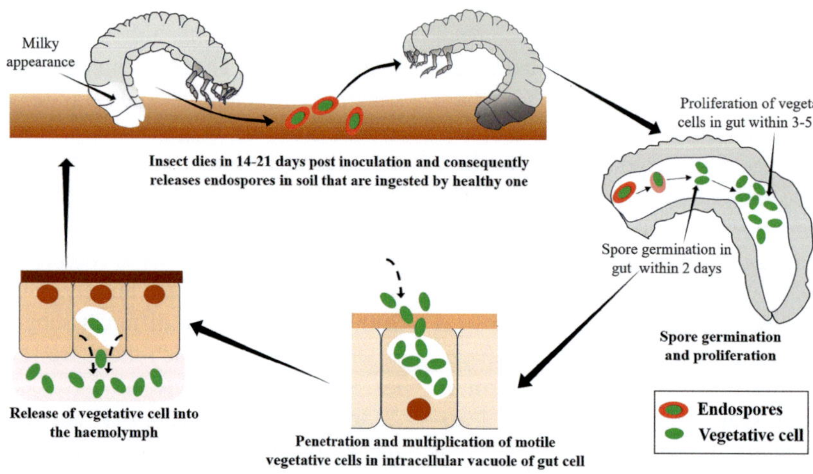

FIGURE 6.1 Mode of action of *Paenibacillus papillae*.

6.5 ADVANTAGES

P. popilliae possesses a very narrow host range (effective against scarab grubs only), so the use of its commercial formulations is advantageous. They are completely safe for man and other vertebrates as they cannot grow at 37°C. They are compatible with other control agents, like entomopathogenic nematodes and chemical insecticides. Moreover, they render persistent and long-lasting control (Deacon, 2013). The bacterium persists in the soil and therefore does not need to be reapplied (Khachatourians, 2009).

6.6 DRAWBACKS

The drawbacks include high *in vivo* production costs and slow action, but the most important one is the lower toxicity to Japanese beetle adults, which also causes obvious and distressing damage. Another disadvantage we have already discussed is the requirement of large treated areas for effective control. Having a narrow host range is also sometimes a disadvantage in its wide adaptability. Farmers must be accurate in identifying the infesting grub species, whether it is the Japanese beetle or not (Deacon, 2013). In some areas of the USA, *P. popilliae* seems to lose its toxicity to Japanese beetles. Only 0.2% of larvae under the experimental area showed milky symptoms compared to 41.5% of disease occurrence in 1946. Also, a field study in Kentucky showed that infection caused by the commercial formulations of *P. popilliae* was moderate (39–44%), and infected grubs consumed almost the same number of roots as the uninfected ones. Researchers concluded that earlier reports of successful management were confined to very high infestations of grubs where other stresses had also become associated due to congestion and thus contributed to increased susceptibility to diseases (Deacon, 1998).

6.7 CONCLUSION

P. popilliae got discovered as a boon to manage the Japanese beetle when the pest was destroying a variety of crops and ornamental plants in the USA and enjoys the status of being the first entomopathogen to get registered in the USA as a microbial agent in 1948 in the brand name of Doom® or Japademic®. Though the bacterium acts slowly, precise attack on the host has made it a good choice for reducing harm to non-targets. Although the inability of in vitro mass production restricts their commercial spread, we hope some effective solution will soon tackle the issue.

REFERENCES

Boucias, D. G., & Pendland, J. C. (1998). *Principles of insect pathology* (1st ed.). Springer. https://doi.org/10.1007/978-1-4615-4915-4

Bulla, L. A., Costilow, R. N., & Sharpe, E. S. (1978). Biology of *Bacillus popilliae*. *Advances in Applied Microbiology*, *23*(C), 1–18. https://doi.org/10.1016/S0065–2164(08)70063–7

Deacon, J. (1998). *Paenibacillus popilliae*. Cornell University. https://biocontrol.entomology.cornell.edu/pathogens/paenibacillus.php

Deacon, J. (2013). *Biological control—Bacillus popilliae*. Institute of Cell and Molecular Biology, The University of Edinburgh. http://archive.bio.ed.ac.uk/jdeacon/microbes/control.htm

Dutky, S. R. (1940). Two new spore-forming bacteria causing milky diseases of Japanese Beetle larvae. *Journal of Agricultural Research*, *61*(1), 57–68. https://naldc.nal.usda.gov/download/IND43969460/PDF

Franken, E., Krieger, L., & Schnetter, W. (1996). *Bacillus popilliae*: A difficult pathogen. *Bulletin OILB/SROP*, *19*(2), 40–45.

Glare, T. R., Jurat-Fuentes, J.-L., & O'Callaghan, M. (2017). Basic and applied research. In L. A. Lacey (Ed.), *Microbial control of insect and mite pests* (pp. 47–67). Elsevier. https://doi.org/10.1016/B978-0-12-803527-6.00004-4

Grady, E. N., MacDonald, J., Liu, L., Richman, A., & Yuan, Z. C. (2016). Current knowledge and perspectives of *Paenibacillus*: A review. *Microbial Cell Factories*, *15*, 203. https://doi.org/10.1186/s12934-016-0603-7

Khachatourians, G. G. (2009). Insecticides, microbial. In M. Schaechter (Ed.), *Encyclopedia of microbiology* (3rd ed., pp. 95–109). Academic Press. https://doi.org/10.1016/B978-012373944-5.00124-3

Klein, M. G., & Jackson, T. A. (1992). Bacterial diseases of scarabs. In T. A. Jackson, & T. R. Glare (Eds.), *Use of pathogens in scarab management* (pp. 43–61). Intercept Ltd.

Koppenhöfer, A. M., & Wu, S. (2017). Microbial control of insect pests of turfgrass. In L. A. Lacey (Ed.), *Microbial control of insect and mite pests: From theory to practice* (pp. 331–341). Academic Press. https://doi.org/10.1016/B978-0-12-803527-6.00022-6

Patel, R., Piper, K., Cockerill, F. R., Steckelberg, J. M., & Yousten, A. A. (2000). The biopesticide *Paenibacillus popilliae* has a vancomycin resistance gene cluster homologous to the enterococcal VanA vancomycin resistance gene cluster. *Antimicrobial Agents and Chemotherapy*, *44*(3), 705. https://doi.org/10.1128/AAC.44.3.705–709.2000

Redmond, C. T., & Potter, D. A. (1995). Lack of Efficacy of in vivo—and putatively in vitro-produced *Bacillus popilliae* against field populations of Japanese beetle (Coleoptera: Scarabaeidae) grubs in Kentucky. *Journal of Economic Entomology*, *88*(4), 846–854. https://doi.org/10.1093/JEE/88.4.846

Rippere, K. E., Tran, M. T., Yousten, A. A., & Hilu, K. H. (1998). *Bacillus popilliae* and *Bacillus lentimorbus*, bacteria causing milky disease in Japanese beetles and related scarab larvae. *International Journal of Systematic Bacteriology*, *48*(2), 395–402. https://doi.org/10.1099/00207713-48-2-395

Splittstoesser, C. M., Kawanishi, C. Y., & Tashiro, H. (1978). Infection of the European chafer, *Amphimallon majalis* by *Bacillus popilliae*: Light and electron microscope observations. *Journal of Invertebrate Pathology*, *31*(1), 84–90. https://doi.org/10.1016/0022–2011(78)90112-X

Zhang, J., Hodgman, T. C., Krieger, L., Schnetter, W., & Schairer, H. U. (1997). Cloning and analysis of the first cry gene from *Bacillus popilliae*. *Journal of Bacteriology*, *179*(13), 4336–4341. https://doi.org/10.1128/JB.179.13.4336–4341.1997

7 Phylum Firmicutes
Other Entomopathogenic Bacteria

ABSTRACT

This chapter deals with several underutilized entomopathogenic bacteria belonging to the phylum Firmicutes. Where most of such bacteria belong to the class Bacilli, a single bacterium is from the class Clostridia. Entomopathogenic nature is reported against several insect pests for these bacteria. Also, some discrete cases of harming beneficial insects, plants, or human beings are reported. This chapter brings all these aspects together to find out the potential entomopathogenic bacteria for possible commercial exploitation soon.

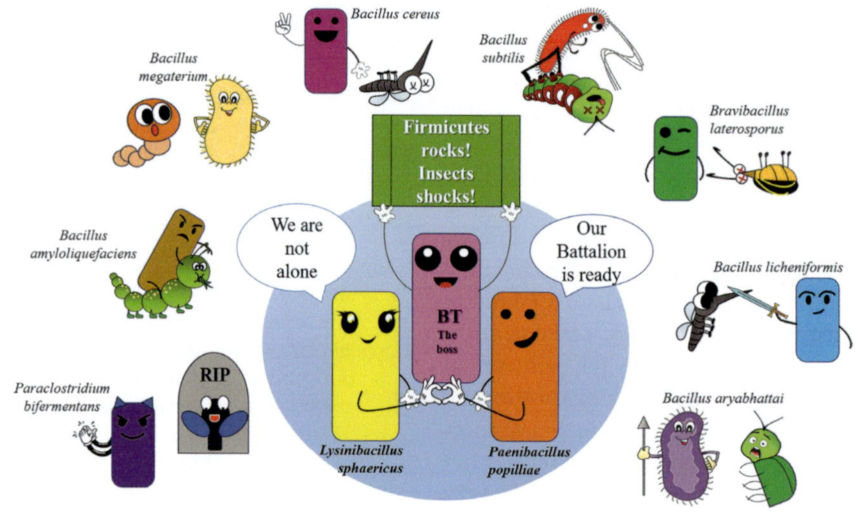

7.1 INTRODUCTION

Phylum Firmicutes is the most important phylum in respect of entomopathogenic bacteria (EPB). There are several famous names in the field of EPB, like *Bacillus thuringiensis* Berliner (Bt), *Lysinibacillus sphaericus* (Meyer and Neide) (Ls), and *Paenibacillus popilliae* Dutky, which have been discussed in previous chapters. Along with these, there are other EPB from class Bacilli, like *P. polymyxa* (Prazmowski), *Brevibacillus laterosporus* (Laubach), *Bacillus cereus* Frankland and Frankland, *B. subtilis* (Ehrenberg), *B. amyloliquefaciens* (Fukumoto), *B. megaterium* de Bary, *B. aryabhattai* Shivaji et al., and *B. licheniformis* (Weigmann). There is also an EPB from the class Clostridia, that is, *Paraclostridium bifarmentans* (Weinberg and Seguin). They are all Gram-positive endospore-forming bacteria effective against different insects.

7.2 *PAENIBACILLUS POLYMYXA*

P. polymyxa is a plant growth–promoting rhizobacterium (PGPR). It has a broad host range, but so far, this organism has not been proven much efficient as a biocontrol agent. In an experiment on the entomopathogenic nature of the *Bacillus* group against lepidopteran and coleopteran insects of the olive tree, it was found that *P. polymyxa* was most efficient amongst all the other *Bacillus* sp. against coleopteran pests *Hylesinus oleiperda* (Fabricius) and *Phloeotribus scarabaeoides* (Bernard). It was also effective against lepidopteran pests like *Prays oleae* Bernard, but the required LC_{50} value is higher (Blibech et al., 2012). However, instances are found where the bacteria helped the pest survive against its entomopathogens, like in the case of a stingless bee (*Melipona scutellaris* Latreille), where the bacteria were found to be present in the larval food and synthesized (L)-(–)-3-phenyllactic acid and fusaricidins, which were active against entomopathogenic fungi (EPF) and *P. larvae* (White) (Menegatti et al., 2018). So this type of dual role can be seen in the case of *P. polymyxa*.

7.3 *BREVIBACILLUS LATEROSPORUS*

B. laterosporus is a rod-shaped, endospore-forming bacterium with broad-spectrum antimicrobial properties. It produces a typical canoe-shaped parasporal body (CSPB) during sporulation that is firmly attached to one side of the spore coat, and this determines its lateral position in

DOI: 10.1201/9781003426882-7

the sporangium. This is the unique morphological character of this species. It is an omnipresent species present in the soil, gemstones, lahar, fresh water, seawater, insect bodies, leaf surfaces, locust beans, compost, milk, cheese, honey, starchy foods, gelatine factory effluents, animal hide, wool, and quails (Ruiu, 2013). It forms flat, smooth (plain) white or beige colonies along with rhizoid margins on the agar surface (Ghazanchyan et al., 2018). The insecticidal activity of different strains of this bacterium has been reported against insects of different orders, including Coleoptera, Lepidoptera, and Diptera, and also against molluscs, nematodes, phytopathogenic bacteria, and fungi. The whole-genome sequencing of *B. laterosporus* reveals that this is potent to produce different toxins, including *Cry* genes, although their role in toxicity has not been demonstrated (Glare et al., 2017). Certain strains that show entomopathogenicity against corn rootworms (*Diabrotrica* spp.) and other coleopteran larvae produce insecticidal secreted proteins (ISPs) that act as binary toxins in the insect midgut and have high homology with vegetative insecticidal proteins (VIPs) produced by Bt (Ruiu, 2015). Some strains (like strains 615 and 921) toxic to mosquitoes produce parasporal inclusion bodies showing similarities to those produced by Bt (Smirnova et al., 1996). Another strain that lacks parasporal crystals is found to be highly toxic to the housefly, *Musca domestica* Linnaeus, and the mode of action indicates histopathological changes in the midgut by disrupting microvillar epithelium (Ruiu, 2015). Spore surface proteins (ExsC, CHRD, CpbA, and CpbB) are also found to be an integral part of a complex set of toxins and virulence factors that *B. laterosporus* engages in its pathogenicity towards houseflies (Marche et al., 2017). *B. laterosporus* isolate was also proved to be highly toxic in comparative bioassays with different Bt isolates against *Tenebrio molitor* Linnaeus (Coleoptera: Tenebrionidae) larvae with a LC_{50} value of 3.388×10^6 spores/mL (Rand & Laing, 2011). Strains of *B. laterosporus* isolated from *Brassica* seeds were found to be effective against insects, including diamondback moth larvae, *Plutella xylostella* (Linnaeus) (Ormskirk et al., 2019; van Zijll de Jong et al., 2016).

Despite all these efficient cases of managing a wide range of pests by *B. laterosporus*, we should not forget that it is also associated with the European foulbrood (EFB) disease of honeybees (Ruiu, 2013). There is also an issue hindering the commercial development; the cause of insect toxicity is not completely understood (Glare & O'Callaghan, 2019). Recently, the bacterium is proved to attack humans also (Curtis et al., 2020). So we must keep these factors in mind before large-scale release of this bacterium in the environment to manage pests.

7.4 *BACILLUS CEREUS*

The majority of *B. cereus* strains that are often connected to insects are symbiotic or saprophytic bacteria found in the digestive tract. Although Bt and *B. cereus* have a lot of genetic similarities, the later lacks the ability to create parasporal crystalline toxins, which reduces its pathogenicity towards insect hosts and encourages saprophytic development. Despite being most prevalent in soil, *B. cereus* is widely distributed and frequently found in the gut flora of several insect species. Isolates of *B. cereus* have been reported to cause natural or induced infections in diverse insects, including larvae of scarab beetles (Selvakumar et al., 2007); flour beetle, *Tribolium castaneum* (Herbst) (Kumari & Neelgund, 1985); spruce budworm, *Choristoneura fumiferana* (Clemens) (Strongman et al., 1997); mosquitoes, *Aedes aegypti* (Linnaeus), *Anopheles stephensi* Liston, and *Culex quinquefasciatus* Say (Ragavendran & Natarajan, 2018); *Trichoplusia ni* (Hübner) (To et al., 1975); and adults of the tsetse fly, *Glossina morsitans* Westwood (Kaaya & Darji, 1989). Production of phospholipase C was linked with *B. cereus* pathogenicity in cockroach, *Leucophaea maderae* (Fabricius) (Rahmet-Alla & Rowley, 1989), whereas production of antibiotics by *B. cereus* in the gut of larvae of *P. xylostella* synergized Bt pathogenicity (Raymond et al., 2008). *B. cereus* subsp. *moritai* was established as a bioinsecticide for the housefly, *M. domestica*. Contrary to Bt, *B. moritai* is not pathogenic to the silkworm, and this will be an advantage to sericultural countries as well as shows no significant effect in mammalian system (Maehashi, 1972). Despite the fact that insecticidal toxins have been found on the spore coat of *B. cereus* isolates, they have significantly lesser activity than analogous toxins from Bt spores (Somerville & Pockett, 1975). The Vip toxins were originally identified in a strain of *B. cereus* pathogenic to larvae of western and northern corn rootworms (*Diabrotica* spp.) (Warren et al., 1996). The Vip toxins from *B. cereus* are similar to the binary toxins from *Clostridium* and are made up of two toxin subunits: the Vip1 toxin, which attaches to the target cell membrane, and the Vip2 toxin, which inhibits actin polymerization and kills the host cell (Han et al., 1999). Both the Vip1 and Vip2 toxins are required for toxicity, and the application of these toxins for the control of insects is restricted by the possibility that they may be harmful to plants or other non-target species while being expressed. A method to create transgenic plants for insect control has been proposed: expression of the Vip2 toxin component in plants as an inactive proenzyme that becomes active once consumed by the rootworm (Jucovic et al., 2008). The binary toxin vip1Ac1-vip2Ae3 showed activity only against *Aphis gossypii* Glover (Hemiptera), but not for colepteran (*T. molitor*, *Holotrichia oblita* [Falderman]), lepidopteran (*Spodoptera exigua* [Hübner], *Helicoverpa armigera* [Hübner], and *Chilo suppressalis* [Walker]), and dipteran (*Cx. quinquefasciatus*) insects (Yu et al., 2011), rendering very target-specific management against aphids. Thermostable and non-proteinaceous exotoxins have also been detected in strains of *B. cereus* (Ohba et al., 1981; Perchat et al., 2005). Cotton boll weevil, *Anthonomus grandis* Boheman, is extremely susceptible to the non-proteinaceous exotoxins, but cotton leafworm, *Spodoptera littoralis* (Boisduval), and black bean aphid, *Aphis fabae* Scopoli, both exhibit reduced activity (Perchat et al., 2005). *B. cereus* vegetative cells can also acquire toxin-encoding plasmids from Bt cells infecting lepidopteran larvae; as a result, *B. cereus* cells can also generate crystalline insecticidal toxins (Yuan et al., 2007). Additionally, they have been discovered to be related to *Rhabditis* (*Oscheius*) sp., an entomopathogenic nematode (EPN), and to have strong antimicrobial properties. The six bioactive compounds isolated were antagonistic to bacterial strains of *B. subtilis*, *Staphylococcus aureus* Rosenbach, *Escherichia coli* (Migula), *Pseudomonas aeruginosa* (Schroeter), and fungus, like *Aspergillus flavus* Link, *Candida albicans* (Robin), *Fusarium oxysporum* (Schlechtendal), *Rhizoctonia solani* Kühn, and *Penicillium expansum* Link, with better performance than fungicide Bavistin® (Kumar et al., 2014a, 2014b). Also, a novel *B. cereus* isolate has been discovered, promoting growth in tomato and inhibiting *Clavibacter michiganensis* (Smith) infection under greenhouse (Solano-Alvarez et al., 2021). These open up scope for managing insects and pathogens with a single agent, making IPM simpler and more effective in both pre- and post-harvest situations.

But development and commercialization of *B. cereus* for microbial insect control have not been considered, owing to the presence of toxins that induce diarrhoea and vomiting in food-borne illnesses in vertebrates. However, the use of *B. cereus* as host for expression of toxins targeting mosquito larvae was proposed because of its amenability to transform and be retained in the mosquito gut (Luxananil et al., 2001).

7.5 BACILLUS SUBTILIS

B. subtilis is a Gram-positive, rod-shaped bacterium that forms heat-resistant, dormant spores (Piggot, 2009). It is a ubiquitous organism, proving its presence in the soil, roots of plants, and aquatic environments. Also, the bacteria are easy to grow and nurture in the laboratory. Although this can grow in the gastrointestinal (GI) tract of animals, it is not considered a human pathogen. In fact, *B. subtilis*, along with other species of *Bacillus*, is also considered a GRAS (generally regarded as safe) organism by the Food and Drug Administration (FDA) (Martinez, 2013). The entomopathogenic nature of *B. subtilis* is rather newly established, and despite having a broad host range, it is not beyond the discussion of pros and cons. It is found to be effective against larvae of different notorious nematodes, like root-knot nematode, *Meloidogyne incognita* (Kofold and White), and cyst nematodes, like *Heterodera cajani* Koshy; *H. zeae* Koshy, Swarup, and Sethi; and *H. avenae* Wollenweber (Gotke & Swarup, 1988). Among insects, it is effective against *Drosophila melanogaster* Meigen (Lazare et al., 1998). It has shown 100% mortality against 97.3% mortality provided by Bt against *S. littoralis* (Abd El-Salam et al., 2011). *B. subtilis* ssp. *subtilis* affected the survival of the housefly larvae by 10–20% (Torres et al., 2022). Crude metabolite of *B. subtilis* was recorded to cause 93.33% mortality of Bihar hairy caterpillar, *Spilosoma obliqua* (Walker), attacking jute (Bhattacharya et al., 2014). However, this bacterium was proved inefficient against dipteran and lepidopteran pests of olive (Blibech et al., 2012). Rather, it induced morbidity in termites (Omoya & Kelly, 2014). However, like *P. polymyxa*, it also antagonises entomopathogens and favours the pest's longevity. In cicadellids and delphacids, the presence of this bacterium in the host cuticle prevents EPF-like *Beauveria bassiana* (Bals.) and *Metarhizium anisopliae* (Metchn.) (Toledo et al., 2015). Also, it has been found to reduce the progeny development in *Heterorhabditis indica* Poinar, Karunakar, and David, an EPN of apple root borer, *Dorysthenes huegelii* (Redtenbacher) (Upadhyay & Mohan, 2021).

7.6 BACILLUS AMYLOLIQUEFACIENS

B. amyloliquefaciens is one of the most common bacterial endophytes infecting many plants (Li et al., 2015). It is a Gram-positive bacterium which produces liquified amylase, hence the name (Ngalimat et al., 2021). The bacterium is mostly effective against several plant-pathogenic fungus (Nam et al., 2016) but also shows effectiveness against several insects. This is effective in deterring the feeding capacity of fall armyworm, *Spodoptera frugiperda* (Smith), in *Hosta* spp. (Li et al., 2015). *B. amyloliquefaciens* AG1 strain showed the best insecticidal activity against larvae of tomato leaf miner, *Tuta absoluta* (Meyrick), with 100% mortality (Ben Khedher et al., 2015). *B. amyloliquefaciens* CBMDLO3 strain is found to reduce the survival of larvae and pupae of housefly, *M. domestica*, and fecundity of female adults is also reduced (Torres et al., 2022). But it also prevents pest management by helping insects like cicadellids and delphacids; it antagonizes the growth of EPF, like *B. bassiana* and *M. anisopliae*, being present in the midgut of the insects (Toledo et al., 2015). So commercialization of this bacterium in the purpose of pest management should consider all these pros and cons.

7.7 BACILLUS MEGATERIUM

B. megaterium is a Gram-positive, aerobic, spore-forming neutralophilic bacteria found in diverse habitats, like soil and rice paddy fields, and embedded in plant ovules and seed tissues but commonly regarded as a soil bacterium. It can utilize different carbon sources and grow at a wide range of temperatures (3°C to 45°C). *B. megaterium* possesses plant growth–promoting activity, including biocontrol ability against different plant pathogenic fungi (Goswami et al., 2018). Within a week, different strains of *B. megaterium* can induce morbidity in 92–100% of green apple aphids, *Aphis pomi* de Geer. England et al. (1997) reported that *B. megaterium* mutants producing a high level of the mosquitocidal binary toxin of Ls were possible to isolate. These mutants were potent to induce mortality in *Cx. quinquefasciatus*, similar to Ls (Aksoy & Ozman-Sullivan, 2008). The bacteria have also been found in the aerobic gut of the codling moth, *Cydia pomonella* (Linnaeus), but they failed to be a highly morbidity-inducing agent (Sevim et al., 2018). *B. megaterium* isolates HMA5 and BM1 caused different abnormal wing shapes and hair loss on the thorax and abdomen when applied on pupa and second instar larvae of fall webworm, *Hyphantria cunea* Drury. Also, pupae treated with the aforementioned strains showed significant reduction in adult emergence rates (1% and 2%, respectively) (Aksoy et al., 2018a). *B. megaterium* is also reported to act as an efficient biocontrol agent against the green shield bug, *Palomena prasina* Linnaeus (Aksoy et al., 2018b). *B. megaterium* TV-3D strain causes 95% mortality of last larval instar of *Cymbalophora rivularis* (Ménétries) attacking forage crops sainfoin, alfalfa, and clover (Tozlu et al., 2022). So it can be utilized as biocontrol agent against various kinds of insect pests. But we have to be cautious about the possible negative effect on plants, as one pathovar, *B. megaterium* pv. *cerealis*, is proved to incite white blotch in wheat (Hosford, Jr., 1982).

7.8 *BACILLUS LICHENIFORMIS*

B. licheniformis is a Gram-positive, endospore-forming bacterium. It can be isolated from plant materials and soil, but there is no report of its pathogenicity for either animals or plants (Veith et al., 2004). So it can be a highly effective entomopathogen, which was first proposed by Ming Yan and team in 2007, when they found that the crystalline inclusions of this bacterium resembles parasporal crystals of Bt (Yan et al., 2007). The bacteria are generally found to be present in poplar pests, like *Paranthrene tabaniformis* Rottemburg (larval stage) and *Nycteola asiatica* (Krulikovskii) (adult stage), and proved to be insect pathogenic in the latter one (Yaman et al., 2017). *B. licheniformis* isolates caused up to 59.2% larval mortality for *P. oleae* and 43.6% for *Palpita unionalis* (Hübner), lepidopteran pests of olive trees. The concerned LC_{50} value against *P. oleae* was 12.5 µg/mL, proving the isolates to be highly toxic and efficient (Blibech et al., 2012). The bacteria have been proven to be an efficient replacement for most used entomopathogenic formulations, like Ls and *Bt* serovar *israelensis* against *Ae. aegypti*, which has started developing resistance against these commercial ones (Rajagopal et al., 2020). However, along with *B. subtilis*, it is also found to suppress the growth of EPN *H. indica*, active against apple root borer (Upadhyay & Mohan, 2021). But it is also found to help in the survival of two famous pests of corn, leafhopper, *Dalbulus maidis* (DeLong and Wolcott) and planthopper, *Delphacodes kuscheli* Fennah, being present in cuticular linings and acting as antagonists of EPF, *B. bassiana* (Toledo et al., 2011). So we have to keep in mind both the positive and negative effects of this entomopathogenic bacterium before commercializing it as a biocontrol agent.

7.9 *BACILLUS ARYABHATTAI*

This is one of the most versatile bacteria, with its presence in cryotubes used to collect air samples (27 to 41 km), soil from sugar cane and rice plantation, rhizosphere of horseweed and *Spartina maritima* (Curtis), dense forest soil, urban tunnel, and deep seawater (Bhattacharyya et al., 2017), and with its diversified activities in plant growth promotion, like modulation of the production of phytohormone (Park et al., 2017) or glyphosate herbicide residue (Venter & Bøhn, 2016). There are some views on merging this species of bacteria with *B. megaterium*, as they share 99.7% homology in genome sequences (Narsing Rao et al., 2019), but this is beyond the scope of discussion of this chapter. Coming to our concern, the entomopathogenic nature is also being explored for this bacterium. It has been isolated with pathogenicity from *Cassida vittata* de Villers in Egyptian sugar beet (Bazazo et al., 2019). A recent publication gave us the news of its pathogenicity against the nematode *Meloidogyne javanica* (Treub) (Antil et al., 2021). So it has the potential to come up with the future of the biocontrol industry.

7.10 *PARACLOSTRIDIUM BIFERMENTANS*

Paraclostridium bifermentans (formerly known as *Clostridium bifermentans*), belonging to class Clostridia, is an obligatory anaerobic, endospore-forming, Gram-positive, catalase-negative bacillus which produces round terminal endospores and is infrequently recorded as infectious to humans (Glare et al., 2017; Valtierra-De-luis et al., 2020). *P. bifermentans* serovar *malaysia*, isolated from mangrove swamp soils of Malaysia for the first time, is effective against mosquitoes and blackflies (de Barjac et al., 1990). Parasporal inclusions having up to 12 'brush-bottle'-like appendages appear 5 hours after the ending of exponential phase and remain attached to the spores when released after sporangium lysis. Larvicidal activity assayed on *An. stephensi* larvae appeared immediately after the end of exponential growth and was maximum between 5 and 8 hours from the end of exponential growth (Charles et al., 1990). Here, the mosquitocidal activity is associated with the production of three major proteins; among them, Cbm71 (67 kDa) shows homology to Bt delta endotoxins (Nicolas et al., 1993). To prove the role of this protein, the encoding gene was cloned and induced to be expressed by transformed Bt, which resulted in toxicity against mosquitoes (Barloy et al., 1996). The strain produces Cry proteins, namely, Cry16A and Cry17A. Other potential toxins are similar to haemolysins, but they cannot cause mortality in mosquitoes individually. However, using the four toxins encoded on a single operon (Cry16A, Cry17A, Cbm17.1, and Cbm17.2), morbidity was found against *Aedes* spp., while *Anopheles* mosquitoes were more susceptible to complete cultures of *P. bifermentans* subsp. *malaysia* (Qureshi et al., 2014). Recently, an unrelated protein from *P. bifermentans*, namely, PMP1 (paraclostridial mosquitocidal protein 1), has been discovered. This is a clostridial BoNT-like neurotoxin which acts through its metalloprotease activity on a neuronal SNARE protein called syntaxin1A. Recombinant PMP1 showed injection toxicity against both larvae and adults of *Anopheles coluzzii* Coetzee and Wilkerson and *Ae. aegypti*, and to adults of *D. melanogaster* (Contreras et al., 2019). It was proved to be more toxic than Bt and Ls to the vector of lymphatic filariasis, *Ae. polynesiensis* Marks (Mercer et al., 1995). The bacterium does not contain plasmids, and inevitably, the toxin genes are present on the chromosome (Park et al., 2006).

Inoculation of the bacterium using diverse routes demonstrated that the bacteria are safe for laboratory mammals and goldfish (Thiery et al., 1992). The bacterium is seemingly non-virulent, but it has been proven to be a causal organism of different diseases in humans, like pneumonia/empyema, septic arthritis, osteomyelitis, soft tissue infection, abdominal infections, and brain injury abscess, bacteraemia, and endocarditis (Edagiz et al., 2015). So both pros and cons should be kept in mind, and different levels of residual tests carried out, before field release. Other strains of *P. bifermentens* are known to be present in anaerobic zones associated with lake sediments.

7.11 CONCLUSION

The entomopathogenic bacteria discussed in this chapter are not explored for commercial utilization for insect pest management, but they have potential scope for that. There are several disadvantages as well associated with several of these bacteria which may hinder their commercial applicability. But critical study of these points will spark the scientists with ideas to overcome them. This was the only objective of the chapter, to discuss the positive and negative sides of these unexplored entomopathogenic bacteria, so that in the near future we can see many alternatives for addressing especially the resistance development issue even against some well-established entomopathogenic bacteria (discussed in Chapter 13).

REFERENCES

Abd El-Salam, A. M. E., Nemat, A. M., & Magdy, A. (2011). Potency of *Bacillus thuringiensis* and *Bacillus subtilis* against the cotton leafworm, *Spodoptera littoralis* (Bosid.) larvae. *Archives of Phytopathology and Plant Protection*, *44*(3), 204–215. https://doi.org/10.1080/03235400902952129

Aksoy, H. M., & Ozman-Sullivan, S. K. (2008). Isolation of *Bacillus megaterium* from *Aphis pomi* (Homoptera: Aphididae) and assessment of its pathogenicity. *Journal of Plant Pathology*, *90*(3), 433–436. http://www.jstor.org/stable/41998537

Aksoy, H. M., Saruhan, I., Kaya, Y., & Ozturk, M. (2018a). Morphological changes caused by *Bacillus megaterium* on adult emergence of fall webworm's pupa, *Hyphantria cunea* (Drury) (Lepidoptera: Erebidae). *Journal of Agricultural Sciences*, *24*, 539–546. https://doi.org/10.15832/ankutbd.349474

Aksoy, H. M., Tuncer, C., Saruhan, I., Erper, I., Ozturk, M., & Akca, I. (2018b). Isolation and characterization of *Bacillus megaterium* isolates from dead pentatomids and their insecticidal activity to *Palomena prasina* nymphs. *Akademik Ziraat Dergisi*, *7*(1), 21–28. https://doi.org/10.29278/azd.440586

Antil, S., Kumar, R., Pathak, D. V., Kumar, A., Panwar, A., Kumari, A., & Kumar, V. (2021). On the potential of *Bacillus aryabhattai* KMT-4 against *Meloidogyne javanica*. *Egyptian Journal of Biological Pest Control*, *31*(1), 67. https://doi.org/10.1186/S41938-021-00417-2

Barloy, F., Delécluse, A., Nicolas, L., & Lecadet, M. M. (1996). Cloning and expression of the first anaerobic toxin gene from *Clostridium bifermentans* subsp. *malaysia*, encoding a new mosquitocidal protein with homologies to *Bacillus thuringiensis* delta-endotoxins. *Journal of Bacteriology*, *178*(11), 3099–3105. https://doi.org/10.1128/JB.178.11.3099-3105.1996

Bazazo, K., Ekram, A., & El-Sheikh, M. (2019). New record of entomopathogenic bacteria, *Bacillus aryabhattai* strain B8w 22, isolated from *Cassida vittata* Vill. and its pathogenicity against this insect in Egyptian sugar beet fields. *Zagazig Journal of Agricultural Research*, *46*(6), 2247–2254. https://doi.org/10.21608/zjar.2019.65077

Ben Khedher, S., Boukedi, H., Kilani-Feki, O., Chaib, I., Laarif, A., Abdelkefi-Mesrati, L., & Tounsi, S. (2015). *Bacillus amyloliquefaciens* AG1 biosurfactant: Putative receptor diversity and histopathological effects on *Tuta absoluta* midgut. *Journal of Invertebrate Pathology*, *132*, 42–47. https://doi.org/10.1016/J.JIP.2015.08.010

Bhattacharyya, C., Bakshi, U., Mallick, I., Mukherji, S., Bera, B., & Ghosh, A. (2017). Genome-guided insights into the plant growth promotion capabilities of the physiologically versatile *Bacillus aryabhattai* strain AB211. *Frontiers in Microbiology*, *8*, 411. https://doi.org/10.3389/fmicb.2017.00411

Bhattacharya, S., Dhar, T., & Senapati, S. K. (2014). Entomopathogenic potentiality of *Bacillus subtilis* against *Spilosoma obliqua* on jute. *Green Farming*, *5*(5), 869–871.

Blibech, I., Ksantini, M., Chaieb, I., Jlassi, B., Rhouma, A., Jaoua, S., & Aifa, S. (2012). Isolation of entomopathogenic *Bacillus* from a biodynamic olive farm and their pathogenicity to lepidopteran and coleopteran insect pests. *Crop Protection*, *31*(1), 72–77. https://doi.org/10.1016/J.CROPRO.2011.09.020

Charles, J. F., Nicolas, L., Sebald, M., & de Barjac, H. (1990). *Clostridium bifermentans* serovar *malaysia*: Sporulation, biogenesis of inclusion bodies and larvicidal effect on mosquito. *Research in Microbiology*, *141*(6), 721–733. https://doi.org/10.1016/0923-2508(90)90066-Y

Contreras, E., Masuyer, G., Qureshi, N., Chawla, S., Dhillon, H. S., Lee, H. L., Chen, J., Stenmark, P., & Gill, S. S. (2019). A neurotoxin that specifically targets *Anopheles* mosquitoes. *Nature Communication*, *10*, 2869. https://doi.org/10.1038/s41467-019-10732-w

Curtis, A. K., Lamb, C., Hassan, W. M., & Foxworth, J. (2020). *Brevibacillus laterosporus* bacteremia in an adult. *Cureus*, *12*(9), e10481. https://doi.org/10.7759/cureus.10481

de Barjac, H., Sebald, M., Charles, J. F., & Cheong, W. H. (1990). *Clostridium bifermentans* serovar *malaysia*, a new anaerobic bacterium toxic to mosquito and blackfly larvae. *Comptes Rendus de l'Academie Des Sciences Serie 3 Sciences de La Vie*, *310*(9), 383–387.

Edagiz, S., Lagace-Wiens, P., Embil, J., Karlowsky, J., & Walkty, A. (2015). Empyema caused by *Clostridium bifermentans*: A case report. *The Canadian Journal of Infectious Diseases & Medical Microbiology*, *26*(2), 105–107. https://doi.org/10.1155/2015/481076

England, D. F., Penfold, R. J., Delaney, S. F., & Rogers, P. L. (1997). Isolation of *Bacillus megaterium* mutants that produce high levels of heterologous protein, and their use to construct a highly mosquitocidal strain. *Current Microbiology*, *35*(2), 71–76. https://doi.org/10.1007/s002849900215

Ghazanchyan, N. L., Kinosyan, M. H., Tadevosyan, P. E., Khachaturyan, N. S., & Afrikian, E. G. (2018). *Brevibacillus laterosporus* as perspective source of new bioinsecticides. *Annals of Agrarian Science*, *16*(4), 413–415. https://doi.org/10.1016/J.AASCI.2018.03.011

Glare, T. R., Jurat-Fuentes, J. L., & O'Callaghan, M. (2017). Basic and applied research. In L. A. Lacey (Ed.), *Microbial control of insect and mite pests* (pp. 47–67). Elsevier. https://doi.org/10.1016/B978-0-12-803527-6.00004-4

Glare, T. R., & O'Callaghan, M. (2019). Microbial biopesticides for control of invertebrates: Progress from New Zealand. *Journal of Invertebrate Pathology*, *165*, 82–88. https://doi.org/10.1016/J.JIP.2017.11.014

Goswami, G., Panda, D., Samanta, R., Boro, R. C., Modi, M. K., Bujarbaruah, K. M., & Barooah, M. (2018). *Bacillus megaterium* adapts to acid stress condition through a network of genes: Insight from a genome-wide transcriptome analysis. *Scientific Reports*, *8*, 16105. https://doi.org/10.1038/s41598-018-34221-0

Gotke, N., & Swarup, G. (1988). On the potential of some bacterial biocides against root-knot and cyst nematodes. *Indian Journal of Nematology*, *18*(1), 152–153.

Han, S., Craig, J. A., Putnam, C. D., Carozzi, N. B., & Tainer, J. A. (1999). Evolution and mechanism from structures of an ADP-ribosylating toxin and NAD complex. *Nature Structural Biology*, *6*(10), 932–936. https://doi.org/10.1038/13300

Hosford, Jr., R. M. (1982). White blotch incited in wheat by *Bacillus megaterium* pv. *cerealis*. *Phytopathology*, *72*(11), 1453–1459. https://doi.org/10.1094/phyto-77-1453

Jucovic, M., Walters, F. S., Warren, G. W., Palekar, N. V., & Chen, J. S. (2008). From enzyme to zymogen: Engineering Vip2, an ADP-ribosyltransferase from *Bacillus cereus*, for conditional toxicity. *Protein Engineering, Design & Selection*, *21*(10), 631–638. https://doi.org/10.1093/protein/gzn038

Kaaya, G. P., & Darji, N. (1989). Mortality in adult tsetse, *Glossina morsitans morsitans*, caused by entomopathogenic bacteria. *Journal of Invertebrate Pathology*, *54*(1), 32–38. https://doi.org/10.1016/0022–2011(89)90136–5

Kumar, S. N., Nambisan, B., Sundaresan, A., Mohandas, C., & Anto, R. J. (2014a). Isolation and identification of antimicrobial secondary metabolites from *Bacillus cereus* associated with a rhabditid entomopathogenic nematode. *Annals of Microbiology*, *64*(1), 209–218. https://doi.org/10.1007/S13213-013-0653-6/TABLES/4

Kumar, S. N., Sreekala, S. R., Chandrasekaran, D., Nambisan, B., & Anto, R. J. (2014b). Biocontrol of *Aspergillus* species on peanut kernels by antifungal diketopiperazine producing *Bacillus cereus* associated with entomopathogenic nematode. *PLoS One*, *9*(8), e106041. https://doi.org/10.1371/journal.pone.0106041

Kumari, S. M., & Neelgund, Y. F. (1985). Preliminary infectivity tests using six bacterial formulations against the red flour beetle, *Tribolium castaneum*. *Journal of Invertebrate Pathology*, *46*(2), 198–199. https://doi.org/10.1016/0022–2011(85)90149–1

Lazare, K., Haubruge, E., Lienard, V., & Gaspar, C. (1998). Use of *Bacillus subtilis* (Cohn) as Insecticide against *Drosophila melanogaster* (Meigen). *Mededelingen—Faculteit Landbouwkundige en Toegepaste Biologische Wetenschappen Universiteit Gent (Belgium)*, *61*(3a), 887–893.

Li, H., Soares, M. A., Torres, M. S., Bergen, M., & White, J. F. (2015). Endophytic bacterium, *Bacillus amyloliquefaciens*, enhances ornamental hosta resistance to diseases and insect pests. *Journal of Plant Interactions*, *10*(1), 224–229. https://doi.org/10.1080/17429145.2015.1056261

Luxananil, P., Atomi, H., Paniym, S., & Imanaka, T. (2001). Isolation of bacterial strains colonizable in mosquito larval guts as novel host cells for mosquito control. *Journal of Bioscience and Bioengineering*, *92*(4), 342–345. https://doi.org/10.1263/JBB.92.342

Maehashi, H. (1972). On the pathogenicity to mammals of the new bacterial insecticide-*Bacillus moritai*. *Japanese Journal of Hygiene*, *27*(2), 267–274. https://doi.org/10.1265/jjh.27.267

Marche, M. G., Mura, M. E., Falchi, G., & Ruiu, L. (2017). Spore surface proteins of *Brevibacillus laterosporus* are involved in insect pathogenesis. *Scientific Reports*, *7*, 43805. https://doi.org/10.1038/srep43805

Martinez, R. M. (2013). Bacillus subtilis. In S. Maloy, & K. Hughes (Eds.), *Brenner's Encyclopedia of genetics* (2nd ed., pp. 246–248). Elsevier. https://doi.org/10.1016/B978-0-12-374984-0.00125-X

Menegatti, C., Da Paixão Melo, D. G., Carrão, D. B., De Oliveira, A. R. M., Do Nascimento, F. S., Lopes, N. P., & Pupo, M. T. (2018). *Paenibacillus polymyxa* associated with the stingless bee *Melipona scutellaris* produces antimicrobial compounds against entomopathogens. *Journal of Chemical Ecology*,*44*(12),1158–1169.https://doi.org/10.1007/s10886-018-1028-z

Mercer, D. R., Nicolas, L., & Thiery, I. (1995). Evaluation of entomopathogenic bacteria against *Aedes polynesiensis*, the vector of lymphatic filariasis in French Polynesia. *Journal of the American Mosquito Control Association*, *11*(4), 485–488.

Nam, H. S., Yang, H. J., Oh, B. J., Anderson, A. J., & Kim, Y. C. (2016). Biological control potential of *Bacillus amyloliquefaciens* KB3 isolated from the feces of *Allomyrina dichotoma* larvae. *The Plant Pathology Journal*, *32*(3), 273–280. https://doi.org/10.5423/PPJ.NT.12.2015.0274

Narsing Rao, M. P., Dong, Z.-Y., Liu, G.-H., Li, L., Xiao, M., & Li, W.-J. (2019). Reclassification of *Bacillus aryabhattai* Shivaji et al. 2009 as a later heterotypic synonym of *Bacillus megaterium* de Bary 1884 (Approved Lists 1980). *FEMS Microbiology Letters*, *366*(22), fnz258. https://doi.org/10.1093/FEMSLE/FNZ258

Ngalimat, M. S., Yahaya, R. S. R., Baharudin, M. M. A. A., Yaminudin, S. M., Karim, M., Ahmad, S. A., & Sabri, S. (2021). A review on the biotechnological applications of the operational group *Bacillus amyloliquefaciens*. *Microorganisms*, *9*(3), 614. https://doi.org/10.3390/microorganisms9030614

Nicolas, L., Charles, J.-F., & de Barjac, H. (1993). *Clostridium bifermentans* serovar *malaysia*: Characterization of putative mosquito larvicidal proteins. *FEMS Microbiology Letters*, *113*(1), 23–28. https://doi.org/10.1111/J.1574–6968.1993.TB06482.X

Ohba, M., Tantichodok, A., & Aizawa, K. (1981). Production of heat-stable exotoxin by *Bacillus thuringiensis* and related bacteria. *Journal of Invertebrate Pathology*, *38*(1), 26–32. https://doi.org/10.1016/0022–2011(81)90030–6

Omoya, F. O., & Kelly, B. A. (2014). Variability of the potency of some selected entomopathogenic bacteria (*Bacillus spp.* and *Serratia spp.*) on termites, *Macrotermes bellicosus* (Isoptera: Termitidae) after exposure to magnetic fields. *International Journal of Tropical Insect Science*, *34*(2), 98–105. https://doi.org/10.1017/S1742758414000253

Ormskirk, M. M., Narciso, J., Hampton, J. G., & Glare, T. R. (2019). Endophytic ability of the insecticidal bacterium *Brevibacillus laterosporus* in Brassica. *PloS One*, *14*(5), e0216341. https://doi.org/10.1371/journal.pone.0216341

Park, H.-W., Federici, B. A., & Sakano, Y. (2006). Inclusion proteins fromother insecticidal bacteria. In J. M. Shively (Ed.), *Inclusions in prokaryotes* (pp. 321–330). https://doi.org/10.1007/3-540-33774-1_13

Park, Y.-G., Mun, B.-G., Kang, S.-M., Hussain, A., Shahzad, R., Seo, C.-W., Kim, A.-Y., Lee, S.-U., Oh, K. Y., Lee, D. Y., Lee, I.-J., & Yun, B.-W. (2017). *Bacillus aryabhattai* SRB02 tolerates oxidative and nitrosative stress and promotes the growth of soybean by modulating the production of phytohormones. *PLoS One*, *12*(3), e0173203. https://doi.org/10.1371/journal.pone.0173203

Perchat, S., Buisson, C., Chaufaux, J., Sanchis, V., Lereclus, D., & Gohar, M. (2005). *Bacillus cereus* produces several nonproteinaceous insecticidal exotoxins. *Journal of Invertebrate Pathology*, *90*(2), 131–133. https://doi.org/10.1016/J.JIP.2005.08.002

Piggot, P. J. (2009). Bacillus subtilis. In M. Schaechter (Ed.), *Encyclopedia of microbiology* (3rd ed., pp. 45–56). Academic Press. https://doi.org/10.1016/B978-012373944-5.00036-5

Qureshi, N., Chawla, S., Likitvivatanavong, S., Lee, H. L., & Gill, S. S. (2014). The cry toxin operon of *Clostridium bifermentans* subsp. *malaysia* is highly toxic to *Aedes* larval mosquitoes. *Applied and Environmental Microbiology*, *80*(18), 5689–5697. https://doi.org/10.1128/AEM.01139-14

Ragavendran, C., & Natarajan, D. (2018). Larval toxicity of entomopathogenic bacteria *Bacillus cereus* against *Aedes aegypti, Anopheles stephensi*, and *Culex quinquefasciatus* mosquitoes. In B. K. Tyagi, & D. Dhanasekaran (Eds.), *Microbial control of vector-borne diseases* (1st ed., pp. 113–135). CRC Press. https://doi.org/10.1201/B22203–7

Rahmet-Alla, M., & Rowley, A. F. (1989). Studies on the pathogenicity of different strains of *Bacillus cereus* for the cockroach, *Leucophaea maderae*. *Journal of Invertebrate Pathology*, *53*(2), 190–196. https://doi.org/10.1016/0022–2011(89)90007–4

Rajagopal, G., Jeyavani, J., & Ilango, S. (2020). Larvicidal and histopathological efficacy of inhabitant pathogenic bacterial strains to reduce the dengue vector competence. *Pest Management Science*, *76*(11), 3587–3595. https://doi.org/10.1002/ps.5892

Rand, N. Du, & Laing, M. D. (2011). Determination of insecticidal toxicity of three species of entomopathogenic spore-forming bacterial isolates against *Tenebrio molitor* L. (Coleoptera: Tenebrionidae). *African Journal of Microbiology Research*, *5*(16), 2222–2228. https://doi.org/10.5897/AJMR11.069

Raymond, B., Lijek, R. S., Griffiths, R. I., & Bonsall, M. B. (2008). Ecological consequences of ingestion of *Bacillus cereus* on *Bacillus thuringiensis* infections and on the gut flora of a lepidopteran host. *Journal of Invertebrate Pathology*, *99*(1), 103–111. https://doi.org/10.1016/J.JIP.2008.04.007

Ruiu, L. (2013). *Brevibacillus laterosporus*, a pathogen of invertebrates and a broad-spectrum antimicrobial species. *Insects*, *4*(3), 476–492. https://doi.org/10.3390/insects4030476

Ruiu, L. (2015). Insect pathogenic bacteria in integrated pest management. *Insects*, *6*(2), 352–367. https://doi.org/10.3390/insects6020352

Selvakumar, G., Mohan, M., Sushil, S. N., Kundu, S., Bhatt, J. C., & Gupta, H. S. (2007). Characterization and phylogenetic analysis of an entomopathogenic *Bacillus cereus* strain WGPSB-2 (MTCC 7182) isolated from white grub, *Anomala dimidiata* (Coleoptera: Scarabaeidae). *Biocontrol Science and Technology*, *17*(5), 525–534. https://doi.org/10.1080/09583150701311663

Sevim, E., Çocar, M., Sezgin, F. M., & Sevim, A. (2018). Aerobic gut bacterial flora of *Cydia pomonella* (L.) (Lepidoptera: Tortricidae) and their virulence to the host. *Egyptian Journal of Biological Pest Control*, *28*, 30. https://doi.org/10.1186/S41938-018-0036-1

Smirnova, T. A., Minenkova, I. B., Orlova, M. V., Lecadet, M. M., & Azizbekyan, R. R. (1996). The crystal-forming strains of *Bacillus laterosporus*. *Research in Microbiology*, *147*(5), 343–350. https://doi.org/10.1016/0923-2508(96)84709-7

Solano-Alvarez, N., Valencia-Hernández, J. A., Rico-García, E., Torres-Pacheco, I., Ocampo-Velázquez, R. V., Escamilla-Silva, E. M., Romero-García, A. L., Alpuche-Solís, Á. G., & Guevara-González, R. G. (2021). A novel isolate of *Bacillus cereus* promotes growth in tomato and inhibits *Clavibacter michiganensis* infection under greenhouse conditions. *Plants*, *10*(3), 506. https://doi.org/10.3390/plants10030506

Somerville, H. J., & Pockett, H. V. (1975). An insect toxin from spores of *Bacillus thuringiensis* and *Bacillus cereus*. *Journal of General Microbiology*, *87*(2), 359–369. https://doi.org/10.1099/00221287-87-2-359

Strongman, D. B., Eveleigh, E. S., van Frankenhuyzen, K., & Royama, T. (1997). The occurrence of two types of entomopathogenic bacilli in natural populations of the spruce budworm, *Choristoneura fumiferana*. *Canadian Journal of Forest Research*, *27*(12), 1922–1927. https://doi.org/10.1139/x97–163

Thiery, I., Hamon, S., Dumanoir, V. C., & de Barjac, H. (1992). Vertebrate safety of *Clostridium bifermentans* serovar *malaysia*, a new larvicidal agent for vector control. *Journal of Economic Entomology*, *85*(5), 1618–1623. https://doi.org/10.1093/JEE/85.5.1618

To, W. N., Gudauskas, R. T., & Harper, J. D. (1975). Pathogenicity of *Bacillus cereus* isolated from *Trichoplusia ni* larvae. *Journal of Invertebrate Pathology*, *26*(1), 135–136. https://doi.org/10.1016/0022–2011(75)90182–2

Toledo, A. V., Alippi, A. M., & De Remes Lenicov, A. M. M. (2011). Growth inhibition of *Beauveria bassiana* by bacteria isolated from the cuticular surface of the corn leafhopper, *Dalbulus maidis* and the planthopper, *Delphacodes kuscheli*, two important vectors of maize pathogens. *Journal of Insect Science*, *11*(1), 29. https://doi.org/10.1673/031.011.0129

Toledo, A. V., López, S., Aulicino, M., de Remes Lenicov, A. M., & Balatti, P. (2015). Antagonism of entomopathogenic fungi by *Bacillus spp.* associated with the integument of cicadellids and delphacids. *International Microbiology*, *18*(2), 91–97. https://doi.org/10.2436/20.1501.01.238

Torres, M. J., Rocha, V. F., Petroselli, G., Villena, R. R., & Audisio, M. C. (2022). Entomopathogenic potential of *Bacillus subtilis* and *Bacillus amyloliquefaciens* strains against *Musca domestica* under controlled conditions. *Entomologia Experimentalis et Applicata*, *170*(7), 584–592. https://doi.org/10.1111/EEA.13186

Tozlu, E., Tozlu, G., & Kotan, R. (2022). Biocontrol of *Cymbalophora rivularis* (Menetries) (Lepidoptera: Erebidae) larvae by entomopathogenic bacteria and fungi. *Egyptian Journal of Biological Pest Control*, *32*, 66. https://doi.org/10.1186/s41938-022-00565-z

Upadhyay, A., & Mohan, S. (2021). *Bacillus subtilis* and *B. licheniformis* isolated from *Heterorhabditis indica* infected apple root borer (*Dorysthenes huegelii*) suppresses nematode production in *Galleria mellonella*. *Acta Parasitologica*, *66*(3), 989–996. https://doi.org/10.1007/S11686-021-00366-8

Valtierra-De-luis, D., Villanueva, M., Berry, C., & Caballero, P. (2020). Potential for *Bacillus thuringiensis* and other bacterial toxins as biological control agents to combat dipteran pests of medical and agronomic importance. *Toxins*, *12*(12), 773. https://doi.org/10.3390/toxins12120773

van Zijll de Jong, E., Roush, T. L., Glare, T. R., & Hampton, J. G. (2016). Discovery of two *Brevibacillus laterosporus* isolates from *Brassica* with insecticidal properties against diamondback moth. *Biocontrol Science and Technology*, *26*(3), 426–431. https://doi.org/10.1080/09583157.2015.1118437

Veith, B., Herzberg, C., Steckel, S., Feesche, J., Maurer, K. H., Ehrenreich, P., Bäumer, S., Henne, A., Liesegang, H., Merkl, R., Ehrenreich, A., & Gottschalk, G. (2004). The complete genome sequence of *Bacillus licheniformis* DSM13, an organism with great industrial potential. *Journal of Molecular Microbiology and Biotechnology*, *7*(4), 204–211. https://doi.org/10.1159/000079829

Venter, H. J., & Bøhn, T. (2016). Interactions between Bt crops and aquatic ecosystems: A review. *Environmental Toxicology and Chemistry*, *35*(12), 2891–2902. https://doi.org/10.1002/ETC.3583

Warren, G. W., Koziel, M. G., Mullins, M. A., Nye, G. J., Carr, B., Desai, N. M., Kostichka, K., Duck, N. B., & Estruch, J. J. (1996). *Novel pesticidal proteins and strains* (WO 96/10083). WIP Organization.

Yaman, M., Erturk, O., Unal, S., & Selek, F. (2017). Isolation and identification of bacteria from four important poplar pests. *Revista Colombiana de Entomología*, *43*(1), 34–37.

Yan, M., Roehrl, M. H., & Wang, J. Y. (2007). Discovery of crystalline inclusions in *Bacillus licheniformis* that resemble parasporal crystals of *Bacillus thuringiensis*. *Canadian Journal of Microbiology*, *53*(9), 1111–1115. https://doi.org/10.1139/W07-076

Yu, X., Liu, T., Liang, X., Tang, C., Zhu, J., Wang, S., Li, S., Deng, Q., Wang, L., Zheng, A., & Li, P. (2011). Rapid detection of vip1-type genes from *Bacillus cereus* and characterization of a novel vip binary toxin gene. *FEMS Microbiology Letters*, *325*(1), 30–36. https://doi.org/10.1111/j.1574-6968.2011.02409.x

Yuan, Y. M., Hu, X. M., Liu, H. Z., Hansen, B. M., Yan, J. P., & Yuan, Z. M. (2007). Kinetics of plasmid transfer among *Bacillus cereus* group strains within lepidopteran larvae. *Archives of Microbiology*, *187*(6), 425–431. https://doi.org/10.1007/S00203-006-0206-5

8 Phylum Proteobacteria

ABSTRACT

This chapter deals with several entomopathogenic bacteria of phylum Proteobacteria, viz., *Serratia entomophila*, *S. proteamaculans*, *S. marcescens*, *Photorhabdus* spp. and *Xenorhabdus* spp., *Yersinia entomophaga*, *Pseudomonas entomophila*, *P. fluorescens*, *P. aeruginosa*, *P. taiwanensis*, *P. cedrina*, *P. putida*, *Rickettsiella*, *Raoultella terrigena* belonging to class Gammaproteobacteria, *Burkholderia* spp. and *Chromobacterium* spp. from class Betaproteobacteria, and finally, *Rickettsia* and *Wolbachia* from class Alphaproteobacteria. A critical analysis of their advantages and drawbacks for mass applicability for insect pest management has been done. Commercial exploitation has been started for some bacteria, yet there is more to accomplish.

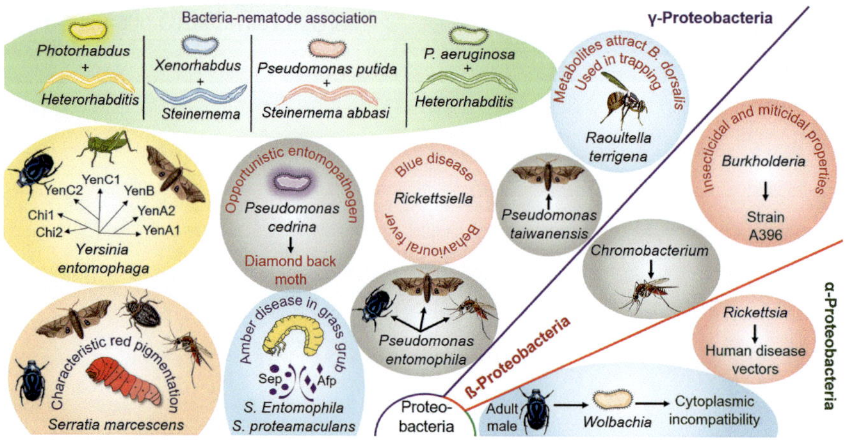

8.1 INTRODUCTION

Proteobacteria is the biggest phylum within the domain of bacteria encompassing all Gram-negative bacteria. There are five classes belonging to this phylum, viz., α-, ß-, γ-, δ-, and ε-proteobacteria; entomopathogenic bacteria are present in only the first three classes. There are many bacteria under the class Gammaproteobacteria discovered with entomopathogenicity, whereas in the classes Beta- and Alphaproteobacteria, only two different genera of bacteria for each have been found to have such entomopathogenic potentiality with varying host ranges (Figure 8.1). They are hardly explored for commercial use but possess high potential.

8.2 CLASS GAMMAPROTEOBACTERIA

8.2.1 *SERRATIA ENTOMOPHILA*

Serratia entomophila Grimont et al. is a Gram-negative, non-spore-forming bacteria (Chattopadhyay & Sen, 2013). Generally, this type of non-spore-forming bacteria can't produce a stable life stage in environment, which limits the mass production for commercial purposes. *S. entomophila* is an exception to this fact, which led to commercialization (strain 154) to control an economically damaging soil-dwelling scarab pest of pasture, the New Zealand grass-grub, *Costelytra zealandica* (White), by causing amber disease. The disease is highly host-specific, affecting only larvae of a single species of scarab (Khachatourians, 2019). These larvae generally feed on decomposed organic matter as well as roots of plants. When these larvae ingest the bacteria, they stop feeding within a couple of days (Glare et al., 2017). The bacterium is believed to bind to the cuticle surface via specialized fimbriae, resulting in colonization at and subsequent clearing of the normal black colour of larval midgut. This results in the characteristic amber coloration, which contributed to the name of the disease. Diseased larvae, also called amber grub or honey grub, can remain active in the soil, starving for several weeks up to one to three months by dissolving their fat bodies, which leads to the weakening of structural tissues and causes lethal septicaemia by allowing the bacteria to enter into the haemocoel (Boucias & Pendland, 1998). The bacterium produces proteases, chitinases, and other tissue-degrading enzymes extracellularly, which adhere to and break the foregut tissue barriers upon ingestion. Two separate gene clusters present in the bacterial plasmid, namely, pADAP, of 155 kb length, have been identified behind this trait (Khachatourians, 2019). Three genes (*sepA*, *sepB*, and *sepC*) are present in the *sep* (*S. entomophila* pathogenicity) gene cluster, which are also detected in plasmids from *S. proteamaculans* (Paine and Stansfield), *S. liquefaciens* (Grimes and Hennerty), and

DOI: 10.1201/9781003426882-8

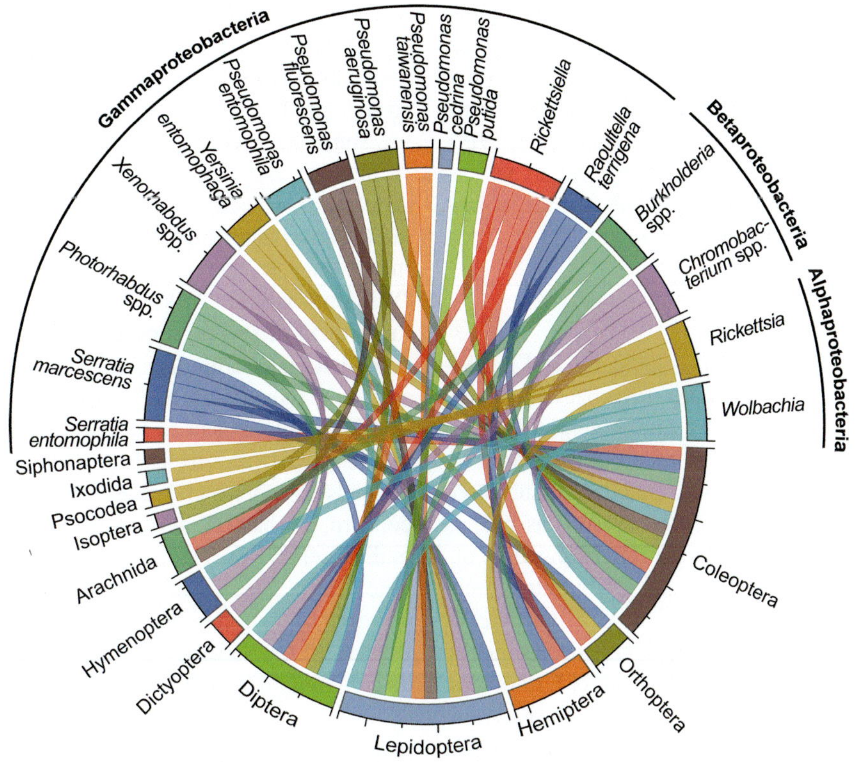

FIGURE 8.1 Classification and host range of different Proteobacteria.

Yersinia frederiksenii Ursing et al. This suggests that they are part of a discrete, horizontally mobile region across different species (Glare et al., 2017; Hurst et al., 2007a). The Sep proteins are similar to the Tc (toxin complex) proteins of *Photorhabdus luminescens* (Thomas and Poinar) and *Xenorhabdus nematophila* (Poinar and Thomas) (Ruiu et al., 2013). *Sep* gene expression in the gut of larvae contributes to gut clearance and amber colouration development upon infection. The second gene cluster, *afp*, has 18 ORFs that code for a novel antifeeding prophage, bullet-shaped toxin-delivery apparatus similar to the R-pyocins of *Pseudomonas aeruginosa* (Schroeter) (Hurst et al., 2007b). For the manifestation of amber disease, both the Sep and Afp virulence factors are essential; Afp causes rapid cessation of feeding upon ingestion, while subsequent induction and release of Sep toxins cause expulsion of gut contents, allowing bacterial colonization at cuticular membranes of the foregut and hindgut, from where constitutive expression of the toxins is maintained (Glare et al., 2017). The AB2 strain of *S. entomophila* shows insecticidal properties against the American bollworm, *Helicoverpa armigera* (Hübner), a foliar pest. Experimental results showed a higher mortality rate when the insect pests were fed with shoot portions of groundnut developed from bacteria-treated seeds. Also, the antifungal property of *S. entomophila* AB2 was found when tested *in vivo* using different parts of the plant developed from bacteria-treated and untreated seeds (Chattopadhyay & Sen, 2013). *S. entomophila* Mor4.1 strain, effective against genera *Phyllophaga* and *Anomala*, was found in Mexico (Glare et al., 2017). All these strains generally produce analogous toxin complexes to those of *Xenorhabdus* spp. and *Photorhabdus* spp. (Ruiu, 2015). Moreover, these cause a reduction in the trypsin activity in the gut, disrupting the normal digestive process (Boucias & Pendland, 1998). Improvements in formulations resulting in expanded shelf-life under ambient conditions have allowed *S. entomophila* to be commercialized (Glare et al., 2017).

8.2.2 Serratia proteamaculans

Serratia proteamaculans (Paine and Stansfield) is a Gram-negative, non-pigmented, common soil bacterium. This is one more causal agent of amber disease of grass-grubs. They are very much similar in the mechanism of their entomopathogenicity, but this bacterium is much faster than *S. entomophila*. *S. entomophila* takes 3–4 months to kill grass-grub, whereas *S. proteamaculans* strain AGR96X causes mortality within 5–12 days of ingestion and successive invasion within the insect haemocoel. AGR96X generates an exclusive variant of the antifeeding prophage (Afp), a cell-free, phage-like organism that is believed to transfer protein toxins to the grass-grub target location, causing feeding activity to cease. The AGR96X Afp, also known as AfpX, has two tail-length termination proteins, which gives it higher length flexibility than other Afp variations. AfpX has dual action against grass-grub (*Costelytra giveni* Coca-Abia & Romero-Samper) and manuka beetle (*Pyronota* sp.) larvae (Hurst et al., 2018). On the other hand, this bacterium possesses remarkable antagonistic traits against plant pathogens (Carter-House et al., 2020; Wang et al., 2014). But this bacterium is also famous for beef spoilage (Manigandan et al., 2018) and has been discovered to cause leaf spot disease in floral crops (Paine & Standsfield, 1919), pointing to its potential risks and requirement of validation of the specificity of strain towards only the target host before exploiting its entomopathogenic nature.

8.2.3 *Serratia marcescens*

Serratia marcescens Bizio is an ecologically and genetically diverse species that includes rhizobacteria, plant endophytes, and insect pests with a high potential for biological control (Grimont & Grimont, 2006). Various *S. marcescens* strains have also been isolated as plant endophytes, promoting the growth of different crops as well as stimulating mercury phytoremediation (Gyaneshwar et al., 2001; Matteoli et al., 2018; Mello et al., 2020; Selvakumar et al., 2008; Zaheer et al., 2016). Furthermore, *S. marcescens* is often identified as a pathogen of several economically significant insect pests belonging to the orders Orthoptera, Hemiptera, Coleoptera, Lepidoptera, and Diptera (Kaaya & Darji, 1989; Niu et al., 2015, 2022; Pineda-Castellanos et al., 2015; Zeng et al., 2020). The bacterium has been found to be effective in the management of scarab beetle, *Phyllophaga blanchardi* (Arrow). Upon oral ingestion and injection bioassays of healthy larvae with the *S. marcescens* isolates, different degrees of antifeeding effect and mortality were observed (Pineda-Castellanos et al., 2015). Strain PS-1, isolated from a moribund striped flea beetle (*Phyllotreta striolata* Fabricius), was found to be lethal to *Bactrocera dorsalis* (Hendel) adults after ingestion. Histological analyses revealed that PS-1 damaged the intestinal epithelium, resulting in cell death within 24 hours. Upon infection, genes encoding the peritrophic matrix constituent were down-regulated, whereas genes involved in lipid and glycan metabolism and the renewal of the gut epithelium, along with genes encoding digestive enzymes and stress response factors, were up-regulated (Zeng et al., 2020). When inoculated in rice seed, it has been found that *S. marcescens* colonizes the roots and shoots, resulting in higher seed germination (9.4–13.3%), root (8.2–36.4%) and shoot lengths (4.1–22.3%), and fresh weights of shoot (19.0–49.0%) and root (26.7–69.3%), along with reduced nymphal growth of brown plant hopper (BPH), *Nilaparvata lugens* (Stål), as compared to plants without inoculation. Rice seeds treated with 10^9 cfu/mL S-JS1 strain of the bacterium resulted in elevated levels of secondary metabolites, which may be associated with *N. lugens* resistance in rice plants (Niu et al., 2022). The bacterium also protects wood from the attack of termite, *Coptotermes curvignathus* Holmgren. Wood treated with strain LGMS 1 at 10^6 cfu mL^{-1} concentration proved more efficient as bait than the higher-concentration applications as it caused a higher mortality rate while still keeping wood very palatable for termites. Wood or soil treated with *S. marcescens* concentration higher than 10^9 cfu mL^{-1} creates a high toxicity and acts as a repellent barrier for termites, but the survival of bacterium on wood after application is less (Chin et al., 2021). Some *S. marcescens* strains produce a red pigment called prodigiosin, which has some larvicidal and pupicidal activity in *Aedes aegypti* (Linnaeus) and *Anopheles stephensi* Liston (Patil et al., 2011; Suryawanshi et al., 2015). Moreover, prodigiosin has antibacterial, antifungal, and trypanolytic properties, which might impact other members of the mosquito microbiota (Williamson et al., 2006). *S. marcescens* VA strain inhibits the growth of several mosquito microbiota isolates using a combination of other secondary metabolites and prodigiosin. In the mosquito, it has been observed that *S. marcescens* VA is highly virulent to larvae in a prodigiosin-dependent manner, while its virulence on adults is lower and largely depends on other metabolites (Heu et al., 2021). But to our concern, the bacterium has been found to be toxic to mammalian cells (Konecka et al., 2019). Also, it causes yellow vine decline disease in Cucurbitaceous plants while being vectored by squash bugs, *Anasa tristis* (DeGeer) (Seebold & Bessin, 2002). So we have to keep this point in mind before its commercial exploitation.

8.2.4 *Photorhabdus* spp. and *Xenorhabdus* spp.

Entomopathogens belonging to these two genera consist of endosymbionts nematicidal properties. The first one is associated with the *Heterorhabditis* entomopathogenic nematodes, while the second one is with the genus *Steinernema*, and both are in obligatory mutualistic relationships with their respective hosts. Members of both the genera are Gram-negative, facultatively anaerobic, rod-shaped bacteria (0.3–2.0 µm by 2.0–10 µm) with peritrichous flagella, can produce intracellular crystalline inclusions, and are resistant to antibiotics like penicillin and beta-lactam. *Xenorhabdus* has distinct primary and secondary phases, having deep blue and red colonies on NBTA medium (Wouts, 1990); *Photorhabdus* colonies are yellow on the same medium (Watson et al., 2018). Unlike *Photorhabdus*, *Xenorhabdus* do not produce pigments due to lack of luciferase gene and are not bioluminescent. Both of them show variation in phase, producing both primary (1°) and secondary phases. This variation observed is similar to the smooth versus rough colony phenotypes observed in various other enteric bacteria. The bacteria are found in the infectious juveniles of host nematodes in the primary phase. Immediately after the host insect dies, most bacteria are in phase I. However, with putrefaction of the host (7–10 days post-mortem), phase transition can be observed, with a higher percentage of free bacteria in phase II (Boucias & Pendland, 1998). They have strong insecticidal activity towards Coleoptera, Dictyoptera, Hymenoptera, and Lepidoptera (Khachatourians, 2019).

The pathogenic action generally involves the release of symbiotic bacteria in the insect haemocoel once the vector nematodes have actively entered the insect body at their juvenile stage via mouth (Forst et al., 1997). The bacteria proliferate by producing various antimicrobial compounds that suppress other microorganisms' growth. Along with these, different enzymes are released from these bacteria that initiate the degradation processes in the haemocoel, thus creating an ideal environment for the growth of the nematode. The bacteria stays unaffected by the host immune system (Silva et al., 2002). Most interestingly, it has been recently discovered that *P. luminescens* has the lowest mutation rate among all bacteria (Pan et al., 2021). In the interaction with the susceptible host, various bacterial virulence factors are involved. In case of *X. budapestensis* Lengyel et al. D43, it produces a 57 kDa insecticidal protein that activates the phenoloxidase cascade and elicits intense immune response in greater wax moth, *Galleria mellonella* (Linnaeus), larvae (Yang et al., 2012). An insecticidal toxins complex (Tc) with a high pest management potential is produced by different *Xenorhabdus* and *Photorhabdus* species. Generally, the Tcs has higher molecular weight and are multi-subunit proteins containing three components, namely, A, B, and C, which are orally active against different insects. Tcs show significant sequence similarities with neurotoxin beta-leptinotarsin isolated from the haemolymph of Colorado potato beetle. While the mode of action is completely not understood, all these components are normally needed to achieve full toxicity. Another insecticidal protein produced by the bacterial *P. luminescens* is represented as *Photorhabdus* insect-related (Pir) proteins, which show similarity to delta endotoxins of *B. thuringiensis* and have been proposed to mimic the juvenile hormone esterases (JHEs) that interfere with insect development regulation. The PirAB binary toxins have oral toxicity against mosquitoes and few caterpillar pests. There are also two more toxins found in this system. One is Mcf1 (makes caterpillars floppy); it promotes programmed cell death, or apoptosis, in great numbers of cells and mimics mammalian BH3 domain-only proteins present in the mitochondrion. The second group of protein is known as PVC (*Photorhabdus* virulence cassettes). These are insecticidal upon injection via an unknown modality (Khachatourians, 2019). Another toxin secreted by *P. luminescens*, Photox, is a single-chain mono-ADP-ribosyltransferase (mART), 46 kDa protein with two domains, which enters a host cell by an unknown mechanism, possibly a type VI secretion system (T6SS), and targets the actin cytoskeleton (Visschedyk et al., 2010). The nematode/bacterial complex is highly virulent and kills its host within 48 hours of infection (Moazami, 2011). To face the immune response of insects, these bacterial endosymbionts evolved different mechanisms in addition to these insecticidal toxins, antibiotics, and various metabolites (Piel, 2010). For instance, PrtA peptidases produced by both of these genera cause effective hydrolysis of denatured proteins, like casein and gelatine. *Photorhabdus* PrtA peptidase selectively cleaved 16 proteins in the haemolymph of tobacco hornworm, *Manduca sexta* (Linnaeus) (Venekei & Marokházi, 2013). It has been shown that to resist the action of insect antimicrobial peptides (AMPs), certain *Photorhabdus* species modify lipopolysaccharides, while *X. nematophila* interferes with the expression mechanisms of host AMPs (Ruiu, 2015). *Photorhabdus* is supposed to produce >20 various small secondary metabolites (Joyce et al., 2011). Subsequently, the nematodes become adults and lay eggs. Eggs hatch into juvenile (J1), which moults into several juvenile stages (J2 to J4). Conversion of J3 to infective juveniles (IJs) or J4 depends on the food quantity/quality and the available space. If scarcity is faced, J3 individuals may become IJs and emerge from the cadaver searching a new host. Or they might moult again to J4 and repeat the cycle (Bastidas et al., 2014). A simplified pictorial description of the mode of action of *Photorhabdus* and *Xenorhabdus* has been given in Figure 8.2.

In addition to *P. luminescens* and *X. nematophila*, the most studied species includes *P. asymbiotica* Fischer-Le Saux et al., *P. temperate* Fischer-Le Saux, *X. beddingii*

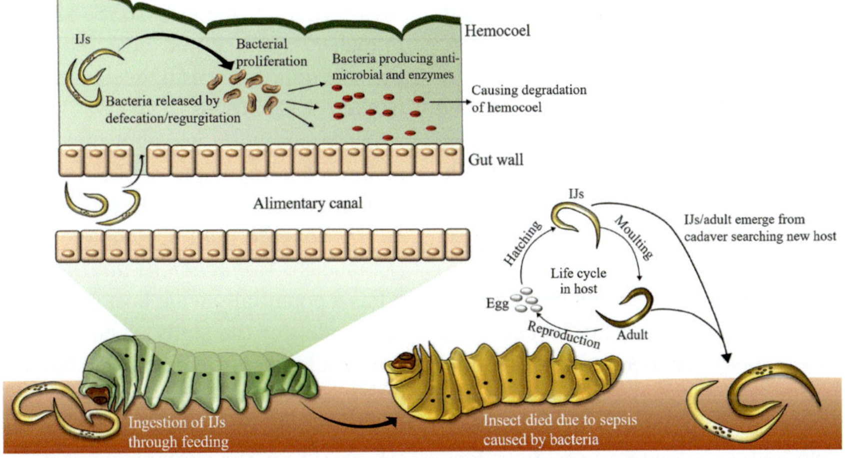

FIGURE 8.2 The mode of action of *Photorhabdus* and *Xenorhabdus* in association with their nematode symbionts to kill the insect hosts.

Akhurst and Boemare, *X. bovienii* (Akhurst), *X. japonica* Nishimura et al., and *X. poinarii* (Akhurst). Along with this, the identification and characterization of new species are continuous. So far, the commercial use of these bacterial endosymbionts is related to the employment of the nematode species with which they are associated (Ruiu, 2015). But we should be careful, as *P. asymbiotica* is pathogenic to humans along with infecting insect hosts (Hapeshi & Waterfield, 2017).

8.2.5 *Yersinia entomophaga*

It is a Gram-negative, rod-shaped, non-spore-forming bacterium. It secretes a multi-subunit toxin complex (Yen-Tc), showing homology with toxin complexes produced by *Photorhabdus* spp. The toxin complex is composed of three protein families, which are A (YenA1, YenA2), B (YenB), and C (YenC1, YenC2), and two chitinases (Chi1 and Chi2) having high endochitinase activity (Hurst et al., 2011). The structure of chitinases has been thoroughly analyzed, and different hypotheses are formulated to explain their role in the complex of toxin. For instance, they may assist in the degradation of the gut peritrophic membrane, opening the way for ABC toxins, or they may play a role in the insects' post-mortem degradation, or they may attach to the toxin complex on the peritrophic membrane. Otherwise, it would be washed through the gut (Ruiu et al., 2013). In the case of *Y. entomophaga* Hurst et al., the study of the 3D structure of the toxin complex showed that the basis of a five-fold symmetric assembly is formed by the subunits YenA1 and YenA2, while a surface-accessible area is formed by subunits B and C and are the main determinants for toxicity. The surface of the TcA scaffold is decorated by the chitinases (Landsberg et al., 2011). These toxins have a broad range of insecticidal activity, including Orthoptera, Coleoptera, and Lepidoptera. Histopathological studies on the effects caused upon ingestion of Yen-Tc revealed that the midgut epithelium of *C. zealandica* progressively disorganized and deteriorated. These effects are reminiscent of those previously reported for other Tc-producing bacteria, like *X. nematophila* and *P. luminescens* (Marshall et al., 2012). After ingestion of bacteria, the larvae of *C. zealandica* exhibit reduced feeding and expulsion of gut contents. By 24 hours, midgut epithelial cells swell and burst, and vesicles are expelled into the midgut lumen. Within 48 hours, the bacteria invade the haemocoel, resulting in death due to septicaemia. By the time of larval death, *Y. entomophaga* multiplies approximately 500-fold in the cadavers (Hurst et al., 2014). This bacterium can also induce mortality within five days in the global pest *Locusta migratoria* (Linnaeus) (McNeill & Hurst, 2018). There is a report regarding promising insecticidal activity of formulations containing *Y. entomophaga* against the pasture pest porina (*Wiseana* spp.) larvae, highlighting its persistence and survival in the field with lethal activity within 7 days, thus suggesting that it can be used as a microbial alternative for the management of porina (Hurst et al., 2020). But the bacterium is also found to be the culprit against acute urinary retention in human beings (Le Guern et al., 2018), so commercialization should be after proper consideration of effects on human and beneficials.

8.2.6 *Pseudomonas entomophila*

P. entomophila Mulet et al. is a ubiquitous Gram-negative aerobic bacterium that colonizes and survives in numerous ecological niches, including soil, water, and plant surfaces (Reimer et al., 2019; Vodovar et al., 2006). It orally infects insect larvae from different orders, determining extensive gut cell rupture. Host–pathogen interactions have been studied in experiments with *Drosophila melanogaster* Meigen (Diptera: Drosophilidae), which highlighted the activation of a specific post-ingestion systemic immune response in both larvae and adults (Vodovar et al., 2005). The complete genome sequencing of *P. entomophila* revealed a specific secretion system and the associated toxins probably responsible for its entomopathogenic properties (Vodovar et al., 2006). *P. entomophila* is the first *Pseudomonas* strain pathogenic in the multicellular organism but is devoid of a T3SS; rather, it follows T1SS and T6SS as secretion mechanisms of effector proteins (Dieppois et al., 2015; Vodovar et al., 2006). The virulence is greatly contributed by Monalysin, a β-barrel pore-forming toxin that damages *Drosophila* tissues and causes necrotic cell death (Nonaka et al., 2020). Natural infection of *Drosophila* larvae by *P. entomophila* induces a global antimicrobial response with preferential induction of the antibacterial peptide gene *Diptericin* (Vodovar et al., 2005). There is a two-component system, GacS/GacA, which plays a key role in *P. entomophila* pathogenicity, controlling many putative virulence factors. AprA, a protease secreted by the bacterium, is found to be important in escaping the fly immune response (Vallet-Gely et al., 2010). The *Pseudomonas virulence factor* (*pvf*) gene cluster encoding a signalling system (Pvf) present in over 500 strains of proteobacteria, including plant and human pathogenic strains, is also present in *P. entomophila* and regulates the expression of approximately 300 genes involved in virulence, T6SS, siderophore transport, and branched-chain amino acid biosynthesis. This system is the sole quorum-sensing mechanism present in the bacteria (Acken & Li, 2023).

P. entomophila is pathogenic for three orders of insects so far, including Diptera, like *Anopheles gambiae* Giles, *D. melanogaster*; Lepidoptera, like *Bombyx mori* (Linnaeus), *G. mellonella*; and Coleoptera, like *Sitophilus oryzae* (Linnaeus), *Monochamus alternatus* Hope, suggesting that it has a potentially wide host range and making it a promising model for the study of host–pathogen interactions and the development of biocontrol agents against insect pests (Dieppois et al., 2015; Wang et al., 2022). It is also reported to be strongly antagonistic to *Xanthomonas citri* subsp. *citri* (Hasse) (*Xcc*), the etiological agent of citrus canker disease (Villamizar et al., 2020), conferring added advantage. But some species of *Drosophila*, like *D. suzukii* (Matsumura), cannot be managed by this bacterium, as the plasmatocytes of this invasive pest cast extracellular traps to combat *P. entomophila*, along with other cell-mediated reactions, such as phagocytosis and the formation of filopodia as a part of the cellular innate immune response. Interestingly, this is very similar to the vertebrate immune system (Carrau et al., 2021). This study elucidates the need for further investigations on insect–pathogen interactions.

8.2.7 *Pseudomonas fluorescens*

P. fluorescens Migula, which is well-known for its plant-beneficial properties, including pathogen and disease suppression, is also recently found to cause lethal infections in insect larvae. Twenty-six (26) different strains belonging to the *P. fluorescens* group exhibit diverse activities toward lepidopteran larvae, ranging from lethal to avirulent. Among them, *P. chlororaphis* (Guignard and Sauvageau) and *P. protegens* Ramette et al. are highly insecticidal, regardless of their origin (animals, plants) (Flury et al., 2016). These bacteria produce Fit toxin, which is similar to Mcf toxin from *Photorhabdus* and *Xenorhabdus* (Ruffner et al., 2015). Chitinase C and phospholipase C are important for causing insect pathogenicity (Flury et al., 2016). It has been found to manage the red spider mites, *Oligonychus coffeae* (Nietner), efficiently in tea (Roobakkumar et al., 2011). *P. fluorescens* KPM-018P strain isolated from tomato leaves can control the larvae of hadda beetles, *Epilachna vigintioctopunctata* (Fabricius). It produces extracellular chitinase, which forms a transparent zone around their colonies due to hydrolyzation of chitin in a minimal medium. Change of incubation temperature turns pale-yellow colonies into red. These characteristics were availed as markers for tracking KPM-018P. The bacteria produced biosurfactants which enabled the bacteria to stabilize and colonize the hydrophobic leaf surface area; they were recovered without any considerable decrease even after a suspension of KPM-018P was sprayed onto leaves. Ingestion of KPM-018P-sprayed leaves by the larvae caused prompt death of these insects to eventually suppress their pupation (Otsu et al., 2010).

But it shows an antagonistic relationship with the entomopathogenic nematode *Heterorhabditis indica* Poinar, Karunakar, and David and its development inside the insect host, *G. mellonella* (Kamra & Mohan, 2011). Also, *P. fluorescens* can cause bacteraemia in humans, with most reported cases being attributed either to transfusion of contaminated blood products or to use of contaminated equipment associated with intravenous infusions. There are several reports identifying it in respiratory samples. There is also an intriguing association between *P. fluorescens* and human disease, in that approximately 50% of Crohn's disease patients develop serum antibodies to *P. fluorescens* (Scales et al., 2014). Though the bacterium is already commercially used as plant growth–promoting rhizobacteria (PGPR) and biofertilizer, we must take care of these negative sides as well.

8.2.8 *Pseudomonas aeruginosa*

P. aeruginosa (Schroeter) is a common encapsulated, Gram-negative, aerobic–facultatively anaerobic, rod-shaped bacterium found widely in nature, soil, and water (Planet, 2023). This is highly effective in the management of tsetse fly, *Glossina morsitans morsitans* Westwood (Kaaya & Darji 1989). When applied to a concentration of 10^9 CFU mL^{-1} on wood, it serves as a slow-acting toxic bait against termite, *C. curvignathus*. However, similar to *Serratia marcescens*, the ability of *P. aeruginosa* to survive on wood is low, which made the bait unable to retain a useful level of toxicity for a long period of time, and frequent reapplication is needed (Chin et al., 2021). This bacterium also has a mutualistic relationship with entomopathogenic nematode *Heterorhabditis bacteriophora* (Poinar) found to attack pupae of *G. mellonella* (He et al., 2019). It has also been isolated from entomopathogenic nematode *H. indica*, having mutualism (Salgado-Morales et al., 2017). Insect consumption and surface application of *P. aeruginosa* resulted in 83.33% and 81.66% mortality of *Trichoplusia ni* (Hübner) larvae, respectively (He et al., 2019). Intra-haemocoelic injection of *P. aeruginosa* in larvae and pre-pupal stage of red palm weevil, *Rhynchophorus ferrugineus* (Olivier), provides effective management of the pest (Banerjee & Dangar, 1995).

But strain BGf-2 of this bacterium is present in the gut of cockroach, *Blattella germanica* (Linnaeus), and protects its host by secreting anti–*Beauveria bassiana* (Bals.) protein. A0A0H2ZK06 showed homology with pyrrolidone carboxyl peptidases (pcps) (Zhang et al., 2020). And most importantly, this bacterium is a well-known human pathogen and can cause infections in the lungs (pneumonia), blood, or other parts of the body after surgery. The bacterium is even coming up with antibiotic resistance in humans via biofilm production (CDC, 2019). Upon root colonization, *P. aeruginosa* forms a biofilm that confers resistance against root-secreted antibiotics in case of *Arabidopsis* and sweet basil (*Ocimum basilicum* L.) and causes plant mortality within 7 days (Walker et al., 2004). So better to be careful of the negative impacts as well before exploiting its entomopathogenicity. Rather, this phenomenon is being used in medical science to study the virulence of human pathogenic *P. aeruginosa* strains on caterpillars of *G. mellonella* as a model organism (Andrejko et al., 2014).

8.2.9 *Pseudomonas taiwanensis*

P. taiwanensis Wang et al. is a Gram-negative, rod-shaped, motile, non-spore-forming bacterium isolated from soil (Wang et al., 2010). It is another entomopathogenic bacterium with broad host range, exhibiting insecticidal activity toward insects like *Plutella xylostella* (Linnaeus), *Spodoptera exigua* Hübner, *Spodoptera litura* (Fabricius), *D. melanogaster*, and *T. ni*. The TccC protein, a component of the toxin complex (Tc), plays an important role in the insecticidal activity of *P. taiwanensis*. TccC protein can be cleaved into two fragments, an N-terminal fragment containing an Rhs-like domain, and a C-terminal fragment containing a Glt symporter domain, and a TraT domain, which might contribute to antioxidative stress activity and defence against macrophagosis, respectively. Natural oral infection by *P. taiwanensis* triggered expression of host programmed cell death–related genes *JNK-2* and *caspase-3* (Chen et al., 2014). It is also effective in antagonizing bacterial leaf blight of rice caused by *Xanthomonas oryzae* pv *oryzae* (Ishiyama) (Chen et al., 2020), adding more benefits of utilizing this bacterium commercially.

8.2.10 *Pseudomonas cedrina*

Along with *P. taiwanensis*, one more Pseudomonad is effective in the management of the diamondback moth, *P. xylostella*, that is, *P. cedrina* Dabboussi et al. (Liu et al., 2019). It is a fluorescent bacterium isolated from the phyllosphere of grasses (Behrendt et al., 2009). Oral administration of *P. cedrina* at levels above 10,000 CFU/mL gave significant mortality to *P. xylostella* larvae. Pathogenicity was also observed through reduced longevity and fecundity in adult females. But to the concern of commercial exploitation, when live bacterial cells were removed, the cultured broth lost its pathogenicity. In response to the bacterial infection, *P. xylostella* expressed antimicrobial and stress-associated genes. A mixture treatment of *P. cedrina* and Bt showed an additive effect on larval mortality of *P. xylostella*. These results indicate that *P. cedrina* is an opportunistic entomopathogen without secretion of toxins (Liu et al., 2019). This bacterium is also associated with tumour of *Pinus patula* Schiede, and the biomass extract is antiproliferative to human solid tumour cell lines (Sánchez-Tafolla et al., 2019). *P. cedrina* was able to induce oxidative foliar photo-necrosis in papaya leaves because of its ability to produce photosensitizing metabolites that generate singlet oxygen in plants it colonizes (Mendoza et al., 2020). So both sides of the coin should be judged properly before commercial exploitation.

8.2.11 *Pseudomonas Putida*

P. putida (Trevisan) is another insect-pathogenic Gram-negative bacterium exhibiting swarming motility. On one hand, it establishes a commensal relationship with plants, and on the other hand, it forms a symbiotic relationship with the entomopathogenic nematode *Steinernema abbasi* Elawad, Ahmad, and Reid (Mahar et al., 2004; Molina et al., 2020). *P. putida* differs from *Xenrorhabdus* spp. because it is aerobic and has one polar flagellum, in contrast to *Xenrorhabdus* spp., which are facultatively anaerobic and have numerous peritrichous flagella. The bacterium is carried in the infective juvenile guts of nematode, which, upon entry into an insect host, releases the bacteria directly into the haemocoel. The bacteria multiply and cause septicaemia to set in, killing the insect within 24–48 hours. They support nematode reproduction by producing nutrients and antimicrobial agents that inhibit the growth of a wide range of other organisms. The bacterium can be grown as free-living organisms under standard laboratory conditions. Insecticidal toxin protein secreted by *P. putida* has been found lethal to insects. *P. putida* has been shown to control beet army worm, *S. exigua*. *P. putida*, via *S. abbasi* and its metabolic secretions, caused the mortality of the pupae of *G. mellonella*. Bacterial cells were recovered from the haemocoel when they were applied to the *G. mellonella* pupae, indicating that bacterial cells can enter the haemocoel even in the absence of nematode vector (Mahar et al., 2004). Strain Ld4 had 100% mortality within 5 days in Colorado potato beetle, *Leptinotarsa decemlineata* (Say) (Muratoğlu et al., 2011). It has also been recently proved that the bacterium can be utilized to remove paraquat, a highly toxic herbicide soluble in water, on corncob biochar as free or immobilized cells (Nguyen et al., 2023). But similar to *P. fluorescens*, *P. putida* are also uncommon causes of skin and soft tissue infections (Singh et al., 2021). So this side also should be taken care of.

8.2.12 *Rickettsiella*

These are Gram-negative, obligate intracellular pathogens with typical bacterial cell walls and no flagella that undergo a developmental cycle by producing various cell phenotypes. The infectious particle is a small, dense rod or disc-shaped cell (0.06–0.6 μm) referred to as an elementary body (EB) that gains entry by inducing endocytosis. These dense particles enlarge within the cell vacuole, forming less-dense reticulate bodies (RB) and intermediate bodies (IB) that undergo replication and convert to the infectious EB phenotype (Jurat-Fuentes & Jackson, 2012). The bacteria were earlier thought to be a part of the genus *Coxiella* only, thus were named as such. It is assumed to have originated from *Coxiella* ~350 years ago (Cordaux et al., 2007). Further, it was renamed as *Rickettsiella* based on its resemblance to *Rickettsia*, belonging to Alphaproteobacteria. All members of *Rickettsiella* are arthropod pathogens. They are highly fastidious and must be propagated within living insect hosts. Transmission of all species occurs by feeding or through wounds. Many *Rickettsiella* extensively replicate in the fat body upon penetration in the alimentary canal. The infective cells are typically small, dense rods which are ingested during feeding and enter the haemocoel by penetrating the midgut epithelium and get entry to the host cells via endocytosis. Once they reach the cell, they produce a pleomorphic form within the cytoplasmic vacuoles, varying from bacteria-like secondary cells to large round rickettsogenic stroma. As the infection progresses, distinctive protein crystals emerge, and the cells transform into miniature rickettsia. When infected cells eventually lyse, masses of rickettsia and crystals are released into the haemolymph, giving infected larvae their white-to-blue hue (Jurat-Fuentes & Jackson, 2012). The isolated entomopathogenic species produce significant visible symptoms and are lethal to insect host. For example, *Rickettsiella popilliae* is the etiological agent of the blue disease in the Japanese beetle, *Popillia japonica* Newman, larvae. This disease initially develops in the fat tissue and produces crystalline bodies as well as the RB, IB, and EB cell phenotypes. The development of these bacteria in the host larvae is lengthy, typical for intracellular bacterial pathogens. Haemocoel injection of bacteria (10^4 cells/larvae) into *P. japonica* larvae results in the expression of blue disease symptoms after 19 days. This disease is also reported in two more scarabaeid, *Phyllophaga anxia* (LeConte) and *P. ephilida* (Say), by the same causal agent (Dutky & Gooden, 1952). *Rickettsiella grylli* Vago and Martoja isolated from crickets within the genus *Gryllus* undergo a developmental cycle similar to that observed with *R. popilliae* (Dutky and Gooden). The infection leads to the swelling of the cricket abdomen, similar to that caused by Iridovirus. When infected with *R. grylli*, some cricket hosts preferentially require a high-temperature habitat, inducing a special kind of fever that suppresses the development of *Rickettsiella*; the phenomenon is known as 'behavioural fever' (Jurat-Fuentes & Jackson, 2012). *R. chirononii* (Weiser), isolated from chironomid hosts, has an unusual disc shape, also seen in the *Rickettsiella* associated with arachnid hosts. The EB of *R. chirononii* does not produce crystalline inclusions during the replicative cycle (Boucias & Pendland, 1998). *Rickettsiella eurygasteris* Lukyanchikov acts as a natural enemy of sunn pest, *Eurygaster integriceps* Puton (Voegele, 1996).

But *Rickettsiella* is also found to reduce the spore formation of the entomopathogenic fungus *Pandora neoaphidis* (Remaudière and Hennebert) in pea aphids (Łukasik et al., 2013). *Rickettsiella* changes the body colour of *Acyrthosiphon pisum* (Harris) from red to green in natural populations by increasing blue-green polycyclic quinones in the body, which consequently results in conferring protection from predators (Tsuchida et al., 2010). The poultry red mite, *Dermanyssus gallinae* DeGeer, is an obligate blood-feeding ectoparasite and a serious threat to the hen egg industry; it uptakes vitamin B1, B2, and B6 from *Rickettsiella* DGE strain (Price et al., 2021), and also, infected hosts often decline over time as a result of rickettsial infections. Affected larvae show visible signs weeks after infection but continue to function normally for more than a month. The larvae stop eating and grow lethargic before dying (Jurat-Fuentes & Jackson, 2012). So we can see that it is not much efficient as an entomopathogen due to its requirement of a long incubation period.

8.2.13 *Raoultella Terrigena*

R. terrigena (Izard et al.) is an environmental bacterium isolated mainly from soil and aquatic environments. This bacterium was earlier classified under the genus *Klebsiella* (Izard et al., 1981). It is also isolated from roots of *Nicotiana tabacum* L. and *Panicum virgatum* L. and is considered to be an endophytic nitrogen-fixing bacterium (Gushgari-Doyle et al., 2021). This bacterium is proven to cause 60% mortality in adults of green shield bug, *Palomena prasina* (Linnaeus) (Ozsahin et al., 2014). Furthermore, *R. terrigena* is detected from the gut of bark beetles *Dendroctonus rhizophagus* Thomas and Bright and *D. valens* (Leconte), exhibiting high acetylene reduction activity (Morales-Jiménez et al., 2013), as well as from the reproductive system of the female oriental fruit fly *B. dorsalis*, and its metabolites are found to be attractive to adults in field trapping assays (Shi et al., 2012). The latter information may be useful in constituting pheromone traps against fruit flies. But the negative side is also there. The quercetin present in *Polygonum viviparum* L. is insecticidal to *S. litura*, but this bacterium, being present in the gut of ghost moth larvae, *Thitarodes xiaojinensis* (Tu.) (Lepidoptera: Hepialidae: Thitarodes), helps its host digest quercetin and survive by feeding on the root of *P. viviparum*. When provided with food, this bacterium helped digest quercetin (one of the insecticidal properties of *Polygonum*) for *S. litura* (Liu et al., 2022).

8.3 CLASS BETAPROTEOBACTERIA

8.3.1 *Burkholderia* Spp.

The bacterial species in the genus *Burkholderia* are ubiquitous organisms in the soil, rhizospheres, insects, fungi, and water (Cordova-Kreylos et al., 2013). Different insect species harbour symbiotic bacteria from the genus *Burkholderia*, mostly associated with specific localized gut regions. In addition to these behavioural mutualistic relationships with insects, *Burkholderia* sp. has also been reported to affect oviposition and fecundity of bean bug, *Riptortus pedestris* (Fabricius) (Kil et al., 2014). The insecticidal properties of strain *B. rinojensis* sp. *nov.* have been reported, which was isolated from soil in Japan. Whole-cell broth cultures of this bacterial strain, A396, show oral toxicity and contact effects against the beet armyworm *S. exigua* and the two-spotted spider mite *Tetranychus urticae* Koch. It is also found to be effective against Colorado potato beetle, *L. decemlineata* (Sabbahi et al., 2022). After heat treatment, insecticidal and miticidal properties are maintained, hence having the potential to become efficient commercial formulations against various chewing, sucking insects and mite pests. Different bacterial metabolites might be involved in the insecticidal action (Cordova-Kreylos et al., 2013).

However, some *Burkholderia* isolates were found to be associated with leaf-cutting ant *Atta sexdens rubropilosa* Forel, where it secretes a potent anti-fungal agent, inhibiting the germination of conidia of the entomopathogenic fungi *B. bassiana*, *Metarhizium anisopliae* (Metschn.), saprophytic fungi, *Verticillium lecanii* (Zimm.), and also of *Escovopsis weberi* Muchovej and Della Lucia, a specialized pathogen of fungus garden of leaf-cutting ants. But significantly, the growth of the ant's mutualist fungus was unaffected, thus proved to help in the survival of the pest (Santos et al., 2004). Sinapigladioside is produced as an antifungal metabolite by the beetle symbiont *Burkholderia gladioli* (Severini), a component spread onto the eggs by female *Lagria* beetles to protect the beetle offspring from pathogens (Dose et al., 2021). Also, some members of *Burkholderia* sp. are established plant and human pathogens, so we should check the pathogenicity of the potent strains towards crops and consumers before releasing them commercially.

8.3.2 *Chromobacterium* spp.

Chromobacterium subtsugae Martin et al. is a Gram-negative Betaproteobacterium developing violet-pigmented colonies on L-agar due to production of violacein, a tryptophan derivative. This was first isolated from a soil sample (Maryland, USA). Along with lethality in diversified insect hosts belonging to the orders Coleoptera, Lepidoptera, and Hemiptera, significant sublethal effects, with special regard to feeding inhibition, can be observed in most cases (Martin et al., 2007). The toxin factors associated with insect mortality were stable to heat, freezing, or pH treatment, rendering characteristics of a good biocontrol agent. It also produces bioactive compounds during the stationary phase. A strain identified as *Chromobacterium* (Csp_P) isolated from the midgut of field-caught *Ae. aegypti* was shown to be able to colonize the insect midgut and display entomopathogenic and anti-pathogen properties. Csp_P activates mosquito immune responses, and the exposure leads to dramatic reduction of the survival larval as well as adult stages. Ingestion of Csp_P significantly reduces the susceptibility of mosquitoes towards the causal organism of malaria, *Plasmodium falciparum* (Welch), and dengue virus infection, thus compromising the ability of mosquitoes to act as vectors (Ramirez et al., 2014). *C. subtsugae* strain PRAA4-1T has been developed into a commercially available product (Grandevo®, Marrone Bio Innovations Inc.) to manage a broad spectrum of chewing and sucking insect species (Ruiu et al., 2013). *Chromobacterium violaceum* Bergonzini kills pyrethroid-resistant mosquito *Anopheles coluzzii* Coetzee and Wilkerson and reduces feeding willingness, egg laying, and hatching of mosquitoes (Gnambani et al., 2020). So the bacterium may come as an important remedy for the problem of mosquitoes and associated fatal human diseases vectored by them, along with the management of various other insect pests. But there are some rare events of medical reports against *C. violacearum* causing human diseases (Parajuli et al., 2016), which should be taken due care of.

8.4 CLASS ALPHAPROTEOBACTERIA

8.4.1 *Rickettsia*

These Gram-negative bacteria are like short-rod (0.3–0.5 µm by 0.8–2.0 µm) intracellular pathogens and obligate that may cycle from invertebrate to vertebrate hosts. *Rickettsia* enters host cells via rickettsial-mediated endocytosis, escapes the endosome prior to its fusion with lysosomes, and replicates in the cytoplasm, producing 100–1,000 progeny bacteria per host cell. *Rickettsia* is highly cytopathic in most cases, causing lysis within 2 to 3 days post-infection. Progeny *Rickettsia* may be released from infected cells within 10 hours post-infection. Members of the genus *Rickettsia* are often causal agents of human disease and are capable of causing widespread epidemics. These are often transmitted from reservoir hosts to humans by various invertebrate vectors (insects, ticks, mites). *Rickettsia prowazekii* da Rocha-Lima, the causal agent of epidemic typhus fever in humans, is transmitted by the human louse *Pediculus humanus* Linnaeus. In the columnar epithelial cells of the louse alimentary tract, *R. prowazekii* undergoes extensive replication. Exfoliated infected cells deposited in faeces serve as an inoculum source for vertebrate infection. The growth of *R. prowazekii* in the midgut of the louse kills the vector within 1 to 2 weeks (Boucias & Pendland, 1998). The infection of *R. prowazekii* is also proved fatal to some ticks, like *Dermacentor andersoni* Stiles, *D. marginatus* (Sulzer), and *D. reticulatus* (Fabricius) (Burgdorfer et al., 1972). The disease-causing agent of murine typhus, *Rickettsia typhi* (Wolbach and Todd), can replicate in the midgut cells of its vector, the rat flea, *Xenopsylla cheopis* (Rothschild). This disease agent has been demonstrated to penetrate the flea's midgut and infect gonadal tissue, resulting in the transovarial transmission of *R. typhi* to progeny fleas (Boucias & Pendland, 1998). *Rickettsia*-infected common bed bug (*Cimex lectularius* Linnaeus) developed more slowly from first instar to adult, and infected females produced fewer fertile eggs (Thongprem et al., 2020). The bacterium was also reported to be present in grass-grub, *C. zealandica*, and chafer, *Odontria* sp., but did not induce sufficient mortality to be used commercially for management (Moore et al., 2010). Sweet potato whiteflies, *Bemisia tabaci* (Gennadius), infected with *Rickettsia* show a decreased rate of death due to *Pseudomonas syringae* van Hall (Hendry et al., 2014). For these reasons, it is still now not utilized in the field.

8.4.2 *Wolbachia*

These bacteria are heterogeneous groups of small rods and cocci, morphologically similar to the *Rickettsia*. *Wolbachia* are endocellular symbionts and parasites of arthropods and nematodes and are not directly pathogenic to either insects or vertebrates but cause alterations in sexuality and the reproductive fitness of insect hosts (Boucias & Pendland, 1998; Brown et al., 2016). The effects of *Wolbachia* infection in *Drosophila* have been pictorially depicted in Figure 8.3. Around 17% of all neotropical insects carry this bacterium (Werren et al., 1995), and 25–75% of all insects are the potential host (Kozek & Rao, 2007). *Wolbachia* is highly fastidious and usually cannot be propagated outside the insect host. It majorly regulates cytoplasmic incompatibility (CI) in insect populations. CI results when a *Wolbachia*-infected adult male is mated with an uninfected adult female. Depending on the *Wolbachia*-infected host, such crosses will block embryogenesis. *Wolbachia* are not present in the sperm of infected males, but their presence in adult males modifies the sperm so that it cannot fertilize the eggs of uninfected females. Crosses between infected males and infected females may result in normal

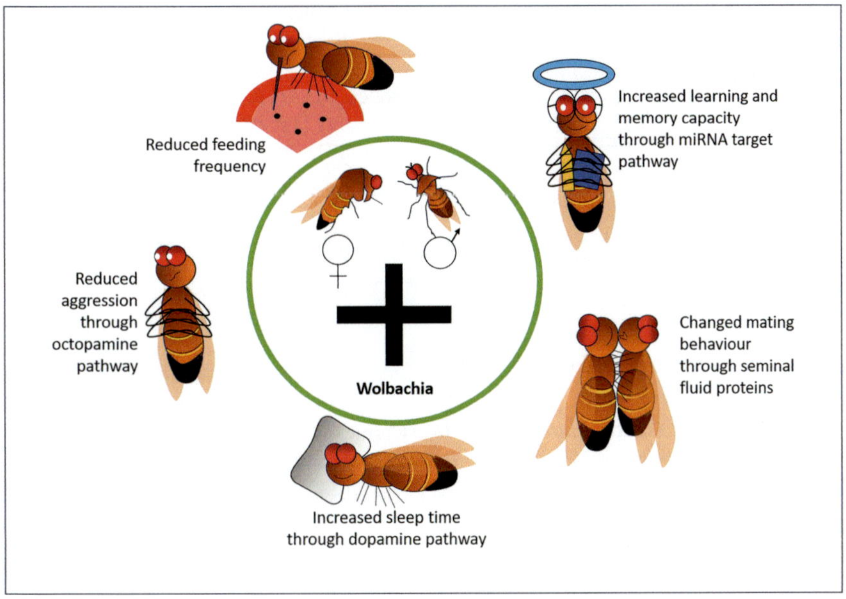

FIGURE 8.3 Behavioural changes in insects due to *Wolbachia* infection in *Drosophila*. **Note:** Reduces appetite but higher survival under starvation in *Drosophila* (Karpova et al., 2023) (also reduces blood-feeding success in *Aedes*; Turley et al., 2009). Reduces aggression of male *Drosophila* (Rohrscheib et al., 2015). Affects sleep behaviour in *Drosophila* (Bi et al., 2018)/Change in post-mating responses (PMR) in *Drosophila* (He et al., 2018). Improves learning and memory capacity (LMC) in *Drosophila* by altering host gene expression in the host nervous system (Bi et al., 2019).

progeny, suggesting that *Wolbachia* in females nullifies their action on the sperm. However, crosses between *Wolbachia*-positive female and male adults result in CI in certain cases. Normally, such bidirectional expression of CI occurs due to the presence of different bacteria in the males and females.

The *Wolbachia* associated with insects are maternally transmitted to offspring via egg cytoplasm. These bacteria are often associated with young nurse cells and oocytes. Several insects of orders Diptera, Hymenoptera, Coleoptera, and Lepidoptera have been reported to harbour *Wolbachia* species expressing the CI trait. The parasitoid wasp genus *Nasonia* also has *Wolbachia* and exhibits either unidirectional or bidirectional incompatibility. *Nasonia* often harbours a double infection of *Wolbachia* groups. Compatible mating occurs when the paternal strain contains an equivalent or a lower bacterial load than that found in its female mate. In addition, to reduce viable progeny, the presence of *Wolbachia* may influence male fertility and the expression of parthenogenesis. For example, the *Wolbachia* associated with the flour beetle, *Tribolium confusum* Duval, increased male fertility. The hymenopteran egg parasitoid *Trichogramma* contains a *Wolbachia* that regulates its sexuality (Boucias & Pendland, 1998). So *Wolbachia*'s CI mechanism can provide us with a very effective pest control strategy.

Except CI mechanism, several mechanisms are followed by *Wolbachia* to maximize their spread by significantly altering the reproductive capabilities of their hosts. Male killing happens when infected males pass away while developing into larvae, which raises the number of newborn infected females. This was first documented in the ladybird *Adalia bipunctata* (Linnaeus) and the butterfly *Acraea encedon* Linnaeus, two innocent insects regarding agricultural aspects (Hurst et al., 1999). Infected men that undergo feminization either become females or become sterile pseudo-females. This is especially common in Lepidoptera species, like the adzuki bean borer, *Ostrinia scapulalis* (Walker), where both male killing and feminization mechanisms are present (Fujii et al., 2001). Infected females also can reproduce without males through a process called parthenogenesis. Some researchers have hypothesized that *Wolbachia* may have impacted parthenogenesis. Some species of the *Trichogramma* parasitoid wasp genus have evolved to reproduce without males as a result of the presence of *Wolbachia*, serving as an example of parthenogenesis driven by *Wolbachia*. The lack of males in this species of wasps may be due to the fact that the same strain of *Wolbachia* has killed so many of them (Knight, 2001). Again, *Wolbachia* is also associated with reproductive distortions of several *Trissolcus* wasps (Hymenoptera: Scelionidae), an important egg parasitoid of the *E. integriceps*, which might become a point of concern (Guz et al., 2012). Also, its compatibility with other biocontrol agents is being checked. *Wolbachia* infections may not hamper and may even improve the success of biological control using entomopathogenic fungi (Zélé et al., 2020). In another experiment, it was found that *Culex quinquefasciatus* Say naturally infected with *Wolbachia pipientis* Hertig *w*PipSJ is less susceptible to the pathogenic action of mosquitocidal bacterial strains of *Bt* subsp. *israelensis*, *Bacillus wiedmannii* biovar *thuringiensis*, and *Lysinibacillus sphaericus* (Meyer and Neide), when compared to non-infected mosquitoes, and is more susceptible to *Bt* subsp. *israelensis* than the other two (Díaz-Nieto et al., 2021). In *Wolbachia*-infected cabbage root fly, *Delia radicum* (Wiedemann), it has been found that *Wolbachia* might antagonize *Erwinia* for being entomopathogenic but would favour *Providencia* and *Serratia* because they might protect the host against chemical plant defences (Ourry et al., 2021).

8.5 CONCLUSION

The phylum Proteobacteria is an assembly of diverse types of entomopathogenic bacteria, starting from endosymbionts to soil bacterium. Along with them, *Photorhabdus*, *Xenorhabdus*, and some other bacteria are also found to be associated with entomopathogenic nematodes and thus help in insect management. There are some commercial products, like Invade™ from *Serratia entomophila* against grass-grub, Majestene® and Venerate® from *Burkholderia* spp. against chewing and sucking insects and mites and nematodes, Grandevo® from *Chromobacterium subtsugae* against chewing and sucking insects and mites (Johnson et al., 2001; Ruiu, 2018), but more scope is there to explore further. We can hope, in the future, we will have many more novel microbial insecticides generated from the bacteria of this phylum.

REFERENCES

Acken, K. A., & Li, B. (2023). *Pseudomonas virulence factor* controls expression of virulence genes in *Pseudomonas entomophila*. *PLoS One*, *18*(5), e0284907. https://doi.org/10.1371/journal.pone.0284907

Andrejko, M., Zdybicka-Barabas, A., & Cytryńska, M. (2014). Diverse effects of *Galleria mellonella* infection with entomopathogenic and clinical strains of *Pseudomonas aeruginosa*. *Journal of Invertebrate Pathology*, *115*(1), 14–25. https://doi.org/10.1016/J.JIP.2013.10.006

Banerjee, A., & Dangar, T. K. (1995). *Pseudomonas aeruginosa*, a facultative pathogen of red palm weevil, *Rhynchophorus ferrugineus*. *World Journal of Microbiology & Biotechnology*, *11*(6), 618–620. https://doi.org/10.1007/BF00361002/METRICS

Bastidas, B., Portillo, E., & San-Blas, E. (2014). Size does matter: The life cycle of *Steinernema* spp. in micro-insect hosts. *Journal of Invertebrate Pathology*, *121*, 46–55. https://doi.org/10.1016/J.JIP.2014.06.010

Behrendt, U., Schumann, P., Meyer, J. M., & Ulrich, A. (2009). *Pseudomonas cedrina* subsp. *fulgida* subsp. *nov*., a fluorescent bacterium isolated from the phyllosphere of grasses; emended description of *Pseudomonas cedrina* and description of *Pseudomonas cedrina* subsp. *cedrina* subsp. *nov*. *International Journal of Systematic and Evolutionary Microbiology*, *59*(6), 1331–1335. https://doi.org/10.1099/ijs.0.005025-0

Bi, J., Sehgal, A., Williams, J. A., & Wang, Y. F. (2018). *Wolbachia* affects sleep behavior in *Drosophila melanogaster*. *Journal of Insect Physiology*, *107*, 81–88. https://doi.org/10.1016/J.JINSPHYS.2018.02.011

Bi, J., Zheng, Y., Wang, R. F., Ai, H., Haynes, P. R., Brownlie, J. C., Yu, X. Q., & Wang, Y. F. (2019). *Wolbachia* infection may improve learning and memory capacity of *Drosophila* by altering host gene expression through microRNA. *Insect Biochemistry and Molecular Biology*, *106*, 47–54. https://doi.org/10.1016/J.IBMB.2018.11.007

Boucias, D. G., & Pendland, J. C. (1998). *Principles of insect pathology* (1st ed.). Springer. https://doi.org/10.1007/978-1-4615-4915-4

Brown, A. M. V., Wasala, S. K., Howe, D. K., Peetz, A. B., Zasada, I. A., & Denver, D. R. (2016). Genomic evidence for plant-parasitic nematodes as the earliest *Wolbachia* hosts. *Scientific Reports*, *6*(1), 34955. https://doi.org/10.1038/srep34955

Burgdorfer, W., Ormsbee, R. A., & Hoogstraal, H. (1972). Ticks as vectors of *Rickettsia prowazeki*-a controversial issue. *The American Journal of Tropical Medicine and Hygiene*, *21*(6), 989–998. https://doi.org/10.4269/ajtmh.1972.21.989

Carrau, T., Thümecke, S., Silva, L. M. R., Perez-Bravo, D., Gärtner, U., Taubert, A., Hermosilla, C., Vilcinskas, A., & Lee, K. Z. (2021). The Cellular innate immune response of the invasive pest insect *Drosophila suzukii* against *Pseudomonas entomophila* involves the release of extracellular traps. *Cells*, *10*(12), 3320. https://doi.org/10.3390/CELLS10123320

Carter-House, D., Chung, J., McDonald, S., Mauck, K., & Stajich, J. E. (2020, September 7). Volatiles from *Serratia marcescens*, *S. proteamaculans*, and *Bacillus subtilis* inhibit growth of *Rhizopus stolonifer* and other fungi. *BioRxiv*, 286443. https://doi.org/10.1101/2020.09.07.286443

CDC. (2019, November 13). *Pseudomonas aeruginosa in healthcare settings*. Centers for Disease Control and Prevention. https://www.cdc.gov/hai/organisms/pseudomonas

Chattopadhyay, P., & Sen, S. K. (2013). Systemic infestation of *Serratia entomophila* AB2 through plant tissue inferred protection against insect pest and fungal pathogens. *African Journal of Microbiology Research*, *7*(21), 2651–2655. https://doi.org/10.5897/AJMR2013.5743

Chen, W. J., Hsieh, F. C., Hsu, F. C., Tasy, Y. F., Liu, J. R., & Shih, M. C. (2014). Characterization of an insecticidal toxin and pathogenicity of *Pseudomonas taiwanensis* against insects. *PLoS Pathogens*, *10*(8), e1004288. https://doi.org/10.1371/journal.ppat.1004288

Chen, W. J., Kuo, T. Y., Chen, C. Y., Hsieh, F. C., Yang, Y. L., Liu, J. R., & Shih, M. C. (2020). Whole genome sequencing and Tn5-Insertion mutagenesis of *Pseudomonas taiwanensis* CMS to probe its antagonistic activity against rice bacterial blight disease. *International Journal of Molecular Sciences*, *21*(s22), 8639. https://doi.org/10.3390/IJMS21228639

Chin, K. L., H'Ng, P. S., Lee, C. L., Wong, W. Z., Go, W. Z., Khoo, P. S., Luqman, A. C., & Ashaari, Z. (2021). Application strategies by selective medium treated with entomopathogenic bacteria *Serratia marcescens* and *Pseudomonas aeruginosa* as potential biocontrol against *Coptotermes curvignathus*. *Royal Society Open Science*, *8*(4), 201311. https://doi.org/10.1098/RSOS.201311

Cordaux, R., Paces-Fessy, M., Raimond, M., Michel-Salzat, A., Zimmer, M., & Bouchon, D. (2007). Molecular characterization and evolution of arthropod-pathogenic *Rickettsiella* bacteria. *Applied and Environmental Microbiology*, *73*(15), 5045–5047. https://doi.org/10.1128/AEM.00378-07

Cordova-Kreylos, A. L., Fernandez, L. E., Koivunen, M., Yang, A., Flor-Weiler, L., & Marrone, P. G. (2013). Isolation and characterization of *Burkholderia rinojensis* sp. *nov*., a non-*Burkholderia cepacia* complex soil bacterium with insecticidal and miticidal activities. *Applied and Environmental Microbiology*, *79*(24), 7669–7678. https://doi.org/10.1128/AEM.02365-13

Díaz-Nieto, L. M., Gil, M. F., Lazarte, J. N., Perotti, M. A., & Berón, C. M. (2021). *Culex quinquefasciatus* carrying *Wolbachia* is less susceptible to entomopathogenic bacteria. *Scientific Reports*, *11*, 1094. https://doi.org/10.1038/s41598-020-80034-5

Dieppois, G., Opota, O., Lalucat, J., & Lemaitre, B. (2015). *Pseudomonas entomophila*: A versatile bacterium with entomopathogenic properties. In J. L. Ramos, J. Goldberg, & A. Filloux (Eds.), *Pseudomonas* (pp. 25–49). Springer. https://doi.org/10.1007/978-94-017-9555-5_2

Dose, B., Niehs, S. P., Scherlach, K., Shahda, S., Flórez, L. V., Kaltenpoth, M., & Hertweck, C. (2021). Biosynthesis of sinapigladioside, an antifungal isothiocyanate from *Bur kholderia* symbionts. *ChemBioChem*, *22*(11), 1920–1924. https://doi.org/10.1002/CBIC.202100089

Dutky, S. R., & Gooden, E. L. (1952). *Coxiella popilliae*, *n. sp.*, a rickettsia causing blue disease of Japanese beetle larvae. *Journal of Bacteriology*, *63*(6), 743–750. https://doi.org/10.1128/JB.63.6.743-750.1952

Flury, P., Aellen, N., Ruffner, B., Péchy-Tarr, M., Fataar, S., Metla, Z., Dominguez-Ferreras, A., Bloemberg, G., Frey, J., Goesmann, A., Raaijmakers, J. M., Duffy, B., Höfte, M., Blom, J., Smits, T. H. M., Keel, C., & Maurhofer, M. (2016). Insect pathogenicity in plant-beneficial pseudomonads: Phylogenetic distribution and comparative genomics. *The ISME Journal*, *10*(10), 2527–2542. https://doi.org/10.1038/ISMEJ.2016.5

Forst, S., Dowds, B., Boemare, N., & Stackebrandt, E. (1997). *Xenorhabdus* and *Photorhabdus* spp.: Bugs that kill bugs. *Annual Review of Microbiology*, *51*, 47–72. https://doi.org/10.1146/ANNUREV.MICRO.51.1.47

Fujii, Y., Kageyama, D., Hoshizaki, S., Ishikawa, H., & Sasaki, T. (2001). Transfection of *Wolbachia* in Lepidoptera: The feminizer of the adzuki bean borer *Ostrinia scapulalis* causes male killing in the Mediterranean flour moth *Ephestia kuehniella*. *Proceedings of the Royal Society of London: Series B: Biological Sciences*, *268*(1469), 855–859. https://doi.org/10.1098/RSPB.2001.1593

Glare, T. R., Jurat-Fuentes, J. L., & O'Callaghan, M. (2017). Basic and applied research. In L. A. Lacey (Ed.), *Microbial control of insect and mite pests* (pp. 47–67). Elsevier. https://doi.org/10.1016/B978-0-12-803527-6.00004-4

Gnambani, E. J., Bilgo, E., Sanou, A., Dabiré, R. K., & Diabaté, A. (2020). Infection of highly insecticide-resistant malaria vector *Anopheles coluzzii* with entomopathogenic bacteria *Chromobacterium violaceum* reduces its survival, blood feeding propensity and fecundity. *Malaria Journal*, *19*, 352. https://doi.org/10.1186/s12936-020-03420-4

Grimont, F., & Grimont, P. A. D. (2006). The genus *Serratia*. In M. Dworkin, S. Falkow, E. Rosenberg, K. H. Schleifer, & E. Stackebrandt (Eds.), *The prokaryotes: A handbook on the biology of bacteria* (3rd ed., pp. 219–244). Springer.

Gushgari-Doyle, S., Schicklberger, M., Li, Y. V., Walker, R., & Chakraborty, R. (2021). Plant growth promotion diversity in switchgrass-colonizing, diazotrophic endophytes. *Frontiers in Microbiology*, *12*, 730440. https://doi.org/10.3389/fmicb.2021.730440

Guz, N., Kocak, E., Emre Akpinar, A., Oktay Gurkan, M., & Neset Kilincer, A. (2012). Wolbachia infection in *Trissolcus* species (Hymenoptera: Scelionidae). *European Journal of Entomology*, *109*(2), 169–174. https://doi.org/10.14411/EJE.2012.022

Gyaneshwar, P., James, E. K., Mathan, N., Reddy, P. M., Reinhold-Hurek, B., & Ladha, J. K. (2001). Endophytic colonization of rice by a diazotrophic strain of *Serratia marcescens*. *Journal of Bacteriology*, *183*(8), 2634–2645. https://doi.org/10.1128/JB.183.8.2634-2645.2001

Hapeshi, A., & Waterfield, N. R. (2017). *Photorhabdus asymbiotica* as an insect and human pathogen. *Current Topics in Microbiology and Immunology*, *402*, 159–177. https://doi.org/10.1007/82_2016_29

He, Y., Qin, Q., DiLegge, M. J., & Vivanco, J. M. (2019). Isolation of *Klebsiella pneumoniae* and *Pseudomonas aeruginosa* from entomopathogenic nematode-insect host relationship to examine bacterial pathogenicity on *Trichoplusia ni*. *Microbial Pathogenesis*, *135*, 103606. https://doi.org/10.1016/J.MICPATH.2019.103606

He, Z., Zhang, H. B., Li, S. T., Yu, W. J., Biwot, J., Yu, X. Q., Peng, Y., & Wang, Y. F. (2018). Effects of *Wolbachia* infection on the postmating response in *Drosophila melanogaster*. *Behavioral Ecology and Sociobiology*, *72*, 146. https://doi.org/10.1007/S00265-018-2561-Z

Hendry, T. A., Hunter, M. S., & Baltrus, D. A. (2014). The facultative symbiont *Rickettsia* protects an invasive whitefly against entomopathogenic *Pseudomonas syringae* strains. *Applied and Environmental Microbiology*, *80*(23), 7161–7168. https://doi.org/10.1128/AEM.02447-14

Heu, K., Romoli, O., Schönbeck, J. C., Ajenoe, R., Epelboin, Y., Kircher, V., Houël, E., Estevez, Y., & Gendrin, M. (2021). The effect of secondary metabolites produced by *Serratia marcescens* on *Aedes aegypti* and its microbiota. *Frontiers in Microbiology*, *12*, 645701. https://doi.org/10.3389/fmicb.2021.645701

Hurst, G. D. D., Jigging, F. M., Von Der Schulenburg, J. H. G., Bertrand, D., West, S. A., Goriacheva, I. I., Zakharov, I. A., Werren, J. H., Stouthamer, R., & Majerus, M. E. N. (1999). Malekilling *Wolbachia* in two species of insect. *Proceedings of the Royal Society of London. Series B: Biological Sciences*, *266*(1420), 735–740. https://doi.org/10.1098/RSPB.1999.0698

Hurst, M. R. H., Beard, S. S., Jackson, T. A., & Jones, S. M. (2007b). Isolation and characterization of the *Serratia entomophila* antifeeding prophage. *FEMS Microbiology Letters*, *270*(1), 42–48. https://doi.org/10.1111/J.1574-6968.2007.00645.X

Hurst, M. R. H., Beattie, A., Jones, S. A., Laugraud, A., van Koten, C., & Harper, L. (2018). *Serratia proteamaculans* srain AGR96X encodes an antifeeding prophage (Tailocin) with activity against grass grub (*Costelytra giveni*) and manuka beetle (*Pyronota* species) larvae. *Applied and Environmental Microbiology*, *84*(10). https://doi.org/10.1128/AEM.02739-17

Hurst, M. R. H., Jones, S. A., Tan, B., Harper, L. A., Jackson, T. A., & Glare, T. R. (2011). The main virulence determinant of *Yersinia entomophaga* MH96 is a broad-host-range toxin complex active against insects. *Journal of Bacteriology*, *193*(8), 1966–1980. https://doi.org/10.1128/JB.01044-10.

Hurst, M. R. H., Jones, S. M., Tan, B., & Jackson, T. A. (2007a). Induced expression of the *Serratia entomophila* Sep proteins shows activity towards the larvae of the New Zealand grass grub *Costelytra zealandica*. *FEMS Microbiology Letters*, *275*(1), 160–167. https://doi.org/10.1111/J.1574-6968.2007.00886.X

Hurst, M. R. H., Swaminathan, J., Wright, D., Hardwick, S., Ferguson, C., Beattie, A., Richards, N., Harper, L., Moss, R., Cave, V., van Koten, C., & McNeill, M. (2020). Development of a *Yersinia entomophaga* bait for control of larvae of the porina moth (*Wiseana spp.*), a pest of New Zealand impro ved grassland systems. *Pest Management Science*, *76*(1), 350–359. https://doi.org/10.1002/PS.5521

Hurst, M. R. H., van Koten, C., & Jackson, T. A. (2014). Pathology of *Yersinia entomophaga* MH96 towards *Costelytra zealandica* (Coleoptera: Scarabaeidae) larvae. *Journal of Invertebrate Pathology*, *115*(1), 102–107. https://doi.org/10.1016/J.JIP.2013.11.004.

Izard, D., Ferragut, C., & Favini, F. (1981). *Klebsiella terrigena*, a new species from soil and water. *International Journal of Systematic Bacteriology*, *31*(2), 116–127. https://doi.org/10.1099/00207713-31-2-116

Johnson, V. W., Pearson, J., & Jackson, T. A. (2001). Formulation of *Serratia entomophila* for biological control of grass grub. *New Zealand Plant Protection*, *54*, 125–127. https://doi.org/10.30843/NZPP.2001.54.3752

Joyce, S. A., Lango, L., & Clarke, D. J. (2011). The regulation of secondary metabolism and mutualism in the insect pathogenic bacterium *Photorhabdus luminescens*. *Advances in Applied Microbiology*, *76*, 1–25. https://doi.org/10.1016/B978-0-12-387048-3.00001-5

Jurat-Fuentes, J. L., & Jackson, T. A. (2012). Bacterial entomopathogens. In F. E. Vega, & H. K. Kaya (Eds.), *Insect pathology* (2nd ed., pp. 265–349). https://doi.org/10.1016/B978-0-12-384984-7.00008-7

Kaaya, G. P., & Darji, N. (1989). Mortality in adult tsetse, *Glossina morsitans morsitans*, caused by entomopathogenic bacteria. *Journal of Invertebrate Pathology*, *54*(1), 32–38. https://doi.org/10.1016/0022-2011(89)90136-5

Kamra, A., & Mohan, S. (2011). Antagonistic effect of *Pseudo monas fluorescens* Migula 1895 on the entomopathogenic nematode, *Heterorhabditis indica* Poinar et al. 1992. *International Journal of Nematology*, *21*(1), 225–229.

Karpova, E. K., Bobrovskikh, M. A., Deryuzhenko, M. A., Shishkina, O. D., & Gruntenko, N. E. (2023). *Wolbachia* effect on *Drosophila melanogaster* lipid and carbohydrate metabolism. *Insects*, *14*(4), 357. https://doi.org/10.3390/insects14040357

Khachatourians, G. G. (2019). Insecticides, microbial. In *Reference module in life sciences*. Elsevier. https://doi.org/10.1016/b978-0-12-809633-8.13066-3

Kil, Y. J., Seo, M. J., Kang, D. K., Oh, S. N., Cho, H. S., Youn, Y. N., Aoki, C., & Yu, Y. M. (2014). Effects of enterobacteria (*Burkholderia* sp.) on development of *Riptortus pedestris*. *Journal of the Faculty of Agriculture, Kyushu University*, *59*(1), 77–84. https://doi.org/10.5109/1434382

Knight, J. (2001). Meet the Herod bug. *Nature*, *412*(6842), 12–14. https://doi.org/10.1038/35083744

Konecka, E., Mokracka, J., Krzyminska, S., & Kaznowski, A. (2019). Evaluation of the pathogenic potential of insecticidal *Serratia marcescens* strains to humans. *Polish Journal of Microbiology*, *68*(2), 185 191. https://doi.org/10.33073/PJM-2019-018

Kozek, W. J., & Rao, R. U. (2007). The discovery of *Wolbachia* in arthropods and nematodes—a historical perspective. *Issues in Infectious Diseases*, *5*, 1–14. https://doi.org/10.1159/000104228

Landsberg, M. J., Jones, S. A., Rothnagel, R., Busby, J. N., Marshall, S. D. G., Simpson, R. M., Lott, J. S., Hankamer, B., & Hurst, M. R. H. (2011). 3D structure of the *Yersinia entomophaga* toxin complex and implications for insecticidal activity. *Proceedings of the National Academy of Sciences*, *108*(51), 20544–20549. https://doi.org/10.1073/PNAS.1111155108.

Le Guern, A. S., Savin, C., Brémont, S., Payro, G., Bon, D., Carniel, E., & Pizarro-Cerdá, J. (2018). First isolation of *Yersinia entomophaga* in human urinary tract. *New Microbes and New Infections*, *26*, 3–7. https://doi.org/10.1016/J.NMNI.2018.08.002

Liu, F. H., Lin, X. L., Kang, Z. W., Tian, H. G., & Liu, T. X. (2019). Isolation and characterization of *Pseudomonas cedrina* infecting *Plutella xylostella* (Lepidoptera: Plutellidae). *Archives of Insect Biochemistry and Physiology*, *102*(3). https://doi.org/10.1002/ARCH.21593

Liu, G., Cao, L., & Han, R. (2022). Plant quercetin degradation by gut bacterium *Raoultella terrigena* of ghost moth *Thitarodes xiaojinensis*. *Frontiers in Microbiology*, *13*, 1079550. https://doi.org/10.3389/fmicb.2022.1079550

Łukasik, P., van Asch, M., Guo, H., Ferrari, J., & Charles, H. (2013). Unrelated facultative endosymbionts protect aphids against a fungal pathogen. *Ecology Letters*, *16*(2), 214–218. https://doi.org/10.1111/ELE.12031

Mahar, A. N., Darban, D. A., Gowen, S. R., Hague, N. G. M., Jan, N. D., Munir, M., & Mahar, A. Q. (2004). Use of entomopathogenic bacterium *Pseudomonas putida* (Enterobacteriaceae) and its secretion against greater wax moth, *Galleria mellonella* pupae. *Journal of Entomology*, *2*(1), 77–85. https://doi.org/10.3923/JE.2005.77.85

Manigandan, V., Karthik, R., Ramachandran, S., & Rajagopal, S. (2018). Chitosan applications in food industry. In A. M. Grumezescu, & A. M. Holban (Eds.), *Handbook of food bioengineering, biopolymers for food design* (pp. 469–491). Elsevier. https://doi.org/10.1016/B978-0-12-811449-0.00015-3

Marshall, S. D. G., Hares, M. C., Jones, S. A., Harper, L. A., Vernon, J. R., Harland, D. P., Jackson, T. A., & Hurst, M. R. H. (2012). Histopathological effects of the Yen-Tc toxin complex from *Yersinia entomophaga* MH96 (Enterobacteriaceae) on the *Costelytra zealandica* (Coleoptera: Scarabaeidae) larval midgut. *Applied and Environmental Microbiology*, *78*(14), 4835–4847. https://doi.org/10.1128/AEM.00431-12.

Martin, P. A. W., Gundersen-Rindal, D., Blackburn, M., & Buyer, J. (2007). *Chromobacterium subtsugae* sp. *nov*., a Betaproteobacterium toxic to colorado potato beetle and other insect pests. *International Journal of Systematic and Evolutionary Microbiology*, *57*, 993–999. https://doi.org/10.1099/ijs.0.64611-0.

Matteoli, F. P., Passarelli-Araujo, H., Reis, R. J. A., Da Rocha, L. O., De Souza, E. M., Aravind, L., Olivares, F. L., & Venancio, T. M. (2018). Genome sequencing and assessment of plant growth-promoting properties of a *Serratia marcescens* strain isolated from vermicompost. *BMC Genomics*, *19*, 750. https://doi.org/10.1186/S12864-018-5130-Y

McNeill, M. R., & Hurst, M. A. (2018, July 1). Yersinia entomophaga, a potential new biopesticide for locusts [Poster presentation]. *New Zealand Plant Protection Conference*.

Mello, I. S., Targanski, S., Pietro-Souza, W., Frutuoso Stachack, F. F., Terezo, A. J., & Soares, M. A. (2020). Endophytic bacteria stimulate mercury phytoremediation by modulating its bioaccumulation and volatilization. *Ecotoxicology and Environmental Safety*, *202*, 110818. https://doi.org/10.1016/J.ECOENV.2020.110818

Mendoza, G., Sánchez-Tafolla, L., & Trigos, Á. (2020). Oxidative foliar photo-necrosis produced by the bacteria *Pseudomonas cedrina*. *Electronic Journal of Biotechnology*, *44*, 14–18. https://doi.org/10.1016/j.ejbt.2020.01.007

Moazami, N. (2011). Biological control. In M. M. Young (Ed.), *Comprehensive biotechnology: Industrial biotechnology and commodity products* (Vol. 3, 2nd ed., pp. 731–739). Academic Press. https://doi.org/10.1016/B978-0-08-088504-9.00235-X

Molina, L., Segura, A., Duque, E., & Ramos, J. L. (2020). The versatility of *Pseudomonas putida* in the rhizosphere environment. *Advances in Applied Microbiology*, *110*, 149–180. https://doi.org/10.1016/BS.AAMBS.2019.12.002

Moore, S. G., Kalmakoff1, J., & Miles1, J. A. R. (2010). An iridescent virus and a rickettsia from the grass grub *Costelytra zealandica* (Coleoptera: Scarabaeidae). *New Zealand Journal of Zoology*, *1*(2), 205–210. https://doi.org/10.1080/03014223.1974.9517828

Morales-Jiménez, J., Vera-Ponce de León, A., García-Domínguez, A., Martínez-Romero, E., Zúñiga, G., & Hernández-Rodríguez, C. (2013). Nitrogen-fixing and uricolytic bacteria associated with the gut of *Dendroctonus rhizophagus* and *Dendroctonus valens* (Curculionidae: Scolytinae). *Microbial Ecology*, *66*(1), 200–210. https://doi.org/10.1007/S00248-013-0206-3

Muratoğlu, H., Demirbağ, Z., & Sezen, K. (2011). *Leptinotarsa decemlineata*'dan izole edilen bir entomopatojenik bakteri, *Pseudomonas putida*. *Turkish Journal of Biology*, *35*(3), 275–282. https://doi.org/10.3906/biy-0902-19

Nguyen, C. T., Nguyen, T. H. H., Tra, V. T., Tungtakanpoung, D., Tran, C. S., Vo, T. K. Q., & Kaewlom, P. (2023). Paraquat removal by free and immobilized cells of *Pseudomonas putida* on corn cob biochar. *Case Studies in Chemical and Environmental Engineering*, *8*, 100376. https://doi.org/10.1016/J.CSCEE.2023.100376

Niu, H., Liu, B., Li, Y., & Guo, H. (2015). Identification of a bacterium isolated from the diseased brown planthopper and determination of its insecticidal activity. *Biocontrol Science and Technology*, *26*(2), 217–226. https://doi.org/10.1080/09583157.2015.1089215

Niu, H., Sun, Y., Zhang, Z., Zhao, D., Wang, N., Wang, L., & Guo, H. (2022). The endophytic bacterial entomopathogen *Serratia marcescens* promotes plant growth and improves resistance against *Nilaparvata lugens* in rice. *Microbiological Research*, *256*, 126956. https://doi.org/10.1016/J.MICRES.2021.126956

Nonaka, S., Salim, E., Kamiya, K., Hori, A., Nainu, F., Asri, R. M., Masyita, A., Nishiuchi, T., Takeuchi, S., Kodera, N., & Kuraishi, T. (2020). Molecular and functional analysis of pore-forming toxin Monalysin from entomopathogenic bacterium *Pseudomonas entomophila*. *Frontiers in Immunology*, *11*, 520. https://doi.org/10.3389/FIMMU.2020.00520

Otsu, Y., Matsuda, Y., Mori, H., Ueki, H., Nakajima, T., Fujiwara, K., Matsumoto, M., Azuma, N., Kakutani, K., Nonomura, T., Sakuratani, Y., Shinogi, T., Tosa, Y., Mayama, S., & Toyoda, H. (2010). Stable phylloplane colonization by entomopathogenic bacterium *Pseudomonas fluorescens* KPM-018P and biological control of phytophagous ladybird beetles *Epilachna vigintioctopunctata* (Coleoptera: Coccinellidae). *Biocontrol Science and Technology*, *14*(5), 427–439. https://doi.org/10.1080/09583150410001683538

Ourry, M., Crosland, A., Lopez, V., Derocles, S. A. P., Mougel, C., Cortesero, A.-M., & Poinsot, D. (2021). Influential insider: *Wolbachia*, an intracellular symbiont, manipulates bacterial diversity in its insect host. *Microorganisms*, *9*(6), 1313. https://doi.org/10.3390/microorganisms9061313

Ozsahin, E., Sezen, K., Demir, I., & Demirbag, Z. (2014). Bacterial isolates from *Palomena prasina* (Hemiptera: Pentatomidae) include potential microbial control agents. *Biocontrol Science and Technology*, *24*(9), 1039–1051. https://doi.org/10.1080/09583157.2014.918584

Paine, S. G., & Standsfield, H. (1919). Studies in bacteriosis. III: A bacterial leaf-spot disease of *Protea cynaroides*, exhibiting a host reaction of possibly bacteriolytic nature. *Annals of Applied Biology*, *6*(1), 27–39. https://doi.org/10.1111/J.1744-7348.1919.TB05299.X

Pan, J., Williams, E., Sung, W., Lynch, M., & Long, H. (2021). The insect-killing bacterium *Photorhabdus luminescens* has the lowest mutation rate among bacteria. *Marine Life Science & Technology*, *3*, 20–27. https://doi.org/10.1007/s42995-020-00060-0

Parajuli, N. P., Bhetwal, A., Ghimire, S., Maharjan, A., Shakya, S., Satyal, D., Pandit, R., & Khanal, P. R. (2016). Bacteremia caused by a rare pathogen—*Chromobacterium violaceum*: A case report from Nepal. *International Journal of General Medicine*, *9*, 441–446. https://doi.org/10.2147/IJGM.S125183

Patil, C. D., Patil, S. V., Salunke, B. K., & Salunkhe, R. B. (2011). Prodigiosin produced by *Serratia marcescens* NMCC46 as a mosquito larvicidal agent against *Aedes aegypti* and *Anopheles stephensi*. *Parasitology Research*, *109*(4), 1179–1187. https://doi.org/10.1007/S00436-011-2365-9

Piel, J. (2010). The chemistry of symbiotic interactions. In H. W. Liu & L. Mander (Eds.), *Comprehensive Natural Products II: Chemistry and Biology: Natural Products Structural Diversity-II Secondary Metabolites: Sources, Structures and Chemical Biology* (Vol. 2, pp. 475–510). Elsevier. https://doi.org/10.1016/B978-008045382-8.00049-6

Pineda-Castellanos, M. L., Rodríguez-Segura, Z., Villalobos, F. J., Hernández, L., Lina, L., & Eugenia Nuñez-Valdez, M. (2015). Pathogenicity of isolates of *Serratia marcescens* towards larvae of the scarab *Phyllophaga blanchardi* (Coleoptera). *Pathogens*, *4*(2), 210–228. https://doi.org/10.3390/pathogens4020210

Planet, P. J. (2023). Pseudomonas aeruginosa. In S. S. Long (Ed.), *Principles and practice of pediatric infectious diseases* (6th ed., pp. 884–889.e2). Elsevier. https://doi.org/10.1016/B978-0-323-75608-2.00155-5

Price, D. R. G., Bartley, K., Blake, D. P., Karp-Tatham, E., Nunn, F., Burgess, S. T. G., & Nisbet, A. J. (2021). A *Rickettsiella* endosymbiont is a potential source of essential B-vitamins for the poultry red mite, *Dermanyssus gallinae*. *Frontiers in Microbiology*, *12*, 2383. https://doi.org/10.3389/FMICB.2021.695346

Ramirez, J. L., Short, S. M., Bahia, A. C., Saraiva, R. G., Dong, Y., Kang, S., Tripathi, A., Mlambo, G., & Dimopoulos, G. (2014). *Chromobacterium* Csp_P reduces malaria and dengue infection in vector mosquitoes and has entomopathogenic and in vitro anti-pathogen activities. *PLoS Pathogens*, *10*(10), e1004398. https://doi.org/10.1371/journal.ppat.1004398

Reimer, L. C., Vetcininova, A., Carbasse, J. S., Söhngen, C., Gleim, D., Ebeling, C., & Overmann, J. (2019). BacDive in 2019: Bacterial phenotypic data for high-throughput biodiversity analysis. *Nucleic Acids Research*, *47*(D1), D631–D636. https://doi.org/10.1093/NAR/GKY879.

Rohrscheib, C. E., Bondy, E., Josh, P., Riegler, M., Eyles, D., van Swinderen, B., Weible, M. W., & Brownlie, J. C. (2015). *Wolbachia* influences the production of octopamine and affects *Drosophila* male aggression. *Applied and Environmental Microbiology*, *81*(14), 4573. https://doi.org/10.1128/AEM.00573-15

Roobakkumar, A., Babu, A., & Kumar, D. V. (2011). *Pseudomonas fluorescens* as an efficient entomopathogen against *Oligo nychus coffeae* Nietner (Acari: Tetranychidae) infesting tea. *Journal of Entomology*, *3*(July), 73–77.

Ruffner, B., Péchy-Tarr, M., Höfte, M., Bloemberg, G., Grunder, J., Keel, C., & Maurhofer, M. (2015). Evolutionary patchwork of an insecticidal toxin shared between plant-associated pseudomonads and the insect pathogens *Photorhabdus* and *Xenorhabdus*. *BMC Genomics*, *16*, 609. https://doi.org/10.1186/s12864-015-1763-2

Ruiu, L. (2015). Insect pathogenic bacteria in integrated pest management. *Insects*, *6*(2), 352–367. https://doi.org/10.3390/insects6020352

Ruiu, L. (2018). Microbial biopesticides in agroecosystems. *Agronomy*, *8*(11), 235. https://doi.org/10.3390/agronomy8110235

Ruiu, L., Satta, A., & Floris, I. (2013). Emerging entomopathogenic bacteria for insect pest management. *Bulletin of Insectology*, *66*(2), 181–186.

Sabbahi, R., Hock, V., Azzaoui, K., Saoiabi, S., & Hammouti, B. (2022). A global perspective of entomopathogens as microbial biocontrol agents of insect pests. *Journal of Agriculture and Food Research*, *10*, 100376. https://doi.org/10.1016/J.JAFR.2022.100376

Salgado-Morales, R., Rivera-Gómez, N., Lozano-Aguirre Beltrán, L. F., Hernández-Mendoza, A., & Dantán-González, E. (2017). Draft genome sequence of a *Pseudomonas aeruginosa* NA04 bacterium isolated from an entomopathogenic nematode. *Genome Announcements*, *5*(36), e00746–17. https://doi.org/10.1128/genomea.00746-17

Sánchez-Tafolla, L., Padrón, J. M., Mendoza, G., Luna-Rodríguez, M., Fernández, J. J., Norte, M., & Trigos, Á. (2019). Antiproliferative activity of biomass extract from *Pseudomonas cedrina*. *Electronic Journal of Biotechnology*, *40*, 40–44. https://doi.org/10.1016/j.ejbt.2019.03.010

Santos, A. V., Dillon, R. J., Dillon, V. M., Reynolds, S. E., & Samuels, R. I. (2004). Ocurrence of the antibiotic producing bacterium *Burkholderia sp.* in colonies of the leaf-cutting ant *Atta sexdens rubropilosa*. *FEMS Microbiology Letters*, *239*(2), 319–323. https://doi.org/10.1016/J.FEMSLE.2004.09.005

Scales, B. S., Dickson, R. P., Lipuma, J. J., & Huffnagle, G. B. (2014). Microbiology, genomics, and clinical significance of the *Pseudomonas fluorescens* species complex, an unappreciated colonizer of humans. *Clinical Microbiology Reviews*, *27*(4), 927–948. https://doi.org/10.1128/CMR.00044-14

Seebold, K. W., & Bessin, R. T. (2002). Yellow vine decline of cucurbits. In *Plant pathology fact sheet*. College of Agriculture, University of Kentucky.

Selvakumar, G., Mohan, M., Kundu, S., Gupta, A. D., Joshi, P., Nazim, S., & Gupta, H. S. (2008). Cold tolerance and plant growth promotion potential of *Serratia marcescens* strain SRM (MTCC 8708) isolated from flowers of summer squash (*Cucurbita pepo*). *Letters in Applied Microbiology*, *46*(2), 171–175. https://doi.org/10.1111/J.1472-765X.2007.02282.X

Shi, Z., Wang, L., & Zhang, H. (2012). Low diversity bacterial community and the trapping activity of metabolites from cultivable bacteria species in the female reproductive system of the oriental fruit fly, *Bactrocera dorsalis* Hendel (Diptera: Tephritidae). *International Journal of Molecular Sciences*, *13*(5), 6266–6278. https://doi.org/10.3390/IJMS13056266

Silva, C. P., Waterfield, N. R., Daborn, P. J., Dean, P., Chilver, T., Au, C. P. Y., Sharma, S., Potter, U., Reynolds, S. E., & Ffrench-Constant, R. H. (2002). Bacterial infection of a model insect: *Photorhabdus luminescens* and *Manduca sexta*. *Cellular Microbiology*, *4*(6), 329–339. https://doi.org/10.1046/J.1462-5822.2002.00194.X

Singh, P., Montano, A., & Bostick, A. (2021). Rapid severe sepsis from *Pseudomonas fluorescens/putida* bacteremia due to skin and soft tissue infection—a case report. *Annals of Medicine and Surgery*, *70*, 102845. https://doi.org/10.1016/J.AMSU.2021.102845

Suryawanshi, R. K., Patil, C. D., Borase, H. P., Narkhede, C. P., Salunke, B. K., & Patil, S. V. (2015). Mosquito larvicidal and pupaecidal potential of prodigiosin from *Serratia marcescens* and understanding its mechanism of action. *Pesticide Biochemistry and Physiology*, *123*, 49–55. https://doi.org/10.1016/J.PESTBP.2015.01.018

Thongprem, P., Evison, S. E. F., Hurst, G. D. D., & Otti, O. (2020). Transmission, tropism, and biological impacts of toxin *Rickettsia* in the common bed bug *Cimex lectularius* (Hemiptera: Cimicidae). *Frontiers in Microbiology*, *11*, 3292. https://doi.org/10.3389/FMICB.2020.608763

Tsuchida, T., Koga, R., Horikawa, M., Tsunoda, T., Maoka, T., Matsumoto, S., Simon, J. C., & Fukatsu, T. (2010). Symbiotic bacterium modifies aphid body color. *Science*, *330*(6007), 1102–1104. https://doi.org/10.1126/SCIENCE.1195463

Turley, A. P., Moreira, L. A., O'Neill, S. L., & McGraw, E. A. (2009). *Wolbachia* infection reduces blood-feeding success in the dengue fever mosquito, *Aedes aegypti*. *PLoS Neglected Tropical Diseases*, *3*(9), e516. https://doi.org/10.1371/journal.pntd.0000516

Vallet-Gely, I., Novikov, A., Augusto, L., Liehl, P., Bolbach, G., Péchy-Tarr, M., Cosson, P., Keel, C., Caroff, M., & Lemaitre, B. (2010). Association of hemolytic activity of *Pseudomonas entomophila*, a versatile soil bacterium, with cyclic lipopeptide production. *Applied and Environmental Microbiology*, *76*(3), 910. https://doi.org/10.1128/AEM.02112-09

Venekei, I., & Marokházi, J. (2013). PrtA peptidases. In N. D. Rawlings, & G. Salvesen (Eds.), *Handbook of proteolytic enzymes* (Vol. 1, pp, 877–882). Academic Press. https://doi.org/10.1016/B978-0-12-382219-2.00184-8

Villamizar, S., Ferro, J. A., Caicedo, J. C., & Alves, L. M. C. (2020). Bactericidal effect of entomopathogenic bacterium *Pseudomonas entomophila* against *Xanthomonas citri* reduces citrus canker disease severity. *Frontiers in Microbiology*, *11*, 1431. https://doi.org/10.3389/fmicb.2020.01431

Visschedyk, D. D., Perieteanu, A. A., Turgeon, Z. J., Fieldhouse, R. J., Dawson, J. F., & Merrill, A. R. (2010). Photox, a novel actin-targeting mono-ADP-ribosyltransferase from *Photorhabdus luminescens*. *Journal of Biological Chemistry*, *285*(18), 13525–13534. https://doi.org/10.1074/jbc.M109.077339

Vodovar, N., Vallenet, D., Cruveiller, S., Rouy, Z., Barbe, V., Acosta, C., Cattolico, L., Jubin, C., Lajus, A., Segurens, B., Vacherie, B., Wincker, P., Weissenbach, J., Lemaitre, B., Médigue, C., & Boccard, F. (2006). Complete genome sequence of the entomopathogenic and metabolically versatile soil bacterium *Pseudomonas entomophila*. *Nature Biotechnology*, *24*(6), 673–679. https://doi.org/10.1038/nbt1212

Vodovar, N., Vinals, M., Liehl, P., Basset, A., Degrouard, J., Spellman, P., Boccard, F., & Lemaitre, B. (2005). *Drosophila* host defense after oral infection by an entomopathogenic *Pseudomonas* species. *Proceedings of the National Academy of Sciences*, *102*(32), 11414–11419. https://doi.org/10.1073/PNAS.0502240102.

Voegele, J. (1996). Review of biological control of sunn pest. In J. G. Morse, & R. H. Miller (Eds.), *Sunn pests and their control in the Near East*. Food and Agriculture Organization of the United Nations.

Walker, T. S., Bais, H. P., Déziel, E., Schweizer, H. P., Rahme, L. G., Fall, R., & Vivanco, J. M. (2004). *Pseudomonas aeruginosa*-plant root interactions. Pathogenicity, biofilm formation, and root exudation. *Plant Physiology*, *134*(1), 320. https://doi.org/10.1104/PP.103.027888

Wang, L.-T., Tai, C.-J., Wu, Y.-C., Chen, Y.-B., Lee, F.-L., & Wang, S.-L. (2010). *Pseudomonas taiwanensis* sp. *nov*., isolated from soil. *International Journal of Systematic and Evolutionary Microbiology*, *60*, 2094–2098. https://doi.org/10.1099/ijs.0.014779-0

Wang, M., Xing, Y., Wang, J., Xu, Y., & Wang, G. (2014). The role of the chi1 gene from the endophytic bacteria *Serratia proteamaculans* 336x in the biological control of wheat take-all. *Canadian Journal of Microbiology*, *60*(8), 533–540. https://doi.org/10.1139/CJM-2014-0212

Wang, S., Han, H., Guo, Y., Carballar-Lejarazú, R., Sheng, L., Liang, G., Hu, X., Wang, R., Zhang, F., & Wu, S. (2022). Isolation and identification of *Pseudomonas entomophila* from *Pinus massoniana* and its insecticidal activity against *Monochamus alternatus*. *Biocontrol Science and Technology*, *32*(2), 163–177. https://doi.org/10.1080/09583157.2021.1977241

Watson, J. R., Wallihan, R., & Antonara, S. (2018). Less commonly encountered enterobacteriaceae. In S. S. Long, C. G. Prober, & M. Fischer (Eds.), *Principles and practice of pediatric infectious diseases* (5th ed., pp. 829–831.e2). Elsevier. https://doi.org/10.1016/B978-0-323-40181-4.00142-0

Werren, J. H., Windsor, D., & Guo, L. (1995). Distribution of *Wolbachia* among neotropical arthropods. *Proceedings of the Royal Society of London. Series B: Biological Sciences*, *262*(1364), 197–204. https://doi.org/10.1098/RSPB.1995.0196

Williamson, N. R., Fineran, P. C., Leeper, F. J., & Salmond, G. P. C. (2006). The biosynthesis and regulation of bacterial prodiginines. *Nature Reviews Microbiology*, *4*(12), 887–899. https://doi.org/10.1038/nrmicro1531

Wouts, W. M. (1990). The primary form of *Xenorhabdus* Species (Enterobacteriaceae: Eubacteriales) may consist of more than one bacterial species. *Nematologica*, *36*, 313–318.

Yang, J., Zeng, H. M., Lin, H. F., Yang, X. F., Liu, Z., Guo, L. H., Yuan, J. J., & Qiu, D. W. (2012). An insecticidal protein from *Xenorhabdus budapestensis* that results in prophenoloxidase activation in the wax moth, *Galleria mellonella*. *Journal of Invertebrate Pathology*, *110*(1), 60–67. https://doi.org/10.1016/J.JIP.2012.02.006

Zaheer, A., Mirza, B. S., Mclean, J. E., Yasmin, S., Shah, T. M., Malik, K. A., & Mirza, M. S. (2016). Association of plant growth-promoting *Serratia spp.* with the root nodules of chickpea. *Research in Microbiology*, *167*(6), 510–520. https://doi.org/10.1016/J.RESMIC.2016.04.001

Zélé, F., Altıntaş, M., Santos, I., Cakmak, I., & Magalhães, S. (2020). Population-specific effect of *Wolbachia* on the cost of fungal infection in spider mites. *Ecology and Evolution*, *10*(9), 3868. https://doi.org/10.1002/ECE3.6015

Zeng, T., Bai, X., Liu, Y. L., Li, J. F., Lu, Y. Y., & Qi, Y. X. (2020). Intestinal responses of the oriental fruit fly *Bactrocera dorsalis* (Hendel) after ingestion of an entomopathogenic bacterium strain. *Pest Management Science*, *76*(2), 653–664. https://doi.org/10.1002/PS.5563

Zhang, F., Yang, C., Zhang, X., Zhu, H., Zhao, D., & Huang, Y. (2020). Isolation of an anti-entomopathogenic fungal protein secreted from *Pseudomonas aeruginosa* BGf-2: An intestinal bacteriam of *Blattella germanica* (L.). *Journal of Invertebrate Pathology*, *173*, 107371. https://doi.org/10.1016/J.JIP.2020.107371

9 Phylum Actinobacteria

ABSTRACT

There are several insecticides of microbial origin. The bacteria from phylum Actinobacteria are famous for this contribution, as they produce diverse types of secondary metabolites which might prove entomopathogenic. From *Streptomyces* sp., we got Avermectin and Emamectin benzoate group of insecticides, which targets GABA receptors in the insect's nervous system. *Saccharopolyspora spinosa* provided us with Spinosad, working in the same mode of action. Both these bacteria also act as EPB by themselves. *Arthrobacter* sp. is one more bacterium with an entomopathogenic nature to several insects. We can look for newer metabolites that can be utilized for preparing new pesticides, along with exploiting these bacteria as EPB.

9.1 INTRODUCTION

With the evolution of insecticides in search of more efficient and targeted effects, newer molecules come up in the scenario—similar way the newer bacteria are also explored in search of specificity and efficiency. Bacteria belonging to the phylum Actinobacteria are known to synthesize diverse kinds of secondary metabolites, which are responsible for conferring entomopathogenicity to some bacteria, whereas some become mutualistic to insects. We have several insecticides, like avermectin, emamectin benzoate, milbamectin, spinosad, etc., which originated from these bacteria. Also in nature, these bacteria are found to play competitive or symbiotic relationships with different insects. Let us discuss the major entomopathogenic bacteria from this phylum, along with the development of several bacterial-origin insecticides.

9.2 *STREPTOMYCES* SPP.

Streptomyces comprise a group of Gram-positive, filamentous, non-motile, aerobic, non-acid-fast, and catalase-positive bacteria, and it looks like a fungus (Sharma et al., 2020). Some species and strains of this genus produce a variety of metabolites that act as potent toxins against insect pests. Among the insecticidal substances produced by *Streptomyces* sp., the first discoveries are flavensomycin (Craveri & Giolitti, 1957), antimycin A (Neft & Farley, 1972), piericidins (Takahashi et al., 1965), macrotetralides (Jizba et al., 1991), and prasinons (Box et al., 1973). Later, it was discovered that avermectins produced by *Streptomyces avermitilis* (Burg et al.) have insecticidal and anthelmintic activities. These macrocyclic lactone derivatives target the gamma-aminobutyric acid (GABA) receptor in the insect's peripheral nervous system. The enhancement of GABA binding creates a cascade of events resulting in neurotransmission inhibition and neuromuscular systems paralysis. Commercial formulations based on avermectins containing 80% avermectin B1a and 20% avermectin B1b, known as abamectin, act as a contact and oral poison and have limited plant translaminar activity. Analogous substances include emamectin benzoate, especially toxic to Lepidoptera, and milbemectin, isolated from *S. hygroscopicus* (Jensen). In addition, diverse *Streptomyces* species produce a variety of other secondary metabolites that have been isolated and characterized so far. Commercialized products have medical and veterinary importance and are very successful against ectoparasites and endoparasites (Ruiu, 2015).

S. griseoplanus Backus et al., *S. bacillaris* (Krassilnikov), and *S. albolongus* Tsukiura et al. were proven to be potent entomopathogens against *Helicoverpa armigera* (Hübner), *Spodoptera litura* (Fabricius), and *Chilo partellus* (Fabricius) (Vijayabharathi et al., 2014). One more interesting side has been recently explored for this bacterium. It shows mutualism with the vespid wasp, *Polybia plebeja* Saussure, a predator of spiders, crickets, cicadas, and other insects. Ionophore antibiotics isolated from wasp-associated

DOI: 10.1201/9781003426882-9

strain M54 were effective against entomopathogenic fungi, *Hirsutella citriformis* Speare, attacking wasps. Peptides like meliponamycins, glycosylated antibiotics like lobophorins, and polycyclic antibiotics have bee-isolated from bee-associated actinobacteria. All these compounds exhibit high activity against *Paenibacillus larvae* (White), the causal agent of American foulbrood (Baranova et al., 2022; Matarrita-Carranza et al., 2017). Similar to human beings using natural products for treatment, honeybees are also found to employ *Streptomyces* sp. from pollen stores to treat American foulbrood (Grubbs et al., 2021). Also, *Streptomyces* sp. protects passalid beetle, *Odontotaenius disjunctus* (Illiger); dung beetle, *Copris tripartitus* Waterhouse; and termites, *Macrotermes barneyi* Light by preventing the entomopathogenic fungi, *Metarhizium anisopliae* (Metchn.), with metabolites (Benndorf et al., 2018; Kim et al., 2021; Li et al., 2021; Pessotti et al., 2021).

S. griseus (Krainsky) also protects forest pests like bark beetle, *Xyleborus saxesenii* (Ratzeburg). Isolated strain XylebKG-1 produces cycloheximide. This well-known antifungal compound strategically inhibits the growth of the parasitic fungus *Nectria* sp., but not of mutualistic *Raffaelea sulphurea* (Batra) (Grubbs et al., 2020). Vicious red imported ants, *Solenopsis invicta* Buren, also prefer to lay eggs in *Streptomyces*-rich soil, to protect the next generation from pathogen attack (Huang et al., 2020). However, avermectins are also toxic against fish, aquatic invertebrates, bees, earthworms (Williamson, 2005), and mammals (Stevens & Breckenridge, 2001), and that is why these insecticides are now considered under the yellow label and have increased considerable demands for it now.

9.3 *SACCHAROPOLYSPORA SPINOSA*

During a screening programme, *S. spinosa* Mertz and Yao was discovered, where the insecticidal activity of isolate A83543 emerged (Mertz & Yao, 1990). Subsequent assays highlighted that the specific compounds isolated from the fermentation broth had broad toxicity, and spinosyns was the generic name given to these isolates. Isolation of a variety of spinosyn analogues has been done, and to investigate the pathway of spinosyn biosynthesis, many studies have been conducted, which led to the characterization of specific gene clusters (Waldron et al., 2000). Since the first experimentations, spinosyns have shown broad-spectrum insecticidal activity against different insect orders, especially Lepidoptera and Diptera (Mertz & Yao, 1990). Mixtures derived from the fermentation of the natural *S. spinosa* were named 'spinosad' and contain two spinosoids, spinosyn A and spinosyn D, as major and minor components, at a ratio of 17:3, respectively. A tetracyclic polyketide aglycone is present in the structure of spinosyn A, and with it, a neutral saccharide substituent is attached (Kirst, 2010). Till now, no one completely understands the mode of action of spinosyns, but it is considered that spinosyns have some unique characteristics compared to other insecticides. It has been demonstrated that spinosyn interacts with gamma-aminobutyric acid (GABA) receptors and nicotinic acetylcholine receptors, leading to neuronal activity disruption. This disruption results in consequent insect paralysis and the death of the insect. Though they have broad-spectrum activity against insects, unlike other insecticides (especially chemical ones), spinosyns have a low risk toward non-target species, including mammals and various aquatic organisms. The continuous research and interest of different industries in this field have led to the discovery of various compounds related to spinosyns produced by another species of *Saccharopolyspora*, that is, *S. pogona* (Ruiu, 2015).

9.4 *ARTHROBACTER* SP.

Arthrobacter, a genus of Gram-positive coryneform bacteria, is frequently found in soil, air, food (such as dairy products), water, plants, and extreme conditions, such as oil, brine, mural paints, clinical specimens, sewage, activated sludge, etc. The majority of *Arthrobacter* bacteria generate pigments in a wide variety of colours, including blue and green (indigoidine, indochrome, and derived salts), orange and yellow (riboflavin, carotenoids), and red (porphyrins, carotenoids) pigments (Sutthiwong et al., 2014). *Arthrobacter gandensis* and *A. gandavensis* were isolated from the wireworm *Agriotes lineatus* (Linnaeus) (Coleoptera: Elateridae), showing 100% mortality 10 days after treatment. This is a serious pest of various vegetables and fruits throughout the world (Danismazoglu et al., 2012).

Several other species under this genus are found to be present as a part of microbial flora in different insect pests. *Arthrobacter pityocampae* sp. *nov.* İnce et al. was isolated from the microbiome of the pine processionary moth, *Thaumetopoea pityocampa* (Denis and Schiffermüller) (Lepidoptera: Thaumetopoidae), one of the most harmful pests of pine species in Mediterranean countries (İnce et al., 2014). *Arthrobacter* spp. are also prevalent in the gut microbiota of the cave beetles *Neobathyscia pasai* Ruffo and *N. mancinii* Jeannel (Coleoptera: Leiodidae) (Latella et al., 2017). Wood-digesting *Arthrobacter* sp. had been isolated from the hind gut of the termite *Reticulitermes hesperus* Banks (Thayer, 1976). They were also found in the microbiome of *Capnodis tenebrionis* (Linnaeus), a serious pest of cultivated stone fruit trees (Barak et al., 2019). So this genus of bacteria acts as a friend of both pests and us simultaneously.

9.5 CONCLUSION

There are several insecticides available in the market, but they are losing their efficiency day by day due to the development of insecticide resistance by several pests. From that perspective, the secondary metabolites from several Actinobacteria that show entomopathogenic properties can act as an illuminator in the search for newer insecticides. All the bacteria under this phylum are rich in secondary metabolites, and a deep dive into them can lead us to discover many such secondary metabolites which can become potent insecticides in the near future. The scope for whether these bacteria can be directly employed as entomopathogens on several pests should also be examined.

REFERENCES

Barak, H., Kumar, P., Zaritsky, A., Mendel, Z., Ment, D., Kushmaro, A., & Ben-Dov, E. (2019). Diversity of bacterial biota in *Capnodis tenebrionis* (Coleoptera: Buprestidae) larvae. *Pathogens*, *8*(1), 4. https://doi.org/10.3390/pathogens8010004

Baranova, A. A., Zakalyukina, Y. V., Ovcharenko, A. A., Korshun, V. A., & Tyurin, A. P. (2022). Antibiotics from insect-associated actinobacteria. *Biology*, *11*(11), 1676. https://doi.org/10.3390/biology11111676

Benndorf, R., Guo, H., Sommerwerk, E., Weigel, C., Garcia-Altares, M., Martin, K., Hu, H., Küfner, M., de Beer, Z. W., Poulsen, M., & Beemelmanns, C. (2018). Natural products from actinobacteria associated with fungus-growing termites. *Antibiotics*, *7*(3), 83. https://doi.org/10.3390/antibiotics7030083

Box, S. J., Cole, M., & Yeoman, G. H. (1973). Prasinons A and B: Potent insecticides from *Streptomyces prasinus*. *Applied Microbiology*, *26*(5), 699–704. https://doi.org/10.1128/AM.26.5.699-704.1973

Craveri, R., & Giolitti, G. (1957). An antibiotic with fungicidal and insecticidal activity produced by Streptomyces. *Nature*, *179*(4573), 1307. https://doi.org/10.1038/1791307a0

Danismazoglu, M., Demir, İ., Sevim, A., Demirbag, Z., & Nalcacioglu, R. (2012). An investigation on the bacterial flora of *Agriotes lineatus* (Coleoptera: Elateridae) and pathogenicity of the flora members. *Crop Protection*, *40*, 1–7. https://doi.org/10.1016/j.cropro.2012.04.012

Grubbs, K. J., May, D. S., Sardina, J. A., Dermenjian, R. K., Wyche, T. P., Pinto-Tomás, A. A., Clardy, J., & Currie, C. R. (2021). Pollen *Streptomyces* produce antibiotic that inhibits the honey bee pathogen *Paenibacillus larvae*. *Frontiers in Microbiology*, *12*, 632637. https://doi.org/10.3389/FMICB.2021.632637

Grubbs, K. J., Surup, F., Biedermann, P. H. W., McDonald, B. R., Klassen, J. L., Carlson, C. M., Clardy, J., & Currie, C. R. (2020). Cycloheximide-producing *Streptomyces* associated with *Xyleborinus saxesenii* and *Xyleborus affinis* fungus-farming ambrosia beetles. *Frontiers in Microbiology*, *11*, 562140. https://doi.org/10.3389/fmicb.2020.562140

Huang, H., Ren, L., Li, H., Schmidt, A., Gershenzon, J., Lu, Y., & Cheng, D. (2020). The nesting preference of an invasive ant is associated with the cues produced by actinobacteria in soil. *PLoS Pathogens*, *16*(9), e1008800. https://doi.org/10.1371/journal.ppat.1008800

İnce, İ. A., Demirbağ, Z., & Katı, H. (2014). *Arthrobacter pityocampae* sp. *nov*., isolated from *Thaumetopoea pityocampa* (Lep., Thaumetopoeidae). *International Journal of Systematic and Evolutionary Microbiology*, *64*(Pt 10), 3384–3389. https://doi.org/10.1099/IJS.0.060731-0

Jizba, J., Sedmera, P., Zima, J., Beran, M., Blumauerová, M., Kandybin, N. V., & Samoukina, G. V. (1991). Macrotetrolide antibiotics produced by *Streptomyces globisporus*. *Folia Microbiologica*, *36*(5), 437–443. https://doi.org/10.1007/BF02884062

Kim, S. H., Park, G., Park, J. S., & Kwon, H. C. (2021). Antifungal *Streptomyces spp*., plausible partners for brood-caring of the dung beetle *Copris tripartitus*. *Microorganisms*, *9*(9), 1980. https://doi.org/10.3390/microorganisms9091980

Kirst, H. A. (2010). The spinosyn family of insecticides: Realizing the potential of batural products research. *Journal of Antibiotics*, *63*(3), 101–111. https://doi.org/10.1038/ja.2010.5.

Latella, L., Castioni, A., Bignotto, L., Salvetti, E., Torriani, S., & Felis, G. E. (2017). Exploring gut microbiota composition of the cave beetles *Neobathyscia pasai* Ruffo, 1950 and *Neobathyscia mancinii* Jeannel, 1924 (Leiodidae: Cholevinae). *Bollettino Del Museo Civico Di Storia Naturale Di Verona*, *41*, 3–24.

Li, J., Sang, M., Jiang, Y., Wei, J., Shen, Y., Huang, Q., Li, Y., & Ni, J. (2021). Polyene-producing *Streptomyces spp*. from the fungus-growing termite *Macrotermes barneyi* exhibit high inhibitory activity against the antagonistic fungus *Xylaria*. *Frontiers in Microbiology*, *12*, 689. https://doi.org/10.3389/FMICB.2021.649962

Matarrita-Carranza, B., Moreira-Soto, R. D., Murillo-Cruz, C., Mora, M., Currie, C. R., & Pinto-Tomas, A. A. (2017). Evidence for widespread associations between neotropical hymenopteran insects and Actinobacteria. *Frontiers in Microbiology*, *8*, 292779. https://doi.org/10.3389/FMICB.2017.02016

Mertz, F. P., & Yao., R. C. (1990). *Saccharopolyspora spinosa* sp. *nov*. isolated from soil collected in a sugar mill rum still. *International Journal of Systematic and Evolutionary Microbiology*, *40*(1), 34–39. https://doi.org/10.1099/00207713-40-1-34

Neft, N., & Farley, T. M. (1972). Conditions influencing antimycin production by a *Streptomyces* species grown in chemically defined medium. *Antimicrobial Agents and Chemotherapy*, *1*(3), 274–276. https://doi.org/10.1128/AAC.1.3.274

Pessotti, R. de C., Hansen, B. L., Reaso, J. N., Ceja-Navarro, J. A., El-Hifnawi, L., Brodie, E. L., & Traxler, M. F. (2021). Multiple lineages of *Streptomyces* produce antimicrobials within passalid beetle galleries across eastern North America. *ELife*, *10*, e65091. https://doi.org/10.7554/ELIFE.65091

Ruiu, L. (2015). Insect pathogenic bacteria in integrated pest management. *Insects*, *6*(2), 352–367. https://doi.org/10.3390/insects6020352

Sharma, V., Sharma, A., Malannavar, A. B., & Salwan, R. (2020). Molecular aspects of biocontrol species of *Streptomyces* in agricultural crops. In V. Sharma, R. Salwan, & L. K. T. Al-Ani (Eds.), *Molecular aspects of plant beneficial microbes in agriculture* (pp. 89–109). Academic Press. https://doi.org/10.1016/B978-0-12-818469-1.00008-0

Stevens, J., & Breckenridge, C. B. (2001). The avermectins: Insecticidal and antiparasitic agents. In R. I. Krieger, & W. C. Krieger (Eds.), *Handbook of pesticide toxicology* (2nd ed., pp. 1157–1167). Academic Press. https://doi.org/10.1016/B978-012426260-7.50059-8

Sutthiwong, N., Fouillaud, M., Valla, A., Caro, Y., & Dufossé, L. (2014). Bacteria belonging to the extremely versatile genus Arthrobacter as novel source of natural pigments with extended hue range. *Food Research International*, *65*(Pt B), 156–162. https://doi.org/10.1016/j.foodres.2014.06.024

Takahashi, N., Suzuki, A., & Tamura, S. (1965). Structure of piericidin A. *Journal of the American Chemical Society*, *87*(9), 2066–2068. https://doi.org/10.1021/ja01087a050

Thayer, D. W. (1976). Facultative wood digesting bacteria from the hind gut of the termite *Reticulitermes hesperus*. *Journal of General Microbiology*, *95*(2), 287–296. https://doi.org/10.1099/00221287-95-2-287

Vijayabharathi, R., Ratna Kumari, B., Sathya, A., Srinivas, V., Abhishek, R., Sharma, H. C., & Gopalakrishnan, S. (2014). Biological activity of entomopathogenic actinomycetes against lepidopteran insects (Noctuidae: Lepidoptera). *Canadian Journal of Plant Science*, *94*, 759–769. https://doi.org/10.4141/CJPS2013-298

Waldron, C., Madduri, K., Crawford, K., Merlo, D. J., Treadway, P., Broughton, M. C., & Baltz, R. H. (2000). A cluster of genes for the biosynthesis of spinosyns, novel macrolide insect control agents produced by *Saccharopolyspora spinosa*. *Antonie van Leeuwenhoek*, *78*, 385–390.

Williamson, K. K. (2005). Avermectins. In P. Wexler (Ed.), *Encyclopedia of toxicology* (2nd ed., pp. 192–194). Elsevier. https://doi.org/10.1016/B0-12-369400-0/00095-8

10 Phylum Tenericutes

ABSTRACT

Tenericutes are unique bacteria which do not have cell wall. Among different genera belonging to this phylum, only *Acholeplasma* and *Spiroplasma* are entomopathogenic. *Acholeplasma* is associated with several pests as well as some beneficial insects too. Same for *Spiroplasma*, with one interesting feature of some of them, to act as sex ratio organisms (SROs), which restrict the production of the male population in the insect it inhabits. Even though the usage of entomopathogenic Tenericutes as such may not be feasible, a thorough understanding of their interaction with insects may lead to the development of some genetically improved versions of these organisms.

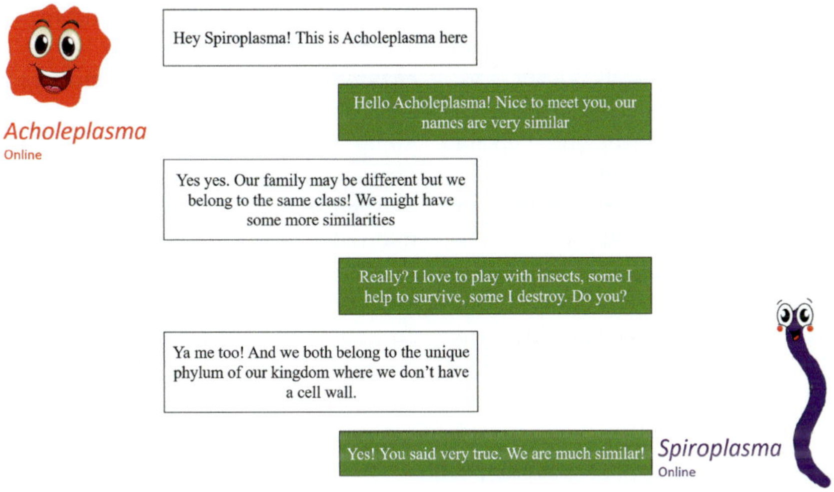

10.1 INTRODUCTION

Compared to the previously mentioned phyla of entomopathogenic bacteria, the phylum Tenericutes is unique among the whole kingdom because this is the only one that does not possess a cell wall. Mollicutes, meaning 'soft wall', is the single class under this phylum. Usually, mollicutes are obligate parasites or commensals of various eukaryotic hosts, like plants, humans, domestic animals, and even insects (Razin et al., 1998). Although members of this class are present in a wide range of habitats, likely due to their small genome size, which demands highly specific requirements that favour their growth, most species show selectivity in host type and tissue (Razin et al., 1998; Waites & Talkington, 2004). This phylum contains members under three groups, viz., the Pneumonia group, which includes genera *Mycoplasma* and *Ureaplasma*; the Spiroplasma group, which includes the genera *Entomoplasma*, *Mesoplasma*, and *Spiroplasma*; and the Anaeroplasma group, which includes the genera *Acholeplasma* and *Phytoplasma* (Davis et al., 2013). Among them, several genera, like *Entomoplasma*, *Mesoplasma*, *Spiroplasma*, *Acholeplasma*, and *Phytoplasma*, are found to be associated with insects. However, only some examples of *Acholeplasma* and *Spiroplasma* are there to be pathogenic to insects (Stülke et al., 2009). In this present chapter, we are going to discuss these two entomopathogenic genera in detail and explore the scope to employ them in pest management.

10.2 *ACHOLEPLASMA*

The *Acholeplasma* are non-motile facultative anaerobes that produce spherical cells and do not require serum or sterol for growth. *Acholeplasma* has been isolated from the gut contents of various tabanid, beetle, honeybee, and lepidopteran hosts. These organisms have been proposed to be classified as strains of *A. entomophilum* Tully et al. (Tully et al., 1988). These organisms, not requiring a sterol source, produce pleomorphic, coccoid cells in vitro, forming colonies that display a typical fried egg morphology.

In addition to being isolated from insects, these strains have also been isolated from flowers. The mycoplasma-like organisms (MLOs) responsible for various diseases of angiosperms (aster yellow MLO, x disease MLO) are phylogenetically related to *Acholeplasma*; these are mostly known as phytoplasma. Insects like leafhoppers and planthoppers disseminate these organisms during feeding excursions (Boucias & Pendland, 1998). From the preceding discussion, we can see that the bacteria are harmful to

DOI: 10.1201/9781003426882-10

crops and beneficial insects like honeybees, so the commercial use of the same is not in reality.

10.3 *SPIROPLASMA*

The *Spiroplasma* is polymorphic, producing both non-motile coccoid and unique helical-shaped cells with linear motors for locomotion. It is the MreB which confers the helical structure of *Spiroplasma*, even in the absence of a cell wall (Madkour & Mayer, 2007). These are fastidious bacteria requiring nutrient-rich medium. The colonies are diffused, showing a fried egg morphology, owing to their motile nature and are 0.1–4 mm in size (Regassa, 2014). *Spiroplasma* is biologically diverse as a group, containing members associated with plants, insects, arachnids, crustaceans, and vertebrate hosts (Boucias & Pendland, 1998, Cisak et al., 2015). Around 5–10% of arthropods possess *Spiroplasma* as their endosymbionts (Ogata et al., 2023). Among insects, they have been isolated from six main insect orders: Hymenoptera, Coleoptera, Diptera, Lepidoptera, Homoptera, and Hemiptera (Regassa, 2014). Certain *Spiroplasma*, which is impossible to culture on cell-free media, has been grown successfully in the presence of insect cells (Boucias & Pendland, 1998). Many *Spiroplasma* species continue to infect their hosts through vertical transmission. For this, *Spiroplasma* employs yolk uptake machinery in *Drosophila* to get into the germline (Herren et al., 2013). But several phylogenetic studies have also supported that some *Spiroplasma* also follow horizontal transmission between unrelated hosts (Ogata et al., 2023). The broad range of different insect hosts/vectors associated with *Spiroplasma* has been graphically represented in Figure 10.1.

Hitherto, a total of 49 serogroups and 15 subgroups of *Spiroplasma* has been reported (Regassa, 2014). Out of them, serogroup I, in addition to containing *Spiroplasma* pathogenic to plants (causal agent of citrus stubborn, corn stunt, etc.), also contains members associated with flower surfaces that replicate in the gut and haemolymph of various insect pollinators (honeybees, bumblebees, butterflies) (Boucias & Pendland, 1998). The type species *S. melliferum* Clark et al. is reported to be pathogenic to honeybees and readily replicates in wax moth larvae. The virulence factors responsible are chitinase utilization enzymes and unique protein clusters with transcriptional regulators and toxins (Regassa, 2014). Members within group II include *Spiroplasma* associated with various *Drosophila* species (Boucias & Pendland, 1998). Certain *Spiroplasma*, like *S. poulsonii* Williamson et al., is referred to as sex ratio organisms, or SROs. When present in the haemolymph of adult females, these bacteria cause the total elimination of the male progeny. The SROs can be transmitted vertically via oogenesis, but the horizontal transfer may also be seen (Regassa, 2014). Though the

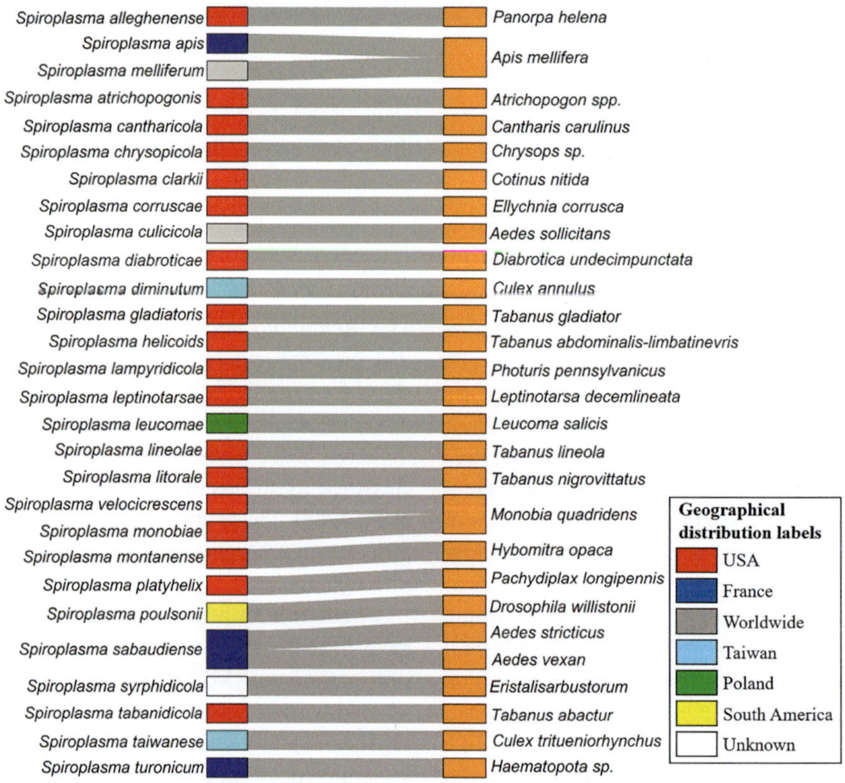

FIGURE 10.1 Representation of the association of different *Spiroplasma* spp. with their respective insect hosts/vectors.

Source: **Cacciola et al. (2017).**

mechanism is not exactly deciphered, one protein named *Sp*AID has been found to be involved in male killing during embryo development via massive apoptosis and neural defects. Male *Drosophila* has single X chromosome, unlike two X chromosomes in females. To compensate for the X-chromosome-linked gene expression in males, dosage compensation mechanism leads hyper-transcription of male X chromosome up to two times than in females, and this is regulated by a protein–RNA complex called the male-specific lethal (MSL) complex, selectively recruited to the male X chromosome. *Sp*AID, with its ANK (ankyrin) repeats and OTU (ovarian tumour deubiquitinase) domain, targets this MSL complex and downstream histone modifications after nuclear localization and disrupts the male X chromosome (Harumoto & Lemaitre, 2018). Such male killing results in female-biased skewed sex ratio, which results in enhanced transmission of *Spiroplasma* from one generation to another by maternal inheritance (Kageyama et al., 2007). This phenomenon successfully prevents the establishment of two parasitic wasps, *Leptopilina heterotoma* (Thomson) and *L. boulardi* (Barbotin et al.), in the *Drosophila* (Xie et al., 2014). On the other hand, *Spiroplasma* increases the susceptibility of *Drosophila* to certain Gram-negative pathogens, like *Enterobacter cloacae* (Jordan). Most interestingly, the activation of systemic defence pathways like Toll and Imd immune pathways increases *Sprioplasma* number in *Drosophila melanogaster* (Herren & Lemaitre, 2011). *S. apis* Mouches et al., which belongs to a serologically different group (group IV) than *S. melliferum*, taken up during nectar gathering, can pass through the gut and cause lethal sepsis in honeybees. *S. leptinotarsae* Hackett et al. attaches to the microvilli of the midgut without causing any pathological impact on the host Colorado potato beetle *Leptinotarsa decemlineata* (Say). *Spiroplasma* also has been frequently detected in the gut and the haemocoel of various firefly beetles (Boucias & Pendland, 1998). *S. floricola* Davis et al. is the agent of lethargy disease of cockchafer, *Melolontha melolontha* (Linnaeus) (Bove, 1997). Some species of *Spiroplasma* have shown pathogenicity against mosquitoes and are expected to be useful vector control tools. Helical replicative forms of *S. taiwanense* Abalain-Colloc et al. were shown to reduce significantly the survival of *Aedes aegypti* (Linnaeus) mosquito larvae and adult males and females of *Ae. aegypti*, *Anopheles stephensi* Liston, and *Culex tritaeniorhynchus* Giles (Abalain-Colloc et al., 1988; Humphery-Smith et al., 1991a; Humphery-Smith et al., 1991b). *S. cantharicola* Whitcomb et al. are pathogenic to the mosquito *Ae. japonicus* (Theobald) (Shimooka et al., 2021). *Pristonchus* is a unique type of entomopathogenic nematode which does not possess a specific type of bacteria but rather harbours a diverse bacterial community of common inhabitants of soil, water, and other insects. *Spiroplasma* is an important member in conferring its entomopathogenicity (Ishaq et al., 2021). It has been found that some *Spiroplasma* produce ribosome-inactivating protein (RIP) related to Shiga-like toxins from *Escherichia coli* (Migula) and confer protection against nematodes, parasitoid wasps, and fungi, which act as natural enemies to their insect hosts (Ballinger & Perlman, 2019; Hamilton et al., 2016). *D. neotestacea* Grimaldi, James, and Jaenike is protected by the *Spiroplasma* from the nematode *Howardula aoronymphium* Welch by depurinating very specifically the 28S rRNA of the nematode, but not of the host (Hamilton et al., 2016). Out of 38 described *Spiroplasma* species, three have been associated with the phloem of the plant (*S. citri* Saglio et al. in citrus, *S. kunkelii* Whitcomb et al. in corn, and *S. phoeniceum* Saillard, Vignault, Bové, Raie, and Tully in periwinkle). Moreover, some species have been related to animal diseases, such as transmissible spongiform encephalopathy (TSE) (a neurodegenerative brain disorder) or cataracts (clouding of eye lens) (Cacciola et al., 2017; Megraud et al., 1983).

So we can see that there is much risk associated with these bacteria on behalf of our crop and beneficial insects, and we should be highly cautious regarding commercial application if it is of utmost need to formulate. But some interesting sites are highlighted here which might be utilized in some modified way for the betterment of human society.

10.4 CONCLUSION

Bacteria, insects, plants, humans we all are integrated parts of nature, same as the interaction among us. Humans, the owner of the thought process, have cultivated plants and defined them as crops, to produce more and more food for their survival. Humans have designated some insects as pests based on the damage they cause to the crop. But the insects feed for their own survival; they do not have any other intention. Same for the bacteria that show mutualism or lethality to pests and/or beneficial insects. Rather than expecting nature to compromise for our benefit, it is us, human beings, who must fit the whole system of nature as per our convenience, keeping in mind that nature should not be affected. The dilemma is more in utilizing Tenericutes because they are simultaneously either causing harm to pests or beneficials or helping them for better survival.

The condition of nature is already disturbed as we started to make nature in our favour with artificial chemicals; consequently, many components of nature are harmed—so are ourselves. So we have to come up with a solution from nature only. The description of the interaction of Tenericutes and insects presented here bears the same motto whether we can find some scope to utilise these agents as an effective management strategy for insect pest management. Though the use of entomopathogenic Tenericutes might not be possible as such, the in-depth knowledge of the mode of interaction with different insects might give us an idea to manipulate or engineer them in a way favourable to all our companions in nature.

REFERENCES

Abalain-Colloc, M. L., Rosen, L., Tully, J. G., Bove, J. M., Chastel, C., & Williamson, D. L. (1988). *Spiroplasma taiwanense sp.nov*. from *Culex tritaeniorhynchus* mosquitoes collected in Taiwan. *International Journal of Systematic Bacteriology*, *38*(1), 103–107. https://doi.org/10.1099/00207713-38-1-103

Ballinger, M. J., & Perlman, S. J. (2019). The defensive *Spiroplasma*. *Current Opinion in Insect Science*, *32*, 36–41. https://doi.org/10.1016/J.COIS.2018.10.004

Boucias, D. G., & Pendland, J. C. (1998). *Principles of insect pathology* (1st ed.). Springer. https://doi.org/10.1007/978-1-4615-4915-4

Bove, J. M. (1997). Spiroplasmas: Infectious agents of plants, arthropods and vertebrates. *Wiener Klinische Wochenschrift*, *109*(14–15), 604–612.

Cacciola, S. O., Bertaccini, A., Pane, A., Furneri, P. M., Cacciola, S. O., Bertaccini, A., Pane, A., & Furneri, P. M. (2017). *Spiroplasma* spp.: A plant, arthropod, animal and human pathogen. *Citrus Pathology*. https://doi.org/10.5772/66481

Cisak, E., Wójcik-Fatla, A., Zając, V., Sawczyn, A., Sroka, J., & Dutkiewicz, J. (2015). *Spiroplasma*—an emerging arthropod-borne pathogen? *Annals of Agricultural and Environmental Medicine*, *22*(4), 589–593. https://doi.org/10.5604/12321966.1185758

Davis, J. J., Xia, F., Overbeek, R. A., & Olsen, G. J. (2013). Genomes of the class erysipelotrichia clarify the firmicute origin of the class Mollicutes. *International Journal of Systematic and Evolutionary Microbiology*, *63*(Pt 7), 2727–2741. https://doi.org/10.1099/ijs.0.048983-0

Hamilton, P. T., Peng, F., Boulanger, M. J., & Perlman, S. J. (2016). A ribosome-inactivating protein in a *Drosophila* defensive symbiont. *Proceedings of the National Academy of Sciences of the United States of America*, *113*(2), 350–355. https://doi.org/10.1073/pnas.1518648113

Harumoto, T., & Lemaitre, B. (2018). Male-killing toxin in a bacterial symbiont of *Drosophila*. *Nature*, *557*(7704), 252–255. https://doi.org/10.1038/s41586-018-0086-2

Herren, J. K., & Lemaitre, B. (2011). *Spiroplasma* and host immunity: Activation of humoral immune responses increases endosymbiont load and susceptibility to certain Gram-negative bacterial pathogens in *Drosophila melanogaster*. *Cellular Microbiology*, *13*(9), 1385–1396. https://doi.org/10.1111/j.1462-5822.2011.01627.x

Herren, J. K., Paredes, J. C., Schüpfer, F., & Lemaitre, B. (2013). Vertical transmission of a Drosophila endosymbiont via cooption of the yolk transport and internalization machinery. *MBio*, *4*(2), e00532–12. https://doi.org/10.1128/mBio.00532-12

Humphery-Smith, I., Grulet, O., & Chastle, C. (1991a). Pathogenicity of *Spiroplasma taiwanense* for larval *Aedes aegypti* mosquitoes. *Medical and Veterinary Entomology*, *5*(2), 229–232. https://doi.org/10.1111/J.1365-2915.1991.TB00545.X

Humphery-Smith, I., Grulet, O., Le Goff, F., & Chastel, C. (1991b). *Spiroplasma* (Mollicutes: Spiroplasmataceae) pathogenic for *Aedes aegypti* and *Anopheles stephensi* (Diptera: Culicidae). *Journal of Medical Entomology*, *28*(2), 219–222. https://doi.org/10.1093/JMEDENT/28.2.219

Ishaq, S. L., Hotopp, A., Silverbrand, S., Dumont, J. E., Michaud, A., MacRae, J. D., Stock, S. P., & Groden, E. (2021). Bacterial transfer from *Pristionchus entomophagus* nematodes to the invasive ant *Myrmica rubra* and the potential for colony mortality in coastal Maine. *I Science*, *24*(6), 102663. https://doi.org/10.1016/j.isci.2021.102663

Kageyama, D., Anbutsu, H., Shimada, M., & Fukatsu, T. (2007). *Spiroplasma* infection causes either early or late male killing in *Drosophila*, depending on maternal host age. *Naturwissenschaften*, *94*(4), 333–337. https://doi.org/10.1007/S00114-006-0195-X

Madkour, M. H. F., & Mayer, F. (2007). Intracellular cytoskeletal elements and cytoskeletons in bacteria. *Science Progress*, *90*(2–3), 73–102. https://doi.org/10.3184/003685007X215913

Megraud, F., Gamon, L. B., & McGarrity, G. J. (1983). Characterization of *Spiroplasma mirum* (suckling mouse cataract agent) in a rabbit lens cell culture. *Infection and Immunity*, *42*(3), 1168–1175. https://doi.org/10.1128/IAI.42.3.1168-1175.1983

Ogata, S., Umemiya-Shirafuji, R., Kusakisako, K., Kakisaka, K., Chatanga, E., Hayashi, N., Taya, Y., Ohari, Y., Pandey, G. S., Abdelbaset, A. E., Qiu, Y., Matsuno, K., Nonaka, N., & Nakao, R. (2023). Investigation of vertical and horizontal transmission of *Spiroplasma* in ticks under laboratory conditions. *Scientific Reports*, *13*(1), 13265. https://doi.org/10.1038/s41598-023-39128-z

Razin, S., Yogev, D., & Naot, Y. (1998). Molecular biology and pathogenicity of mycoplasmas. *Microbiology and Molecular Biology Reviews*, *62*(4), 1094–1156. https://doi.org/10.1128/mmbr.62.4.1094-1156.1998

Regassa, L. B. (2014). The family Spiroplasmataceae. In E. Rosenberg, E. F. DeLong, S. Lory, E. Stackebrandt, & F. Thompson (Eds.), *The Prokaryotes* (pp. 551–567). Springer. https://doi.org/10.1007/978-3-642-30120-9_389

Shimooka, M., Sakurai, Y., Muramatsu, Y., & Uchida, L. (2021). Isolation and characterization of mosquito-associated *Spiroplasma cantharicola* from *Aedes japonicus* collected in Hokkaido, Japan. *Insects*, *12*(12), 1056. https://doi.org/10.3390/insects12121056

Stülke, J., Eilers, H., & Schmidl, S. R. (2009). *Mycoplasma* and *Spiroplasma*. In M. Schaechter (Ed.), *Encyclopedia of Microbiology* (3rd ed., pp. 208–219). Academic Press. https://doi.org/10.1016/B978-012373944-5.00229-7

Tully, J. G., Rose, D. L., Carle, P., Bové, J. M., Hackett, K. J., & Whitcomb., R. F. (1988). *Acholeplasma entomophilum* sp. *nov*. from gut contents of a wide range of host insects. *International Journal of Systematic and Evolutionary Microbiology*, *38*(2), 164–167. https://doi.org/10.1099/00207713-38-2-164

Waites, K. B., & Talkington, D. F. (2004). *Mycoplasma pneumoniae* and its role as a human pathogen. *Clinical Microbiology Reviews*, *17*(4), 697–728. https://doi.org/10.1128/cmr.17.4.697-728.2004

Xie, J., Butler, S., Sanchez, G., & Mateos, M. (2014). Male killing *Spiroplasma* protects *Drosophila melanogaster* against two parasitoid wasps. *Heredity*, *112*, 399–408. https://doi.org/10.1038/hdy.2013.118

11 Genetic Engineering of Entomopathogenic Bacteria for Sustainable Pest Management

ABSTRACT

Undoubtedly, entomopathogenic bacteria (EPBs) play a very crucial role in modern-day insect pest management programmes. However, there are certain factors that limit what they can do. To that end, genetic engineering has proven a viable way for enhancing their performances. *Bacillus thuringiensis* (Bt), which is the most researched EPB, has been genetically improved to increase its toxicity, host spectrum, photostability, etc. Isolation of toxin genes from one strain and incorporating them into another gives rise to recombinant strains that have enhanced toxicity against different insects, with the ability to colonize and persist for a longer time in the target area. In addition, the development of Bt transgenic crops by transferring the desired toxin genes from Bt into the crops, either through *Agrobacterium tumefaciens*–mediated gene transfer or by using the 'gene gun' method, has been proven highly effective in controlling various insects. Other than Bt, different other EPBs, such as *Lysinibacillus sphaericus*, *Bacillus cereus*, *Xenorhabdus*, *Photorhabdus*, *Serratia*, *Pseudomonas*, etc., have also been genetically engineered to enhance their activity.

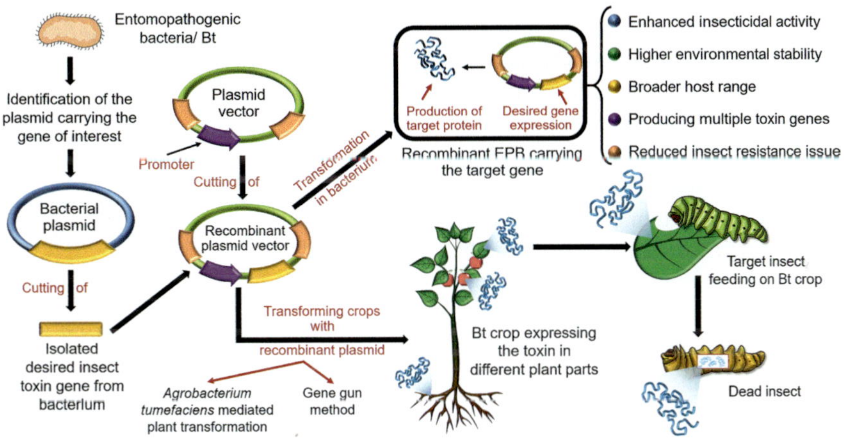

11.1 INTRODUCTION

Entomopathogenic bacteria (EPB) are highly effective against multiple insect pests from different insect orders. For many years, these EPBs have been successfully utilized in agriculture as a tool in integrated pest management (IPM), as well as for domestic pest control. Efficiency, host specificity, environmental safety, safety to human health and other non-target organisms, etc. are certain features that make EPBs the most widely adopted bioinsecticide. However, certain factors limit the effectiveness of these bacteria. In our previous chapters, we have discussed such factors restraining the efficiency of EPBs. For example, in the case of *Bacillus thuringiensis* Berliner (Bt), one of the major concerns is its sunlight-mediated inactivation (Leong et al., 1980; Pusztai et al., 1991), which makes it less persistent in the environment and reduces its activity under field conditions. Therefore, use of Bt becomes expensive due to the requirement of repeated application. Similarly, different EPB strains contain certain toxins that are specific to a particular group of insects. That's why controlling other insect pest groups requires different protection measures. This narrow host range also sometimes reduces the acceptability of EPBs. However, scientific advancement resolved these problems. In this aspect, genetic engineering (GE) has played a crucial role. GE of EPB natural strains has promising impacts on improving the efficacy and cost-effectiveness of EPB-based products. The three main objectives of engineering various EPB strains are (1) to increase the toxicity towards a particular pest, (2) to broaden the target pest range, and (3) to delay the onset of pest resistance by including toxins that bind to different sites or have different modes of action (Sansinenea et al., 2010). In the last few decades, through GE, a number of recombinant EPB strains have been developed, showing higher activity, environmental stability, broader host range, etc., as compared

DOI: 10.1201/9781003426882-11

to the naturally occurring strains. In this chapter, we have focused on such GE-related aspects of different EPBs to enhance their effectiveness.

11.2 GENETIC IMPROVEMENT RELATED TO BT

To date, the most widely studied and utilized EPB is Bt. Therefore, researchers have focused most of their genetic improvement efforts on Bt to expand its host range, increase its activity, maintain environmental stability, and create Bt transgenic crops.

11.2.1 Toxin Gene Isolation from Bt

The first step to achieve the aforementioned objectives is to isolate and sequence the toxin gene and then obtain the amino acid sequence of the toxin. While comparing the amino acid sequence, all proteins showed a common toxin domain. It was found that the whole gene sequence is always not necessary for the toxin to have its insecticidal activity. A portion of the gene, a chemically synthesized coding sequence, or in some cases, the whole gene can be used for further manipulation (Boucias & Pendland, 1998). Most of the Cry and Cyt proteins are encoded by the plasmid of Bt (Sansinenea, 2012). A toxin-producing strain can be conjugated with a strain that lacks insecticidal activity to test whether the toxin gene is present in plasmid or in chromosomal DNA. If the latter strain acquires the ability to synthesize the insecticidal toxin, then the toxin gene is most likely to be present in a plasmid, because the transfer of chromosomal DNA during conjugation is rare. To isolate the protoxin-encoding plasmid DNA sequence, total cellular DNA can be isolated and separated into plasmid and chromosomal DNA using caesium chloride gradient centrifugation. The sucrose gradient is further used to fractionate the plasmids based on different sizes. As an example, *Bt* subsp. *kurstaki* (Btk), which is one of the widely used Bts, contains an insecticidal protoxin gene on one of the total seven plasmids it possesses (i.e. plasmids of 2, 7.4, 7.8, 8.2, 14.4, 45, and 71 kilobase [kb] in length) (Alvarez & Loto, 2012). To determine which Btk plasmid carries the protoxin gene, the plasmid DNA sample is divided into three fractions by sucrose gradient centrifugation, that is, small (2.0 kb), medium (7.4, 7.8, 8.2, and 14.4 kb), and large (45 and 71 kb). The fraction with the small plasmid (2.0 kb) is generally discarded as it is too small to encode a protein equivalent to the 130 kDa (kilodalton) protoxin. The rest of the plasmid fractions are each partially digested with the restriction enzyme Sau3AI and then ligated into the BamHI site of plasmid pBR322. These clone banks are transformed into *Escherichia coli* (Migula) and then screened immunologically using the following procedure:

- Colonies are transferred from agar plates to a nitrocellulose membrane.
- The transferred colonies are partially lysed with organic solvents.
- All available sites on the membrane to which primary and secondary antibodies could bind are blocked by treating the membrane with bovine serum albumin.
- The bovine serum albumin–treated membranes are treated with rabbit antiserum that contains antibodies against the insecticidal toxin.
- The membranes are washed to remove unbound antibodies and then treated with ^{25}I-labeled *Streptococcus aureus* Rosenbach protein A, which binds to the Fc portion of the bound antibodies.
- Spots on the membrane analogous to colonies that actively synthesize the insecticidal toxin are visualized by autoradiography. By this method, the 71 kb plasmid of Btk was found to encode the toxin gene

(Boucias & Pendland, 1998)

However, recent techniques are based on polymerase chain reaction (PCR); 16S–23S rRNA internal transcribed spacer (ITS) regions of most of the Bt isolates are found to be highly conserved. The *Cry1* gene family is most abundant, followed by the *Cry2* gene family. Nowadays, toxin genes can be amplified with especially designed primers, then cloned and sequenced for novelty judgment from other existing strains (Jayakumar & Kaur, 2013).

Usually, Bt δ-endotoxin genes are expressed during the sporulation phase only. Therefore, it might be advantageous to have the toxin gene transcribed and translated during the vegetative growth phase. This would permit the toxin to be synthesized by a continuous fermentation process, which decreases the cost of toxin production (Boucias & Pendland, 1998). However, cloning of the δ-endotoxin genes downstream from the tetracycline resistance gene promoter resulted in the toxin crystals being expressed continuously throughout the growth cycle (in both sporulation and vegetative phases) in Bt. This condition allows the production of δ-endotoxin at higher rates and also allows the production of δ-endotoxin in Bt strains, which are unable to produce δ-endotoxin due to defected sporulation (Li & Yu, 2012; Mettus & Macaluso, 1990).

11.2.2 Genetically Improved Bt with Expanded Host Range

Initially, Bt was reported to infect nearly 525 species of insects belonging to 13 insect orders around the world (Jayaraj, 1985). Since then, the host range of Bt is constantly expanded along with an increase in its commercial interest. One approach to achieve the objective of expanding the host range is the pyramiding of appropriate toxin genes. In order to enhance the insecticidal efficiency and expand the host range of Bt, efforts are made by transferring the Bt toxin genes into nonhomologous Bt isolates to combine δ-endotoxins, showing either additive or synergistic effects (Arora & Shera, 2014). Fusion of multiple Bt toxins has higher effectiveness against different insects. In this way, hybrid toxins were made by exchanging a fragment of 451–623 amino acid residues of domain III of Cry1Aa with Cry1Ac, which resulted in altered receptor binding properties in Gypsy moth midgut (Lee et al., 1995). Cloned Bt genes from *Bt* subsp. *tenebrionis* (Btt) were introduced into native isolates of *Bt* subsp. *israelensis* (Bti) and Btk using the electroporation method, which ultimately increased their host spectrum. As expected, the developed organisms showed a high level of toxicity towards beetles and mosquitoes. However, the hybrid was also found to possess an additional low level of toxicity against the cabbage butterfly, *Pieris brassicae* (Linnaeus), unlike the parent subspecies *israelensis* or donor *tenebrionis*, which were non-toxic to this insect (Crickmore et al., 1990).

Using the technique of stacking multiple toxin genes from Bt, different commercial products have been developed, showing a higher response against various groups of insect pests. Lereclus et al. (1992) conceived a novel approach of in vivo genetic recombination and showed the expansion in the Bt host range. For achieving this genetic recombination, they inserted sequence IS232 to deliver *Cry3A* gene from Btt into an isolate (Btk) producing CryIA toxin. This resulted in the development of a strain that showed insecticidal activity against both coleopteran and lepidopteran insects. The transconjugant known as 'Foil' is a bioinsecticide product marketed by Ecogen Inc. and applied to potatoes, brinjals, and maize, showing toxicity against both orders (Carlton & Gawron-Burke, 1993). Ecogen Inc. also produced the commercial product Condor® (strain EG2348) by transferring self-transmissible Cry1A plasmids via conjugation from a strain of *Bt* subsp. *aizawai* to a host strain of Btk. The combined insecticidal property functioned well against the targeted lepidopteran pests (Gawron-Burke & Baum, 1991). Later on, the company also produced an rDNA variant (ECX9399) of EG2348 which showed higher toxicity against the fall armyworm, *Spodoptera frugiperda* (J. E. Smith), as compared to the parent strain (All et al., 1994). Similarly, Agree® and Design®, two commercial products, were the result of conjugation between strains of Btk and *Bt* subsp. *aizawai* (Arora & Shera, 2014). Another type of bioinsecticide, named GCSC-BtA, based on 'Germany–China Scientific Cooperation' research, was also developed through the conjugation of toxin proteins from Bt with *Streptomyces avermitilis* (Burg et al.)–produced toxin, that is, abamectin (Liu & Sengonca, 2003). This conjugated bioinsecticide possesses a broader host range with rapid killing efficiency against multiple insect pests and shows superior performance than the Bt crystal or abamectin alone (Zhu et al., 2006).

Different strains of *Bt* subsp. *canadiensis*, *colmeneri*, *entomocidus*, *israelensis*, *kenyea*, *konkukian*, *kurstaki*, and *pakistani* harbour genes encoding different types of chitinase enzymes (Karabörklü et al., 2018). Overexpression of Bt homolog chitinases proved to enhance the insecticidal activities of Bt against different insect pests, such as *Ephestia kuehniella* Zeller (Driss et al., 2011), *Drosophila melanogaster* Meigen and *Galleria mellonella* (Linnaeus) (Ozgen et al., 2013), *Helicoverpa armigera* (Hübner) (Hu et al., 2009), *Lymantria dispar* (Linnaeus) (Lertcanawanichakul et al., 2004), *Plutella xylostella* (Linnaeus) (Wiwat et al., 2000), *S. frugiperda* (González-Ponce et al., 2019; Hu et al., 2009), *Tuta absoluta* (Meyrick) (Atia et al., 2019), etc. That is why different chitinase (*chi*) genes (e.g. *chiB* and *chiC*) have been identified and characterized from a wide variety of microorganisms. Incorporating these genes into Bt resulted in recombinant strains with significantly enhanced insecticidal activity (Lertcanawanichakul et al., 2004; Okay et al., 2008; Ozgen et al., 2013).

Although expanding the host range of Bt through the production of hybrid toxins has proven to be effective against various insect pests, however, while expanding this host range of Bt, the potential activity of the hybrid toxins on beneficial arthropods needs to be monitored carefully.

11.2.3 Genetically Improved Bt with Enhanced Photostability

The exposure of Bt spores or toxins to sunlight, specifically to ultraviolet (UV) light, causes rapid degradation of spores and toxins either by directly damaging the DNA or by producing reactive oxygen-derived free radicals, which ultimately limits the field use of Bt-based insecticides (Pusztai et al., 1991; Zhang et al., 2008). In this case, melanin shows promising activity, as it absorbs the radiation and consequently provides photoprotection to Bt-based biopesticides (Sansinenea & Ortiz, 2015). A mutant (Bt-m) of Btk that can resist UV irradiation through the production of melanin was obtained by inducing mutagenesis using ethyl methanesulphonate (EMS). These melanin-producing mutants show higher larval mortality to *H. armigera* and potato tuber moth, *Phthorimaea operculella* (Zeller), than their parents (Mohamed et al., 2010; Saxena et al., 2002). TD841 is another recombinant strain of Bt that can express both *mel* (from *B. cereus* 799) and *CryIA* genes and shows higher activity against *H. armigera*, with 9.7-fold enhanced UV resistance as compared to its parental strain after 4-hour UV irradiation (Zhang et al., 2008).

Sanchis et al. (1999) constructed an asporogenic recombinant sporulation-deficient Bt strain, SigK$^-$ (expressing *CryIC/Ab* gene), in which the crystals remained encapsulated within the cells and got protected from UV degradation. This new strain also showed a broader host spectrum and greater potency than the parent strain. Btt strain NB-176 is a gamma-irradiated mutant and high-yielding descendant of the strain NB-125, containing two to three copies of Cry3A. This mutant strain is used to formulate the commercial product Novodor® (by Abbott Laboratories) with enhanced field activity through the production of large thromboid crystals made of Cry 3A toxin (EFSA, 2013).

11.2.4 Enhancing Bt Activity through Alternate Delivery Systems

By using alternate ways of delivering the toxin into the target insects, it is also possible to enhance the insecticidal activity of Bt. This is achieved by cloning the Bt genes in epiphytic, endophytic, and/or aquatic bacteria and eukaryotic plants, which enhances the environmental persistency of Bt toxins through a self-propagating host that can colonize the insect feeding site and produce the crystal proteins continuously (Arora & Shera, 2014).

Lampel et al. (1994) have engineered a bacterial endophyte to enhance insecticidal activity against European corn borer, *Ostrinia nubilalis* (Hübner), by introducing the *Cry1Ac* gene from Btk HD-73 into the chromosome of an endophytic bacterium, *Clavibacter xyli* subsp. *cynodontis* (Cxc), that can colonize the vascular system of different plants. Similarly, the endophytic bacterium *Bacillus megatarium* de Bary strain RS-1, which can colonize the cotton leaf, was used as the host for *Cry1A* genes from Btk HD-1 (Bora et al., 1994). Two nodule-forming endophytes, *Rhizobium leguminosarum* (Frank) and *R. meliloti* Dangeard, were also modified with the *Cry3A* gene from Btt to control coleopteran insects *Sitona lineatus* (Linnaeus) and *S. hispidulus* (Fabricius), respectively (Bezdicek et al., 1991). Recombinants with improved colonizing ability in the tomato phylloplane and an LC_{50} value comparable to the donor strain against *T. absoluta* were developed by inserting the *Cry1Ab* gene from Bt strain LM-466 into the naturally occurring strains of *B. subtilis* (Ehrenberg) and *B. licheniformis* (Weigmann) (Theoduloz et al., 2003). In a study, a recombinant bacterial endophyte, *Burkholderia pyrrocinia* strain JKSH007, was developed through the introduction of *Cry218* gene of Bt to control lepidopteran larvae (Li et al., 2017). Scientists from Madurai Kamraj University of Tamil Nadu isolated a strain, RSl, which was identified as *Bacillus* sp. and capable of colonizing efficiently on cotton phyllosphere. They introduced the *Cry1Aa* gene from Bt into that *Bacillus* sp. through conjugal transfer. The conjugant showed a prolonged period of colonization on cotton leaves and protected the plants from the attack of *H. armigera* for more than 30 days (Narayanan, 2002).

One of the major concerns regarding Bt-based products is their lack of stability in soil. That is why insects attacking the plant roots are completely inaccessible to these Bt-based insecticides. Therefore, introducing the toxin gene from Bt to a bacterium that colonizes close to the roots can solve this problem. The introduction of the engineered bacteria into the soil will allow the bacterium to colonize the soil and continuously release the toxin close to the roots. This will also mitigate the need for continuous spraying. Thus, the Btk insecticidal toxin has been introduced into the chromosomal DNA of *Pseudomonas fluorescens* (Trevisan) that colonizes corn roots and releases the Bt toxin beneath the soil to control the root-feeding insects (Sanchis, 2012). Nambiar et al. (1990) cloned a toxin gene from Bti and transferred it into the pigeon pea nodule-forming bacterium *Bradyrhizobium* spp. through conjugative mobilization. The modified organism provided protection against the larvae of root nodule–feeding dipteran insect, *Rivellia angulata* (Hendel). The Bt crystal protein Cry5B has lethal effects on nematodes. Cloning of genes encoding the Cry5B toxin and a truncated version of it in *Lactococcus lactis* (Lister) proved to have the potential as an anthelmintic that could be delivered orally against parasitic nematode infections (Durmaz et al., 2016).

The major limitation of using Bti in mosquito control programmes is the rapid sedimentation of the crystals and spores below the feeding zone of mosquito larvae, causing Bt to fail to interact with the target insect. This problem has been solved by using an alternate delivery system to increase the persistency of toxins in the aquatic environment. So in those cases, the toxin genes (*Cry4A*, *Cry4B*, *Cry11A*) from Bti can be incorporated into any other suitable vectors, such as *Anabaena* (Lluisma et al., 2001), *Caulobacter crescentus* Poindexter (Boussiba et al., 2000; Thanabalu et al., 1992a), *Agmenellum quadruplicatum* (Menegh.) (Angsuthanasombat & Panyim, 1989; Murphy & Stevens, 1992), *Asticcacaulis excentricus* Poindexter (Armengol et al., 2005; Zheng et al., 2007), etc., which are predominantly found near the feeding zone of mosquito larvae and provide prolonged mosquito control. *Beauveria bassiana* (Bals.) Vuill. is an entomopathogenic fungi having slow action against mosquitoes. However, a recombinant *B. bassiana* strain, *Bb*-Cyt2Ba, expressing Cyt2Ba toxin from Bt, shows significantly enhanced virulence against mosquitoes (Deng et al., 2019).

11.2.5 Process of Developing Transgenic Bt Crops

When one or multiple isolated toxin genes of Bt are incorporated in plant genomes, they can synthesize the specific toxin in the plant system as well. These crops are known as Bt crops, capable of resisting insect attacks via these toxin genes. As these special types of plants are developed through genetic modifications, they are known as genetically modified (GM) crops, transgenic crops, or genetically engineered (GE) crops (Abbas, 2018). The first ever Bt crop was Bt tobacco, developed by the Plant Genetic Systems biotech company (now part of Bayer CropScience) through the introduction of a toxin gene, *bt2*, from Bt strain Berliner 1715 into the tobacco plant to control the tobacco hornworm, *Manduca sexta* (Linnaeus) (Vaeck et al., 1987). From then to now, this Bt crop technology has made a huge growth, and currently, a wide variety of Bt crops, including Bt cotton, soybean, rice, maize, potato, brinjal, etc., is available.

For the transfer of Bt toxin genes to the plant genome, we need some genetic vectors. The most common method of gene transfer is via the utilization of one soil bacterium, *Agrobacterium tumefaciens* (Smith and Townsend), which is peculiar to have a tumour-inducing gene in its plasmid (thus known as the Ti plasmid). The Ti plasmid has three significant regions: the origin of replication, the virulence region, and the T-DNA (transferred DNA) region, possessing auxin, cytokinin, and opine (octopine [ocs], mannopine [mas], and agropine [ags]) synthesis genes (Gelvin, 2003). The origin of replication is responsible for the initiation of plasmid replication via the rolling circle model (Ruiz-Masó et al., 2015). The virulence region comprises seven major loci (*virA*, *virB*, *virC*, *virD*, *virE*, *virF*, and *virG*), responsible for transferring the T-DNA from bacteria to the plant genome (Hwang et al., 2017). T-DNA, which is the major responsible area for causing tumours, is our major interest, as the Bt toxin gene will be inserted in place of this region (Gelvin, 2003). The T-DNA region has a left border and a right border and does not possess a unique restriction site. As the Ti plasmid is big and often has a low copy number, it is problematic to be isolated and cloned. Most of the researchers use a binary vector method to get over these problems. This technique carries a T-DNA region comprising a cloning site for the desired gene and a selectable marker for plants on a broad-host range replicon with an ori site for *Agrobacterium* as well as *E. coli* and a selectable marker for bacteria, and the *vir* genes required for T-DNA transfer are located on the disarmed Ti plasmid, which is called the helper plasmid (Hwang et al., 2017). The Bt toxin gene of interest as well as the Ti plasmid should be restricted with the same restriction enzyme to create complementary sticky ends for ligation. Competent cells are prepared via treatment with calcium chloride combined with heat shock, or 10% glycerol method, and transformation is done via freeze-thaw method or electroporation, where the bacterial cell wall becomes weak and of neutral charge so that the bacteria can accept the recombinant plasmid and do not repel it due to negative charge in both nucleic acid and cell wall (Das & Dash, 2015). The transformed bacterial cells are selected based on antibiotic resistance or blue-white screening (Grooms, 2019) (details of these methods are not described here, as these are beyond the scope of the book). The modified *A. tumefaciens* cells containing the plasmid with the introduced gene are mixed with plant cells or cut pieces of plants, such as leaves or stems (explants). Some of the cells take up the T-DNA (AUDA-NEPAD, 2020). The *Agrobacterium* inoculation can also be done via pinprick method either in stem or germinated seeds, or agroinfiltration via syringe (leaf injection), vacuum infiltration, or root or floral dipping as well (Amal et al., 2020; Hwang et al., 2017; Kaur & Devi, 2019; Zia et al., 2011). The *A. tumefaciens* inserts the Bt toxin gene into one of the chromosomes of the plant to form the GM (or transgenic) cells.

However, this method of gene transfer is only applicable to all dicot plants, along with a few monocot plants and callus-forming plants (Gelvin, 2003; Su et al., 2023). For other plants, some other methods should be taken, although their success rates are lower. Particle bombardment, sometimes known as a 'gene gun', is the second most used technique for transferring DNA. In this procedure, tiny particles coated with DNA molecules are fired into the cell using a biological approach (Dohm, 2003). Using lasers to temporarily breach the cell membrane and allow DNA to enter the cytoplasm is known as 'laser-induced DNA delivery' (Heinemann et al., 2022). Chemical treatments are used in chemical-mediated transfer to enable cell membrane permeation (Lalani & Misra, 2011). Additional techniques include sonication, microinjection, silicon carbide fibres, liposomes, etc. (Khan, 2009).

Only a small percentage of the plant cells carry the desired gene after transformation, and most of the time, selectable marker genes that provide resistance to either antibiotics or herbicides are employed to encourage the growth of the transformed cells in callus culture in comparison to the non-transformed cells (Manimaran et al., 2011). In this technique, the resistance genes are delivered to the plant cells together with the gene(s) that impart desirable characteristics by inserting them into the vector. Only the transformed cells expressing the selectable marker gene will survive when the cells are exposed to the antibiotic or herbicide. Tissue culture techniques are then used to regenerate the transformed cells into whole plants.

Plant transformation verification involves confirming that the gene was successfully introduced and is regularly inherited. Analyses are conducted to ascertain the quantity of inserted copies, the integrity of the copies, and the possibility of unintentional consequences arising from gene interference caused by the insertion. To ensure that a gene is functioning, tests of gene expression are carried out, which include assessing the trait of interest and producing messenger RNA and/or protein (AUDA-NEPAD, 2020). The second phase of the procedure involves conducting multi-location, multi-year assessment trials in field and greenhouse settings to assess the transgene's overall performance and impacts. Evaluations of food and environmental safety are conducted in coordination with plant performance testing (Arora et al., 2019).

There are several regulatory agencies to be supplied with information about the marker genes and whether or not they will be present in the created GM plant. Before the tests are carried out, the developer must provide the regulatory authorities with comprehensive information on the genes of interest, promoters, marker genes, vectors, and transformation strategies. The developer provides the regulators with well-presented methods and findings to ascertain if the gene was inserted, how many copies were inserted, whether the copies are undamaged, and whether the insertion does not interfere with normal plant function and gene expression (AUDA-NEPAD, 2020). Currently, the Genetic Engineering Approval Committee (GEAC) under the Ministry of Environment, Forest, and Climate Change approves genetically modified organisms in India. The government of India has some other agencies that enforce laws and standards for transgenic crops, including the Department of Biotechnology (DBT), the Ministry of Science and Technology, and the Review Committee on Genetic Manipulation (RCGM) (Pandey & Warrier, 2015). The Biotechnology Regulatory Authority of India (BRAI) Bill was introduced in 2013 for regulating biotechnology-related activities, such as research, transportation, import, and use of GM crops. The BRAI would also monitor field trials and protect the health and safety of humans and animals, as well as the environment. But as of October 2018, the Bill has expired, and a new and relaunched version was anticipated (Ajaykumar, 2024).

The Bt gene can be added to the male, female, or both parental lines that are being utilized to develop hybrids. From the standpoint of seed testing, it is necessary to investigate variations in the toxin expression levels in the plants and seeds of Bt hybrids when either the male parent or the female parent, or both, is utilized as the Bt parent. The government has created regulations for the management of Bt hybrid seed quality in response to the rising demand for these hybrids. *Apomixis* is a phenomenon reported in agricultural plants that allows for seed formation without meiosis or fertilization. For breeders, it is a very important tool for fixing genes and multiplying heterotic genetic combinations. Apomixes aid in the stabilization of hybrid vigour after a desired combination has been chosen and developed. The hybrid seed may be grown and cared for in the same way as a regular variety. As Bt cotton seeds are too costly, apomictic Bt hybrids will prove to be cost-effective because they don't require seeds to be bought every season. This hybrid development approach is also known as the 'one-line method'. Apomictic line development is ongoing in cotton, and efforts are being made to investigate the possibility of developing apomictic Bt hybrids (Santhy et al., 2008). We have discussed these Bt crops briefly in Chapter 12, the following chapter.

11.3 GENETIC IMPROVEMENT RELATED TO *LYSINIBACILLUS SPHAERICUS* (LS)

To date, most GE-related works have been done to improve the efficiency of toxins produced by Bt, either in commercial formulation or by incorporating the toxin gene into the crops. Very little importance has been given to the genetic improvement of other EPBs. Hitherto, in the case of mosquitocidal bacteria *Lysinibacillus sphaericus* (Meyer and Neide) (Ls), some studies have already been conducted to improve its efficiency through GE.

As mentioned in Chapter 5, the highly toxic strains of Ls used to produce binary (Bin) toxin, which is a heteromorphic protein consisting of 41.9 kDa (BinA) and 51.4 kDa (BinB) active peptides required at an equimolar proportion to show activity against mosquito larvae (Broadwell et al., 1990; Nicolas et al., 1993). On the other hand, low-toxic Ls strains, such as SSII-1, produce a 100 kDa mosquitocidal (Mtx) toxin instead of producing binary toxins (Thanabalu et al., 1991). However, the purified, recombinant 97 kDa Mtx toxin, derived after the deletion of the putative signal sequence from the 100 kDa protein, has an LC_{50} value comparable to the binary toxins of Ls against larvae of *Culex quinquefasciatus* Say (Thanabalu et al., 1992b). Thanabalu and Porter (1995) found that the lower toxicity of this 100 kDa Mtx toxin in strain SSII-1 during sporulation phase is probably due to the degradation of the toxin, synthesized during vegetative growth phase, by the action of intracellular and extracellular proteases produced during sporulation phase. That is why they expressed the *mtx* gene in a protease-negative Ls strain 1693, using a native promoter, and found high levels of 100 kDa toxin expression in both vegetative and sporulated cells of strain 1693, making the recombinant strain as toxic as the natural, highly toxic strains of Ls. Thus, they suggested the protease-negative strain as a potential vehicle for delivering *mtx* and other protease-susceptible toxin-producing genes to mosquito larvae. By using such protease-deficient bacteria, the delivery of toxins to the targeted insects could be enhanced, and thus, genetically modified Ls will be of enhanced efficacy against the targeted aquatic Dipteran insects (Harrison & Bonning, 2000).

As stated in Chapter 5, Ls strains have higher persistency in the environment but show moderate toxicity to *Anopheles* mosquitoes and no toxicity to *Aedes* mosquitoes. Apart from that, different *Culex* mosquitoes have also developed resistance against Ls strains. On the other hand, strains of Bti are highly toxic to both *Anopheles* and *Aedes* mosquitoes but do not persist in the environment for longer periods. These issues have been addressed by developing recombinant Ls strains carrying toxins from Bti. A recombinant strain of Ls was constructed by integrating the dipteran-active *Cry11A1* gene from Bti into Ls strain 2297 via homologous recombination. Thus, the recombinant produces two toxins and shows longer environmental persistence in the aquatic feeding zones of various mosquitoes, unlike donor Bti, with higher activity against *Aedes* and *Anopheles* mosquitoes, which are less affected by the parent strain, Ls (Poncet et al., 1997). Similarly, Servant et al. (1999) have introduced the *Cry11A* gene from Bti and the *Cry11Ba* gene from *Bt* subsp. *jegathesan*, separately and in combination, into Ls strain 2297. Unlike the parent strain, the recombinant strains showed toxicity to *Aedes aegypti* (Linnaeus) larvae, with the strain containing Cry11Ba having higher toxicity. Moreover, the recombinant strains also partially overcame the resistance development in *Cx. pipiens* Linnaeus and *Cx. quinquefasciatus* to the Ls parent strain. Likewise, the recombinant Bti IPS-82 strain incorporated with binary toxin genes from Ls strain 2362 demonstrated a 21- and 32-fold increase in effectiveness against fourth instar *Cx. quinquefasciatus* larvae as compared to the parent strains, respectively. Additionally, it successfully addressed the problem of resistance development in insects against the original strain (Park et al., 2005).

Synergism between the toxins is an excellent way to enhance the insecticidal activity. Cai et al. (2007) discovered that chitinase can enhance the activity of the binary toxin against *Cx. quinquefasciatus* by converting Ls strain 2297 with the chitinase gene *chiAC* from Bti, resulting in a 4,297-fold greater toxicity than the original strain. Transforming an acrystalliferous strain of Bt with the *mtx 1* gene from Ls strain SSII-1 individually showed moderate toxicity against both binary toxin-susceptible and resistant *Cx. quinquefasciatus* larvae. On the other hand, transforming the Bt acrystalliferous strain with the *mtx 1* gene in combination with the cytolytic toxin gene, *cyt1Aa*, from

Bti showed a very low LC_{50} value and enhanced the Mtx1 activity to target mosquito larvae, indicating that Cyt1Aa synergizes the activity of Mtx1 toxin (Zhang et al., 2006).

Similar to Bt, an alternate way of delivering the binary toxin from Ls into mosquitoes is also an effective way of enhancing the insecticidal activity and stability of Ls. A study found that *E. coli* recombinants carrying recombinant plasmids harbouring *Cry* genes (*Cry4A*, *Cry4B*, and *Cry11A*) from Bti and binary toxin genes from Ls had broad-spectrum action against all *Aedes*, *Anopheles*, and *Culex* larvae (Tanapongpipat et al., 2003). There are some species of bacteria having a close relationship with mosquito larvae and are able to colonize rapidly in the guts of mosquito larvae. Luxananil et al. (2003) isolated such two strains of *Bacillus cereus* Frankland and Frankland, Ae10 and Cx5, from the guts of mosquito larvae and transformed them with a pBS373 recombinant plasmid that harboured binary toxin genes from strain 2297. The very high toxicity of these recombinant *B. cereus* strains against the larvae of *Cx. quinquefasciatus*, combined with their close association with mosquito larvae and ability to recolonize in the guts of mosquito larvae, made them a promising alternate delivery system of binary toxins.

C. crescentus is a Gram-negative bacterium widely distributed in freshwater ecosystems and most commonly found at or near the water surface. Transforming the *C. crescentus* strain CB 15 with a recombinant plasmid harbouring the *mtx 1* gene from Ls strain SSII-1 and the *bin* gene from strain 2297 resulted in transformants that were very active against *Culex* mosquitoes (Yuan, 2002). Similarly, different groups of cyanobacteria, like *Anabaena*, are also widely available on the surface of different aquatic ecosystems and have a close association with the mosquitoes. Transforming *Anabaena* with the binary toxin genes from Ls also serves as an efficient way of delivering Ls toxins into target hosts with high larvicidal activity (Xu et al., 2000; Xudong et al., 1993). Introduction of the binary toxin gene from Ls strain 2362 and the *Cry11Bb* gene from *Bt* subsp. *medellin* strain CIB 163–131 into *A. excentricus* yielded recombinant strains that were able to express both toxins and remained in the top water column for a longer period than the parent strains (Romero et al., 2001).

Apart from the mosquitocidal activity of Ls, Abdel-Salam et al. (2018) enhanced the nematicidal activity of this bacterium against root-knot nematode (RKN), *Meloidogyne incognita* (Kofoid and White), through protoplast fusion technique for Ls strain Amira and *B. amyloliquefaciens* subsp. *plantarum* strain SA5. The fusants were more efficient in killing J2 of RKN and produced a greater amount of chitinase, which is positively correlated with the nematicidal activity, providing better results in shoot length and the fresh and dry weights over the control.

11.4 GENETIC IMPROVEMENT RELATED TO OTHER *BACILLUS*

Similar to Bt, cloning of the δ-endotoxin genes into the downstream region of a tetracycline resistance gene promoter resulted in the toxic crystal proteins being expressed continuously throughout the growth cycle (in both sporulation and vegetative phases) in *B. megaterium* (Mettus & Macaluso, 1990). Different species of *Bacillus* also acted as a potential host for delivering the toxins from Bt and Ls to the target insects. Different strains of *B. cereus* are used as an adequate host, in which binary toxin genes from Ls and *Cry* genes from Bti have been introduced to enhance the mosquito larvicidal activity as well as prevent the mosquito population from developing resistance against parent strains. Conjugal transfer of pBtoxis plasmid from Bti into *B. cereus* strain showed higher toxicity against dipteran insects (Hu et al., 2005). Similarly, transferring a recombinant plasmid, pBS373, containing binary toxin genes from Ls strain 2297, into Ae10 and Cx5 strains of *B. cereus* caused the production of 51 kDa toxin protein in both the recombinant strains, posing high toxicity against larvae of *C. quinquefasciatus* with recolonizing ability in mosquito gut (Luxananil et al., 2003). Transforming the *Paenibacillus polymyxa* (Prazmowski) strain NMO10 in order to combine the insecticidal properties of the *Cry1C* gene from Bt and the growth-promoting properties of *P. polymyxa* demonstrated higher toxicity against lepidopteran insects than Bt (Ibrahim et al., 2014). Similarly, different other species of *Bacillus*, such as *B. subtilis* and *B. licheniformis* (Theoduloz et al., 2003) and *B. velezensis* (Roh et al., 2009), are also genetically modified to carry the toxin genes from Bt into the target insects.

11.5 GENETIC IMPROVEMENT RELATED TO *XENORHABDUS* AND *PHOTORHABDUS*

Xenorhabdus nematophilus (Poinar and Thomas) and *Photorhabdus luminescens* (Thomas and Poinar) are two important entomopathogenic bacteria that are engaged in symbiotic relationships with entomopathogenic nematodes, *Steinernema* and *Heterorhabditis*, respectively. *X. nematophilus* strain PMFI296 produces *xpt* genes, responsible for insecticidal activity (Morgan et al., 2001). Expression of *xptA1*, *xptA2*, *xptB1*, and *xptC1* genes from *X. nematophilus* strain PMFI296 in *E. coli*, either individually or in combination, against *Pieris rapae* (Linnaeus), *P. brassicae*, and *Heliothis virescens* (Fabricius), revealed how a combination of three toxins should be administered together to show the insecticidal activity (Sergeant et al., 2003). Lee et al. (2004) identified and cloned a novel toxin gene (*tccC1/xptB1*) from a strain of this bacterium, which was isolated from a Korea-specific entomopathogenic nematode, *Steinernema glaseri* (Steiner) MK. In *E. coli*, this toxin gene expressed successfully, and the recombinant toxin protein showed high efficiency, with 80% larval mortality of *G. mellonella* within 2 days. Apart from that, Khandelwal et al. (2004a) isolated and purified a 17 kDa pilin subunit from *X. nematophilus* cell surface, which showed cytotoxicity to *H. armigera* larval haemocytes. Later on, they cloned the 537-bp DNA fragment encoding the pilin protein from *X. nematophilus* into *E. coli* strain

BL21(DE3). The recombinant toxin showed oral toxicity against fourth and fifth instar *H. armigera* larvae by causing serious damage to the midgut epithelial membrane (Khandelwal et al., 2004b).

Unlike *X. nematophilus*, the toxin complex (Tc) of *P. luminescens* consists of three components, termed TcA, TcB, and TcC (Azizoglu et al., 2020). Although the direct use of *Photorhabdus* as an insecticide is limited, the recombinant toxins from *Photorhabdus* expressed in alternate hosts have been found to have oral insecticidal activity against various insect pests (Azizoglu et al., 2020). Waterfield et al. (2001) showed that higher oral toxicity to *M. sexta* requires the combination of three genes, *tcdA*, *tcdB*, and *tccC*, all on the same plasmid when expressed in *E. coli*. Liu et al. (2003) genetically transformed *Arabidopsis thaliana* (L.) plants to express *tcdA* gene, encoding a 283 kDa protein, toxin A, from *P. luminescens*. These transgenic plants were found to be toxic to *M. sexta*, representing this toxin as an alternative to Bt toxins for deployment in transgenic crops. Apart from that, the isolated toxin A from these transgenic plants also inhibited the growth of the Southern corn rootworm, *Diabrotica undecimpunctata howardi* Barber (Liu et al., 2003). However, the transgenic plants expressed TcdA in low levels. That is why, to enhance the activity of the toxin TcdA, it requires potentiation by any of these two pairs of gene products, TcdB1 and TccC1 or TcdB2 and TccC3 (Waterfield et al., 2005a). *Enterobacter cloacae* (Jordan) is a bacteria found in the gut of the Formosan subterranean termite, *Coptotermes formosanus* Shiraki. Transforming this bacterium with genes (*tcdA1* and *tcdB1*) expressing orally insecticidal proteins from *P. luminescens* subsp. *laumondii* strain TT01 showed a significantly higher toxicity against *C. formosanus*, and the recombinant bacteria also transferred between donor workers and recipient workers (Zhao et al., 2008). In addition, loci *plu* 4093-*plu* 4092 and *plu* 4437-*plu* 4436 of *P. luminescens* strain TT01 encoded '*Photorhabdus* insect-related' (Pir) proteins (PirA and PirB), showing oral toxicity against both mosquito and moth larvae (Duchaud et al., 2003; Waterfield et al., 2005b). Injecting the recombinant *E. coli* strain EC100 expressing both PirA (45 kDa) and PirB (14 kDa) toxin into the haemocoel of *G. mellonella* larvae kills the larvae within 72 hours. However, individually, none of these toxins show insecticidal activity (Waterfield et al., 2005b).

11.6 GENETIC IMPROVEMENT RELATED TO *SERRATIA*

Serratia entomophila Grimont et al. is also an important entomopathogenic bacteria, effective against many insects causing amber disease. This bacterium contains a 115 kb plasmid, pADAP, which is required for causing amber disease (Glare et al., 1993). Hurst et al. (2000) defined a 16.9 kb region in this pADAP plasmid, encoding three genes, *sepA*, *sepB*, and *sepC*, that is sufficient to confer disease-causing ability to the grass-grub, *Costelytra zealandica* (White), on pADAP-cured *S. entomophila* strains and on *E. coli* strains. Furthermore, products of these genes resemble components of the insecticidal toxin complexes of *P. luminescens*. Later on, Hurst et al. (2004) constructed a 47 kb clone through deletion analysis of pADAP plasmid and subsequent sequence data. Expression of this clone either in *E. coli* or *Serratia* background produced strong antifeeding activity and caused rapid death in infected larvae of *C. zealandica*.

Apart from that, different species of *Serratia* are able to produce high levels of chitinases. Overexpression of *chiA*, *chiB*, and *chiC* genes from *S. marcescens* Bizio in *E. coli* shows high chitinolytic activity against *Malacosoma neustria* (Linnaeus) (47–66%) and *H. armigera* (45–80%) (Danişmazoğlu et al., 2015). Furthermore, transferring chitinase-encoding genes (*chiA*, *chiB*) from strains of *S. marcescens* to Btk and Bti strains and anti-Coleopteran Bt strain 3023 yielded recombinants that express higher levels of chitinase along with the corresponding Bt toxins, showing greater insecticidal activity than the parental strains (Okay et al., 2008; Ozgen et al., 2013).

11.7 GENETIC IMPROVEMENT RELATED TO *PSEUDOMONAS*

Pseudomonas spp. are omnipotent bacteria having antagonistic activity against plant pathogens and also possess insecticidal activity. Genetic engineering has proven to enhance the activity of this bacteria against different insects. Downing et al. (2000) isolated a strain of *P. fluorescens* from sugarcane phylloplane and transformed it with the *Cry1Ac7* gene of Bt strain 234 either with or without *tac* promoter. The recombinants, under the control of the *tac* promoter, resulted in higher mortality in sugarcane bore, *Eldana saccharina* Walker, larvae. In addition, a combination of this recombinant *P. fluorescens* with *P. fluorescens* carrying the *chiA* gene from *S. marcescens*, under the con trol of the *tac* promoter, resulted in enhanced toxic effects on *E. saccharina* larvae as compared to individual expression of the toxins (Downing et al., 2000). The recombinant *Pseudomonas* sp. carrying gene encoding the Cry9Aa toxin from *Bt* subsp. *galleriae* showed effectiveness against *G. mellonella* and pine processionary moth, *Thaumetopoea pityocampa* (Denis and Schiffermüller) (Alberghini et al., 2005, 2006). Like *Photorhabdus*, some *Pseudomonas* spp., such as strains of insect-associated *P. entomophila* and unassociated *P. fluorescens*, and *P. syringae* pv. *tomato* also harbour insecticidal *Tc* genes (Liu et al., 2010; Vodovar et al., 2006). Heterologous expression of the *tccC* gene from a novel species, *P. taiwanensis*, in *E. coli* resulted in recombinants that caused 60% mortality in *D. melanogaster* within 72 hours (Liu et al., 2010).

11.8 CONCLUSION

About 125 years have elapsed since the discovery of Bt's entomopathogenic potential against silkworm. The ability of Bt to produce various toxins specific to different target insects makes it the most widely adopted bioinsecticide. Its host range has also widened, and its stability has also enhanced through genetic improvement. In addition, the development of transgenic Bt crops is something that changed the crop protection scenario. Currently, different Bt crops occupy a good area coverage and are accepted by many. However, most of the genetic engineering–related efforts have been made for Bt and its toxins, whereas the other entomopathogenic bacteria remain quite unexplored in this aspect. One reason for this might be that the Bt toxins can be used to control a broad range of agriculturally important as well as household pests, whereas the other EPBs usually have some limited uses. Apart from that, developing Bt-based products is also easy and cheap. For now, some research has already been conducted to enhance the activity of Ls, *Xenorhabdus*, *Photorhabdus*, *Serratia*, *Pseudomonas*, etc., through GE, but that is not enough. That is why scope is there to explore those unexplored ones and unleash their potential through genetic improvement.

REFERENCES

Abbas, M. S. T. (2018). Genetically engineered (modified) crops (*Bacillus thuringiensis* crops) and the world controversy on their safety. *Egyptian Journal of Biological Pest Control*, *28*(1), 52. https://doi.org/10.1186/s41938-018-0051-2

Abdel-Salam, M. S., Ameen, H. H., Soliman, G. M., Elkelany, U. S., & Asar, A. M. (2018). Improving the nematicidal potential of *Bacillus amyloliquefaciens* and *Lysinibacillus sphaericus* against the root-knot nematode *Meloidogyne incognita* using protoplast fusion technique. *Egyptian Journal of Biological Pest Control*, *28*(1), 31. https://doi.org/10.1186/s41938-018-0034-3

Ajaykumar, S. (2024). *Laboratory biosafety in India: In search of a sound regulatory framework* (Issue No. 705). Observer Research Foundation. https://www.orfonline.org/public/uploads/posts/pdf/20240408185232.pdf

Alberghini, S., Filippini, R., Marchetti, E., Dindo, M. L., Shevelev, A. B., Battisti, A., & Squartini, A. (2005). Construction of a *Pseudomonas* sp. derivative carrying the *cry9Aa* gene from *Bacillus thuringiensis* and a proposal for new standard criteria to assess entomocidal properties of bacteria. *Research in Microbiology*, *156*(5–6), 690–699. https://doi.org/10.1016/j.resmic.2005.02.003

Alberghini, S., Filippini, R., Shevelev, A. B., Squartini, A., & Battisti, A. (2006). Extended plant protection by an epiphytic *Pseudomonas* sp. derivative carrying the *cry9Aa* gene from *Bacillus thuringiensis galleriae* against the pine processionary moth *Thaumetopoea pityocampa*. *Biocontrol Science and Technology*, *16*(7), 709–715. https://doi.org/10.1080/09583150600699960

All, J. N., Stancil, J. D., Johnson, T. B., & Gouger, R. (1994). A genetically-modified *Bacillus thuringiensis* product effective for control of the fall armyworm (Lepidoptera: Noctuidae) on corn. *Florida Entomologist*, *77*(4), 437–440. https://doi.org/10.2307/3495697

Alvarez, A., & Loto, F. V. (2012). Characterization and biological activity of *Bacillus thuringiensis* isolates that are potentially useful in insect pest control. In G. A. Lameed (Ed.), *Biodiversity enrichment in a diverse world* (pp. 133–153). IntechOpen.

Amal, T. C., Karthika, P., Dhandapani, G., Selvakumar, S., & Vasanth, K. (2020). A simple and efficient *Agrobacterium*-mediated in planta transformation protocol for horse gram (*Macrotyloma uniflorum* Lam. Verdc.). *Journal of Genetic Engineering & Biotechnology*, *18*(1), 9. https://doi.org/10.1186/S43141-020-00023-Z

Angsuthanasombat, C., & Panyim, S. (1989). Biosynthesis of 130-kilodalton mosquito larvicide in the cyanobacterium *Agmenellum quadruplicatum* PR-6. *Applied and Environmental Microbiology*, *55*(9), 2428–2430. https://doi.org/10.1128/aem.55.9.2428–2430.1989

Armengol, G., Guevara, O. E., Orduz, S., & Crickmore, N. (2005). Expression of the *Bacillus thuringiensis* mosquitocidal toxin Cry11Aa in the aquatic bacterium *Asticcacaulis excentricus*. *Current Microbiology*, *51*, 430–433. https://doi.org/10.1007/s00284-005-0166-z

Arora, R., & Shera, P. S. (2014). Genetic improvement of biocontrol agents for sustainable pest management. In K. Sahayaraj (Ed.), *Basic and applied aspects of biopesticides* (pp. 255–285). Springer. https://doi.org/10.1007/978-81-322-1877-7_15

Arora, S., Chimata, M. K., & Ahuja, V. (2019). *Handbook for seed inspectors: Genetically modified seeds and regulations*. Ministry of Environment, Forest and Climate Change.

Atia, H. H., Kamal, M. I., El Sayed, A. I., El-Adl, A. M., & Zaied, K. A. (2019). Additive effect of chitinase genes in transconjugants of entomopathogens *Bacillus thuringiensis* to improve biological control of *Tuta absoluta*. *Journal of Agricultural Chemistry & Biotechnology*, *10*(8), 159–170. https://doi.org/10.21608/JACB.2019.57723

AUDA-NEPAD. (2020). *Process of developing GM crops*. African Union Development Agency. https://www.nepad.org/content/process-of-developing-gm-crops

Azizoglu, U., Jouzani, G. S., Yilmaz, N., Baz, E., & Ozkok, D. (2020). Genetically modified entomopathogenic bacteria, recent developments, benefits and impacts: A review. *Science of the Total Environment*, *734*, 139169. https://doi.org/10.1016/j.scitotenv.2020.139169

Bezdicek, D. F., Quinn, M. A., & Kahn, M. (1991). *Genetically engineering Bacillus thuringiensis genes into Rhizobia to control nodule feeding insects* (Report No. 9131200). USDA CRIS.

Bora, R. S., Murty, M. G., Shenbagarathai, R., & Sekar, V. (1994). Introduction of a lepidopteran-specific insecticidal crystal protein gene of *Bacillus thuringiensis* subsp. *kurstaki* by conjugal transfer into a *Bacillus megaterium* strain that persists in the cotton phyllosphere. *Applied and Environmental Microbiology*, *60*(1), 214–222. https://doi.org/10.1128/aem.60.1.214-222.1994

Boucias, D. G., & Pendland, J. C. (1998). *Principles of insect pathology*. Springer. https://doi.org/10.1016/0022-2011(67)90149-8.

Boussiba, S., Wu, X-Q., Ben-Dov, E., Zarka, A., & Zaritsky, A. (2000). Nitrogen-fixing cyanobacteria as gene delivery system for expressing mosquitocidal toxins of *Bacillus thuringiensis* ssp. *israelensis*. *Journal of Applied Phycology*, *12*, 461–467. https://doi.org/10.1023/A:1008114929490

Broadwell, A. H., Baumann, L., & Baumann, P. (1990). Larvicidal properties of the 42 and 51 kilodalton *Bacillus sphaericus* proteins expressed in different bacterial hosts: Evidence for a binary toxin. *Current Microbiology*, *21*, 361–366.

Cai, Y., Yan, J., Hu, X., Han, B., & Yuan, Z. (2007). Improving the insecticidal activity against resistant *Culex quinquefasciatus* mosquitoes by expression of chitinase gene *chiAC* in *Bacillus sphaericus*. *Applied and Environmental*

Microbiology, *73*(23), 7744–7746. https://doi.org/10.1128/AEM.01510-07

Carlton, B. C., & Gawron-Burke, C. (1993). Genetic improvement of *Bacillus thuringiensis* for biopesticide development. In L. Kim (Ed.), *Advanced engineered pesticides* (pp. 43–62). Marcel Dekker, Inc.

Crickmore, N., Bone, E. J., & Ellar, D. J. (1990). Genetic manipulation of *Bacillus thuringiensis*: Towards an improved pesticide. *Aspects of Applied Biology*, *24*, 17–24.

Danişmazoğlu, M., Demir, I., Sezen, K., Muratoğlu, H., & Nalçacioğlu, R. (2015). Cloning and expression of chitinase A, B, and C (*chiA*, *chiB*, *chiC*) genes from *Serratia marcescens* originating from *Helicoverpa armigera* and determining their activities. *Turkish Journal of Biology*, *39*(1), 78–87. https://doi.org/10.3906/biy-1404-31

Das, S., & Dash, H. R. (2015). Cloning and transformation. In S. Das, & H. R. Dash (Eds.), *Microbial biotechnology—a laboratory manual for bacterial systems* (pp. 35–72). Springer. https://doi.org/10.1007/978-81-322-2095-4_2

Deng, S. Q., Zou, W. H., Li, D. L., Chen, J. T., Huang, Q., Zhou, L. J., Tian, X. X., Chen, Y. J., & Peng, H. J. (2019). Expression of *Bacillus thuringiensis* toxin Cyt2Ba in the entomopathogenic fungus *Beauveria bassiana* increases its virulence towards *Aedes* mosquitoes. *PLoS Neglected Tropical Diseases*, *13*(7), e0007590. https://doi.org/10.1371/journal.pntd.0007590

Dohm, A. (2003). Biotechnologies for breeding | Genetic transformation. In A. V. Roberts (Ed.), *Encyclopedia of rose science* (pp. 15–25). Academic Press. https://doi.org/10.1016/B0-12-227620-5/00008-2

Downing, K. J., Leslie, G., & Thomson, J. A. (2000). Biocontrol of the sugarcane borer *Eldana saccharina* by expression of the *Bacillus thuringiensis cry1Ac7* and *Serratia marcescens chiA* genes in sugarcane-associated bacteria. *Applied and Environmental Microbiology*, *66*(7), 2804–2810. https://doi.org/10.1128/AEM.66.7.2804-2810.2000

Driss, F., Rouis, S., Azzouz, H., Tounsi, S., Zouari, N., & Jaoua, S. (2011). Integration of a recombinant chitinase into *Bacillus thuringiensis* parasporal insecticidal crystal. *Current Microbiology*, *62*, 281–288. https://doi.org/10.1007/s00284-010-9704-4

Duchaud, E., Rusniok, C., Frangeul, L., Buchrieser, C., Givaudan, A., Taourit, S., Bocs, S., Boursaux-Eude, C., Chandler, M., Charles, J. F., Dassa, E., Derose, R., Derzelle, S., Freyssinet, G., Gaudriault, S., Médigue, C., Lanois, A., Powell, K., Siguier, P., Vincent, R., . . . Kunst, F. (2003). The genome sequence of the entomopathogenic bacterium *Photorhabdus luminescens*. *Nature Biotechnology*, *21*(11), 1307–1313. https://doi.org/10.1038/nbt886

Durmaz, E., Hu, Y., Aroian, R. V., & Klaenhammer, T. R. (2016). Intracellular and extracellular expression of *Bacillus thuringiensis* crystal protein Cry5B in *Lactococcus lactis* for use as an anthelminthic. *Applied and Environmental Microbiology*, *82*(4), 1286–1294. https://doi.org/10.1128/AEM.02365-15

EFSA (European Food Safety Authority). (2013). Conclusion on the peer review of the pesticide risk assessment of the active substance *Bacillus thuringiensis* ssp. *tenebrionis* strain NB-176. *EFSA Journal*, *11*(1), 3024–3059. https://doi.org/10.2903/j.efsa.2013.3024

Gawron-Burke, C., & Baum, J. A. (1991). Genetic manipulation of *Bacillus thuringiensis* insecticidal crystal protein genes in bacteria. In J. K. Setlow (Ed.), *Genetic engineering: Principles and methods* (pp. 237–263). Springer. https://doi.org/10.1007/978-1-4615-3760-1_11

Gelvin, S. B. (2003). *Agrobacterium*-mediated plant transformation: The biology behind the "gene-jockeying" tool. *Microbiology and Molecular Biology Reviews*, *67*(1), 16–37. https://doi.org/10.1128/mmbr.67.1.16-37.2003

Glare, T. R., Corbett, G. E., & Sadler, T. J. (1993). Association of a large plasmid with amber disease of the New Zealand grass grub, *Costelytra zealandica*, caused by *Serratia entomophila* and *Serratia proteamaculans*. *Journal of Invertebrate Pathology*, *62*(2), 165–170. https://doi.org/10.1006/jipa.1993.1091

González-Ponce, K. S., Casados-Vázquez, L. E., Lozano-Sotomayor, P., Bideshi, D. K., Del Rincón-Castro, M. C., & Barboza-Corona, J. E. (2019). Expression of ChiA74Δ sp and its truncated versions in *Bacillus thuringiensis* HD1 using a vegetative promoter maintains the integrity and toxicity of native Cry1A toxins. *International Journal of Biological Macromolecules*, *124*, 80–87. https://doi.org/10.1016/j.ijbiomac.2018.11.173

Grooms, K. (2019, June 21). *Selecting the right colony: The answer is there in blue and white*. Promega Connections. https://www.promegaconnections.com/selecting-the-right-colony-the-answer-is-there-in-blue-and-white/

Harrison, R. L., & Bonning, B. C. (2000). Genetic engineering of biocontrol agents for insects. In J. E. Rechcigl, & N. A. Rechcigl (Eds.), *Biological and biotechnological control of pests* (pp. 243–280). Lewis Publishers.

Heinemann, D., Zabic, M., Terakawa, M., & Boch, J. (2022). Laser-based molecular delivery and its applications in plant science. *Plant Methods*, *18*, 82. https://doi.org/10.1186/S13007-022-00908-9

Hu, S. B., Liu, P., Ding, X. Z., Yan, L., Sun, Y. J., Zhang, Y. M., Li, W. P., & Xia, L. Q. (2009). Efficient constitutive expression of chitinase in the mother cell of *Bacillus thuringiensis* and its potential to enhance the toxicity of Cry1Ac protoxin. *Applied Microbiology and Biotechnology*, *82*, 1157–1167. https://doi.org/10.1007/s00253-009-1910-2

Hu, X., Hansen, B. M., Yuan, Z., Johansen, J. E., Eilenberg, J., Hendriksen, N. B., Smidt, L., & Jensen, G. B. (2005). Transfer and expression of the mosquitocidal plasmid pBtoxis in *Bacillus cereus* group strains. *FEMS Microbiology Letters*, *245*(2), 239–247. https://doi.org/10.1016/j.femsle.2005.03.012

Hurst, M. R., Glare, T. R., & Jackson, T. A. (2004). Cloning *Serratia entomophila* antifeeding genes—a putative defective prophage active against the grass grub *Costelytra zealandica*. *Journal of Bacteriology*, *186*(15), 5116–5128. https://doi.org/10.1128/jb.186.15.5116-5128.2004

Hurst, M. R., Glare, T. R., Jackson, T. A., & Ronson, C. W. (2000). Plasmid-located pathogenicity determinants of *Serratia entomophila*, the causal agent of amber disease of grass grub, show similarity to the insecticidal toxins of *Photorhabdus luminescens*. *Journal of Bacteriology*, *182*(18), 5127–5138. https://doi.org/10.1128/jb.182.18.5127-5138.2000

Hwang, H.-H., Yu, M., & Lai, E.-M. (2017). *Agrobacterium*-mediated plant transformation: Biology and applications. *The Arabidopsis Book*, *15*(15), e0186. https://doi.org/10.1199/tab.0186

Ibrahim, N. A. G. A., Hussien, A. I., Hatem, A. E. S., Aldebis, H. K., & Vargas-Osuna, E. (2014). Persistence of the transformed *Paenibacillus polymyxa* expressing CRY1C in the plant leaves and its effect on chlorophyll and carotenoid. *Life Science Journal*, *11*, 433–442.

Jayakumar, S., & Kaur, S. (2013). Occurrence of *cry* genes in *Bacillus thuringiensis* (Bt) isolates recovered from phylloplanes of crops growing in the New Delhi region of India and toxicity towards diamond-back moth (*Plutella xylostella*). *Journal of Biological Sciences*, *13*(6), 463–473. https://doi.org/10.3923/JBS.2013.463.473.

Jayaraj, S. (1985). *Microbial control and pest management*. Tamil Nadu Agricultural University.

Karabörklü, S., Azizoglu, U., & Azizoglu, Z. B. (2018). Recombinant entomopathogenic agents: A review of biotechnological approaches to pest insect control. *World Journal of Microbiology and Biotechnology*, *34*, 14. https://doi.org/10.1007/s11274-017-2397-0

Kaur, R. P., & Devi, S. (2019). In planta transformation in plants: A review. *Agricultural Reviews*, *40*(3), 159–174. https://doi.org/10.18805/ag.r-1597

Khan, K. H. (2009). Gene transfer technologies in plants: Roles in improving crops. *Recent Research in Science and Technology*, *1*(3), 116–123.

Khandelwal, P., Bhatnagar, R., Choudhury, D., & Banerjee, N. (2004a). Characterization of a cytotoxic pilin subunit of *Xenorhabdus nematophila*. *Biochemical and Biophysical Research Communications*, *314*(4), 943–949. https://doi.org/10.1016/j.bbrc.2003.12.187

Khandelwal, P., Choudhury, D., Birah, A., Reddy, M. K., Gupta, G. P., & Banerjee, N. (2004b). Insecticidal pilin subunit from the insect pathogen *Xenorhabdus nematophila*. *Journal of Bacteriology*, *186*(19), 6465–6476. https://doi.org/10.1128/jb.186.19.6465-6476.2004

Lalani, J., & Misra, A. (2011). Gene delivery using chemical methods. In A. Mishra (Ed.), *Challenges in delivery of therapeutic genomics and proteomics* (pp. 127–206). Elsevier. https://doi.org/10.1016/B978-0-12-384964-9.00004-9

Lampel, J. S., Canter, G. L., Dimock, M. B., Kelly, J. L., Anderson, J. J., Uratani, B. B., Foulke, J. S. Jr., Turner, J. T., & Turner, J. T. (1994). Integrative cloning, expression, and stability of the *cryIA(c)* gene from *Bacillus thuringiensis* subsp. *kurstaki* in a recombinant strain of *Clavibacter xyli* subsp. *cynodontis*. *Applied and Environmental Microbiology*, *60*(2), 501–508. https://doi.org/10.1128/aem.60.2.501-508.1994

Lee, M. K., Young, B. K., & Dean, D. H. (1995). Domain III exchanges of *Bacillus thuringiensis* CryIA toxins affect binding to different gypsy moth midgut receptors. *Biochemical and Biophysical Research Communications*, *216*(1), 306–312. https://doi.org/10.1006/BBRC.1995.2625

Lee, P. J., Ahn, J. Y., Kim, Y. H., Kim, S. W., Kim, J. Y., Park, J. S., & Lee, J. (2004). Cloning and heterologous expression of a novel insecticidal gene (*tccC1*) from *Xenorhabdus nematophilus* strain. *Biochemical and Biophysical Research Communications*, *319*(4), 1110–1116. https://doi.org/10.1016/j.bbrc.2004.04.203

Leong, K. L. H., Cano, R. J., & Kubinski, A. M. (1980). Factors affecting *Bacillus thuringiensis* total field persistence. *Environmental Entomology*, *9*(5), 593–599. https://doi.org/10.1093/ee/9.5.593

Lereclus, D., Vallade, M., Chaufaux, J., Arantes, O., & Ramband, S. (1992). Expansion of insecticidal host range of *Bacillus thuringiensis* by in vivo genetic recombination. *Biotechnology*, *10*, 418–421. https://doi.org/10.1038/nbt0492-418

Lertcanawanichakul, M., Wiwat, C., Bhumiratana, A., & Dean, D. H. (2004). Expression of chitinase-encoding genes in *Bacillus thuringiensis* and toxicity of engineered *B. thuringiensis* subsp. *aizawai* toward *Lymantria dispar* larvae. *Current Microbiology*, *48*, 175–181. https://doi.org/10.1007/s00284-003-4119-0

Li, L., & Yu, Z. (2012). Genetically modified *Bacillus thuringiensis* biopesticides. In E. Sansinenea (Ed.), *Bacillus thuringiensis biotechnology* (pp. 231–258). Springer. https://doi.org/10.1007/978-94-007-3021-2_13

Li, Y., Wu, C., Xing, Z., Gao, B., & Zhang, L. (2017). Engineering the bacterial endophyte *Burkholderia pyrrocinia* JK-SH007 for the control of lepidoptera larvae by introducing the *cry218* genes of *Bacillus thuringiensis*. *Biotechnology & Biotechnological Equipment*, *31*(6), 1167–1172. https://doi.org/10.1080/13102818.2017.1379361

Liu, B., & Sengonca, C. (2003). Conjugation of δ-endotoxin from *Bacillus thuringiensis* with abamectin of *Streptomyces avermitilis* as a new type of biocide, GCSC-BtA, for control of agricultural insect pests. *Journal of Pest Science*, *76*, 44–49. https://doi.org/10.1046/j.1439-0280.2003.03010.x

Liu, D., Burton, S., Glancy, T., Li, Z. S., Hampton, R., Meade, T., & Merlo, D. J. (2003). Insect resistance conferred by 283-kDa *Photorhabdus luminescens* protein TcdA in *Arabidopsis thaliana*. *Nature Biotechnology*, *21*(10), 1222–1228. https://doi.org/10.1038/nbt866

Liu, J. R., Lin, Y. D., Chang, S. T., Zeng, Y. F., & Wang, S. L. (2010). Molecular cloning and characterization of an insecticidal toxin from *Pseudomonas taiwanensis*. *Journal of Agricultural and Food Chemistry*, *58*(23), 12343–12349. https://doi.org/10.1021/jf103604r

Lluisma, A., Karmacharya, N., Zarka, A., Ben-Dov, E., Zaritsky, A., & Boussiba, S. (2001). Suitability of *Anabaena* PCC7120 expressing mosquitocidal toxin genes from *Bacillus thuringiensis* subsp. *israelensis* for biotechnological application. *Applied Microbiology and Biotechnology*, *57*, 161–166. https://doi.org/10.1007/s002530100776

Luxananil, P., Tanapongpipat, S., Promdonkoy, B., Atomi, H., Imanaka, T., & Panyim, S. (2003). Expression of binary toxin genes in the mosquito-colonizable bacteria, *Bacillus cereus*, leads to high toxicity against *Culex quinquefasciatus* larvae. *Current Microbiology*, *47*, 372–375. https://doi.org/10.1007/s00284-002-4026-9

Manimaran, P., Ramkumar, G., Sakthivel, K., Sundaram, R. M., Madhav, M. S., & Balachandran, S. M. (2011). Suitability of non-lethal marker and marker-free systems for development of transgenic crop plants: Present status and future prospects. *Biotechnology Advances*, *29*(6), 703–714. https://doi.org/10.1016/J.BIOTECHADV.2011.05.019

Mettus, A. M., & Macaluso, A. (1990). Expression of *Bacillus thuringiensis* delta-endotoxin genes during vegetative growth. *Applied and Environmental Microbiology*, *56*(4), 1128–1134. https://doi.org/10.1128/aem.56.4.1128-1134.1990

Mohamed, H. A. L. A., Abd El-Fatah, M. R., Sabbour, M. M., El-Sharkawey, A. Z., & El-sayed, R. S. (2010). Genetic modification of *Bacillus thuringiensis* var. *kurstaki* hd-73 to overproduce melanin, UV resistance and their insecticidal potentiality against potato tuber moth. *International Journal of Academic Research*, *2*(6), 73–83.

Morgan, J. A. W., Sergeant, M., Ellis, D., Ousley, M., & Jarrett, P. (2001). Sequence analysis of insecticidal genes from *Xenorhabdus nematophilus* PMFI296. *Applied and Environmental Microbiology*, *67*(5), 2062–2069. https://doi.org/10.1128/AEM.67.5.2062-2069.2001

Murphy, R. C., & Stevens Jr., S. E. (1992). Cloning and expression of the *cryIVD* gene of *Bacillus thuringiensis* subsp. *israelensis* in the cyanobacterium *Agmenellum quadruplicatum* PR-6 and its resulting larvicidal activity. *Applied and Environmental Microbiology*, *58*(5), 1650–1655. https://doi.org/10.1128/aem.58.5.1650-1655.1992

Nambiar, P. T. C., Ma, S. W., & Iyer, V. N. (1990). Limiting an insect infestation of nitrogen-fixing root nodules of the pigeon pea (*Cajanus cajan*) by engineering the expression of an entomocidal gene in its root nodules. *Applied and Environmental Microbiology*, *56*(9), 2866–2869. https://doi.org/10.1128/aem.56.9.2866-2869.1990

Narayanan, E. S. (2002). Microbial control of insect pests: Role of genetic engineering and tissue culture. In O. Koul, & G. S. Dhaliwal (Eds.), *Microbial biopesticides* (pp. 117–179). Taylor & Francis.

Nicolas, L., Nielsen-Leroux, C., Charles, J. F., & Delécluse, A. (1993). Respective role of the 42-and 51-kDa components of the *Bacillus sphaericus* toxin overexpressed in *Bacillus thuringiensis*. *FEMS Microbiology Letters*, *106*(3), 275–279. https://doi.org/10.1111/j.1574-6968.1993.tb05976.x

Okay, S., Tefon, B. E., Özkan, M., & Özcengiz, G. (2008). Expression of chitinase A (*chiA*) gene from a local isolate of *Serratia marcescens* in Coleoptera-specific *Bacillus thuringiensis*. *Journal of Applied Microbiology*, *104*(1), 161–170. https://doi.org/10.1111/j.1365-2672.2007.03570.x

Ozgen, A., Sezen, K., Demir, I., Demirbag, Z., & Nalcacioglu, R. (2013). Molecular characterization of chitinase genes from a local isolate of *Serratia marcescens* and their contribution to the insecticidal activity of *Bacillus thuringiensis* strains. *Current Microbiology*, *67*, 499–504. https://doi.org/10.1007/s00284-013-0395-5

Pandey, H. K., & Warrier, R. (2015). *Regulatory framework for GE plants in India*. Ministry of Environment, Forest and Climate Change.

Park, H. W., Bideshi, D. K., Wirth, M. C., Johnson, J. J., Walton, W. E., & Federici, B. A. (2005). Recombinant larvicidal bacteria with markedly improved efficacy against *Culex* vectors of West Nile virus. *American Journal of Tropical Medicine and Hygiene*, *72*(6), 732–738.

Poncet, S., Bernard, C., Dervyn, E., Cayley, J., Klier, A., & Rapoport, G. (1997). Improvement of *Bacillus sphaericus* toxicity against dipteran larvae by integration, via homologous recombination, of the Cry11A toxin gene from *Bacillus thuringiensis* subsp. *israelensis*. *Applied and Environmental Microbiology*, *63*(11), 4413–4420. https://doi.org/10.1128/AEM.63.11.4413-4420.1997.

Pusztai, M., Fast, P., Gringorten, L., Kaplan, H., Lessard, T., & Carey, P. R. (1991). The mechanism of sunlight-mediated inactivation of *Bacillus thuringiensis* crystals. *Biochemical Journal*, *273*(1), 43–47. https://doi.org/10.1042/bj2730043

Roh, J. Y., Liu, Q., Choi, J. Y., Wang, Y., Shim, H. J., Xu, H. G., Choi, G. J., Kim, J. C., & Je, Y. H. (2009). Construction of a recombinant *Bacillus velezensis* strain as an integrated control agent against plant diseases and insect pests. *Journal of Microbiology and Biotechnology*, *19*(10), 1223–1229. https://doi.org/10.4014/jmb.0902.065

Romero, M., Gil, F. M., & Orduz, S. (2001). Expression of mosquito active toxin genes by a Colombian native strain of the gram-negative bacterium *Asticcacaulis excentricus*. *Memórias do Instituto Oswaldo Cruz*, *96*, 257–263. https://doi.org/10.1590/S0074-02762001000200021

Ruiz-Masó, J. A., Machón, C., Bordanaba-Ruiseco, L., Espinosa, M., Coll, M., & del Solar, G. (2015). Plasmid rolling-circle replication. In M. E. Tolmasky, & J. C. Alonso (Eds.), *Plasmids: Biology and impact in biotechnology and discovery* (pp. 45–69). Wiley. https://doi.org/10.1128/9781555818982.ch4

Sanchis, V. (2012). Genetic improvement of Bt strains and development of novel biopesticides. In E. Sansinenea (Ed.), *Bacillus thuringiensis biotechnology* (pp. 215–228). Springer. https://doi.org/10.1007/978-94-007-3021-2_12

Sanchis, V., Gohar, M., Chaufaux, J., Arantes, O., Meier, A., Agaisse, H., Cayley, J., & Lereclus, D. (1999). Development and field performance of a broad-spectrum nonviable asporogenic recombinant strain of *Bacillus thuringiensis* with greater potency and UV resistance. *Applied and Environmental Microbiology*, *65*(9), 4032–4039. https://doi.org/10.1128/AEM.65.9.4032-4039.1999

Sansinenea, E. (2012). Discovery and description of *Bacillus thuringiensis*. In E. Sansinenea (Ed.), *Bacillus thuringiensis biotechnology* (pp. 3–18). Springer. https://doi.org/10.1007/978-94-007-3021-2_1

Sansinenea, E., & Ortiz, A. (2015). Melanin: A photoprotection for *Bacillus thuringiensis* based biopesticides. *Biotechnology Letters*, *37*, 483–490. https://doi.org/10.1007/s10529-014-1726-8

Sansinenea, E., Vázquez, C., & Ortiz, A. (2010). Genetic manipulation in *Bacillus thuringiensis* for strain improvement. *Biotechnology Letters*, *32*, 1549–1557. https://doi.org/10.1007/S10529-010-0338-1

Santhy, V., Khadi, B. M., Singh, P., Vijaya Kumari, P. R., Deshmukh, R. K., & Vishwanathan, A. (2008). *Hybrid seed production in cotton*. Central Institute for Cotton Research.

Saxena, D., Ben-Dov, E., Manasherob, R., Barak, Z. E., Boussiba, S., & Zaritsky, A. (2002). A UV tolerant mutant of *Bacillus thuringiensis* subsp. *kurstaki* producing melanin. *Current Microbiology*, *44*, 25–30. https://doi.org/10.1007/s00284-001-0069-6

Sergeant, M., Jarrett, P., Ousley, M., & Morgan, J. A. W. (2003). Interactions of insecticidal toxin gene products from *Xenorhabdus nematophilus* PMFI296. *Applied and Environmental Microbiology*, *69*(6), 3344–3349. https://doi.org/10.1128/AEM.69.6.3344-3349.2003

Servant, P., Rosso, M. L., Hamon, S., Poncet, S., Delécluse, A., & Rapoport, G. (1999). Production of Cry11A and Cry11Ba toxins in *Bacillus sphaericus* confers toxicity towards *Aedes aegypti* and resistant *Culex* populations. *Applied and Environmental Microbiology*, *65*(7), 3021–3026. https://doi.org/10.1128/AEM.65.7.3021-3026.1999

Su, W., Xu, M., Radani, Y., & Yang, L. (2023). Technological development and application of plant genetic transformation. *International Journal of Molecular Sciences*, *24*(13), 10646. https://doi.org/10.3390/IJMS241310646

Tanapongpipat, S., Luxananil, P., Promdonkoy, B., Chewawiwat, N., Audtho, M., & Panyim, S. (2003). A plasmid encoding a combination of mosquito-larvicidal genes from *Bacillus thuringiensis* subsp. *israelensis* and *Bacillus sphaericus* confers toxicity against a broad range of mosquito larvae when expressed in gram-negative bacteria. *FEMS Microbiology Letters*, *228*(2), 259–263. https://doi.org/10.1016/S0378-1097(03)00780–8

Thanabalu, T., Hindley, J., & Berry, C. (1992b). Proteolytic processing of the mosquitocidal toxin from *Bacillus sphaericus* SSII-1. *Journal of Bacteriology*, *174*(15), 5051–5056. https://doi.org/10.1128/jb.174.15.5051-5056.1992

Thanabalu, T., Hindley, J., Brenner, S., Oei, C., & Berry, C. (1992a). Expression of the mosquitocidal toxins of *Bacillus sphaericus* and *Bacillus thuringiensis* subsp. *israelensis* by recombinant *Caulobacter crescentus*, a vehicle for biological control of aquatic insect larvae. *Applied and Environmental Microbiology*, *58*(3), 905–910. https://doi.org/10.1128/AEM.58.3.905-910.1992.

Thanabalu, T., Hindley, J., Jackson-Yap, J., & Berry, C. (1991). Cloning, sequencing, and expression of a gene encoding a 100-kilodalton mosquitocidal toxin from *Bacillus sphaericus* SSII-1. *Journal of Bacteriology*, *173*(9), 2776–2785. https://doi.org/10.1128/jb.173.9.2776-2785.1991

Thanabalu, T., & Porter, A. G. (1995). Efficient expression of a 100-kilodalton mosquitocidal toxin in protease-deficient recombinant *Bacilus sphaericus*. *Applied and Environmental Microbiology*, *61*(11), 4031–4036. https://doi.org/10.1128/aem.61.11.4031-4036.1995

Theoduloz, C., Vega, A., Salazar, M., González, E., & Meza-Basso, L. (2003). Expression of a *Bacillus thuringiensis* δ-endotoxin *cry1Ab* gene in *Bacillus subtilis* and *Bacillus licheniformis* strains that naturally colonize the phylloplane of tomato plants (*Lycopersicon esculentum*, Mills). *Journal of Applied Microbiology*, *94*(3), 375–381. https://doi.org/10.1046/j.1365-2672.2003.01840.x

Vaeck, M., Reynaerts, A., Hofte, H., Jansens, S., DeBeuckleer, M., Dean, C., Zabeau, M., Van Montagu, M., & Leemans, J. (1987). Transgenic plants protected from insect attack. *Nature*, *327*, 33–37. https://doi.org/10.1038/328033a0

Vodovar, N., Vallenet, D., Cruveiller, S., Rouy, Z., Barbe, V., Acosta, C., Cattolico, L., Jubin, C., Lajus, A., Segurens, B., Vacherie, B., Wincker, P., Weissenbach, J., Lemaitre, B., Me´digue, C., & Boccard, F. (2006). Complete genome sequence of the entomopathogenic and metabolically versatile soil bacterium *Pseudomonas entomophila*. *Nature Biotechnology*, *24*(6), 673–679. https://doi.org/10.1038/nbt1212

Waterfield, N., Dowling, A., Sharma, S., Daborn, P. J., Potter, U., & Ffrench-Constant, R. H. (2001). Oral toxicity of *Photorhabdus luminescens* W14 toxin complexes in *Escherichia coli*. *Applied and Environmental Microbiology*, *67*(11), 5017–5024. https://doi.org/10.1128/AEM.67.11.5017-5024.2001

Waterfield, N., George Kamita, S., Hammock, B. D., & Ffrench-Constant, R. (2005b). The *Photorhabdus* Pir toxins are similar to a developmentally regulated insect protein but show no juvenile hormone esterase activity. *FEMS Microbiology Letters*, *245*(1), 47–52. https://doi.org/10.1016/j.femsle.2005.02.018

Waterfield, N., Hares, M., Yang, G., Dowling, A., & Ffrench-Constant, R. (2005a). Potentiation and cellular phenotypes of the insecticidal toxin complexes of *Photorhabdus* bacteria. *Cellular Microbiology*, *7*(3), 373–382. https://doi.org/10.1111/j.1462-5822.2004.00467.x

Wiwat, C., Thaithanun, S., Pantuwatana, S., & Bhumiratana, A. (2000). Toxicity of chitinase-producing *Bacillus thuringiensis* ssp. *kurstaki* HD-1 (G) toward *Plutella xylostella*. *Journal of Invertebrate Pathology*, *76*(4), 270–277. https://doi.org/10.1006/jipa.2000.4976

Xu, X., Yan, G., Kong, R., Liu, X., & Yu, L. (2000). Analysis of expression of the binary toxin genes from *Bacillus sphaericus* in *Anabaena* and the potential in mosquito control. *Current Microbiology*, *41*, 352–356. https://doi.org/10.1007/s002840010148

Xudong, X., Renqiu, K., & Yuxiang, H. (1993). High larvicidal activity of intact recombinant cyanobacterium *Anabaena* sp. PCC 7120 expressing Gene 51 and Gene 42 of *Bacillus sphaericus* sp. 2297. *FEMS Microbiology Letters*, *107*(2–3), 247–250. https://doi.org/10.1111/j.1574-6968.1996.tb08543.x

Yuan, Z. (2002). *Bacillus sphaericus*: Mechanism and application as a mosquito larvicide. In R. K. Upadhyay (Ed.), *Advances in microbial control of insect pests* (pp. 41–69). Springer. https://doi.org/10.1007/978-1-4757-4437-8_3

Zhang, B., Liu, M., Yang, Y., & Yuan, Z. (2006). Cytolytic toxin Cyt1Aa of *Bacillus thuringiensis* synergizes the mosquitocidal toxin Mtx1 of *Bacillus sphaericus*. *Bioscience, Biotechnology, and Biochemistry*, *70*(9), 2199–2204. https://doi.org/10.1271/bbb.60140

Zhang, J. T., Yan, J. P., Zheng, D. S., Sun, Y. J., & Yuan, Z. M. (2008). Expression of *mel* gene improves the UV resistance of *Bacillus thuringiensis*. *Journal of Applied Microbiology*, *105*(1), 151–157. https://doi.org/10.1111/j.1365-2672.2008.03729.x

Zhao, R., Han, R., Qiu, X., Yan, X., Cao, L., & Liu, X. (2008). Cloning and heterologous expression of insecticidal-protein-encoding genes from *Photorhabdus luminescens* TT01 in *Enterobacter cloacae* for termite control. *Applied and Environmental Microbiology*, *74*(23), 7219–7226. https://doi.org/10.1128/AEM.00977-08

Zheng, D., Valdez-Cruz, N. A., Armengol, G., Sevrez, C., Munoz-Olaya, J. M., Yuan, Z., Orduz, S., & Crickmore, N. (2007). Co-expression of the mosquitocidal toxins Cyt1Aa and Cry11Aa from *Bacillus thuringiensis* subsp. *israelensis* in *Asticcacaulis excentricus*. *Current Microbiology*, *54*, 58–62. https://doi.org/10.1007/s00284-006-0352-7

Zhu, Y., Liu, B., & Sengonca, C. (2006). Efficiency of GCSC-BtA, as a new type of biocide, on different agricultural arthropod pests and its side-effect on some predators. *Mitteilungen der Deutsche Gesellschaft fur Allgemeine und Angewandte Entomologie*, *15*, 309–313.

Zia, M., Arshad, W., Bibi, Y., Nisa, S., & Chaudhary, M. F. (2011). Does Agro-injection to soybean pods transform embryos? *Plant OMICS*, *4*(7), 384–390.

12 An Overview of Bt Crops and Their Safety

ABSTRACT

When it comes to crop protection, Bt crops are an integral part of modern day pest management programmes. Since 1987, with the first successful insertion of Bt genes into commercial tobacco cultivars to generate transgenic plants with greater resistance to insect herbivores, this Bt technology has made remarkable progress in the last few decades. Researchers have already developed several transgenic Bt hybrids for maize, cotton, rice, potato, soybean, brinjal, and other crops, demonstrating a high level of resistance against multiple insect pests. Progress has been made by developing Bt hybrids that can express multiple Bt genes for higher effectiveness and can tackle insect resistance issues. However, there is a lot of controversy surrounding Bt crops. Commercialization of these Bt crops confronts greater challenges due to public concerns about the safety of this crop technology on the environment, humans, and other non-target organisms, along with different political, social, and industrial issues. Although, under certain circumstances, these Bt crops may show some negative effects, the consequences will undoubtedly be far less severe than those caused by the use of chemical pesticides.

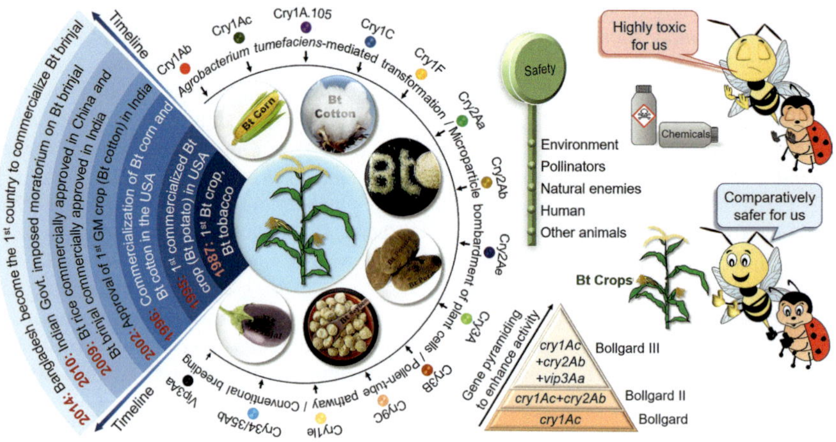

12.1 INTRODUCTION

In our previous chapters, we have discussed how successful the entomopathogenic bacteria (EPB) are in controlling different insect pests. Undoubtedly, among different EPBs, the foremost runner is the *Bacillus thuringiensis* Berliner (Bt). Since the discovery of its insecticidal potential, Bt has been extensively studied and widely used as a successful bioinsecticide for many years. As the science makes progress, the field of Bt also gets expanded. Through genetic engineering (GE), the host range of Bt, its environmental stability, the efficacy on the target host, etc. have been improved (Arora & Shera, 2014; Carlton & Gawron-Burke, 1993). However, this Bt technology took momentum after the development of Bt crops. Over the last few decades, the field of recombinant DNA technology has made remarkable progress by developing crops incorporated with novel genes to enhance their yield attributes, stability, and protection from different stressors. Similarly, progress was made by developing insect-resistant transgenic crops that were mainly based on the integration of toxic protein-encoded bacterial genes, especially from Bt. Through the successful introduction of toxin genes from different Bt subspecies, a wide range of Bt crops have been developed. In the previous chapter (Chapter 11), we have briefly discussed these genetic improvement–related approaches. Expression of specific toxin genes by the Bt crops confers high levels of resistance against target pests. To date, these Bt crops are in use throughout the world, protecting the agricultural yield from a wide range of insect pests. This field of Bt crops is still going through a lot of genetic advancement, developing new transgenic lines with enhanced insect-resistant ability. In this present chapter, we have focused on the following two aspects: (1) an overview of the major Bt crops against different insect pests, and (2) the safety status of these Bt crops.

DOI: 10.1201/9781003426882-12

12.2 GLOBAL STATUS OF GENETICALLY MODIFIED (GM) CROPS

Although the first ever genetically modified (GM) crop (Flavr Savr™ tomato by Calgene Company in the USA) was commercialized in 1994 (James & Krattiger, 1996), cultivation of GM crops only took momentum in 1996 (James, 1998). At that time, the total global area under transgenic crops was 1.70 mi ha (million hectares) (James, 1998), which was later increased to 185.60 mi ha in 2020 (Brookes, 2022), indicating an unprecedented 110-fold increase in its area coverage. According to a report, in 2023, the total area under GM crops reached 206.30 mi ha with 27 GM crop–cultivating countries (Agbioinvestor, 2024). This makes biotech crops the fastest-adopted crop technology in the history of modern agriculture. In 2020, due to GM crops, direct global farm income was benefitted with $18.80 billion, and since 1996, farm revenue was also increased by $261.30 billion (Brookes, 2022). The USA has the largest area under transgenic crops, which is near about 36% of the global transgenic crop area, followed by Brazil (30%) (Brookes, 2022). These two countries, along with Argentina, Canada, and India, cover 91% of the total biotech crop area (Brookes, 2022; Dionglay, 2021). Among different biotech crops, GM soybean has the largest share (50%), followed by maize (33%), cotton (12%), and canola (5%) (Brookes, 2022). Among the Asia-Pacific countries, India has the largest area (12.1 mi ha) under GM crops, followed by China (2.8 mi ha) and Pakistan (2.3 mi ha), and in most of this area, only GM cotton has grown (Agbioinvestor, 2024).

12.3 A BRIEF OVERVIEW ON BT CROPS

Coming to Bt crops, the first Bt crop that was approved for commercial use by the United States Environmental Protection Agency (US EPA) in the USA in 1995 was Bt potato, NewLeaf® by Monsanto (NatureMark) Company (Shelton et al., 2002). Thereafter, huge progress was made in the field of Bt crops, and different Bt crops gained much popularity among growers. However, the area coverage for herbicide-tolerant (HT) GM crops (60%) is more than insect-resistant (IR) GM crops (40%) (Brookes, 2022). All this progress was initiated during the early 1980s, as the genes encoding δ-endotoxins were successfully cloned. In 1987, the first ever Bt transgenic crop was developed by the Plant Genetic Systems biotech company (now part of Bayer CropScience) from Belgium through the introduction of a toxin gene, *bt2*, from Bt strain Berliner 1715 into the tobacco plant to control the tobacco hornworm, *Manduca sexta* (Linnaeus) (Vaeck et al., 1987). Another attempt was made by Barton et al. (1987) as they transformed the tobacco cv. Havana 425 with the toxin gene from *Bt* var. *kurstaki* (Btk) HD-1 and the transformants showed resistance to *M. sexta*. In the same year, Fischhoff et al. (1987) developed transgenic tomato by incorporating truncated genes from Btk HD-1 to confer tolerance to lepidopteran pests. These were some of the earliest studies on the development of Bt crops. Since then, Bt genes have been incorporated into a wide range of crop species and commercialized worldwide. In what follows, we discuss some of the successful transgenic Bt crops in brief. In Figure 12.1, we have presented

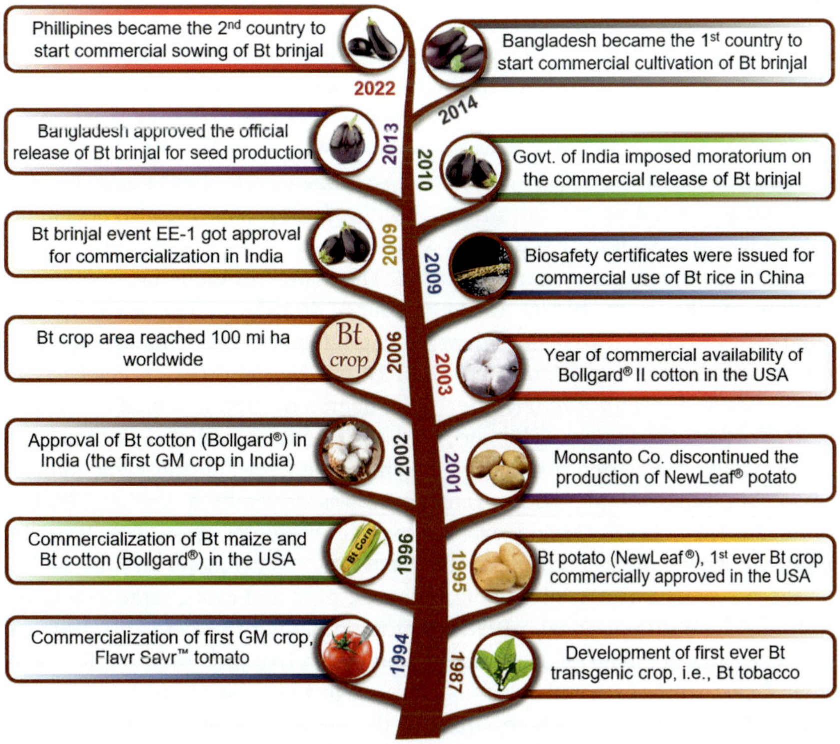

FIGURE 12.1 Timeline on some of the historic events related to the development and commercialization of Bt crops.

a brief timeline on some of the important historical events related to the development and commercialization of transgenic Bt crops.

12.3.1 Maize

The first ever Bt maize was developed in 1991 by Barnason et al. (1991), containing the *Cry1A* gene, and was highly efficient in controlling European corn borer (ECB), *Ostrinia nubilalis* (Hübner). Later on, Koziel et al. (1993) developed transgenic maize by transferring truncated *Cry1A(b)* gene from Btk HD-1 using the cauliflower mosaic virus (CaMV) 35S promoter and a combination of PEPC (phosphoenolpyruvate carboxylase) and a pollen-specific promoter. These Bt transgenic lines even showed tolerance to second-generation ECB, exhibiting less stalk tunnelling (Armstrong et al., 1995) and giving better results even in stress conditions (Traore et al., 2000). After commercialization in the USA in 1996 (James, 1997), Bt maize events Bt11 (trade name Agrisure™ CB/LL) from Novartis Seeds (now merged as Syngenta) and MON810 (trade name YieldGard™) from Monsanto Co., expressing *Cry1A(b)* gene and having 'YieldGard technology' (Ostlie et al., 1997), showed the most promising effect on ECB. In recent times, these two events have been highly utilized for crossing with other events to develop different Bt maize varieties. During the initial period, another Bt maize event, 176 (trade names NaturGard™ by Mycogen and KnockOut™ by Novartis Seeds), was also used in many studies (Ostlie et al., 1997). However, its inability to control second-generation ECB (Archer et al., 2000), low level of expression of Bt proteins in fresh leaves (Benedict & Ring, 2004), and lower-level control than MON810 and Bt11 events (Ostlie et al., 1997) make event 176 less utilized in modern-day Bt maize development programmes.

Although maize hybrids containing these events were originally developed to control ECB, they also offered potential to reduce the damage caused by the fall army worm, *Spodoptera frugiperda* (JE Smith); Asian corn borer, *Ostrinia furnacalis* (Guenée); corn earworm, *Helicoverpa zea* (Boddie); maize stem borer, *Chilo partellus* (Swinhoe); maize stalk borer, *Busseola fusca* (Fuller); southwestern corn borer, *Diatraea grandiosella* Dyar; Mediterranean corn borer, *Sesamia nonagrioides* (Lefebvre); bollworm, *Helicoverpa armigera* (Hübner), etc. (Buntin et al., 2001; Erasmus & Van den Berg, 2014; González-Núñez et al., 2000; Singh et al., 2005; Tan et al., 2011; Williams et al., 1997, 1998) (except event 176, [Buntin et al., 2004]). However, the effectiveness of Cry1Ab protein against these insects was partial (Huang, 2021; Pérez-Hedo et al., 2012; Williams et al., 1998). That was why it required the development of Bt maize hybrids expressing other Cry toxins. Currently, different Bt maize hybrids have been developed, expressing Cry1Ac (source Btk; event DBT418; trade name Bt Xtra™ from Monsanto Co.) (Chen et al., 2016), Cry1F (source *Bt* var. *aizawai*; event TC1507; trade name Herculex™ I from Dow AgroSciences) (Siebert et al., 2008), combination of Cry2Ab2 (source *Bt* subsp. *kumamotoensis*) and Cry1A.105 (synthetic three-domain chimera of Cry1Ab, Cry1Ac, and Cry1F) (event MON89034; trade name YieldGard™ VT Pro™ from Monsanto Co.) (Jalali et al., 2015), Cry9C (source *Bt* subsp. *tolworthi*; event CBH354 from AgrEvo) (Archer et al., 2000), Vip3Aa20 (source *Bt* strain AB88; event MIR162; trade name Agrisure™ Viptera from Syngenta Seeds) (Burkness et al., 2010), Cry1Ie (source *Bt* subsp. *thuringiensis* HD-2) (Song et al., 2003; Zhang et al., 2013), and Cry1C (Du et al., 2014) proteins, showing high efficiency against these non-target pests of maize. Apart from that, Bt maize events MON863 and MON88017 (trade names YieldGard™ Rootworm and YieldGard VT Rootworm/RR2™, respectively, from Monsanto Co.) expressing Cry3Bb1 protein from *Bt* var. *tenebrionis* and event DAS-59122-7 (trade name Herculex™ RW from Dow AgroSciences) expressing Cry34/35Ab1 protein from *Bt* strain PS149B1 have also been developed to resist the infestation caused by the destructive coleopteran pest of maize, that is, corn rootworm, *Diabrotica virgifera virgifera* Leconte, and other species of *Diabrotica* (Hitchon et al., 2015; Meissle et al., 2011; Vaughn et al., 2005).

However, with rising concerns of insect resistance to Bt, pyramided maize hybrids have been developed, showing more effectiveness than single-protein maize hybrids (Burkness et al., 2010; Huang, 2015; Jiang et al., 2016). Bt maize hybrid expressing four Bt genes, for example, PowerCore™ Ultra expressing Cry1F, Cry1A.105, Cry2Ab2, and Vip3Aa20 proteins (from Dow AgroSciences, now Corteva AgroScience) (Marques et al., 2019; Moscardini et al., 2020), was found to be highly effective against multiple target pests, with fewer chances of resistance development in insects. Even maize varieties expressing five (*Cry1Ab*, *eCry3.1Ab*, *mCry3A*, *Cry1Fa2*, and *Vip3Aa20*) and six (*Cry2Ab2*, *Cry1F*, *Cry1A.105*, *Cry34Ab1*, *Cry35Ab1*, and *Cry3Bb1*) Bt genes have also been introduced (Xiao & Wu, 2019).

There are different factors, such as transgenic background, transformation event, promoter used, environmental factors, etc., influencing the level of expression of Bt proteins in different plant tissues that ultimately set the degree of suppression of target pests by the transgenic plants. It was found that the expression of the *Cry1A(b)* gene in the leaves of transgenic maize was higher when the gene was driven by a combination of PEPC and a pollen-specific promoter. However, transgenic lines developed using the 35S promoter showed higher gene expression in leaves, pith, and roots, representing the constitutive nature of the promoter (Koziel et al., 1993). The levels of expression of Cry1Ab protein in events MON810 and Bt11 range from 3 to 10 µg/g in fresh leaf tissues, with event Bt11 being at the top end and event MON810 at the bottom end of the range (Icoz & Stotzky, 2008). However, in grains, the protein expression is much lower (Bøhn et al., 2008), resulting in good control against leaf-feeding insects than grain-attacking ones (Benedict & Ring, 2004).

12.3.2 Cotton

Bt cotton is one of the most successful examples of developing Bt transgenic crops for insect pest control. Considerable progress is achieved by introducing cotton cultivars expressing Bt genes for insect pest resistance. In 1996, Bt cotton expressing the *CrylAc* gene was first commercialized in the USA (Bollgard® [event MON531]) and Australia (Ingard® [event MON757]) by Monsanto Co. and approved in China in the next year (Xiao & Wu, 2019). Although in India, field trials on Bt cotton began in 1997 (Qaim & Zilberman, 2003), the tale of Bt cotton was started after its approval in 2002 with three transgenic cotton hybrids, namely, Bollgard-MECH-12, Bollgard-MECH-162, and Bollgard-MECH-184. MAHYCO (Maharashtra Hybrid Seeds Company Pvt. Ltd., Jalna, India) was the company that introduced this technology in India under license from Monsanto Co., USA (Kranthi et al., 2005).

In the USA, Australia, and China, the introduction of Bt cotton was a very necessary step at that point of time, as during that period, tobacco budworm, *Heliothis* (=*Chloridea*) *virescens* (Fabricius), and pink bollworm, *Pectinophora gossypiella* (Saunders), in the USA (Leonard et al., 1988; Sparks et al., 1993), and American bollworm, *H. armigera*, in China (Wei, 1992) and Australia (Forrester et al., 1993), became resistant to many of the conventional insecticides. The introduction of Bt cotton efficiently solved this problem. In one of the earliest studies, Benedict et al. (1996) genetically modified the commercial cotton cultivar Coker 312 to express *crylAb* and *crylAc* genes. These transgenic lines provided a high degree of resistance against *H. virescens*, *Trichoplusia ni* (Hübner), *Spodoptera exigua* (Hübner), and *H. zea* by significantly reducing the percent injury in flower buds and capsels and provided a significantly higher lint yield than non-transgenic Coker 312 lines (Benedict et al., 1996). This study was one of the earliest studies to demonstrate the superiority of Bt cotton over the commercial non-Bt cultivars in both pest suppression and yield aspects. However, the Cry1Ac toxin was reported to be about 30-fold less toxic to *H. armigera* as compared to *H. virescens* (Liao et al., 2002), which initially created obstacles to achieve the level of efficacy from these Bt crops in regions with high *H. armigera* infestation (Downes & Mahon, 2012). To date, various cotton events have already been developed expressing single Cry toxins, such as events MON531, MON1076, MON757, 3006-210-23, 31707, 31803, Event1 (expressing *CrylAc* gene), 281-24-236 (*CrylF*), GFM (*CrylA*), GK12 (*CrylAb-Ac*), GHB119 (*Cry2Ae*), MLS9124 (*CrylC*), T304-40 (*CrylAb*), etc.

The level of expression of *CrylAc* gene greatly varies in different tissues of Bollgard hybrids, which is usually highest in upper-canopy leaves, followed by lower-canopy leaves, and lowest in mid-canopy leaves. The Cry1Ac concentration is much lower in other plant parts, like squares, flowers, ovaries, raw seeds, boll bracts, and loculi walls, compared to leaves and lowest in boll rind (Kranthi et al., 2005). A very low concentration of Cry toxin in Bt cotton pollen (Bajaj et al., 2023) reduces the chances of toxin exposure to the pollinators. The expression of *CrylAc* gene also reduces as the crop gets matured and sets bolls (Greenplate, 1999; Kranthi et al., 2005; Olsen et al., 2005). That is why, at the later growth stages, plants become more susceptible to insect pest attack (Greenplate et al., 2003). Sometimes the toxin is highly expressed in root tissues, which is a non-feeding zone for lepidopteran pests (Singh et al., 2016). Moreover, *H. zea* was found not to be controlled satisfactorily during crop blooming stage, due to the lower concentration of Cry protein on the reproductive parts, like flowers, where this insect used to feed during the blooming stage (Benedict & Ring, 2004).

To overcome such shortcomings and insect resistance to single-toxin Bt cotton, scientists came up with the second generation of Bt cotton hybrids, expressing multiple *Cry* genes with higher efficiency in pest control and having potentiality against a broad range of lepidopteran pests. That is why Monsanto Co. has introduced Bollgard® II in 2002 (US EPA, 2002), a second-generation Bt cotton developed by inserting a new gene, *Cry2Ab*, from Btk into the already-existing Bollgard® variety, DP50B (event MON531), expressing the *CrylAc* gene, thus creating the new pyramided gene event, MON15985 (Icoz & Stotzky, 2008; US EPA, 2002). Bollgard® II cotton expresses the newly inserted Cry2Ab toxin at a significantly higher level than the level of Cry1Ac toxin expressed in both Bollgard® and Bollgard® II cotton (Adamczyk et al., 2001; Greenplate et al., 2003). This Cry2Ab toxin governs the larger toxic effect in Bollgard® II (Greenplate et al., 2003) and thus controls the lepidopteran pests more efficiently than do the single-gene Bollgard® varieties (Adamczyk et al., 2001, 2008; Stewart et al., 2001). Moreover, Cry2Ab and Cry1Ac have limited homology (Charles et al., 2000), making it difficult for insects to develop cross-resistance (Tabashnik et al., 2002). Laboratory and field studies revealed that mortality rates in *H. zea*, *S. frugiperda*, *S. exigua*, and *T. ni* increased from 84.2 to 92.2%, 16.1 to 100%, 50.1 to 94.9%, and 1.2 to 97.4%, respectively, in Bollgard® II (Benedict & Ring, 2004). After Monsanto Co., Dow AgroSciences also developed a double-stacked cotton variety, WideStrike® (MXB-13 line), expressing both *CrylC* and *CrylF* genes and approved by the US EPA in October 2004 (US EPA, 2004). WideStrike® also possesses good control against lepidopteran pests, and for some pests, it performs better than Bollgard® II (Adamczyk & Gore, 2004; Adamczyk et al., 2008).

More recently, companies have also developed triple-stacked Bt cotton varieties, such as Bollgard® III (expressing Cry1Ac, Cry2Ab2, Vip3A(a) by Monsanto Co.), WideStrike® 3 (expressing Cry1Ac, Cry1F, Vip3Aa19 by Corteva Agriscience), and TwinLink® Plus (expressing Cry1Ab, Cry2Ae, Vip3Aa19 by BASF), which conferred a high level of resistance to lepidopteran pests and are an efficient solution to delay the resistance development in insects (Brown et al., 2019; Reisig & Goldsworthy, 2023). Advancement is made by developing cotton varieties showing multiple traits, such as insect-resistant and herbicide-tolerant (IR-HT) cotton varieties. For example, Roundup Ready™ Bollgard™ is an IR-HT cotton variety developed by Monsanto Co. posing both traits (EFSA, 2012).

12.3.3 Rice

The story of Bt rice dates back to 1993, when Japanese scientists Fujimoto et al. (1993) first introduced a truncated Bt gene, *Cry1A(b)*, from Btk into a *japonica* rice cultivar. This transgenic rice plant, with a maximum Cry1A(b) expression of 0.05% of total soluble leaf protein, provided a high degree of resistance to the striped stem borer, *Chilo suppressalis* (Walker), and rice leaf folder, *Cnaphalocrocis medinalis* (Guenée). Later on, another effort was made by transforming the *indica* rice breeding line IR58 through the introduction of the *Cry1A(b)* gene, which showed 100% mortality in *C. suppressalis* and rice yellow stem borer (YSB), *Scirpophaga incertulas* (Walker), and inhibited feeding by leaf folders, *C. medinalis* and *Marasmia patnalis* Bradley (Wünn et al., 1996). After *Cry1A(b)*, *indica* rice IR64 transformed with *Cry1Ac* gene was also developed, showing a high level of toxicity against YSB (Nayak et al., 1997). In another study, *indica* rice cvs. Basmati 370 and M7 were genetically modified to express *Cry2A* gene, and transformed plants showed protein expression of up to 5% of the total leaf protein, showing high resistance to YSB and leaf folder (Maqbool et al., 1998). Some other early works on the development of transgenic Bt rice lines with high efficacy against different target pests were conducted by Cheng et al. (1998), Ghareyazie et al. (1997), Khanna and Raina (2002), Tu et al. (1998), and Wu et al. (1997).

Later on, research programmes were initiated to develop multiple *Cry* gene–stacked Bt rice varieties. Maqbool and Christou (1999) first reported the double *Cry* gene–stacked Bt rice by transforming the *indica* rice cvs. Basmati 370 and M7 with *Cry1Ac* and *Cry2A* genes. The double gene–stacked Bt Shanyou 63 rice variety (expressing Cry1Ab and Cry1Ac toxins) showed a high level of protection against *S. incertulas* and *C. medinalis*, with very little to no evidence of deadheart, whitehead, and leaf folder damage (Tu et al., 2000). Similarly, Xu et al. (2018) also developed transgenic rice expressing the fusion of the Cry1Ab and Vip3A proteins. However, these *Cry1A*, *Cry2A*, and *Vip3A* genes are toxic only to lepidopteran pests and do not pose any threat to coleopteran pests (Li et al., 2015). That is why progress was made to develop Bt rice against coleopteran pests. Lee et al. (2013) developed transgenic rice lines Btt8R and Btt12R by transforming the *japonica* rice cv. Nakdong with *Cry3A* gene. These transgenic rice lines significantly reduced the growth rate and population of rice water weevil, *Lissorhoptrus oryzophilus* Kuschel. Recently, Cry64Ba and Cry64Ca toxins were also found to be highly effective against rice planthoppers *Laodelphax striatellus* (Fallén) and *Sogatella furcifera* (Horváth) (Liu et al., 2018a) and can be genetically engineered to develop transgenic Bt rice plants effective against hemipteran pests.

In Bt rice, higher expression of Cry proteins is usually found in main stems, grain husks, leaves, and leaf sheaths as compared to roots (Hu et al., 2024; Wang et al., 2014; Wu et al., 2002). In grains, the concentration of Cry proteins is very low, specifically at the maturation stage of the crop (Wu et al., 2002). According to different studies, toxin concentration in Bt rice is higher either at the booting stage (Wu et al., 2002), at the filling stage (Hu et al., 2024), or at the seedling to tillering stages (Wang et al., 2014) of the crop. Protein expression is usually lower at the heading stage (Wang et al., 2014; Wu et al., 2002), although it varies with transgenic variety. However, Breitler et al. (2004) did not find any influence of developmental stages in toxin accumulation. The level of *Cry* gene expression also depends on the promoter used to drive Bt genes, where the expression is higher for the PEPC promoter than of 35SP, Actin-1, and pith-specific promoters (Datta et al., 1998).

In India, field tests on Bt rice were initiated in 2001 (High et al., 2004). In China, the Ministry of Agriculture has issued the biosafety certificates for the commercial production of Bt rice lines Huahui 1 and Bt Shanyou 63 in the Hubei Province in 2009, with an expectation that China will employ the transgenic Bt technology in commercial production of rice (Li et al., 2016). However, due to different political reasons, the commercialization of Bt rice in China has been postponed, and hitherto, it is not commercially grown in China (Jin et al., 2019; Li et al., 2016). Apart from these two Bt rice lines, T1C-19 and T2A-1 are other two important rice lines that were developed in China, conferring resistance at different stages of crop growth and exhibiting insecticidal activity to a variety of insects (Xiao & Wu, 2019). Currently, various universities and research institutes from China, along with the International Rice Research Institute (IRRI), and other universities from different countries are working on the development of transgenic rice by inserting different *Cry* genes singly or in combination into conventional rice cultivars. There is no doubt that the development of transgenic Bt rice lines can provide good season-long control against different lepidopteran pests (Benedict & Ring, 2004).

12.3.4 Potato

Peferoen et al. (1990) first reported GM potatoes expressing *Cry1Ab5* gene that provided protection from potato tuber moth (PTM), *Phthorimaea operculella* (Zeller). An improved expression of this toxin gene was obtained by modifying the gene structure (van Rie et al., 1994). That is why Jansens et al. (1995) introduced a native truncated *Cry1Ab* gene (*bt884*) and a modified truncated *Cry1Ab* gene (*Cry1Ab6*) into different potato lines. Here, unlike *Cry1Ab6* gene–expressing plants, the size and number of leaf tunnels in the potato lines expressing *bt884* gene had increased with time. In addition, expression of *Cry1Ab6* gene also resulted in almost 100% PTM larval mortality, with a storage period of potato tubers up to 7 months. To date, a number of potato cultivars have been transformed by introducing Bt toxin genes, like *Cry1Ia1* (previously *Cry5*) (Douches et al., 1998, 2002), *Cry1Ab* (Cañedo et al., 1999; Kumar et al., 2010), *Cry1Ac9* (Davidson et al., 2004; Meiyalaghan et al., 2006a), *Cry1Ba1*, *Cry1Ca5* (Meiyalaghan et al., 2006b), *Cry9Aa2* (Meiyalaghan et al., 2006a, 2011), modified *Cry9Aa2* (Jacobs et al., 2009), pyramided *Cry1Ac9*, *Cry9Aa2* (Meiyalaghan et al., 2010), etc., that confer moderate to high level of resistance to PTM. Further advancement was made by transforming the potato cultivar 'Spunta' to express both *Cry5* and *PVY^{O}cp* (potato virus Y^{o} coat protein) genes. These transgenic lines resist both PTM and PVY infections more efficiently than non-transgenic lines, with an almost-equivalent efficiency to the individual gene-expressed transgenic lines (Li et al., 1999).

For the Colorado potato beetle (CPB), *Leptinotarsa decemlineata* (Say), which is a serious pest of potato worldwide, Cry3A proteins possess high insecticidal activity (Krieg et al., 1983; Slaney et al., 1992). That's why Perlak et al. (1993) first genetically improved the Russet Burbank potato plants by inserting the *Cry3A* gene. The Newleaf® potato exemplifies the most prevalent Cry3Aa potato cultivar. These Cry3A potato lines provide complete protection and 100% mortality of CPB larvae (Reed et al., 2001; Salehian et al., 2021). Apart from Cry3A transformants, transgenic potato plants expressing *Cry3B* gene also com pletely inhibit egg production by CPB females (Arpaia et al., 2000). Haffani et al. (2000) also made an effort by introducing *Cry3Ca1* gene (potent against CPB) into transgenic potato lines. But in the transformants, this gene was very poorly expressed. In addition, Naimov et al. (2001) constructed a Cry1Ia/Cry1Ba hybrid toxin with increased activity against CPB.

Bt potato was the first GM crop to be commercialized anywhere and approved by US EPA in 1995 (Shelton et al., 2002). However, the market for Bt potatoes was not that stable, unlike other Bt crops. The NewLeaf® potato variety from Monsanto Co., which was approved in the USA in 1995 and marketed in the USA, Romania, and Canada (Shelton et al., 2002), provided a significant reduction in pesticide usage and cost of production to the growers (Kaniewski & Thomas, 2004). However, due to raising concerns on Newleaf® potato (Kadoić Balaško et al., 2020), Monsanto Co. ultimately discontinued its sales in 2001 (Shelton et al., 2002).

12.3.5 Soybean

Soybean is one of the most successful examples of implementing genetic engineering into the crop. According to a report, out of a total of 111 mi ha of global soybean-farmed area, biotech soybean occupies 92.1 mi ha (i.e. about 83% of total area), and the majority of this (>80%) is Bt soybean (Bengyella et al., 2018). Moreover, in 2020, GM soybeans contributed the largest share (50%) among the major GM crops in terms of total cultivated area (Brookes, 2022).

The first successful attempt to develop Bt soybeans expressing a native *Cry1Ab* gene was made by Parrott et al. (1994). Thereafter, Stewart et al. (1996a) genetically modified the soybean cultivar 'Jack' to express the Bt gene *Cry1Ac*, providing protection from the corn earworm, *H. zea*; soybean looper, *Chrysodeixis (=Pseudoplusia) includens* (Walker); tobacco budworm, *H. virescens*; and velvet-bean caterpillar, *Anticarsia gemmatalis* Hübner. Even this Bt soybean showed 100% mortality in all six larval instars of *H. armigera* (Azambuja et al., 2015). To date, a number of Bt soybean cultivars have been developed expressing several Bt genes, like *Cry1Ab*, *Cry1Ac*, *Cry1A.105*, *Cry1A2*, *Cry1B2*, *Cry1AF*, *Cry1F842*, *Cry2Ab2*, *Cry8*-like genes, *Cry14Ab*, and *Vip3A* (Bengyella et al., 2018; Chen et al., 2021; Kahn et al., 2021; Parrott et al., 1994; Qin et al., 2019; Stewart et al., 1996a).

Advancements were made in the field of GM soybeans by developing stacked-trait cultivars. For example, in 2016, Monsanto Co. developed and registered a double-event stacked GM soybean, MON87751 × MON87701, through conventional breeding between MON87751 (expressing Cry1A.105 and Cry2Ab2 toxins) and MON87701 (expressing Cry1Ac toxin), showing a high level of resistance against different lepidopteran pests (US EPA, 2016). Similarly, transgenic soybean event DAS-81419-2 (from Conkesta™ technology), expressing both Cry1Ac and Cry1F proteins, was found to control both primary and secondary lepidopteran pests of soybean efficiently (Marques et al., 2017). In addition, MON87701 × MON89788 (trade name Intacta™ Roundup Ready™ 2 Pro) is an event of IR-HT soybean, producing Cry1Ac + EPSPS (5-enolpyruvylshikimate-3-phosphate synthase for glyphosate resistance) proteins. This event shows high level of control against *A. gemmatalis*, *C. includens*, *H. armigera*, and *H. virescence* (Bernardi et al., 2012, 2014a; Bortolotto et al., 2014; Horikoshi et al., 2021a; Yu et al., 2013). However, these soybean lines are less efficient in resisting *Agrotis ypsilon* (Rottemberg) and *Spodoptera* spp. (Bernardi et al., 2014b; Yu et al., 2013).

A four-event pyramided GM soybean (MON87751 × MON87701 × MON87708 × MON89788) was also developed by Monsanto Co. through conventional breeding that expresses Cry1A.105, Cry2Ab2, and Cry1Ac Bt proteins, along with HT traits (EFSA et al., 2019), and approved by the Brazilian National Biosafety Technical Commission (CTNBio) in 2018 (CTNBio, 2018) and the European Food Safety Authority (EFSA) in 2019 (EFSA et al., 2019). This four-event stacked GM soybean was proved to be very effective against *A. gemmatalis*, *C. includens*, and *H. armigera* (Bacalhau et al., 2020). However, while managing *Spodoptera* species, it had provided effective control only against *S. cosmioides* (Walker) and *S. albula* (Walker), with reasonable control against *S. eridania* (Stoll), but was unable to control *S. frugiperda* (Godoy et al., 2022).

However, secondary pest outbreaks have been reported as a consequence of excessive use of Bt soybeans. A 10-year study on Cry1Ac soybeans used in Argentina revealed that the population levels of non-target stink bugs (*Diceraeus furcatus* [Fabricius], *Edessa meditabunda* [Fabricius], *Nezara viridula* [Linnaeus], and *Piezodorus guildinii* [Westwood]) and soybean weevils (*Rhyssomatus subtilis* Fiedler, *Sternechus subsignatus* Boehman, and *Promecops* sp.) have increased after the introduction of Bt soybeans in the country. These pests were found to be more abundant in the Bt soybeans than non-Bt ones; that might be due to the higher competition with targeted lepidopteran insects on non-Bt crops (Jerez et al., 2023).

Although Bt soybeans have earned much popularity around the world, however, in India, farmers are mostly reliant on the production of conventional high-yielding soybeans (Bengyella et al., 2018), and as per ISAAA data, out of five soybean events approved in India, only one (i.e. MON87701 × MON89788) has Bt traits in it (ISAAA, n.d.).

12.3.6 Brinjal

In the history of Bt crops in India, Bt brinjal kept its mark because of a lot of controversy that has been raised around it. In India, it was the first GM vegetable crop that reached the stage of approval for commercialization and was just on the way to be consumed by humans. MAHYCO, which is the leading seed company from India, developed this transgenic brinjal (Kumar et al., 2011). Kumar et al. (1998) were the first to develop transgenic brinjal by inserting the *Cry1Ab* gene from Bt into the brinjal cv. Pusa Purple Long. These transgenic plants showed significant insecticidal activity against the most devastating pest of brinjal, that is, fruit and shoot borer (FSB), *Leucinodes orbonalis* Guenée. Later on, transgenic brinjal transformed with the *Cry1Ac* gene was developed by Pal et al. (2009). These transgenic plants produced significant levels of the toxin, resulting in high mortality of FSB larvae. Furthermore, Rai et al. (2013) transformed the brinjal cv. Kashi Taru to express *Cry1Aa3* gene, which also showed high larval mortality in brinjal shoots and fruits. To date, a few more experiments have been conducted to develop more transgenic Bt brinjal varieties expressing *Cry2Ab* (Narendran et al., 2006), *Cry1Fa1* (Shrivastava et al., 2011), and *Cry2Aa* (Yadav et al., 2023) genes and conferring a high level of resistance against brinjal FSB.

According to several reports, in India, the creation of Bt brinjal dates back to 2000, and after a series of greenhouse evolution, safety tests, multi-location field trials, etc. (Kumar et al., 2011), Bt brinjal (expressing Cry1Ac toxin) finally got approval from the Genetic Engineering Approval Committee (GEAC) for commercialization on October 14, 2009 (Shelton, 2010). But some scientists, farmers, and anti-GMO activists raised concerns on the commercialization of Bt brinjal. As a result, on February 9, 2010, the government of India officially announced that it would take more time for the commercial release of Bt brinjal (TOI, 2010). However, on October 30, 2013, Bangladesh took a historic decision by approving the official release of four Bt brinjal varieties for seed production, and commercial sowing began in 2014. Thus, Bangladesh not only became the first country to start cultivation of Bt brinjal (Choudhary et al., 2014); it was also the first developing country to approve a genetically engineered food crop (Shelton et al., 2019). After Bangladesh, the Philippines also approved Bt brinjal on October 18, 2022, and became the second country to allow the commercial release of transgenic brinjal (Bugnosen, 2022).

In Table 12.1 we have presented brief details on different transgenic crops, introduced toxin gene, gene incorporation method, corresponding event, trade name, and their target pests.

12.4 SAFETY OF BT CROPS

According to different biotech companies, GM crops that contain Bt genes are completely safe to use. That is because of the unique mode of action of this bacteria (Bravo et al., 2007; Gill, 1995), making it effective only against insects, specifically the lepidopteran ones. However, in order to get a safety status, these transgenic crops need to be safe for the environment without causing any pollution, ingestion of these incorporated toxin by humans and/or other animals must not cause any toxic effect to them, and it must not cause any mortality to non-target beneficial organisms. To date, many studies have already been conducted to assess the safety status of these transgenic Bt crops.

12.4.1 Safety to the Environment

These Bt crops contain toxic proteins in all plant parts, including leaves, pollen, grains, fruits, and roots (Deng et al., 2019). These Bt proteins are directly secreted through root exudates and enter into the soil ecosystem (Liu et al., 2018b; Miethling-Graff et al., 2010). Moreover, in the case of Bt potatoes, the concentration of proteins in the root is higher (Saxena et al., 2004). A large amount of dispersed pollen from these Bt crops falls onto the ground and reaches the soil (Baudrot et al., 2021). Even all the plant residue ultimately returns to the soil and mixes the Bt proteins with soil (Liu et al., 2021). In soil, these Bt proteins get rapidly adsorbed and tightly bound to soil constituents, like clay, due to the influence of humic acid functional groups, which protect the Bt toxins from degradation and keep their insecticidal activity intact (Crecchio & Stotzky, 1998). So there is a possibility that the accumulation of these Bt toxins may cause toxic effects on soil bacteria, fungi, protozoa, archaea, nematodes, algae, and earthworms. However, different studies revealed that growing Bt crops does not cause any significant changes in the soil microbial community when compared with the corresponding non-Bt crops (Li et al., 2018b; Qi et al., 2018; Rui et al., 2005; Saxena & Stotzky, 2001; Xie et al., 2016). Rather, these Bt toxins undergo a rapid decline in their concentration without causing any significant changes in the diversity and community structure of soil microbes (Zhaolei et al., 2018), and no trace of the Bt

TABLE 12.1
Transgenic Crops Carrying Bt Genes for Insect Resistance

Transgenic Crop	Transferred Gene	Event Name	Method of Trait Introduction	Trade Name	Target Insect Pest(s)	References
Alfalfa	*Cry1C*	—	*Agrobacterium tumefaciens*-mediated plant transformation	—	*Spodoptera littoralis* (Biosduval) *Spodoptera exigua* (Hübner)	Strizhov et al. (1996)
	Cry3a	—	Do	—	*Hypera postica* (Gyllenhal)	Tohidfar et al. (2013)
Brinjal	*Cry1Ac*	Event EE1	Do	EARI Bt Begun-1, -2, -3 and -4	*Leucinodes orbonalis* Guenée	Prodhan et al. (2018)
	Cry2Aa	—	Do	—	*Leucinodes orbonalis* Guenée	Yadav et al. (2023)
Broccoli	*Cry1C*	—	Do	—	*Plutella xylostella* (Linnaeus) *Trichoplusia ni* (Hübner) *Pieris rapae* (Linnaeus)	Cao et al. (1999)
	Cry1Ab	—	Do	—	*Plutella xylostella* (Linnaeus)	Viswakarma et al. (2004)
	Cry1Aa	—	Do	—	*Plutella xylostella* (Linnaeus)	Kumar et al. (2018)
Canola	*Cry1Ac*	—	Do	—	*Helicoverpa zea* (Boddie) *Plutella xylostella* (Linnaeus) *Trichoplusia ni* (Hübner) *Spodoptera exigua* (Hübner)	Stewart et al. (1996b)
Cauliflower	*Cry9Aa*	—	Do	—	*Plutella xylostella* (Linnaeus)	Kuvshinov et al. (2001)
Chickpea	*Cry1Ac*	—	Microparticle bombardment of plant cells/tissue	—	*Helicoverpa armigera* (Hübner)	Kar et al. (1997)
	Cry2Aa	—	*A. tumefaciens*-mediated plant transformation	—	*Helicoverpa armigera* (Hübner)	Acharjee et al. (2010)
Cowpea	*Cry1Ab*	AAT709A	Do	—	*Maruca vitrata* (Fabricius)	Addae et al. (2020)
	Cry2Aa	—	Do	—	*Maruca vitrata* (Fabricius)	Kumar et al. (2021)
	Vip3Ba	—	Do	—	*Maruca vitrata* (Fabricius)	Bett et al. (2017)
Cotton	*Cry1Ac*	MON531, MON1076, MON757, 3006-210-23, 31707, 31803	Do	Bollgard™ (MON1076, MON757, MON531) Ingard™ (MON531)	Different Lepidopteran cotton pests	Pyke (2003); Rahman et al. (2021); Torres et al. (2010)
		Event1	Microparticle bombardment of plant cells/tissue	JK 1		
	Cry1Ac/Cry2Ab2	MON15985	Do	Bollgard™ II	Different Lepidopteran cotton pests	Rahman et al. (2021); Torres et al. (2010)
	Cry1Ab	COT67B (IR67B), COT69D, T303-3, T304-40	*A. tumefaciens*-mediated plant transformation	—	*Heliothis virescens* (Fabricius)	O'Reilly et al. (2006)
					Helicoverpa zea (Boddie) *Pectinophora gossypiella* (Saunders)	
					Spodoptera exigua (Hübner) *Spodoptera frugiperda* (J.E. Smith)	Adamczyk and Mahaffey (2008)

(Continued)

TABLE 12.1 (*Continued*)
Transgenic Crops Carrying Bt Genes for Insect Resistance

Transgenic Crop	Transferred Gene	Event Name	Method of Trait Introduction	Trade Name	Target Insect Pest(s)	References
	Cry2Ae	GHB119	Do	—	*Spodoptera frugiperda* (J.E. Smith)	Yang et al. (2016)
	Cry2Ab2	MON15947	Do	—	Lepidopteran cotton pests	—
	Cry1F	281-24-236				
	Vip3Aa	COT102 (IR102), COT202, COT203	Do	—	*Heliothis virescens* (Fabricius) *Helicoverpa zea* (Boddie) *Spodoptera exigua* (Hübner) *Spodoptera frugiperda* (J.E. Smith)	Adamczyk and Mahaffey (2008)
	mCry51Aa2 (now Mpp51Aa2)	MON88702	Do	ThryvOn™	*Lygus* spp. *Frankliniella* spp.	Akbar et al. (2019); Yates-Stewart et al. (2023)
	Cry1C	MLS 9124	Do	—	*Earias vittella* (Fabricius) *Helicoverpa armigera* (Hübner)	Hallad et al. (2014); Kamatham et al. (2021)
	Cry1Ab-Ac	GK12, GFM Cry1A	Pollen-tube pathway (PTP)		*Pectinophora gossypiella* (Saunders) *Spodoptera litura* (Fabricius)	Hallad et al. (2014); Kamatham et al. (2021); Zhang et al. (2006);
	Cry1A/cpTI	SGK321	Do	—	*Ostrinia furnacalis* Guenée *Helicoverpa armigera* (Hübner) *Aphis gossypii* Glover *Empoasca flavescens* (Fabricius) *Sylepta derogata* (Fabricius)	He et al. (2004 Hongxu et al. (2004) Wu et al. (2012)
	Cry1Ac/ Cry2Ab2/Vip3Aa	COT102 × MON15985	Conventional breeding	Bollgard™ III	Different Lepidopteran cotton pests	Rahman et al. (2021)
	Cry1Ab/vip3Aa	COT102 × COT67B	Do	VIPCOT™	Different Lepidopteran cotton pests	Rahman et al. (2021); Torres et al. (2010)
	Cry1Ac/Cry1F	MXB-13	Do	WideStrike™	Different Lepidopteran cotton pests	Rahman et al. (2021); Torres et al. (2010)
	Cry1Ab/Cry2Ae	T304-40 × GHB119	Do	TwinLink™	Different Lepidopteran cotton pests	Rahman et al. (2021); USDA (2011)
Groundnut	*Cry1AcF*	—	*A. tumefaciens*-mediated plant transformation	—	*Spodoptera litura* (Fabricius)	Keshavareddy et al. (2013)
	Cry1Ec	—	Do	—	*Spodoptera litura* (Fabricius)	Tiwari et al. (2008)
	Cry1X	—	Do	—	*Helicoverpa armigera* (Hübner) *Spodoptera litura* (Fabricius)	Entoori et al. (2008)

Maize	*Cry1Ab*	MON810, Bt11, event 176	Microparticle bombardment of plant cells/tissue	YieldGard™, MaizeGard™ (MON810) Agrisure™ (Bt11) NaturGard™, KnockOut™ (176)	*Busseola fusca* (Fuller) *Chilo partellus* (Swinhoe) *Diatraea grandiosella* Dyar *Diatraea saccharalis* (Fabricius) *Ostrinia furnacalis* Guenée *Ostrinia nubilalis* (Hübner) *Sesamia calamistis* Hampson *Sesamia nonagrioides* (Lefèbvre) *Helicoverpa zea* (Boddie) *Spodoptera frugiperda* (J.E. Smith)	Van Rensburg (1999) Archer et al. (2000) Farias et al. (2013) Alcantara et al. (2011) Ostlie et al. (1997) Van den Berg and Van Wyk (2007) Farinós et al. (2011) Buntin (2008)
	Cry1Ac	DBT418	Do	Bt Xtra™	*Helicoverpa armigera* (Hübner) *Ostrinia furnacalis* Guenée	Chen et al. (2016)
	Cry1F	TC1507	Do	Herculex™ I, Herculex™ CB	*Ostrinia nubilalis* (Hübner) *Ostrinia furnacalis* Guenée *Sesamia nonagrioides* (Lefèbvre) *Diatraea grandiosella* Dyar *Diatraea saccharalis* (Fabricius) *Elasmopalpus lignosellus* (Zeller) *Helicoverpa zea* (Boddie) *Spodoptera exigua* (Hübner) *Spodoptera frugiperda* (J.E. Smith)	CFIA (2002) Thompson et al. (2010) Farinós et al. (2012) Siebert et al. (2012) Siebert et al. (2008)
	Cry9C	CBH354, CBH351	Do	Starlink™ Maize (CBH351)	*Diatraea grandiosella* Dyar *Ostrinia nubilalis* (Hübner)	Archer et al. (2000)
	Cry1C	—	Do	—	*Ostrinia furnacalis* Guenée	Du et al. (2014)
	Cry1A.105/Cry2Ab2	MON89034	*A. tumefaciens*-mediated plant transformation	YieldGard™ VT Pro™	*Diatraea grandiosella* Dyar *Diatraea saccharalis* (Fabricius) *Elasmopalpus lignosellus* (Zeller) *Helicoverpa zea* (Boddie) *Spodoptera exigua* (Hübner) *Spodoptera frugiperda* (J.E. Smith)	Siebert et al. (2012)
	Cry1B.868/Cry1Da_7	MON95379	Do	—	*Spodoptera frugiperda* (J.E. Smith) *Diatraea saccharalis* (Fabricius)	Horikoshi et al. (2021b) Horikoshi et al. (2022)

(*Continued*)

TABLE 12.1 (Continued)
Transgenic Crops Carrying Bt Genes for Insect Resistance

Transgenic Crop	Transferred Gene	Event Name	Method of Trait Introduction	Trade Name	Target Insect Pest(s)	References
	Vip3Aa20	MIR162	Do	Agrisure™ Viptera	*Spodoptera frugiperda* (J.E. Smith)	Eghrari et al. (2022)
	Cry1Ab/*Cry2Aj* *Cry1Ab*/*Vip3DA*	N30 V3	Do	—	*Ostrinia nubilalis* (Hübner)	Chang et al. (2013)
	Cry3Bb1	MON863	Microparticle bombardment of plant cells/tissue	YieldGard™ Rootworm, MaxGard™	*Diabrotica virgifera virgifera* LeConte	Clark et al. (2012)
		MON88017	*A. tumefaciens*-mediated plant transformation	YieldGard™ VT™ Rootworm™	*Diabrotica virgifera virgifera* LeConte	Meissle et al. (2011)
	Cry34Ab1/*Cry35Ab1*	59122	Do	Herculex™ RW	*Diabrotica virgifera virgifera* LeConte	Hitchon et al. (2015)
	mCry3A	MIR604	Do	Agrisure™ RW	*Diabrotica virgifera virgifera* LeConte	Hibbard et al. (2011)
	eCry3.1Ab	5307	Do	Agrisure® Duracade™	*Diabrotica virgifera virgifera* LeConte	
	Cry1Ie	IE09S034	Do	—	*Ostrinia furnacalis* Guenée *Helicoverpa armigera* (Hübner)	Zhang et al. (2013)
	Cry1AcM	—	Do	—	*Ostrinia furnacalis* Guenée	Sun et al. (2015)
	Cry1Ah	—	Do	—	*Ostrinia furnacalis* Guenée	Dai et al. (2014)
Poplar	*Cry1Ac*	Bt poplar, Poplar 12	*A. tumefaciens*-mediated plant transformation	—	*Apochemia cineraius* Erschoff *Lymantria dispar* (Linnaeus)	Wang et al. (1996)
	Cry3Aa	—	Do	—	*Chrysomela tremula* Fabricius	Génissel et al. (2003)
Potato	*Cry1Ab*	—	Do	—	*Phthorimaea operculella* (Zeller)	Jansens et al. (1995)
	Cry1Ac9	—	Do	—	*Phthorimaea operculella* (Zeller)	Davidson et al. (2004)
	Cry1Ia1 (or *Cry5*)	—	Do	—	*Phthorimaea operculella* (Zeller) *Leptinotarsa decemlineata* (Say)	Douches et al. (1998)
	Cry1Ba1 *Cry1Ca5*	—	Do	—	*Phthorimaea operculella* (Zeller)	Meiyalaghan et al. (2006b)
	Cry3A	BT06, BT10, BT12, etc.	Do	Newleaf®	*Leptinotarsa decemlineata* (Say)	Perlak et al. (1993)
	Cry9Aa2	—	Do	—	*Phthorimaea operculella* (Zeller)	Meiyalaghan et al. (2006a)
Rice	*Cry1Ab*	—	Do Electroporation Microparticle bombardment of plant cells/tissue	—	*Chilo suppressalis* (Walker) *Cnaphalocrocis medinalis* (Guenée) *Marasmia patnalis* Bradley *Scirpophaga incertulas* (Walker)	Fujimoto et al. (1993); Liu et al. (2016); Wünn et al. (1996)
	Cry1Ac	—	Do	—	*Scirpophaga incertulas* (Walker)	Nayak et al. (1997)

	Cr*y1Ab*/**Cr***y1Ac*	GM Shanyou 63, Huahui-1/TT51-1	Microparticle bombardment of plant cells/tissue (Huahui1)	Bt Shanyou 63, Huahui1	*Chilo suppressalis* (Walker) *Cnaphalocrocis medinalis* (Guence) *Scirpophaga incertulas* (Walker)	Tu et al. (2000); Wang et al. (2016)
	C*ry2A*	T2A-1	Do *A. tumefaciens*-mediated plant transformation	—	*Chilo suppressalis* (Walker) *Cnaphalocrocis medinalis* (Guence) *Scirpophaga incertulas* (Walker)	Chen et al. (2005); Maqbool et al. (1998); Wang et al. (2016)
	C*ry1C*	TC1-19	Do	—	*Chilo suppressalis* (Walker) *Cnaphalocrocis medinalis* (Guence)	Wang et al. (2016); Ye et al. (2009)
	C*ry9C*	—	Do	—	*Chilo suppressalis* (Walker) *Scirpophaga incertulas* (Walker)	Chen et al. (2008)
	C*ry1Ab/Vip3H*	—	Do	—	*Chilo suppressalis* (Walker) *Sesamia inferens* (Walker)	Chen et al. (2010)
	C*ry1Ac*/**C***ry1I*-like	—	Do	—	*Chilo suppressalis* (Walker) *Cnaphalocrocis medinalis* (Guence)	Yang et al. (2014a)
	C*ry1Ac*/**C***pTI*	—	Microparticle bombardment of plant cells/tissue	—	*Chilo suppressalis* (Walker) *Cnaphalocrocis medinalis* (Guence) *Scirpophaga incertulas* (Walker) *Sesamia inferens* (Walker)	Liu et al. (2016)
Sorghum	**C***ry1Ac*	—	Do	—	*Chilo partellus* (Swinhoe)	Girijashankar et al. (2005)
Soybean	**C***ry1Ab*	—	Do	—	*Anticarsia gemmatalis* Hübner	Dufourmantel et al. (2005)
	C*ry1Ac*	MON87701	Do *A. tumefaciens*-mediated plant transformation	—	*Anticarsia gemmatalis* Hübner *Helicoverpa zea* (Boddie) *Heliothis virescens* (Fabricius) *Chrysodeixis* (=*Pseudoplusia*) *includens* (Walker)	EFSA (2011); Stewart et al. (1996a)
	C*ry1A.105*/**C***ry2Ab2*	MON87751	Do	—	*Anticarsia gemmatalis* Hübner *Chrysodeixis includens* (Walker) *Helicoverpa armigera* (Hübner)	Bacalhau et al. (2020); EFSA et al. (2018)
	C*ry1Ac*/**C***ry1F*	DAS-81419-2	Do	Conkesta™	*Anticarsia gemmatalis* Hübner *Chrysodeixis includens* (Walker) *Heliothis virescens* (Fabricius) *Spodoptera cosmioides* (Walker)	Marques et al. (2016)
					Elasmopalpus lignosellus (Zeller) *Agrotis ipsilon* (Hufnagel) *Helicoverpa armigera* (Hübner)	Marques et al. (2017)

(*Continued*)

TABLE 12.1 (*Continued*)
Transgenic Crops Carrying Bt Genes for Insect Resistance

Transgenic Crop	Transferred Gene	Event Name	Method of Trait Introduction	Trade Name	Target Insect Pest(s)	References
	Cry8-like	—	Do	—	*Holotrichia parallela* Motschulsky	Qin et al. (2019)
Sugarcane	*Cry1Ab*	CTC20BT	Electroporation	—	*Diatraea saccharalis* (Fabricius)	Arencibia et al. (1997)
			A. tumefaciens-mediated plant transformation	—	*Chilo curicilius* Dudgeon	Singh et al. (2011)
	Cry1Ac	CTC-91087-6	Do	—	*Sesamia cretica* Lederer	Dessoky et al. (2021)
Tobacco	*Cry1Ab*, *Cry1Ac*	—	Do	—	*Helicoverpa armigera* (Hübner) *Manduca sexta* (Linnaeus)	Barton et al. (1987); Sohail et al. (2012)
	Cry2a5	—	Do	—	*Helicoverpa armigera* (Hübner)	Selvapandiyan et al. (1998)
	Cry1C	—	Do	—	*Spodoptera littoralis* (Biosduval) *Spodoptera exigua* (Hübner)	Strizhov et al. (1996)
	Cry2Ah1	—	Do	—	*Helicoverpa armigera* (Hübner)	Li et al. (2018a)
	Cry9Aa2	—	Microparticle bombardment of plant cells/tissue	—	*Phthorimaea operculella* (Zeller)	Chakrabarti et al. (2006)
Tomato	*Cry1Ac*	5345	*A. tumefaciens*-mediated plant transformation	—	*Helicoverpa zea* (Boddie) *Manduca sexta* (Linnaeus) *Keiferia lycopersicella* (Walsingham)	Delannay et al. (1989)

Note: Some of the information has been collected from the website of International Service for the Acquisition of Agri-Biotech Applications (ISAAA): www.isaaa.org/gmapprovaldatabase/default.asp.

proteins remains after one year of crop residue incorporation (Zhang et al., 2019). Here, different soil factors, like soil properties, water content, biota, pH, temperature, etc., influence the degradation of Bt proteins in soil (Feng et al., 2011; Helassa et al., 2011; Yaoyu et al., 2006; Zhang et al., 2015). Moreover, Bt proteins, which move to aquatic ecosystems from soil through surface runoff, have been filtered in the soil before reaching the water body and degraded over time by different aquatic organisms (Liu et al., 2021).

12.4.2 Safety to Insect Pollinators

These Bt crops contain toxin genes that are specific to their target insect groups. Usually, they are mostly effective against lepidopteran pests and, to some extent, against coleopteran pests (due to the Cry3 toxin group). So now one question may arise: Is there any possibility that they cause any toxic effect to non-target beneficial arthropods, such as pollinators, parasitoids, predators, and other economically beneficial arthropods?

Usually, these Bt crops contain toxin genes in pollen at varying concentrations. So chances are there that the pollen from these crops may have the potential to cause lethal effects on pollinating insects. However, target specificity of these toxin genes makes them completely safe for hymenopteran pollinators, such as honeybees, solitary bees, wasps, ants, etc. (Malone & Burgess, 2009). Feeding on pollen containing Cry toxins does not show any significant effect on survival, weight, pollen consumption, learning capacity, hypopharyngeal gland development, and activity of detoxification and midgut enzymes of honeybees (Duan et al., 2008; Han et al., 2010, 2012; Rose et al., 2007). Rather, it is possible that collection of Bt pollen by honeybees can be used to control lepidopteran pests, such as the greater wax moth, *Galleria mellonella* (Linnaeus), which causes significant damage to honeybee colonies (Hanley et al., 2003). In addition to hymenopteran pollinators, many studies suggest that pollinating beetles, like *Haptoncus luteolus* (Erichson) and *Astylus atromaculatus* Blanchard, are also not affected by Bt pollen (Chen et al., 2011; Niu et al., 2018; Pierre & Hofs, 2010). However, these studies were conducted using Bt crops expressing lepidopteran-specific Cry toxins. So the possibility is there that the transgenic crops expressing Cry3 toxins can show toxicity to these pollinating beetles. Nectar and resin, which are the other products collected by the pollinators from crops, usually contain very low amounts of transgene products, posing no threat to the pollinators (Groot & Dicke, 2002; Malone & Pham-Delègue, 2001; Paula et al., 2014).

Apart from hymenopterans and coleopterans, different butterflies and moths are also comprised of a good fraction of pollinating insects. As most of the Bt toxins are toxic to lepidopterans, there is a chance that pollinating butterflies and moths can also get affected by these Bt crops. Concern regarding the impact of Bt crops on these non-target arthropods arose after the publication of Losey et al. (1999), who raised their concern about the potential risk of Bt maize pollen to the monarch butterfly, *Danaus plexippus* (Linnaeus). This butterfly is a good pollinator, and its larvae usually feed on milkweed plants (*Asclepias* sp.). Here, they found detrimental effects on monarch butterfly larvae when fed with milkweed leaves dusted with the pollen of Bt maize. Based on this, they assumed that the Bt maize pollen could be deposited on other plants near the Bt maize fields, and different non-target organisms consuming those plants might ingest the Bt pollen, which could be a threat to their survival. Later on, Hansen Jesse and Obrycki (2000) reported the natural deposition of Bt maize pollen on milkweed plants in Bt maize fields, causing higher rates of mortality (20 ± 3% at 48 hours of feeding) of monarch butterfly larvae. However, the level of toxicity to monarch butterfly depends on the Bt crop event. For example, pollen from Bt maize events Bt11 and Mon810 show negligible effect on survivorship of the larvae, whereas event 176 shows significantly lower survivorship and weight gain by the larvae (Stanley-Horn et al., 2001). Similar to the monarch butterfly, the peacock butterfly, *Aglais io* (Linnaeus), was also found to get affected while feeding on leaves of nettle plants (*Urtica* sp.) contaminated with Bt maize pollen (Felke et al., 2010).

However, long-range dispersal of pollen is only possible for the crops, like maize and rice, having anemophilous modes of pollination (Kanya et al., 2009; Luna et al., 2001). The total cultivated area of these Bt crops is not that much expanded (Agbioinvestor, 2024). So the chances of non-target lepidopteran larvae feeding on other plants to get exposed to the pollen from these Bt crops are also less. This considers a negligible impact of these Bt pollen on these non-target lepidopteran pollinators (Sears et al., 2001). Apart from that, unlike the adult hymenopterans and coleopterans, the adult butterflies feed only on nectar and do not consume pollen, preventing the Bt toxin from entering the adult gut. Bt crops like maize and rice don't have any nectaries, and their pollen is considered as a medium-quality food for pollinators (Sabugosa-Madeira et al., 2007). These crops also do not require insect pollinators for pollination, and the presence of their pollen on the pollinator's body might be due to contamination (Saha et al., 2023; Silici & Gökceoglu, 2007), where the electrostatic charge of body hairs of the pollinators attracts these anemophilous pollen (Vaknin et al., 2000; Zakon, 2016). All these allow the pollinators to collect a small amount of pollen from these Bt crops. Moreover, for many transgenic Bt hybrids, the expression of toxin in pollen is less (Bajaj et al., 2023; Sears et al., 2001; Székács et al., 2010), which could not cause mortality to the insects. Apart from that, even if the pollen from the Bt crops cause adverse effects on the population of non-target lepidopteran pollinators under certain circumstances, it can still be considered to cause minimal impacts when compared with habitat loss and non-judicious widespread application of chemical pesticides throughout the ecosystem (Pimentel & Raven, 2000).

12.4.3 Safety to Natural Enemies

According to many reports, transgenic Bt crops do not show any adverse effects on non-target natural enemies of different insect pests (Dahi, 2013; Li et al., 2008; Moar et al., 2002; Romeis et al., 2012). In a study, it was found that the protein expressed in Bt crops can be transmitted to the predatory arthropods through the consumption of herbivorous prey. But these toxins do not affect the predator biology (Zhao et al., 2013). Moreover, adoption of Bt crops also increases the abundance of these natural enemies. This is simply correlated with pesticide use. Usually, in Bt crops, the usage of chemical pesticides is very minimal, which in turn increases the abundance of beneficial arthropods. In different studies, it was found that large-scale adoption of Bt crops reduced the insecticide spray and thereby increased the abundance and potentiality of biological control agents, which in turn significantly reduced the population of other key and sporadic pests of the crop (Lu et al., 2012; Moussa et al., 2018; Naranjo, 2011).

However, some authors reported the toxic effects of Cry toxins on natural enemies (Hilbeck et al., 1998, 1999; Schmidt et al., 2009). Moreover, an indirect effect of Bt crops on natural enemies might be there, specifically in the tritrophic level of interaction between crop, herbivorous pest, and pest's natural enemies. It is probable that widespread adoption of Bt crops, which suppresses the regional population of target pests, will result in a regional loss of parasitoids unique to those pests. Usually, specialist parasitoids are more vulnerable to loss as compared to the target pests since they can only reproduce in the presence of their target hosts. So suppression of target pests can also cause loss in their specialist parasitoids (Sisterson & Tabashnik, 2005). Moreover, the parasitoids may fail to complete their development due to the premature mortality of the host feeding on the Bt crops (Schuler et al., 2004). A significantly higher mortality and delayed development were noticed in *Chrysoperla carnea* (Stephens) when fed with larvae of *Spodoptera littoralis* (Boisduval) reared on Bt maize (Dutton et al., 2002). Similarly, crops expressing Cry3 toxins have some negative impacts on the coleopteran predators, even if the toxin entered into the predator's body through the herbivorous prey. For example, a negative impact was noticed on the survival of carabid predators, *Agonum muelleri* (Herbst) and *Poecilus lucublandus* (Say), provided with ECB larvae, and coccinellid predator, *Harmonia axyridis* (Pallas), provided with aphids, where both the host insects were fed with or collected from Cry3Bb *Bt* maize (Stephens et al., 2012).

12.4.4 Safety to Other Non-Target Insects

The mulberry silkworm, *Bombyx mori* (Linnaeus), is another economically important insect that is also highly susceptible to Bt toxins. However, in different studies, it was found that Bt crops show negligible to no lethal or sublethal effects on the *silkworm larvae under field conditions* (Niu et al., 2013; Yang et al., 2014b). No negative impact was noticed when *B. mori* larvae were fed with mulberry leaves covered with Bt rice pollen (expressing Cry1C and Cry2A protein) for 3 days at a pollen density of 1,800 pollen grains/cm^2 mulberry leaf, which is about 20 times higher than the naturally occurring pollen density when mulberry trees are present near rice fields (i.e. 93 grains/cm^2 leaf) (Yang et al., 2014b). Apart from that, Bt crops are mostly effective against lepidopteran and coleopteran pests. So continuous cultivation of Bt crops sometimes causes higher infestation by non-target secondary sucking insect pests, which later earned the major pest status for the crop (Jerez et al., 2023; Lu, 2010).

12.4.5 Safety to Human and Other Mammals

In 2001, the EPA reassessed the four registered Bt crops that had been approved for use in agriculture since 1995, with one of the objectives being to look for their potential impacts on human health. These crops included Bt potato (Cry3A), Bt cotton (Cry1Ac), and Bt maize (Cry1Ab, Cry1F). The assessment confirmed that the proteins were unstable in the presence of digestive fluids, with no persistency in the digestive system, and degraded in gastric fluid within 7 minutes. Apart from that, the Bt proteins were demonstrated to become inactive during food processing or when exposed to heat. EPA did not find any apparent acute oral toxicity while treating mice with a pure preparation of the protein at a dose ranged from 3,280 to 5,000 mg/kg of body weight. Thus, EPA concluded that the registered Bt crops are safe for use (Mendelsohn et al., 2003). Later on, Wang et al. (2002) also reported no adverse effects (in terms of behaviour, feed utilization rate, and weight gain) of Bt rice KMD1 (containing *Cry1Ab* gene) on rats when treated with a dosage of ≤64 g/kg of body weight.

However, in one of the earliest studies by the researchers from the Committee for Independent Information and Research on Genetic Engineering (CRIIGEN), France, depicted were dose-specific significant changes in weight, signs of hepatorenal toxicity, increased triglyceride levels, diminished urine phosphorus and sodium excretions in rats fed with MON863 maize that contained Cry3Bb1 toxin (Séralini et al., 2007). However, based on that data, they were not able to conclude the safety status of Bt maize MON863. Later on, they reported induced signs of hepatorenal alterations in rats when fed with MON810 maize producing Cry1Ab toxin (de Vendômois et al., 2009). These studies initially indicated the potential risk of Bt toxins on mammals. Thereafter, a study was conducted by testing the effects of Cry1Ab and Cry1Ac toxins on the human embryonic kidney cell line 293. Here, Cry1Ab was found to cause cell death. According to this study, exposure to Cry1Ab caused plasma membrane alterations and respiration inhibition, which Cry1Ac was unable to do (Mesnage et al., 2013). However, this toxicity appeared at a relatively higher concentration (100 ppm) of the toxin as compared to the toxin concentrations in the GM plants (i.e. 1–20 ppm). Mesnage et al. (2013) also revealed the possibility of Bt toxins to show combined side effects in the presence of other pesticide residues specific to GM crops. Concern was also raised when Aris and Leblanc (2011) detected traces of Cry1Ab toxin from the blood of both pregnant and non-pregnant women in Canada. However, the level of concentration was very low. Apart from Bt crops, another possible indirect way for Bt toxin to enter the human body is the consumption of meat from herbivorous animals raised on Bt crops. Chowdhury et al. (2003a, 2003b) reported the presence of trace amounts of Cry1Ab toxin from Bt11 maize in the gastrointestinal contents of pigs and calves, but the toxin was not transferred to peripheral blood and degraded quickly. Anilkumar et al. (2010) also reported no detrimental effects on sheep fed with Bt cotton for 3 months.

According to the characteristics and mechanism of Bt endotoxin, it is possible to say that the Bt crops do not have any adverse effect on human health. Even if there is some probability to cause harm to human health, that will highly depend on the choice of toxin gene, transformation process, transformed event, and level of expression of toxin gene. Moreover, the task of determining the level of threat that Bt crops possess to human health is difficult. The human diet is highly complex, consisting of a maximum proportion of non-Bt foods, with Bt foods comprising a small fraction. This makes it difficult to trace the acute or chronic toxicity of consuming Bt foods. Even indirect risk can be there while consuming meat of herbivorous animals raised on Bt crops. That is why a lot of controversy is there regarding the harmful effects of Bt crops on the natural ecosystem, which may take time to resolve. We all know that the introduction of every new technology has some potential risks. But the major consideration is whether the new technology possesses more risk than the existing practice or not. While considering the negative effects of Bt crops on the environment, humans, and other non-target organisms, it is surely a much safer option than the use of chemical pesticides.

12.5 CONCLUSION

The first article documenting the effective insertion of Bt genes into commercial cultivars to generate transgenic plants with greater resistance to insect herbivores marks the completion of about four decades. Since then, crop science has made considerable progress by producing several Bt transgenic crops that express a variety of Bt toxins, revolutionizing crop protection, particularly in the field of integrated pest management (IPM). Adoption of these Bt transgenic crops in global agriculture has led to significant reductions in insecticide usage, and also reduced production costs. It does not stop here. Progress is still being made to generate novel Bt hybrids with stacked genes that can address insect resistance challenges. However, one worry is human consciousness. People may not accept eating a bacterial gene through a food crop. Apart from that, different political, social, and industrial issues are also there, restricting the Bt crop technology from reaching its full-fledged potential. That is why a lot of controversy arose surrounding the Bt crops, making their commercialization a challenging job. A number of Bt hybrids have already been developed that have not been commercialized yet due to different issues. Sometimes, questions have been raised pointing out the safety status of Bt crops. Even if we assume that, under certain circumstances, these Bt crops would have some negative effects on the environment, humans, and other non-target organisms, the consequences will undoubtedly be far less severe than those caused by the use of chemical pesticides. Even after all those controversies, it is hard to argue against this technology's contribution to the fight against insect pests, which has improved global agriculture. Therefore, in order to fully utilize this technology for the benefit of mankind, a solution is necessary to resolve the debate.

REFERENCES

Acharjee, S., Sarmah, B. K., Kumar, P. A., Olsen, K., Mahon, R., Moar, W. J., Moore, A., & Higgins, T. J. V. (2010). Transgenic chickpeas (*Cicer arietinum* L.) expressing a sequence-modified cry2Aa gene. *Plant Science*, *178*(3), 333–339. https://doi.org/10.1016/j.plantsci.2010.02.001

Adamczyk Jr., J. J., Bew, K., Adams, L. C., & Hardee, D. D. (2001). Evaluation of Bollgard II (cv. DP50BII) in the mississippi delta: Field efficacy against various Lepidoptera while profiling season-long expression of Cry1Ac and Cry2Ab. *Proceedings of the Beltwide Cotton Conference*, *2*, 835–837.

Adamczyk Jr., J. J., & Gore, J. (2004). Laboratory and field performance of cotton containing Cry1Ac, Cry1F, and both Cry1Ac and Cry1F (Widestrike®) against beet armyworm and fall armyworm larvae (Lepidoptera: Noctuidae). *Florida Entomologist*, *87*(4), 427–432. https://doi.org/10.1653/0015-4040(2004)087[0427:LAFPOC]2.0.CO;2

Adamczyk Jr., J. J., Greenberg, S., Armstrong, J. S., Mullins, W. J., Braxton, L. B., Lassiter, R. B., & Siebert, M. W. (2008). Evaluations of Bollgard®, Bollgard II®, and Widestrike® technologies against beet and fall armyworm larvae (Lepidoptera: Noctuidae). *Florida Entomologist*, *91*(4), 531–536. https://doi.org/10.1653/0015-4040-91.4.531

Adamczyk Jr., J. J., & Mahaffey, J. S. (2008). Efficacy of Vip3A and Cry1Ab transgenic traits in cotton against various lepidopteran pests. *Florida Entomologist*, *91*(4), 570–575. https://doi.org/10.1653/0015-4040-91.4.570

Addae, P. C., Ishiyaku, M. F., Tignegre, J. B., Ba, M. N., Bationo, J. B., Atokple, I. D., Abudulai, M., Dabiré-Binso, C. L., Traore, F., Saba, M., Umar, M. L., Adazebra, G. A., Onyekachi, F. N., Nemeth, M. A., Huesing, J. E., Beach, L. R., Higgins, T. J. V., Hellmich, R. L., & Pittendrigh, B. R. (2020). Efficacy of a cry1Ab gene for control of *Maruca vitrata* (Lepidoptera: Crambidae) in cowpea (Fabales: Fabaceae). *Journal of Economic Entomology*, *113*(2), 974–979. https://doi.org/10.1093/jee/toz367

Agbioinvestor. (2024, February). *Global GM crop area review*. Agbioinvestor GM Monitor. Retrieved July 22, 2024, from https://gm.agbioinvestor.com

Akbar, W., Gowda, A., Ahrens, J. E., Stelzer, J. W., Brown, R. S., Bollman, S. L., Greenplate, J. T., Gore, J., Catchot, A.

L., Lorenz, G., Stewart, S. D., Kerns, D. L., Greene, J. K., Toews, M. D., Herbert, D. A., Reisig, D. D., Sword, G. A., Ellsworth, P. C., Godfrey, L. D., & Clark, T. L. (2019). First transgenic trait for control of plant bugs and thrips in cotton. *Pest Management Science*, *75*(3), 867–877. https://doi.org/10.1002/ps.5234

Alcantara, E., Estrada, A., Alpuerto, V., & Head, G. (2011). Monitoring Cry1Ab susceptibility in Asian corn borer (Lepidoptera: Crambidae) on Bt corn in the Philippines. *Crop Protection*, *30*(5), 554–559. https://doi.org/10.1016/j.cropro.2010.12.019

Anilkumar, B., Reddy, A. G., Kalakumar, B., Rani, M. U., Anjaneyulu, Y., Raghunandan, T., Reddy, R., Jyothi, K., & Gopi, K. S. (2010). Sero-biochemical studies in sheep fed with Bt cotton plants. *Toxicology International*, *17*(2), 99–101. https://doi.org/10.4103/0971-6580.72680

Archer, T. L., Schuster, G., Patrick, C., Cronholm, G., Bynum Jr., E. D., & Morrison, W. P. (2000). Whorl and stalk damage by European and Southwestern corn borers to four events of *Bacillus thuringiensis* transgenic maize. *Crop Protection*, *19*(3), 181–190. https://doi.org/10.1016/S0261-2194(99)00092-7

Arencibia, A., Vázquez, R. I., Prieto, D., Téllez, P., Carmona, E. R., Coego, A., Hernández, L., De la Riva, G. A., & Selman-Housein, G. (1997). Transgenic sugarcane plants resistant to stem borer attack. *Molecular Breeding*, *3*, 247–255. https://doi.org/10.1023/A:1009616318854

Aris, A., & Leblanc, S. (2011). Maternal and fetal exposure to pesticides associated to genetically modified foods in eastern townships of Quebec, Canada. *Reproductive Toxicology*, *31*(4), 528–533. https://doi.org/10.1016/j.reprotox.2011.02.004

Armstrong, C. L., Parker, G. B., Pershing, J. C., Brown, S. M., Sanders, P. R., Duncan, D. R., Stone, T., Dean, D. A., DeBoer, D. L., Hart, J., Howe, A. R., Morrish, F. M., Pajeau, M. E., Petersen, W. L., Reich, B. J., Rodriguez, R., Santino, C. G., Sato, S. J., Schuler, W., Sims, S. R., . . . Fromm, M. E. (1995). Field evaluation of European corn borer control in progeny of 173 transgenic corn events expressing an insecticidal protein from *Bacillus thuringiensis*. *Crop Science*, *35*(2), 550–557. https://doi.org/10.2135/cropsci1995.0011183X003500020045x

Arora, R., & Shera, P. S. (2014). Genetic improvement of biocontrol agents for sustainable pest management. In K. Sahayaraj (Ed.), *Basic and applied aspects of biopesticides* (pp. 255–285). Springer New Delhi. https://doi.org/10.1007/978-81-322-1877-7_15

Arpaia, S., De Marzo, L., Di Leo, G. M., Santoro, M. E., Mennella, G., & van Loon, J. J. (2000). Feeding behaviour and reproductive biology of Colorado potato beetle adults fed transgenic potatoes expressing the *Bacillus thuringiensis* Cry3B endotoxin. *Entomologia Experimentalis et Applicata*, *95*(1), 31–37. https://doi.org/10.1046/j.1570-7458.2000.00638.x

Azambuja, R., Degrande, P. E., Santos, R. O. D., Souza, E. P. D., & Gomes, C. E. C. (2015). Effect of Bt soybean on larvae of *Helicoverpa armigera* (Hubner) (Lepidoptera: Noctuidae). *Journal of Agricultural Science*, *7*(8), 90–94. https://dx.doi.org/10.5539/jas.v7n8p90

Bacalhau, F. B., Dourado, P. M., Horikoshi, R. J., Carvalho, R. A., Semeão, A., Martinelli, S., Berger, G. U., Head, G. P., Salvadori, J. R., & Bernardi, O. (2020). Performance of genetically modified soybean expressing the Cry1A.105, Cry2Ab2, and Cry1Ac proteins against key lepidopteran pests in Brazil. *Journal of Economic Entomology*, *113*(6), 2883–2889. https://doi.org/10.1093/JEE/TOAA236

Bajaj, K., Chhuneja, P. K., Singh, J., Choudhary, A., & Mohindru, B. (2023). Assessment of Cry1Ac and Cry2Ab proteins in pollen of various transgenic cotton cultivars and effect on *Apis mellifera* L. colonies' development. *Journal of Entomological Research*, *47*(2), 245–252. https://doi.org/10.5958/0974-4576.2023.00045.2

Barnason, A., Armstrong, C., Dean, D., Deaton, R., Fischhoff, D., Fromm, M., Horsch, R., LaVallee, B., Maher, G., Peterson, W., & Simms, S. (1991). Production of transgenic corn plants resistant to the European corn borer. In *Agronomy abstracts* (p. 192). ASA.

Barton, K., Whitely, H., & Yang, N. S. (1987). *Bacillus thuringiensis* δ-endotoxin in transgenic *Nicotiana tabacum* provides resistance to lepidopteran insects. *Plant Physiology*, *85*(4), 1103–1109. https://doi.org/10.1104/pp.85.4.1103

Baudrot, V., Walker, E., Lang, A., Stefanescu, C., Rey, J. F., Soubeyrand, S., & Messéan, A. (2021). When the average hides the risk of Bt-corn pollen on non-target Lepidoptera: Application to *Aglais io* in Catalonia. *Ecotoxicology and Environmental Safety*, *207*, 111215. https://doi.org/10.1016/j.ecoenv.2020.111215

Benedict, J. H., & Ring, D. R. (2004). Transgenic crops expressing Bt proteins: Current status, challenges and outlook. In O. Koul, & G. S. Dhaliwal (Eds.), *Transgenic crop protection: Concepts and strategies* (pp. 15–84). Science Publishers, Inc.

Benedict, J. H., Sachs, E. S., Altman, D. W., Deaton, W. R., Kohel, R. J., Ring, D. R., & Berberich, S. A. (1996). Field performance of cottons expressing transgenic CryIA insecticidal proteins for resistance to *Heliothis virescens* and *Helicoverpa zea* (Lepidoptera: Noctuidae). *Journal of Economic Entomology*, *89*(1), 230–238. https://doi.org/10.1093/jee/89.1.230

Bengyella, L., Yekwa, E. L., Iftikhar, S., Nawaz, K., Jose, R. C., Fonmboh, D. J., Tambo, E., & Roy, P. (2018). Global challenges faced by engineered *Bacillus thuringiensis* Cry genes in soybean (*Glycine max* L.) in the twenty-first century. *3 Biotech*, *8*(11), 464. https://doi.org/10.1007/s13205-018-1484-8

Bernardi, O., Dourado, P. M., Carvalho, R. A., Martinelli, S., Berger, G. U., Head, G. P., & Omoto, C. (2014a). High levels of biological activity of Cry1Ac protein expressed on MON 87701×MON 89788 soybean against *Heliothis virescens* (Lepidoptera: Noctuidae). *Pest Management Science*, *70*(4), 588–594. https://doi.org/10.1002/ps.3581

Bernardi, O., Malvestiti, G. S., Dourado, P. M., Oliveira, W. S., Martinelli, S., Berger, G. U., Head, G. P., & Omoto, C. (2012). Assessment of the high-dose concept and level of control provided by MON 87701× MON 89788 soybean against *Anticarsia gemmatalis* and *Pseudoplusia includens* (Lepidoptera: Noctuidae) in Brazil. *Pest Management Science*, *68*(7), 1083–1091. https://doi.org/10.1002/ps.3271

Bernardi, O., Sorgatto, R. J., Barbosa, A. D., Domingues, F. A., Dourado, P. M., Carvalho, R. A., Martinelli, S., Head, G. P., & Omoto, C. (2014b). Low susceptibility of *Spodoptera cosmioides*, *Spodoptera eridania* and *Spodoptera frugiperda* (Lepidoptera: Noctuidae) to genetically-modified soybean expressing Cry1Ac protein. *Crop Protection*, *58*, 33–40. https://doi.org/10.1016/j.cropro.2014.01.001

Bett, B., Gollasch, S., Moore, A., James, W., Armstrong, J., Walsh, T., Harding, R., & Higgins, T. J. (2017). Transgenic cowpeas (*Vigna unguiculata* L. Walp) expressing *Bacillus thuringiensis* Vip 3Ba protein are protected against the *Maruca* pod borer (*Maruca vitrata*). *Plant Cell, Tissue and Organ*

Culture (PCTOC), *131*, 335–345. https://doi.org/10.1007/s11240-017-1287-3

Bøhn, T., Primicerio, R., Hessen, D. O., & Traavik, T. (2008). Reduced fitness of *Daphnia magna* fed a Bt-transgenic maize variety. *Archives of Environmental Contamination and Toxicology*, *55*(4), 584–592. https://doi.org/10.1007/S00244-008-9150-5

Bortolotto, O. C., Bueno, A. F., Braga, K., Barbosa, G. C., & Sanzovo, A. (2014). Biological characteristics of *Heliothis virescens* fed with Bt-soybean MON 87701×MON 89788 and its conventional isoline. *Anais da Academia Brasileira de Ciências*, *86*(2), 973–980. https://doi.org/10.1590/0001-3765201420130495

Bravo, A., Gill, S. S., & Soberón, M. (2007). Mode of action of *Bacillus thuringiensis* Cry and Cyt toxins and their potential for insect control. *Toxicon*, *49*(4), 423–435. https://doi.org/10.1016/j.toxicon.2006.11.022

Breitler, J. C., Vassal, J. M., Del Mar Catala, M., Meynard, D., Marfà, V., Melé, E., Royer, M., Murillo, I., Segundo, B. S., Guiderdoni, E., & Messeguer, J. (2004). Bt rice harbouring cry genes controlled by a constitutive or wound-inducible promoter: Protection and transgene expression under mediterranean field conditions. *Plant Biotechnology Journal*, *2*(5), 417–430. https://doi.org/10.1111/j.1467-7652.2004.00086.x

Brookes, G. (2022). *GM crops: Global socio-economic and environmental impacts 1996–2020*. PG Economics Ltd.

Brown, S., Walker, W., Cole, C., & Kerns, D. L. (2019). Efficacy and field performance of Bt cotton in Louisiana. In S. Boyd, M. Huffman, L. Krogman, & A. Sarkissian (Eds.), *Proceedings of the 2019 beltwide cotton conferences, New Orleans, LA, USA* (pp. 8–10). National Cotton Council of America.

Bugnosen, Z. J. (2022, October 26). *Philippines approves Bt eggplant for commercial cultivation*. ISAAA. https://www.isaaa.org/blog/entry/default.asp?BlogDate=10/26/2022

Buntin, G. D. (2008). Corn expressing Cry1Ab or Cry1F endotoxin for fall armyworm and corn earworm (Lepidoptera: Noctuidae) management in field corn for grain production. *Florida Entomologist*, *91*(4), 523–530. https://doi.org/10.1653/0015-4040-91.4.523

Buntin, G. D., All, J. N., Lee, R. D., & Wilson, D. M. (2004). Plant-incorporated *Bacillus thuringiensis* resistance for control of fall armyworm and corn earworm (Lepidoptera: Noctuidae) in corn. *Journal of Economic Entomology*, *97*(5), 1603–1611. https://doi.org/10.1603/0022-0493-97.5.1603

Buntin, G. D., Lee, R. D., Wilson, D. M., & McPherson, R. M. (2001). Evaluation of YieldGard transgenic resistance for control of fall armyworm and corn earworm (Lepidoptera: Noctuidae) on corn. *Florida Entomologist*, *84*(1), 37–42. https://doi.org/10.2307/3496660

Burkness, E. C., Dively, G., Patton, T., Morey, A. C., & Hutchison, W. D. (2010). Novel Vip3A *Bacillus thuringiensis* (Bt) maize approaches high-dose efficacy against *Helicoverpa zea* (Lepidoptera: Noctuidae) under field conditions: Implications for resistance management. *GM Crops*, *1*(5), 337–343. https://doi.org/10.4161/gmcr.1.5.14765

Cañedo, V., Benavides, J., Golmirzaie, A., Cisneros, F., Ghislain, M., & Lagnaoui, A. (1999). Assessing Bt transformed potatoes for potato tuber moth, *Phthorimaea operculella* (Zeller) management. In *Impact on a changing world, program report 1997–98* (pp. 161–169). International Potato Centre (CIP).

Cao, J., Tang, J. D., Strizhov, N., Shelton, A. M., & Earle, E. D. (1999). Transgenic broccoli with high levels of *Bacillus thuringiensis* Cry1C protein control diamondback moth larvae resistant to Cry1A or Cry1C. *Molecular Breeding*, *5*, 131–141. https://doi.org/10.1023/A:1009619924620

Carlton, B. C., & Gawron-Burke, C. (1993). Genetic improvement of *Bacillus thuringiensis* for bioinsecticide development. In L. Kim (Ed.), *Advanced engineered biopesticides* (pp. 43–61). CRC Press.

CFIA. (2002, October). *DD2002–41: Determination of the safety of Dow AgroSciences Canada Inc. and pioneer hibred international's insect resistant and glufosinate—Ammonium tolerant corn (Zea mays L.) line 1507*. Canadian Food Inspection Agency. https://inspection.canada.ca/en/plant-varieties/plants-novel-traits/approved-under-review/decision-documents/dd2002-41.

Chakrabarti, S. K., Lutz, K. A., Lertwiriyawong, B., Svab, Z., & Maliga, P. (2006). Expression of the *cry9Aa2* Bt gene in tobacco chloroplasts confers resistance to potato tuber moth. *Transgenic Research*, *15*, 481–488. https://doi.org/10.1007/s11248-006-0018-z

Chang, X., Liu, G. G., He, K. L., Shen, Z. C., Peng, Y. F., & Ye, G. Y. (2013). Efficacy evaluation of two transgenic maize events expressing fused proteins to Cry1Ab-susceptible and-resistant *Ostrinia furnacalis* (Lepidoptera: Crambidae). *Journal of Economic Entomology*, *106*(6), 2548–2556. https://doi.org/10.1603/EC13100

Charles, J., Delecluse, A., & Nielsen-LeRoux, C. (2000). *Entomopathogenic bacteria: From laboratory to field application*. Springer Dordrecht. https://doi.org/10.1007/978-94-017-1429-7

Chen, D., Moar, W. J., Jerga, A., Gowda, A., Milligan, J. S., Bretsynder, E. C., Rydel, T. J., Baum, J. A., Semeao, A., Fu, X., Guzov, V., Gabbert, K., Head, G. P., & Haas, J. A. (2021). *Bacillus thuringiensis* chimeric proteins Cry1A.2 and Cry1B.2 to control soybean lepidopteran pests: New domain combinations enhance insecticidal spectrum of activity and novel receptor contributions. *PLoS One*, *16*(6), e0249150. https://doi.org/10.1371/JOURNAL.PONE.0249150

Chen, H., Tang, W., Xu, C., Li, X., Lin, Y., & Zhang, Q. (2005). Transgenic indica rice plants harboring a synthetic *cry2A** gene of *Bacillus thuringiensis* exhibit enhanced resistance against lepidopteran rice pests. *Theoretical and Applied Genetics*, *111*, 1330–1337. https://doi.org/10.1007/s00122-005-0062-8

Chen, H., Zhang, G., Zhang, Q., & Lin, Y. (2008). Effect of transgenic *Bacillus thuringiensis* rice lines on mortality and feeding behavior of rice stem borers (Lepidoptera: Crambidae). *Journal of Economic Entomology*, *101*(1), 182–189. https://doi.org/10.1093/jee/101.1.182

Chen, H. X., Yang, R., Yang, W., Zhang, L., Camara, I., Dong, X. H., Liu, Y. Q., & Shi, W. P. (2016). Efficacy of Bt maize producing the Cry1Ac protein against two important pests of corn in China. *Environmental Science and Pollution Research*, *23*, 21511–21516. https://doi.org/10.1007/s11356-016-7340-7

Chen, L., Cui, J., Ma, W., Niu, C., & Lei, C. (2011). Pollen from Cry1Ac/CpTI-transgenic cotton does not affect the pollinating beetle *Haptoncus luteolus*. *Journal of Pest Science*, *84*, 9–14. https://doi.org/10.1007/s10340-010-0319-0

Chen, Y., Tian, J. C., Shen, Z. C., Peng, Y. F., Hu, C., Guo, Y. Y., & Ye, G. Y. (2010). Transgenic rice plants expressing a fused protein of Cry1Ab/Vip3H has resistance to rice stem borers under laboratory and field conditions. *Journal of Economic Entomology*, *103*(4), 1444–1453. https://doi.org/10.1603/EC10014

Cheng, X., Sardana, R., Kaplan, H., & Altosaar, I. (1998). *Agrobacterium*-transformed rice plants expressing synthetic *cryIA (b)* and *cryIA (c)* genes are highly toxic to striped stem borer and yellow stem borer. *Proceedings of the National Academy of Sciences*, *95*(6), 2767–2772. https://doi.org/10.1073/pnas.95.6.2767

Choudhary, B., Nasiruddin, K. M., & Gaur, K. (2014). *The status of commercialized BT brinjal in Bangladesh* (Brief No. 47). ISAAA.

Chowdhury, E. H., Kuribara, H., Hino, A., Sultana, P., Mikami, O., Shimada, N., Guruge, K. S., Saito, M., & Nakajima, Y. (2003a). Detection of corn intrinsic and recombinant DNA fragments and Cry1Ab protein in the gastrointestinal contents of pigs fed genetically modified corn Bt11. *Journal of Animal Science*, *81*(10), 2546–2551. https://doi.org/10.2527/2003.81102546x

Chowdhury, E. H., Shimada, N., Murata, H., Mikami, O., Sultana, P., Miyazaki, S., Yoshioka, M., Yamanaka, N., Hirai, N., & Nakajima, Y. (2003b). Detection of Cry1Ab protein in gastrointestinal contents but not visceral organs of genetically modified Bt11-fed calves. *Veterinary and Human Toxicology*, *45*(2), 72–75.

Clark, T. L., Frank, D. L., French, B. W., Meinke, L. J., Moellenbeck, D., Vaughn, T. T., & Hibbard, B. E. (2012). Mortality impact of MON863 transgenic maize roots on western corn rootworm larvae in the field. *Journal of Applied Entomology*, *136*(10), 721–729. https://doi.org/10.1111/j.1439-0418.2012.01709.x

Crecchio, C., & Stotzky, G. (1998). Insecticidal activity and biodegradation of the toxin from *Bacillus thuringiensis* subsp. *kurstaki* bound to humic acids from soil. *Soil Biology and Biochemistry*, *30*(4), 463–470. https://doi.org/10.1016/S0038-0717(97)00147-8

CTNBio. (2018). *Commercial release of genetically modified insect-resistant and herbicide-tolerant soy containing genetically modified events MON87701, MON89788 and MON87708* (Technical Opinion No. 5832/2018). Comissão Técnica Nacional de Biossegurança.

Dahi, H. F. (2013). Assessment of the effects of transgenic Bt cotton that contains Cry1Ac and Cry2Ab toxins on the abundance of non-target organisms community. *Journal Nature and Science (JNSCI)*, *11*(2), 117–122.

Dai, J., Li, X., Zhu, L., Wang, H., Zhang, J., He, K., Lang, Z., & Huang, D. (2014). Molecular detection and agronomic traits analysis of insect-resistant transgenic maize harboring Bt *cry1Ah* gene. *Biotechnology Bulletin*, *5*, 63–68.

Datta, K., Vasquez, A., Tu, J., Torrizo, L., Alam, M. F., Oliva, N., Abrigo, E., Khush, G. S., & Datta, S. K. (1998). Constitutive and tissue-specific differential expression of the *cryIA (b)* gene in transgenic rice plants conferring resistance to rice insect pest. *Theoretical and Applied Genetics*, *97*, 20–30. https://doi.org/10.1007/s001220050862

Davidson, M. M., Takla, M. F. G., Jacobs, J. M. E., Butler, R. C., Wratten, S. D., & Conner, A. J. (2004). Transformation of potato (*Solanum tuberosum*) cultivars with a *cry1Ac9* gene confers resistance to potato tuber moth (*Phthorimaea operculella*). *New Zealand Journal of Crop and Horticultural Science*, *32*(1), 39–50. https://doi.org/10.1080/01140671.2004.9514278

Delannay, X., LaVallee, B. J., Proksch, R. K., Fuchs, R. L., Sims, S. R., Greenplate, J. T., Marrone, P. G., Dodson, R. B., Augustine, J. J., Layton, J. J., & Fischhoff, D. A. (1989). Field performance of transgenic tomato plants expressing the *Bacillus thuringiensis* var. *kurstaki* insect control protein. *Bio/Technology*, *7*(12), 1265–1269. https://doi.org/10.1038/nbt1289–1265

Deng, J., Wang, Y., Yang, F., Liu, Y., & Liu, B. (2019). Persistence of insecticidal Cry toxins in Bt rice residues under field conditions estimated by biological and immunological assays. *Science of the Total Environment*, *679*, 45–51. https://doi.org/10.1016/j.scitotenv.2019.05.026

Dessoky, E. S., Ismail, R. M., Elarabi, N. I., Abdelhadi, A. A., & Abdallah, N. A. (2021). Improvement of sugarcane for borer resistance using *Agrobacterium* mediated transformation of *cry1Ac* gene. *GM Crops & Food*, *12*(1), 47–56. https://doi.org/10.1080/21645698.2020.1809318

de Vendômois, J. S., Roullier, F., Cellier, D., & Séralini, G. E. (2009). A comparison of the effects of three GM corn varieties on mammalian health. *International Journal of Biological Sciences*, *5*(7), 706–726. https://doi.org/10.7150/ijbs.5.706

Dionglay, C. (2021, May 19). *Five countries produce 91% of the world's biotech crops*. ISAAA. https://www.isaaa.org/blog/entry/default.asp?BlogDate=5/19/2021

Douches, D. S., Li, W., Zarka, K., Coombs, J., Pett, W., Grafius, E., & El-Nasr, T. (2002). Development of Bt-cry5 insect-resistant potato lines' Spunta-G2' and 'Spunta-G3'. *HortScience*, *37*(7), 1103–1107. https://doi.org/10.21273/HORTSCI.37.7.1103

Douches, D. S., Westedt, A. L., Zarka, K., Schroeter, B., & Grafius, E. J. (1998). Potato transformation to combine natural and engineered resistance for controlling tuber moth. *HortScience*, *33*(6), 1053–1056.

Downes, S., & Mahon, R. (2012). Successes and challenges of managing resistance in *Helicoverpa armigera* to Bt cotton in Australia. *GM Crops & Food*, *3*(3), 228–234. https://doi.org/10.4161/gmcr.20194

Du, D., Geng, C., Zhang, X., Zhang, Z., Zheng, Y., Zhang, F., Lin, Y., & Qiu, F. (2014). Transgenic maize lines expressing a *cry1C** gene are resistant to insect pests. *Plant Molecular Biology Reporter*, *32*, 549–557. https://doi.org/10.1007/s11105-013-0663-3

Duan, J. J., Marvier, M., Huesing, J., Dively, G., & Huang, Z. Y. (2008). A meta-analysis of effects of Bt crops on honey bees (Hymenoptera: Apidae). *PLoS One*, *3*(1), e1415. https://doi.org/10.1371/journal.pone.0001415

Dufourmantel, N., Tissot, G., Goutorbe, F., Garcon, F., Muhr, C., Jansens, S., Pelissier, B., Peltier, G., & Dubald, M. (2005). Generation and analysis of soybean plastid transformants expressing *Bacillus thuringiensis* Cry1Ab protoxin. *Plant Molecular Biology*, *58*, 659–668. https://doi.org/10.1007/s11103-005-7405-3

Dutton, A., Klein, H., Romeis, J., & Bigler, F. (2002). Uptake of Bt-toxin by herbivores feeding on transgenic maize and consequences for the predator *Chrysoperla carnea*. *Ecological Entomology*, *27*(4), 441–447. https://doi.org/10.1046/j.1365–2311.2002.00436.x

EFSA Panel on Genetically Modified Organisms (GMO). (2011). Scientific opinion on application (EFSA-GMO-BE-2010-79) for the placing on the market of insect resistant genetically modified soybean MON 87701 for food and feed uses, import and processing under regulation (EC) No 1829/2003 from Monsanto. *EFSA Journal*, *9*(7), 2309. https://doi.org/10.2903/j.efsa.2011.2309

EFSA Panel on Genetically Modified Organisms (GMO). (2012). Scientific opinion on applications EFSA-GMO-UK-2005-09 and EFSA-GMO-RX-MON531× MON1445 for the placing on the market of food and feed produced from or containing ingredients produced from insect-resistant and herbicide-tolerant genetically modified cotton MON 531× MON 1445, and for the renewal of authorisation of existing products produced from cotton MON 531× MON 1445, both under regulation (EC) No 1829/2003 from Monsanto. *EFSA Journal*, *10*(3), 2608. https://doi.org/10.2903/j.efsa.2012.2608

EFSA Panel on Genetically Modified Organisms (GMO), Naegeli, H., Birch, A. N., Casacuberta, J., De Schrijver, A., Gralak, M. A., Jones, H., Manachini, B., Messéan, A., Nielsen, E.

E., Nogué, F., Robaglia, C., Rostoks, N., Sweet, J., Tebbe, C., Visioli, F., Wal, J. M., Álvarez, F., Ardizzone, M., Dumont, A. F., . . . Paraskevopoulos, K. (2018). Assessment of genetically modified soybean MON 87751 for food and feed uses under regulation (EC) No 1829/2003 (application EFSA-GMO-NL-2014-121). *EFSA Journal*, *16*(8), e05346. https://doi.org/10.2903/j.efsa.2018.5346

EFSA Panel on Genetically Modified Organisms (GMO), Naegeli, H., Bresson, J.-L., Dalmay, J.-L., Dewhurst, I. C., Epstein, M. M., Firbank, L. G., Guerche, P., Hejatko, J., Moreno, F. J., Mullins, E., Nogué, F., Rostoks, N., Sánchez Serrano, J. J., Savoini, G., Veromann, E., Veronesi, F., Álvarez, F., Ardizzone, M., De Sanctis, G., . . . Raffaello, T. (2019). Assessment of genetically modified soybean MON 87751 × MON 87701 × MON 87708 × MON 89788 for food and feed uses, under regulation (EC) No 1829/2003 (application EFSA-GMO-NL-2016-128). *EFSA Journal*, *17*(11), e05847. https://doi.org/10.2903/j.efsa.2019.5847

Eghrari, K., Oliveira, S. C., Nascimento, A. M., Queiroz, B., Fatoretto, J., de Souza, B. H. S., Fernandes, O. A., & Môro, G. V. (2022). The implications of homozygous vip3Aa20-and cry1Ab-maize on *Spodoptera frugiperda* control. *Journal of Pest Science*, *95*(1), 115–127. https://doi.org/10.1007/s10340-021-01362-7

Entoori, K., Sreevathsa, R., Arthikala, M. K., Kumar, P. A., Kumar, A. R. V., Madhusudhan, B., & Makarla, U. (2008). A chimeric *cry1X* gene imparts resistance to *Spodoptera litura* and *Helicoverpa armigera* in the transgenic groundnut. *Eurasian Journal of BioSciences*, *2*, 53–65.

Erasmus, A., & Van den Berg, J. (2014). Effect of Bt-maize expressing Cry1Ab toxin on non-target coleoptera and lepidoptera pests of maize in South Africa. *African Entomology*, *22*(1), 167–179.

Farias, J. R., Costa, E. C., Guedes, J. V., Arbage, A. P., Neto, A. P., Bigolin, M., & Pinto, F. F. (2013). Managing the sugarcane borer, *Diatraea saccharalis*, and corn earworm, *Helicoverpa zea*, using Bt corn and insecticide treatments. *Journal of Insect Science*, *13*(1), 109. https://doi.org/10.1673/031.013.10901

Farinós, G. P., Andreadis, S. S., De la Poza, M., Mironidis, G. K., Ortego, F., Savopoulou-Soultani, M., & Castañera, P. (2011). Comparative assessment of the field-susceptibility of *Sesamia nonagrioides* to the Cry1Ab toxin in areas with different adoption rates of Bt maize and in Bt-free areas. *Crop Protection*, *30*(7), 902–906. https://doi.org/10.1016/j.cropro.2011.03.011

Farinós, G. P., De la Poza, M., Ortego, F., & Castañera, P. (2012). Susceptibility to the Cry1F toxin of field populations of *Sesamia nonagrioides* (Lepidoptera: Noctuidae) in mediterranean maize cultivation regions. *Journal of Economic Entomology*, *105*(1), 214–221. https://doi.org/10.1603/EC11213

Felke, M., Langenbruch, G. A., Feiertag, S., & Kassa, A. (2010). Effect of Bt-176 maize pollen on first instar larvae of the Peacock butterfly (*Inachis io*) (Lepidoptera; Nymphalidae). *Environmental Biosafety Research*, *9*(1), 5–12. https://doi.org/10.1051/ebr/2010006

Feng, Y., Ling, L., Fan, H., Liu, Y., Tan, F., Shu, Y., & Wang, J. (2011). Effects of temperature, water content and pH on degradation of Cry1Ab protein released from Bt corn straw in soil. *Soil Biology and Biochemistry*, *43*(7), 1600–1606. https://doi.org/10.1016/j.soilbio.2011.04.011

Fischhoff, D. A., Bowdish, K. S., Perlak, F. J., Marrone, P. G., McCormick, S. M., Niedermeyer, J. G., Dean, D. A., Kusano-Kretzmer, K., Mayer, F. J., Rochester, D. E., Rogers, S. G., & Fraley, R. T. (1987). Insect tolerant transgenic tomato plants. *Bio/Technology*, *5*, 807–812.

Forrester, N. W., Cahill, M., Bird, L. J., & Layland, J. K. (1993). Management of pyrethroid and endosulfan resistance in *Helicoverpa armigera* (Lepidoptera: Noctuidae) in Australia. *Bulletin of Entomological Research: Supplement Series*, R1–132.

Fujimoto, H., Itoh, K., Yamamoto, M., Kyozuka, J., & Shimamoto, K. O. (1993). Insect resistant rice generated by introduction of a modified δ-endotoxin gene of *Bacillus thuringiensis*. *Bio/Technology*, *11*(10), 1151–1155. https://doi.org/10.1038/nbt1093–1151

Génissel, A., Leplé, J. C., Millet, N., Augustin, S., Jouanin, L., & Pilate, G. (2003). High tolerance against *Chrysomela tremulae* of transgenic poplar plants expressing a synthetic cry3Aa gene from *Bacillus thuringiensis* ssp *tenebrionis*. *Molecular Breeding*, *11*, 103–110. https://doi.org/10.1023/A:1022453220496

Ghareyazie, B., Alinia, F., Menguito, C. A., Rubia, L. G., de Palma, J. M., Liwanag, E. A., Cohen, M. B., Khush, G. S., & Bennett, J. (1997). Enhanced resistance to two stem borers in an aromatic rice containing a synthetic *cryIA (b)* gene. *Molecular Breeding*, *3*, 401–414. https://doi.org/10.1023/A:1009695324100

Gill, S. S. (1995). Mechanism of action of *Bacillus thuringiensis* toxins. *Memórias do Instituto Oswaldo Cruz*, *90*, 69–74. https://doi.org/10.1590/S0074–02761995000100016

Girijashankar, V., Sharma, H. C., Sharma, K. K., Swathisree, V., Prasad, L. S., Bhat, B. V., Royer, M., Secundo, B. S., Narasu, M. L., Altosaar, I., & Seetharama, N. (2005). Development of transgenic sorghum for insect resistance against the spotted stem borer (*Chilo partellus*). *Plant Cell Reports*, *24*, 513–522. https://doi.org/10.1007/s00299-005-0947-7

Godoy, D. N., Pretto, V. E., Weschenfelder, M. A., de Almeida, P. G., Wendt, A. D. F., Palharini, R. B., Reis, A. C., Horikoshi, R. J., Dourado, S. M., Head, G., & Bernardi, O. (2022). Managing *Spodoptera* species (Lepidoptera: Noctuidae) found in Brazilian soybean fields with Bt soybean and insecticides. *Agronomy*, *12*(11), 2864. https://doi.org/10.3390/agronomy12112864

González-Núñez, M., Ortego, F., & Castañera, P. (2000). Susceptibility of Spanish populations of the corn borers *Sesamia nonagrioides* (Lepidoptera: Noctuidae) and *Ostrinia nubilalis* (Lepidoptera: Crambidae) to a *Bacillus thuringiensis* endotoxin. *Journal of Economic Entomology*, *93*(2), 459–463. https://doi.org/10.1603/0022-0493-93.2.459

Greenplate, J. T. (1999). Quantification of *Bacillus thuringiensis* insect control protein Cry1Ac over time Bollgard cotton fruit and terminals. *Journal of Economic Entomology*, *92*, 1377–1383. https://doi.org/10.1093/jee/92.6.1377

Greenplate, J. T., Mullins, J. W., Penn, S. R., Dahm, A., Rech, B. J., Osborn, J. A., Rahn, P. R., Ruscheke, R. L., & Shapple, Z. W. (2003). Partial characterization of cotton plants expressing two toxin proteins from *Bacillus thuringiensis*: Relative toxin contribution, toxin interaction, and resistance management. *Journal of Applied Entomology*, *127*(6), 340–347. https://doi.org/10.1046/j.1439–0418.2003.00766.x

Groot, A. T., & Dicke, M. (2002). Insect-resistant transgenic plants in a multi-trophic context. *The Plant Journal*, *31*(4), 387–406. https://doi.org/10.1046/j.1365–313X.2002.01366.x

Haffani, Y. Z., Overney, S., Yelle, S., Bellemare, G., & Belzile, F. J. (2000). Premature polyadenylation contributes to the poor expression of the *Bacillus thuringiensis cry3Ca1* gene in transgenic potato plants. *Molecular and General Genetics MGG*, *264*, 82–88. https://doi.org/10.1007/s004380000282

Hallad, A. V., Udikeri, S. S., Patil, S. B., Biradar, D. P., & Khadi, B. M. (2014). Characterization of resistance to all bollworms and *Spodoptera litura* (Fab.) in different Bt transgenic events of cotton. *International Journal of Current Microbiology and Applied Sciences*, *3*(3), 594–600.

Han, P., Niu, C. Y., Biondi, A., & Desneux, N. (2012). Does transgenic Cry1Ac+ CpTI cotton pollen affect hypopharyngeal gland development and midgut proteolytic enzyme activity in the honey bee *Apis mellifera* L. (Hymenoptera, Apidae)?. *Ecotoxicology*, *21*, 2214–2221. https://doi.org/10.1007/s10646-012-0976-2

Han, P., Niu, C. Y., Lei, C. L., Cui, J. J., & Desneux, N. (2010). Use of an innovative T-tube maze assay and the proboscis extension response assay to assess sublethal effects of GM products and pesticides on learning capacity of the honey bee *Apis mellifera* L. *Ecotoxicology*, *19*, 1612–1619. https://doi.org/10.1007/s10646-010-0546-4

Hanley, A. V., Huang, Z. Y., & Pett, W. L. (2003). Effects of dietary transgenic Bt corn pollen on larvae of *Apis mellifera* and *Galleria mellonella*. *Journal of Apicultural Research*, *42*(4), 77–81. https://doi.org/10.1080/00218839.2003.11101097

Hansen Jesse, L. C., & Obrycki, J. J. (2000). Field deposition of Bt transgenic corn pollen: Lethal effects on the monarch butterfly. *Oecologia*, *125*, 241–248. https://doi.org/10.1007/s004420000502

He, K., Wang, Z., Bai, S., Zheng, L., & Wang, Y. (2004). Field efficacy of transgenic cotton containing single and double toxin genes against the Asian corn borer (Lep., Pyralidae). *Journal of Applied Entomology*, *128*(9–10), 710–715. https://doi.org/10.1111/j.1439–0418.2004.00919.x

Helassa, N., M'charek, A., Quiquampoix, H., Noinville, S., Déjardin, P., Frutos, R., & Staunton, S. (2011). Effects of physicochemical interactions and microbial activity on the persistence of Cry1Aa Bt (*Bacillus thuringiensis*) toxin in soil. *Soil Biology and Biochemistry*, *43*(5), 1089–1097. https://doi.org/10.1016/j.soilbio.2011.01.030

Hibbard, B. E., Frank, D. L., Kurtz, R., Boudreau, E., Ellersieck, M. R., & Frederick Odhiambo, J. (2011). Mortality impact of Bt transgenic maize roots expressing eCry3. 1Ab, mCry3A, and eCry3. 1Ab plus mCry3A on western corn rootworm larvae in the field. *Journal of Economic Entomology*, *104*(5), 1584–1591. https://doi.org/10.1603/EC11186

High, S. M., Cohen, M. B., Shu, Q. Y., & Altosaar, I. (2004). Achieving successful deployment of Bt rice. *Trends in Plant Science*, *9*(6), 286–292. https://doi.org/10.1016/j.tplants.2004.04.002

Hilbeck, A., Baumgartner, M., Fried, P. M., & Bigler, F. (1998). Effects of transgenic *Bacillus thuringiensis* corn-fed prey on mortality and development time of immature *Chrysoperla carnea* (Neuroptera: Chrysopidae). *Environmental Entomology*, *27*(2), 480–487. https://doi.org/10.1093/ee/27.2.480

Hilbeck, A., Moar, W. J., Pusztai-Carey, M., Filippini, A., & Bigler, F. (1999). Prey-mediated effects of Cry1Ab toxin and protoxin and Cry2A protoxin on the predator *Chrysoperla carnea*. *Entomologia Experimentalis et Applicata*, *91*(2), 305–316. https://doi.org/10.1046/j.1570–7458.1999.00497.x

Hitchon, A. J., Smith, J. L., French, B. W., & Schaafsma, A. W. (2015). Impact of the Bt corn proteins Cry34/35Ab1 and Cry3Bb1, alone or pyramided, on western corn rootworm (Coleoptera: Chrysomelidae) beetle emergence in the field. *Journal of Economic Entomology*, *108*(4), 1986–1993. https://doi.org/10.1093/jee/tov125

Hongxu, Z., Jianying, G., & Fanghao, W. (2004). Effect of transgenic Cry1Ac+ CpTI cotton (SGK321) on population dynamics of pests and their natural enemies. *Acta Entomologica Sinica*, *47*(4), 538–542.

Horikoshi, R. J., Dourado, P. M., Berger, G. U., de S. Fernandes, D., Omoto, C., Willse, A., Martinelli, S., Head, G. P., & Corrêa, A. S. (2021a). Large-scale assessment of lepidopteran soybean pests and efficacy of Cry1Ac soybean in Brazil. *Scientific Reports*, *11*(1), 15956. https://doi.org/10.1038/s41598-021-95483-9

Horikoshi, R. J., Ferrari, G., Dourado, P. M., Climaco, J. I., Vertuan, H. V., Evans, A., Pleau, M., Morrell, K., José, M. O. M. A., Anderson, H., Martinelli, S., Ovejero, R. F. L., Berger, G. U., & Head, G. (2022). MON 95379 Bt maize as a new tool to manage sugarcane borer (*Diatraea saccharalis*) in South America. *Pest Management Science*, *78*(8), 3456–3466. https://doi.org/10.1002/ps.6986

Horikoshi, R. J., Vertuan, H., de Castro, A. A., Morrell, K., Griffith, C., Evans, A., Tan, J., Asiimwe, P., Anderson, H., José, M. O. M. A., Dourado, P. M., Berger, G., & Head, G. (2021b). A new generation of Bt maize for control of fall armyworm (*Spodoptera frugiperda*). *Pest Management Science*, *77*(8), 3727–3736. https://doi.org/10.1002/ps.6334

Hu, Y., Tian, C., Feng, Y., Ma, W., Zhang, Y., Yang, Q., & Zhang, X. (2024). Transgenic early japonica rice: Integration and expression characterization of stem borer resistance Bt gene. *Gene*, *927*, 148753. https://doi.org/10.1016/j.gene.2024.148753

Huang, F. N. (2015). Resistance management for Bt maize and above-ground lepidopteran targets in the USA: From single gene to pyramided traits. In A. Bravo, M. Soberón, & Y. Gao (Eds.), *Bt resistance: Haracterization and strategies for GM crops producing Bacillus thuringiensis toxins* (pp. 173–185). CABI.

Huang, F. N. (2021). Resistance of the fall armyworm, *Spodoptera frugiperda*, to transgenic *Bacillus thuringiensis* Cry1F corn in the Americas: Lessons and implications for Bt corn IRM in China. *Insect Science*, *28*(3), 574–589. https://doi.org/10.1111/1744–7917.12826

Icoz, I., & Stotzky, G. (2008). Fate and effects of insect-resistant Bt crops in soil ecosystems. *Soil Biology and Biochemistry*, *40*(3), 559–586. https://doi.org/10.1016/j.soilbio.2007.11.002

ISAAA. (n.d.). *GM Crop Events Approved in India*. Retrieved July 14, 2024, from https://www.isaaa.org/gmapprovaldatabase/approvedeventsin/default.asp?CountryID=IN

Jacobs, J. M., Takla, M. F., Docherty, L. C., Frater, C. M., Markwick, N. P., Meiyalaghan, S., & Conner, A. J. (2009). Potato transformation with modified nucleotide sequences of the *cry9Aa2* gene improves resistance to potato tuber moth. *Potato Research*, *52*, 367–378. https://doi.org/10.1007/s11540-009-9140-7

Jalali, S. K., Yadavalli, L., Ojha, R., Kumar, P., Sulaikhabeevi, S. B., Sharma, R., Nair, R., Kadanur, R. C., Kamath, S. P., & Komarlingam, M. S. (2015). Baseline sensitivity of maize borers in India to the *Bacillus thuringiensis* insecticidal proteins Cry1A. 105 and Cry2Ab2. *Pest Management Science*, *71*(8), 1082–1090. https://doi.org/10.1002/ps.3888

James, C. (1997). *Global status of transgenic crops in 1997* (Briefs No. 5). ISAAA.

James, C. (1998). *Global review of commercialized transgenic crops: 1998* (Briefs No. 8). ISAAA.

James, C., & Krattiger. A. F. (1996). *Global review of the field testing and commercialization of transgenic plants, 1986 to 1995: The first decade of crop biotechnology* (Briefs No. 1). ISAAA.

Jansens, S., Cornelissen, M., Clerçq, R., de Reynaerts, A., & Peferoen, M. (1995). *Phthorimaea operculella* (Lepidoptera Gelechiidae) resistance in potato by expression of *Bacillus thuringiensis* Cry1A(b) insecticidal crystal protein. *Journal of Economic Entomology*, *88*, 1469–1476. https://doi.org/10.1093/jee/88.5.1469

Jerez, P. G. P., Hill, J. G., Pereira, E. J., Alzogaray, R. A., & Vera, M. T. (2023). Ten years of Cry1Ac Bt soybean use in Argentina: Historical shifts in the community of target and non-target pest insects. *Crop Protection*, *170*, 106265. https://doi.org/10.1016/j.cropro.2023.106265

Jiang, F., Zhang, T., Bai, S., Wang, Z., & He, K. (2016). Evaluation of Bt corn with pyramided genes on efficacy and insect resistance management for the Asian corn borer in China. *PLoS One*, *11*(12), e0168442. https://doi.org/10.1371/journal.pone.0168442

Jin, Y., Drabik, D., Heerink, N., & Wesseler, J. (2019). The cost of postponement of Bt rice commercialization in China. *Frontiers in Plant Science*, *10*, 1226. https://doi.org/10.3389/fpls.2019.01226

Kadoić Balaško, M., Mikac, K. M., Bažok, R., & Lemic, D. (2020). Modern techniques in Colorado potato beetle (*Leptinotarsa decemlineata* Say) control and resistance management: History review and future perspectives. *Insects*, *11*(9), 581. https://doi.org/10.3390/insects11090581

Kahn, T. W., Duck, N. B., McCarville, M. T., Schouten, L. C., Schweri, K., Zaitseva, J., & Daum, J. (2021). A *Bacillus thuringiensis* Cry protein controls soybean cyst nematode in transgenic soybean plants. *Nature Communications*, *12*(1), 3380. https://doi.org/10.1038/s41467-021-23743-3

Kamatham, S., Munagapati, S., Manikanta, K. N., Vulchi, R., Chadipiralla, K., Indla, S. H., & Allam, U. S. (2021). Recent advances in engineering crop plants for resistance to insect pests. *Egyptian Journal of Biological Pest Control*, *31*, 120. https://doi.org/10.1186/s41938-021-00465-8

Kaniewski, W. K., & Thomas, P. E. (2004). The potato story. *AgBioForum*, *7*(1&2), 41–46.

Kanya, J. I., Kinyamario, J. I., Amugune, N. O., & Hauser, T. P. (2009). Dispersal distance of rice (*Oryza Sativa* L.) pollen at the Tana River delta in the coast province, Kenya. *African Journal of Biotechnology*, *8*(10), 2265–2270.

Kar, S., Basu, D., Das, S., Ramkrishnan, N. A., Mukherjee, P., Nayak, P., & Sen, S. K. (1997). Expression of *cryIA (c)* gene of *Bacillus thuringiensis* in transgenic chickpea plants inhibits development of pod-borer (*Heliothis armigera*) larvae. *Transgenic Research*, *6*, 177–185. https://doi.org/10.1023/A:1018433922766

Keshavareddy, G., Rohini, S., Ramu, S. V., Sundaresha, S., Kumar, A. R. V., Kumar, P. A., & Udayakumar, M. (2013). Transgenics in groundnut (*Arachis hypogaea* L.) expressing *cry1AcF* gene for resistance to *Spodoptera litura* (F.). *Physiology and Molecular Biology of Plants*, *19*, 343–352. https://doi.org/10.1007/s12298-013-0182-6

Khanna, H. K., & Raina, S. K. (2002). Elite *indica* transgenic rice plants expressing modified Cry1Ac endotoxin of *Bacillus thuringiensis* show enhanced resistance to yellow stem borer (*Scirpophaga incertulas*). *Transgenic Research*, *11*, 411–423. https://doi.org/10.1023/A:1016378606189

Koziel, M. G., Beland, G. L., Bowman, C., Carozzi, N. B., Crenshaw, R., Crossland, L., Dawson, J., Desai, N., Hill, M., Launis, K., Lewis, K., Maddox, D., McPherson, K., Meghii, M. R., Merlin, E., Rhodes, R., Warren, G. W., Wright, M., & Evola, S. V. (1993). Field performance of elite transgenic maize plants expressing an insecticidal protein derived from *Bacillus thuringiensis*. *Bio/Technology*, *11*(2), 194–200. https://doi.org/10.1038/nbt0293–194

Kranthi, K. R., Naidu, S., Dhawad, C. S., Tatwawadi, A., Mate, K., Patil, E., Bharose, A. A., Behere, G. T., Wadaskar, R. M., & Kranthi, S. (2005). Temporal and intra-plant variability of Cry1Ac expression in Bt-cotton and its influence on the survival of the cotton bollworm, *Helicoverpa armigera* (Hübner) (Noctuidae: Lepidoptera). *Current Science*, *89*(2), 291–298.

Krieg, A. V., Huger, A. M., Langenbruch, G. A., & Schnetter, W. (1983). *Bacillus thuringiensis* var. *tenebrionis*: ein neuer, gegenüber Larven von Coleopteren wirksamer Pathotyp. *Zeitschrift für angewandte Entomologie*, *96*(1–5), 500–508. https://doi.org/10.1111/j.1439–0418.1983.tb03704.x

Kumar, A., Jaiwal, R., Sreevathsa, R., Chaudhary, D., & Jaiwal, P. K. (2021). Transgenic cowpea plants expressing Bacillus thuringiensis Cry2Aa insecticidal protein imparts resistance to *Maruca vitrata* legume pod borer. *Plant Cell Reports*, *40*, 583–594. https://doi.org/10.1007/s00299-020-02657-2

Kumar, M., Chimote, V., Singh, R., Mishra, G. P., Naik, P. S., Pandey, S. K., & Chakrabarti, S. K. (2010). Development of Bt transgenic potatoes for effective control of potato tuber moth by using *cry1Ab* gene regulated by GBSS promoter. *Crop Protection*, *29*(2), 121–127. https://doi.org/10.1016/j.cropro.2009.11.001

Kumar, P. A., Gambhir, G., Gaur, A., Sharma, K. C., Thakur, A. K., & Srivastava, D. K. (2018). Development of transgenic broccoli with *cryIAa* gene for resistance against diamondback moth (*Plutella xylostella*). *3 Biotech*, *8*, 299. https://doi.org/10.1007/s13205-018-1316-x

Kumar, P. A., Mandaokar, A., Sreenivasu, K., Chakrabarti, S. K., Bisaria, S., Sharma, S. R., Kaur, S., & Sharma, R. P. (1998). Insect-resistant transgenic brinjal plants. *Molecular Breeding*, *4*, 33–37. https://doi.org/10.1023/A:1009694016179

Kumar, S., Misra, A., Verma, A. K., Roy, R., Tripathi, A., Ansari, K. M., Das, M., & Dwivedi, P. D. (2011). Bt Brinjal in India: A long way to go. *GM Crops*, *2*(2), 92–98. https://doi.org/10.4161/gmcr.2.2.16335

Kuvshinov, V., Koivu, K., Kanerva, A., & Pehu, E. (2001). Transgenic crop plants expressing synthetic *cry9Aa* gene are protected against insect damage. *Plant Science*, *160*(2), 341–353. https://doi.org/10.1016/S0168–9452(00)00398–8

Lee, J. H., Shin, K. S., Suh, S. C., Rhim, S. L., Lee, Y. H., Lim, M. H., Woo, H. J., Qin, Y., & Cho, H. S. (2013). CryIIIA toxin gene expression in transgenic rice confers resistance to rice water weevil. *Plant Cell, Tissue and Organ Culture (PCTOC)*, *115*, 243–252. https://doi.org/10.1007/s11240-013-0356-5

Leonard, R. B., Graves, J. B., Sparks, T. C., & Pavloff, A. M. (1988). Variation in resistance of field populations of tobacco budworm and bollworm (Lepidoptera: Noctuidae) to selected insecticides. *Journal of Economic Entomology*, *81*(6), 1521–1528. https://doi.org/10.1093/jee/81.6.1521

Li, P., Li, Y., Shi, J., Yu, Z., Pan, A., Tang, X., & Ming, F. (2018b). Impact of transgenic Cry1Ac+ CpTI cotton on diversity and dynamics of rhizosphere bacterial community of different root environments. *Science of the Total Environment*, *637*, 233–243. https://doi.org/10.1016/j.scitotenv.2018.05.013

Li, S., Wang, Z., Zhou, Y., Li, C., Wang, G., Wang, H., Zhang, J., Liang, G., & Lang, Z. (2018a). Expression of *cry2Ah1* and two domain II mutants in transgenic tobacco confers high resistance to susceptible and Cry1Ac-resistant cotton bollworm. *Scientific Reports*, *8*(1), 508. https://doi.org/10.1038/s41598-017-19064-5

Li, W., Zarka, K. A., Douches, D. S., Coombs, J. J., Pett, W. L., & Grafius, E. J. (1999). Coexpression of potato PVY^{O} coat protein and *cryV-Bt* genes in potato. *Journal of the American Society for Horticultural Science*, *124*(3), 218–223. https://doi.org/10.21273/JASHS.124.3.218

Li, Y., Hallerman, E. M., Liu, Q., Wu, K., & Peng, Y. (2016). The development and status of Bt rice in China. *Plant Biotechnology Journal*, *14*(3), 839–848. https://doi.org/10.1111/pbi.12464

Li, Y., Meissle, M., & Romeis, J. (2008). Consumption of Bt maize pollen expressing Cry1Ab or Cry3Bb1 does not harm adult green lacewings, *Chrysoperla carnea* (Neuroptera: Chrysopidae). *PLoS One*, *3*(8), e2909. https://doi.org/10.1371/journal.pone.0002909

Li, Y., Zhang, X., Chen, X., Romeis, J., Yin, X., & Peng, Y. (2015). Consumption of Bt rice pollen containing Cry1C or Cry2A does not pose a risk to *Propylea japonica* (Thunberg) (Coleoptera: Coccinellidae). *Scientific Reports*, *5*(1), 7679. https://doi.org/10.1038/srep07679

Liao, C., Heckel, D. G., & Akhurst, R. (2002). Toxicity of *Bacillus thuringiensis* insecticidal proteins for *Helicoverpa armigera* and *Helicoverpa punctigera* (Lepidoptera: Noctuidae), major pests of cotton. *Journal of Invertebrate Pathology*, *80*(1), 55–63. https://doi.org/10.1016/S0022–2011(02)00035–6

Liu, J., Liang, Y. S., Hu, T., Zeng, H., Gao, R., Wang, L., & Xiao, Y. H. (2021). Environmental fate of Bt proteins in soil: Transport, adsorption/desorption and degradation. *Ecotoxicology and Environmental Safety*, *226*, 112805. https://doi.org/10.1016/j.ecoenv.2021.112805

Liu, L., Wu, L., & Eickhorst, T. (2018b). Accumulation of Cry1Ab/Ac proteins released from transgenic Bt-rice in the rhizosphere of a paddy soil. *Rhizosphere*, *6*, 39–46. https://doi.org/10.1016/j.rhisph.2018.02.002

Liu, Q., Hallerman, E., Peng, Y., & Li, Y. (2016). Development of Bt rice and Bt maize in China and their efficacy in target pest control. *International Journal of Molecular Sciences*, *17*(10), 1561. https://doi.org/10.3390/ijms17101561

Liu, Y., Wang, Y., Shu, C., Lin, K., Song, F., Bravo, A., Soberón, M., & Zhang, J. (2018a). Cry64Ba and Cry64Ca, two ETX/MTX2-type *Bacillus thuringiensis* insecticidal proteins active against hemipteran pests. *Applied and Environmental Microbiology*, *84*(3), e01996–e02017. https://doi.org/10.1128/AEM.01996-17

Losey, J. E., Rayor, L. S., & Carter, M. E. (1999). Transgenic pollen harms monarch larvae. *Nature*, *399*(6733), 214–214. https://doi.org/10.1038/20338

Lu, Y., Wu, K., Jiang, Y., Guo, Y., & Desneux, N. (2012). Widespread adoption of Bt cotton and insecticide decrease promotes biocontrol services. *Nature*, *487*(7407), 362–365. https://doi.org/10.1038/nature11153

Lu, Y., Wu, K., Jiang, Y., Xia, B., Li, P., Feng, H., Wyckhuys, K. A. G., & Guo, Y. (2010). Mirid bug outbreaks in multiple crops correlated with wide-scale adoption of Bt cotton in China. *Science*, *328*(5982), 1151–1154. https://doi.org/10.1126/science.1187881

Luna, V. S., Figueroa, M. J., Baltazar, M. B., Gomez, L. R., Townsend, R., & Schoper, J. B. (2001). Maize pollen longevity and distance isolation requirements for effective pollen control. *Crop Science*, *41*(5), 1551–1557. https://doi.org/10.2135/cropsci2001.4151551x

Malone, L. A., & Burgess, E. P. J. (2009). Impact of genetically modified crops on pollinators. In N. Ferry, & A. M. R. Gatehouse (Eds.), *Environmental impact of genetically modified crops* (pp. 199–222). CABI.

Malone, L. A., & Pham-Delègue, M. H. (2001). Effects of transgene products on honey bees (*Apis mellifera*) and bumblebees (*Bombus* sp.). *Apidologie*, *32*(4), 287–304. https://doi.org/10.1051/apido:2001130

Maqbool, S. B., & Christou, P. (1999). Multiple traits of agronomic importance in transgenic indica rice plants: Analysis of transgene integration patterns, expression levels and stability. *Molecular Breeding*, *5*, 471–480. https://doi.org/10.1023/A:1009634226797

Maqbool, S. B., Husnain, T., Riazuddin, S., Masson, L., & Christou, P. (1998). Effective control of yellow stem borer and rice leaf folder in transgenic rice *indica* varieties Basmati 370 and M 7 using the novel δ-endotoxin *cry2A Bacillus thuringiensis* gene. *Molecular Breeding*, *4*, 501–507. https://doi.org/10.1023/A:1009660315970

Marques, L. H., Castro, B. A., Rossetto, J., Silva, O. A. B. N., Moscardini, V. F., Zobiole, L. H. S., Santos, A. C., Valverde-Garcia, P., Babcock, J. M., Rule, D. M., & Fernandes, O. A. (2016). Efficacy of soybean's event DAS-81419-2 expressing Cry1F and Cry1Ac to manage key tropical lepidopteran pests under field conditions in Brazil. *Journal of Economic Entomology*, *109*(4), 1922–1928. https://doi.org/10.1093/jee/tow153

Marques, L. H., Santos, A. C., Castro, B. A., Moscardini, V. F., Rosseto, J., Silva, O. A. N., & Babcock, J. M. (2019). Assessing the efficacy of *Bacillus thuringiensis* (Bt) pyramided proteins Cry1F, Cry1A. 105, Cry2Ab2, and Vip3Aa20 expressed in Bt maize against lepidopteran pests in Brazil. *Journal of Economic Entomology*, *112*(2), 803–811. https://doi.org/10.1093/jee/toy380

Marques, L. H., Santos, A. C., Castro, B. A., Moscardini, V. F., Rossetto, J., Silva, O. A. N., Zobiole, L. H. S., Valverde-Garcia, P., Babcock, J. M., Storer, N. P., Rule, D. M., & Fernandes, O. A. (2017). Field evaluation of soybean transgenic event DAS-81419-2 expressing Cry1F and Cry1Ac proteins for the control of secondary lepidopteran pests in Brazil. *Crop Protection*, *96*, 109–115. https://doi.org/10.1016/j.cropro.2017.02.014

Meissle, M., Hellmich, R. L., & Romeis, J. (2011). Impact of Cry3Bb1-expressing Bt maize on adults of the western corn rootworm, *Diabrotica virgifera virgifera* (Coleoptera: Chrysomelidae). *Pest Management Science*, *67*(7), 807–814. https://doi.org/10.1002/ps.2117

Meiyalaghan, S., Barrell, P. J., Jacobs, J. M., & Conner, A. J. (2011). Regeneration of multiple shoots from transgenic potato events facilitates the recovery of phenotypically normal lines: Assessing a *cry9Aa2* gene conferring insect resistance. *BMC Biotechnology*, *11*, 93. https://doi.org/10.1186/1472-6750-11-93

Meiyalaghan, S., Jacobs, J. M. E., Butler, R. C., Wratten, S. D., & Conner, A. J. (2006a). Expression of *cry1Ac9* and *cry9Aa2* genes under a potato light-inducible Lhca 3 promoter in transgenic potatoes for tuber moth resistance. *Euphytica*, *147*, 297–309. https://doi.org/10.1007/s10681-005-9012-4

Meiyalaghan, S., Jacobs, J. M. E., Butler, R. C., Wratten, S. D., & Conner, A. J. (2006b). Transgenic potato lines expressing *cry1Ba1* or *cry1Ca5* genes are resistant to potato tuber moth. *Potato Research*, *49*, 203–216. https://doi.org/10.1007/s11540-007-9017-6

Meiyalaghan, S., Pringle, J. M., Barrell, P. J., Jacobs, J. M., & Conner, A. J. (2010). Pyramiding transgenes for potato tuber moth resistance in potato. *Plant Biotechnology Reports*, *4*(4), 293–301. https://doi.org/10.1007/s11816-010-0145-0

Mendelsohn, M., Kough, J., Vaituzis, Z., & Matthews, K. (2003). Are Bt crops safe?. *Nature Biotechnology*, *21*(9), 1003–1009. https://doi.org/10.1038/nbt0903–1003

Mesnage, R., Clair, E., Gress, S., Then, C., Székács, A., & Séralini, G. E. (2013). Cytotoxicity on human cells of Cry1Ab and Cry1Ac Bt insecticidal toxins alone or with a glyphosate-based herbicide. *Journal of Applied Toxicology*, *33*(7), 695–699. https://doi.org/10.1002/jat.2712

Miethling-Graff, R., Dockhorn, S., & Tebbe, C. C. (2010). Release of the recombinant Cry3Bb1 protein of Bt maize MON88017 into field soil and detection of effects on the diversity of rhizosphere bacteria. *European Journal of Soil Biology*, *46*(1), 41–48. https://doi.org/10.1016/j.ejsobi.2009.10.003

Moar, W. J., Eubanks, M., Freeman, B., Turnipseed, S., Ruberson, J., & Head, G. (2002). Effects of Bt cotton on biological control agents in the southeastern United States. In R. G. Van Driesche (Ed.), *Proceedings of the first international symposium on biological control of arthropods* (pp. 275–277). U.S. Department of Agriculture Forest Service.

Moscardini, V. F., Marques, L. H., Santos, A. C., Rossetto, J., Silva, O. A., Rampazzo, P. E., & Castro, B. A. (2020). Efficacy of *Bacillus thuringiensis* (Bt) maize expressing Cry1F, Cry1A. 105, Cry2Ab2 and Vip3Aa20 proteins to manage the fall armyworm (Lepidoptera: Noctuidae) in Brazil. *Crop Protection*, *137*, 105269. https://doi.org/10.1016/j.cropro.2020.105269

Moussa, S., Baiomy, F., Abouzaid, K., Nasr, M., Moussa, E. M., & Kamel, E. A. (2018). Potential impact of host pest fed on Bt-modified corn on the development of *Chrysoperla carnea* (Stephens) (Neuroptera: Chrysopidae). *Egyptian Journal of Biological Pest Control*, *28*, 23. https://doi.org/10.1186/s41938-017-0018-8

Naimov, S., Weemen-Hendriks, M., Dukiandjiev, S., & de Maagd, R. A. (2001). *Bacillus thuringiensis* delta-endotoxin Cry1 hybrid proteins with increased activity against the Colorado potato beetle. *Applied and Environmental Microbiology*, *67*(11), 5328–5330. https://doi.org/10.1128/AEM.67.11.5328–5330.2001

Naranjo, S. E. (2011). Impacts of Bt transgenic cotton on integrated pest management. *Journal of Agricultural and Food Chemistry*, *59*(11), 5842–5851. https://doi.org/10.1021/jf102939c

Narendran, M., Lalwani, S., Girhepuje, P., Shirale, D., Mustafa, G., Gandhi, R., Parimi, S., Char, B. R., & Zehr, U. B. (2006). Production of marker-free insect resistant transgenic brinjal (*Solanum melongena* L.) carrying *cry2Ab* gene. In M. I. Chadha, G. Kuo, & C. L. L. Gowda (Eds.), *I international conference on indigenous vegetables and legumes. Prospectus for fighting poverty, hunger and malnutrition* (pp. 535–538). https://doi.org/10.17660/ActaHortic.2007.752.100

Nayak, P., Basu, D., Das, S., Basu, A., Ghosh, D., Ramakrishnan, N. A., Ghosh, M., & Sen, S. K. (1997). Transgenic elite indica rice plants expressing CryIAc∂-endotoxin of *Bacillus thuringiensis* are resistant against yellow stem borer (*Scirpophaga incertulas*). *Proceedings of the National Academy of Sciences*, *94*(6), 2111–2116. https://doi.org/10.1073/pnas.94.6.2111

Niu, L., Ma, Y., Mannakkara, A., Zhao, Y., Ma, W., Lei, C., & Chen, L. (2013). Impact of single and stacked insect-resistant Bt-cotton on the honey bee and silkworm. *PLoS One*, *8*(9), e72988. https://doi.org/10.1371/journal.pone.0072988

Niu, L., Tian, Z., Liu, H., Zhou, H., Ma, W., Lei, C., & Chen, L. (2018). Transgenic Bt cotton expressing Cry1Ac/Cry2Ab or Cry1Ac/EPSPS does not affect the plant bug *Adelphocoris suturalis* or the pollinating beetle *Haptoncus luteolus*. *Environmental Pollution*, *234*, 788–793. https://doi.org/10.1016/j.envpol.2017.11.047

Olsen, K. M., Daly, J. C., Holt, H. E., & Finnegan, E. J. (2005). Season-long variation in expression of Cry1Ac gene and efficacy of *Bacillus thuringiensis* toxin in transgenic cotton against *Helicoverpa armigera* (Lepidoptera: Noctuidae). *Journal of Economic Entomology*, *98*(3), 1007–1017. https://doi.org/10.1603/0022-0493-98.3.1007

O'Reilly, D., Mascarenhas, V., Green, M., Townley, A., Potts, A., & McCaffery, A. (2006). Determination of the high dose status of cotton events expressing Cry1Ab. In *Proceedings of the 2006 Beltwide cotton conference* (pp. 1428–1436). National Cotton Council of America.

Ostlie, K. R., Hutchison, W. D., & Hellmich, R. L. (1997). *Bt-corn & European corn borer: Long term success through resistance management*. University of Nebraska–Lincoln.

Pal, J. K., Singh, M., Rai, M., Satpathy, S., Singh, D. V., & Kumar, S. (2009). Development and bioassay of Cry1Ac-transgenic eggplant (*Solanum melongena* L.) resistant to shoot and fruit borer. *The Journal of Horticultural Science and Biotechnology*, *84*(4), 434–438. https://doi.org/10.1080/14620316.2009.11512545

Parrott, W. A., All, J. N., Adang, M. J., Bailey, M. A., Boerma, H. R., & Stewart, C. N. (1994). Recovery and evaluation of soybean plants transgenic for a *Bacillus thuringiensis* var. *kurstaki* insecticidal gene. *In Vitro Cellular & Developmental Biology—Plant*, *30*(3), 144–149. https://doi.org/10.1007/BF02632204

Paula, D. P., Andow, D. A., Timbó, R. V., Sujii, E. R., Pires, C. S., & Fontes, E. M. (2014). Uptake and transfer of a Bt toxin by a Lepidoptera to its eggs and effects on its offspring. *PLoS One*, *9*(4), e95422. https://doi.org/10.1371/journal.pone.0095422

Peferoen, M., Jansens, S., Reynaerts, A., & Leemans, J. (1990). Potato plants with engineered resistance against insect attack. In W. R. Belknap, M. E. Vayda, & W. D. Park (Eds.), *The molecular and cellular biology of the potato* (pp. 193–204). CABI.

Pérez-Hedo, M., López, C., Albajes, R., & Eizaguirre, M. (2012). Low susceptibility of non-target Lepidopteran maize pests to the Bt protein Cry1Ab. *Bulletin of Entomological Research*, *102*(6), 737–743. https://doi.org/10.1017/S0007485312000351

Perlak, F. J., Stone, T. B., Muskopf, Y. N., Peterson, L. J., Parker, G. B., McPherson, S. A., Wyman, J., Love, S., Reed, G., Biever, D., & Fischoff, D. A. (1993). Genetically improved potatoes: Protection from damage by Colorado potato beetles. *Plant Molecular Biology*, *22*, 313–321. https://doi.org/10.1007/BF00014938

Pierre, J., & Hofs, J. L. (2010). *Astylus atromaculatus* (Coleoptera: Melyridae): Abundance and role in pollen dispersal in Bt and non-Bt cotton in South Africa. *Environmental Entomology*, *39*(5), 1523–1531. https://doi.org/10.1603/EN09142

Pimentel, D. S., & Raven, P. H. (2000). Bt corn pollen impacts on nontarget Lepidoptera: Assessment of effects in nature. *Proceedings of the National Academy of Sciences*, *97*(15), 8198–8199. https://doi.org/10.1073/pnas.97.15.8198

Prodhan, M. Z. H., Hasan, M. T., Chowdhury, M. M. I., Alam, M. S., Rahman, M. L., Azad, A. K., Hossain, M. J., Naranjo, S.

E., & Shelton, A. M. (2018). Bt eggplant (*Solanum melongena* L.) in Bangladesh: Fruit production and control of eggplant fruit and shoot borer (*Leucinodes orbonalis* Guenee), effects on non-target arthropods and economic returns. *PLoS One*, *13*(11), e0205713. https://doi.org/10.1371/journal.pone.0205713

Pyke, B. A. (2003). The performance of Bt transgenic (INGARD®) cotton in Autralia over six seasons. In A. Swanepoel (Ed.), *Cotton Production for the New Millennium—Proceedings of the World Cotton Conference-3* (pp. 1274–1280). Agricultural Research Council, Institute for Industrial Crops.

Qaim, M., & Zilberman, D. (2003). Yield effects of genetically modified crops in developing countries. *Science*, *299*(5608), 900–902. https://doi.org/10.1126/science.108060

Qi, X., Liu, B., Wu, H., Song, Q., Jiang, J., Bu, Y., Rui, J., Zou, B., & Zhou, G. (2018). Bacterial communities under long-term conventional and transgenic cotton farming systems using V3-V5 and V5-V9 of 16S rDNA. *Ecotoxicology and Environmental Safety*, *164*, 618–628. https://doi.org/10.1016/j.ecoenv.2018.08.038

Qin, D., Liu, X. Y., Miceli, C., Zhang, Q., & Wang, P. W. (2019). Soybean plants expressing the *Bacillus thuringiensis cry8*-like gene show resistance to *Holotrichia parallela*. *BMC Biotechnology*, *19*(1), 66. https://doi.org/10.1186/S12896-019-0563-1/TABLES/3

Rahman, M., Zafar, Y., & Zhang, T. (2021). *Cotton precision breeding*. Springer Cham. https://doi.org/10.1007/978-3-030-64504-5

Rai, N. P., Rai, G. K., Kumar, S., Kumari, N., & Singh, M. (2013). Shoot and fruit borer resistant transgenic eggplant (*Solanum melongena* L.) expressing *cry1Aa3* gene: Development and bioassay. *Crop Protection*, *53*, 37–45. https://doi.org/10.1016/j.cropro.2013.06.005

Reed, G. L., Jensen, A. S., Riebe, J., Head, G., & Duan, J. J. (2001). Transgenic Bt potato and conventional insecticides for Colorado potato beetle management: Comparative efficacy and non-target impacts. *Entomologia experimentalis et applicata*, *100*(1), 89–100. https://doi.org/10.1046/j.1570–7458.2001.00851.x

Reisig, D., & Goldsworthy, E. (2023). Yield and efficacy of foliar chlorantraniliprole (Vantacor) against bollworms in non-Bt, Bollgard II, WideStrike 3, TwinLink Plus, and Bollgard 3 cotton, 2021. *Arthropod Management Tests*, *48*(1), tsad080. https://doi.org/10.1093/amt/tsad080

Romeis, J., Álvarez-Alfageme, F., & Bigler, F. (2012). Putative effects of Cry1Ab to larvae of *Adalia bipunctata*-reply to Hilbeck et al. (2012). *Environmental Sciences Europe*, *24*, 18. https://doi.org/10.1186/2190-4715-24-18

Rose, R., Dively, G. P., & Pettis, J. (2007). Effects of Bt corn pollen on honey bees: Emphasis on protocol development. *Apidologie*, *38*(4), 368–377. https://doi.org/10.1051/apido:2007022

Rui, Y. K., Yi, G. X., Zhao, J., Wang, B. M., Li, Z. H., Zhai, Z. X., He, Z. P., & Li, Q. X. (2005). Changes of Bt toxin in the rhizosphere of transgenic Bt cotton and its influence on soil functional bacteria. *World Journal of Microbiology and Biotechnology*, *21*, 1279–1284. https://doi.org/10.1007/s11274-005-2303-z

Sabugosa-Madeira, B., Abreu, I., Ribeiro, H., & Cunha, M. (2007). Bt transgenic maize pollen and the silent poisoning of the hive. *Journal of Apicultural Research*, *46*(1), 57–58. https://doi.org/10.1080/00218839.2007.11101367

Saha, S., Nath, R., Dey, A., Laskar, N., & Kundu, A. (2023). Melissopalynological analysis of *Apis* honey from northern districts of West Bengal. *Grana*, *62*(5–6), 352–368. https://doi.org/10.1080/00173134.2023.2273542

Salehian, H., Rahnama, H., Dezhsetan, S., & Babaei, S. (2021). Constitutive expression of *cry3A* gene in transgenic potato plants for resistance to Colorado Potato Beetle (CPB). *Potato Research*, *64*(4), 667–679. https://doi.org/10.1007/s11540-021-09500-5

Saxena, D., Stewart, C. N., Altosaar, I., Shu, Q., & Stotzky, G. (2004). Larvicidal Cry proteins from *Bacillus thuringiensis* are released in root exudates of transgenic *B. thuringiensis* corn, potato, and rice but not of *B. thuringiensis* canola, cotton, and tobacco. *Plant Physiology and Biochemistry*, *42*(5), 383–387. https://doi.org/10.1016/j.plaphy.2004.03.004

Saxena, D., & Stotzky, G. (2001). *Bacillus thuringiensis* (Bt) toxin released from root exudates and biomass of Bt corn has no apparent effect on earthworms, nematodes, protozoa, bacteria, and fungi in soil. *Soil Biology and Biochemistry*, *33*(9), 1225–1230. https://doi.org/10.1016/S0038–0717(01)00027-X

Schmidt, J. E., Braun, C. U., Whitehouse, L. P., & Hilbeck, A. (2009). Effects of activated Bt transgene products (Cry1Ab, Cry3Bb) on immature stages of the ladybird *Adalia bipunctata* in laboratory ecotoxicity testing. *Archives of Environmental Contamination and Toxicology*, *56*, 221–228. https://doi.org/10.1007/s00244-008-9191-9

Schuler, T. H., Denholm, I., Clark, S. J., Stewart, C. N., & Poppy, G. M. (2004). Effects of Bt plants on the development and survival of the parasitoid *Cotesia plutellae* (Hymenoptera: Braconidae) in susceptible and Bt-resistant larvae of the diamondback moth, *Plutella xylostella* (Lepidoptera: Plutellidae). *Journal of Insect Physiology*, *50*(5), 435–443. https://doi.org/10.1016/j.jinsphys.2004.03.001

Sears, M. K., Hellmich, R. L., Stanley-Horn, D. E., Oberhauser, K. S., Pleasants, J. M., Mattila, H. R., Siegfried, B. D., & Dively, G. P. (2001). Impact of Bt corn pollen on monarch butterfly populations: A risk assessment. *Proceedings of the National Academy of Sciences*, *98*(21), 11937–11942. https://doi.org/10.1073/pnas.211329998

Selvapandiyan, A., Reddy, V. S., Kumar, P. A., Tewari, K. K., & Bhatnagar, R. K. (1998). Transformation of *Nicotiana tabacum* with a native *cry1Ia5* gene confers complete protection against *Heliothis armigera*. *Molecular Breeding*, *4*, 473–478. https://doi.org/10.1023/A:1009696027993

Séralini, G. E., Cellier, D., & de Vendomois, J. S. (2007). New analysis of a rat feeding study with a genetically modified maize reveals signs of hepatorenal toxicity. *Archives of Environmental Contamination and Toxicology*, *52*, 596–602. https://doi.org/10.1007/s00244-006-0149-5

Shelton, A. M. (2010). The long road to commercialization of Bt brinjal (eggplant) in India. *Crop Protection*, *29*(5), 412–414. https://doi.org/10.1016/j.cropro.2010.02.016

Shelton, A. M., Hossain, M. J., Paranjape, V., Prodhan, M. Z., Azad, A. K., Majumder, R., Sarwer, S. H., & Hossain, M. A. (2019). Bt brinjal in Bangladesh: The first genetically engineered food crop in a developing country. *Cold Spring Harbor Perspectives in Biology*, *11*(10), a034678. https://doi.org/10.1101/cshperspect.a034678

Shelton, A. M., Zhao, J. Z., & Roush, R. T. (2002). Economic, ecological, food safety, and social consequences of the development of Bt transgenic plants. *Annual Review of Entomology*, *47*, 845–881. https://doi.org/10.1146/annurev.ento.47.091201.145309

Shrivastava, D., Dalal, M., Nain, V., Sharma, P. C., & Kumar, P. A. (2011). Targeted Integration of *Bacillus thuringiensis*-Endotoxin cry1Fa1 in Brinjal (*Solanum melongena* L.). *Current Trends in Biotechnology and Pharmacy*, *5*(2), 1149–1156.

Siebert, M. W., Babock, J. M., Nolting, S., Santos, A. C., Adamczyk Jr., J. J., Neese, P. A., King, J. E., Jenkins, J. N., McCarty, J., Lorenz, G. M., Fromme, D. D., & Lassiter, R. B. (2008). Efficacy of Cry1F insecticidal protein in maize and cotton for control of fall armyworm (Lepidoptera: Noctuidae). *Florida Entomologist*, *91*(4), 555–565. https://doi.org/10.1653/0015-4040-91.4.555

Siebert, M. W., Nolting, S. P., Hendrix, W., Dhavala, S., Craig, C., Leonard, B. R., Stewart, S. D., All, J., Musser, F. R., Buntin, G. D., & Samuel, L. (2012). Evaluation of corn hybrids expressing Cry1F, Cry1A. 105, Cry2Ab2, Cry34Ab1/Cry35Ab1, and Cry3Bb1 against southern United States insect pests. *Journal of Economic Entomology*, *105*(5), 1825–1834. https://doi.org/10.1603/EC12155

Silici, S., & Gökceoglu, M. (2007). Pollen analysis of honeys from mediterranean region of Anatolia. *Grana*, *46*(1), 57–64. https://doi.org/10. 1080/00173130601138783.

Singh, A. K., Paritosh, K., Kant, U., Burma, P. K., & Pental, D. (2016). High expression of Cry1Ac protein in cotton (*Gossypium hirsutum*) by combining independent transgenic events that target the protein to cytoplasm and plastids. *PLoS One*, *11*(7), e0158603. https://doi.org/10.1371/journal.pone.0158603

Singh, R. K., Channappa, R. K. Deeba, F., Nagaraj, N. J., Sukavaneaswaran, M., K., & Manjunath, T. M. (2005). Tolerance of Bt corn (MON 810) to maize stem borer, *Chilo partellus* (Lepidoptera: Pyralidae). *Plant Cell Reports*, *24*, 556–560. https://doi.org/10.1007/s00299-005-0942-z

Singh, R. K., Khan, M. S., Yadav, S., Singh, M. R., Madhok, H. L., Srivastava, S., & Singh, J. (2011). *Agrobacterium*-mediated genetic transformation of subtropical sugarcane variety with Cry1Ab gene. *National Academy Science Letters*, *34*(3&4), 103–111.

Sisterson, M. S., & Tabashnik, B. E. (2005). Simulated effects of transgenic Bt crops on specialist parasitoids of target pests. *Environmental Entomology*, *34*(4), 733–742. https://doi.org/10.1603/0046-225X-34.4.733

Slaney, A. C., Robbins, H. L., & English, L. (1992). Mode of action of *Bacillus thuringiensis* toxin CryIIIA: An analysis of toxicity in *Leptinotarsa decemlineata* (Say) and *Diabrotica undecimpunctata howardi* Barber. *Insect Biochemistry and Molecular Biology*, *22*(1), 9–18. https://doi.org/10.1016/0965–1748(92)90094-U

Sohail, M. N., Karimi, S. M., Asad, S., Mansoor, S., Zafar, Y., & Mukhtar, Z. (2012). Development of broad-spectrum insect-resistant tobacco by expression of synthetic *cry1Ac* and *cry2Ab* genes. *Biotechnology Letters*, *34*, 1553–1560. https://doi.org/10.1007/s10529-012-0923-6

Song, F., Zhang, J., Gu, A., Wu, Y., Han, L., He, K., Chen, Z., Yao, J., Hu, Y., Li, G., & Huang, D. (2003). Identification of *cry1I*-type genes from *Bacillus thuringiensis* strains and characterization of a novel *cry1I*-type gene. *Applied and Environmental Microbiology*, *69*(9), 5207–5211. https://doi.org/10.1128/AEM.69.9.5207–5211.2003

Sparks, T. C., Graves, J. B., & Leonard, B. R. (1993). Insecticide resistance and the tobacco budworm: Past, present and future. *Reviews in Pesticide Toxicology (USA)*, *2*, 149–183.

Stanley-Horn, D. E., Dively, G. P., Hellmich, R. L., Mattila, H. R., Sears, M. K., Rose, R., Hansen Jesse, L. C., Losey, J. E., Obrycki, J. J., & Lewis, L. (2001). Assessing the impact of Cry1Ab-expressing corn pollen on monarch butterfly larvae in field studies. *Proceedings of the National Academy of Sciences*, *98*(21), 11931–11936. https://doi.org/10.1073/pnas.211277798

Stephens, E. J., Losey, J. E., Allee, L. L., DiTommaso, A., Bodner, C., & Breyre, A. (2012). The impact of Cry3Bb Bt-maize on two guilds of beneficial beetles. *Agriculture, Ecosystems & Environment*,*156*,72–81.https://doi.org/10.1016/j.agee.2012.05.002

Stewart Jr., C. N., Adang, M. J., All, J. N., Boerma, H. R., Cardineau, G., Tucker, D., & Parrott, W. A. (1996a). Genetic transformation, recovery, and characterization of fertile soybean transgenic for a synthetic *Bacillus thuringiensis cryIAc* gene. *Plant Physiology*, *112*(1), 121–129. https://doi.org/10.1104/PP.112.1.121

Stewart Jr., C. N., Adang, M. J., All, J. N., Raymer, P. L., Ramachandran, S., & Parrott, W. A. (1996b). Insect control and dosage effects in transgenic canola containing a synthetic *Bacillus thuringiensis cryIAc* gene. *Plant Physiology*, *112*(1), 115–120. https://doi.org/10.1104/pp.112.1.115

Stewart, S. D., Adamczyk Jr., J. J., Knighten, K. S., & Davis, F. M. (2001). Impact of Bt cottons expressing one or two insecticidal proteins of *Bacillus thuringiensis* Berliner on growth and survival of noctuid (Lepidoptera) larvae. *Journal of Economic Entomology*, *94*(3), 752–760. https://doi.org/10.1603/0022-0493-94.3.752

Strizhov, N., Keller, M., Mathur, J., Koncz-Kálmán, Z., Bosch, D., Prudovsky, E., Schell, J., Sneh, B., Koncz, C., & Zilberstein, A. (1996). A synthetic *cryIC* gene, encoding a *Bacillus thuringiensis* δ-endotoxin, confers *Spodoptera* resistance in alfalfa and tobacco. *Proceedings of the National Academy of Sciences*, *93*(26), 15012–15017. https://doi.org/10.1073/pnas.93.26.15012

Sun, Y., Liu, X., Li, L., Guan, Y., & Zhang, J. (2015). Breeding of transgenic maize with resistance to the Asian corn borer (*Ostrinia furnacalis*) and tolerance to glyphosate. *Journal of Agricultural Biotechnology*, *23*, 52–60.

Székács, A., Lauber, É., Juracsek, J., & Darvas, B. (2010). Cry1Ab toxin production of MON 810 transgenic maize. *Environmental Toxicology and Chemistry*, *29*(1), 182–190. https://doi.org/10.1002/etc.5

Tabashnik, B. E., Dennehy, T. J., Sims, M. A., Larkin, K., Head, G. P., Moar, W. J., & Carriere, Y. (2002). Control of resistant pink bollworm (*Pectinophora gossypiella*) by transgenic cotton that produces *Bacillus thuringiensis* toxin Cry2Ab. *Applied and Environmental Microbiology*, *68*(8), 3790–3794. https://doi.org/10.1128/AEM.68.8.3790-3794.2002

Tan, S. Y., Cayabyab, B. F., Alcantara, E. P., Ibrahim, Y. B., Huang, F., Blankenship, E. E., & Siegfried, B. D. (2011). Comparative susceptibility of *Ostrinia furnacalis*, *Ostrinia nubilalis* and *Diatraea saccharalis* (Lepidoptera: Crambidae) to *Bacillus thuringiensis* Cry1 toxins. *Crop Protection*, *30*(9), 1184–1189. https://doi.org/10.1016/j.cropro.2011.05.009

Thompson, G. D., Dalmacio, S. C., Criador, A. I., Alvarez, E. R., & Hechanova, R. F. (2010). Field performance of TC1507 transgenic corn hybrids against Asian corn borer in the Philippines. *Philippine Agricultural Scientist*, *93*(4), 375–383.

Tiwari, S., Mishra, D. K., Singh, A., Singh, P. K., & Tuli, R. (2008). Expression of a synthetic *cry1EC* gene for resistance against *Spodoptera litura* in transgenic peanut (*Arachis hypogaea* L.). *Plant Cell Reports*, *27*, 1017–1025. https://doi.org/10.1007/s00299-008-0525-x

Tohidfar, M., Zare, N., Jouzani, G. S., & Eftekhari, S. M. (2013). *Agrobacterium*-mediated transformation of alfalfa (*Medicago sativa*) using a synthetic *cry3a* gene to enhance resistance against alfalfa weevil. *Plant Cell, Tissue and Organ Culture (PCTOC), 113*, 227–235. https://doi.org/10.1007/s11240-012-0262-2

TOI. (2010, February 9). Govt. says no to Bt brinjal for now. *The Times of India*. https://timesofindia.indiatimes.com/india/govt-says-no-to-bt-brinjal-for-now/articleshow/5552403.cms

Torres, J. B., Ruberson, J. R., & Whitehouse, M. (2010). Transgenic cotton for sustainable pest management: A review. In E. Lichtfouse (Ed.), *Organic farming, pest control and remediation of soil pollutants* (pp. 15–53). Springer Dordrecht. https://doi.org/10.1007/978-1-4020-9654-94

Traore, S. B., Carlson, R. E., Pilcher, C. D., & Rice, M. E. (2000). Bt and non-Bt maize growth and development as affected by temperature and drought stress. *Agronomy Journal, 92*(5), 1027–1035. https://doi.org/10.2134/agronj2000.9251027x

Tu, J., Datta, K., Alam, M. F., Fan, Y., Khush, G. S., & Datta, S. K. (1998). Expression and function of a hybrid Bt toxin gene in transgenic rice conferring resistance to insect pest. *Plant Biotechnology, 15*(4), 195–203. https://doi.org/10.5511/plantbiotechnology.15.195

Tu, J., Zhang, G., Datta, K., Xu, C., He, Y., Zhang, Q., Khush, G. S., & Datta, S. K. (2000). Field performance of transgenic elite commercial hybrid rice expressing *Bacillus thuringiensis* δ-endotoxin. *Nature Biotechnology, 18*(10), 1101–1104. https://doi.org/10.1038/80310

USDA. (2011). *Determination of plant pest risk for insect-resistant and glufosinate ammonium-tolerant (TwinLink™) cotton, Gossypium hirsutum, events T304–40 x GHB119* (Petition 08–340–01p). United States Department of Agriculture.

US EPA. (2002). *Bacillus thuringiensis Cry2Ab2 protein and the genetic material necessary for its production in cotton (006487) fact sheet*. U.S. Environmental Protection Agency.

US EPA. (2004). *EPA conditionally registers new Bt cotton product*. U.S. Environmental Protection Agency. https://www.epa.gov/archive/epapages/newsroom_archive/newsreleases/49f29ccb2e5a2357852570350056ef75.html

US EPA. (2016). *Notice of pesticide registration*. U.S. Environmental Protection Agency. https://www3.epa.gov/pesticides/chem_search/ppls/000524-00627-20160406.pdf

Vaeck, M., Reynaerts, A., Hofte, H., Jansens, S., DeBeuckleer, M., Dean, C., Zabeau, M., Van Montagu, M., & Leemans, J. (1987). Transgenic plants protected from insect attack. *Nature, 327*, 33–37. https://doi.org/10.1038/328033a0

Vaknin, Y., Gan-Mor, S., Bechar, A., Ronen, B., & Eisikowitch, D. (2000). The role of electrostatic forces in pollination. In A. Dafni, M. Hesse, & E. Pacini (Eds.), *Pollen and pollination* (pp. 133–142). Springer Vienna. https://doi.org/10.1007/978-3-7091-6306-1_7

Van den Berg, J., & Van Wyk, A. (2007). The effect of Bt maize on *Sesamia calamistis* in South Africa. *Entomologia Experimentalis et Applicata, 122*(1), 45–51. https://doi.org/10.1111/j.1570-7458.2006.00492.x

Van Rensburg, J. B. J. (1999). Evaluation of Bt.-Transgenic maize for resistance to the stem borers *Busseola fusca* (Fuller) and *Chilo partellus* (Swinhoe) in South Africa. *South African Journal of Plant and Soil, 16*(1), 38–43.

van Rie, J., Jansens, S., & Reynaerts, A. (1994). Engineered resistance against potato tuber moth. In G. M. Zehnder, M. L. Powelson, R. K. Jansson, & K. V. Raman (Eds.), *Advances in potato pest biology and management* (pp. 499–508). The American Phytopathological Society press.

Vaughn, T., Cavato, T., Brar, G., Coombe, T., DeGooyer, T., Ford, S., Groth, M., Howe, A., Johnson, S., Kolacz, K., Pilcher, C., Purcell, J., Romano, C., English, L., & Pershing, J. (2005). A method of controlling corn rootworm feeding using a *Bacillus thuringiensis* protein expressed in transgenic maize. *Crop Science, 45*(3), 931–938. https://doi.org/10.2135/cropsci2004.0304

Viswakarma, N., Bhattacharya, R. C., Chakrabarty, R., Bhat, S. R., Kirti, P. B., Shastri, N. V., & Chopra, V. L. (2004). Insect resistance of transgenic broccoli ('Pusa Broccoli KTS-1') expressing a synthetic *cryIA (b)* gene. *The Journal of Horticultural Science and Biotechnology, 79*(2), 182–188. https://doi.org/10.1080/14620316.2004.11511746

Wang, G., Castiglione, S., Chen, Y., Li, L., Han, Y., Tian, Y., Gabriel, D. W., Han, Y., Mang, K., & Sala, F. (1996). Poplar (*Populus nigra* L.) plants transformed with a *Bacillus thuringiensis* toxin gene: Insecticidal activity and genomic analysis. *Transgenic Research, 5*, 289–301. https://doi.org/10.1007/BF01968939

Wang, Y. N., Ke, K. Q., Li, Y. H., Han, L. Z., Liu, Y. M., Hua, H. X., & Peng, Y. F. (2016). Comparison of three transgenic Bt rice lines for insecticidal protein expression and resistance against a target pest, *Chilo suppressalis* (Lepidoptera: Crambidae). *Insect Science, 23*(1), 78–87. https://doi.org/10.1111/1744-7917.12178

Wang, Y. N., Zhang, L., Li, Y., Liu, Y., Han, L., Zhu, Z., Wang, F., & Peng, Y. (2014). Expression of Cry1Ab protein in a marker-free transgenic Bt rice line and its efficacy in controlling a target pest, *Chilo suppressalis* (Lepidoptera: Crambidae). *Environmental Entomology, 43*(2), 528–536. https://doi.org/10.1603/EN13254

Wang, Z. H., Wang, Y., Cui, H. R., Xia, Y. W., Altosaar, I., & Shu, Q. Y. (2002). Toxicological evaluation of transgenic rice flour with a synthetic *cry1Ab* gene from *Bacillus thuringiensis*. *Journal of the Science of Food and Agriculture, 82*(7), 738–744. https://doi.org/10.1002/jsfa.1105

Wei, C. (1992). Investigation on the resistance level of the cotton bollworm *Heliothis armigera* to four kinds of pyrethroid pesticides in China in 1991. *Resistant Pest Management, 4*(1), 17–21.

Williams, W. P., Buckley, P. M., Sagers, J. B., & Hanten, J. A. (1998). Evaluation of transgenic corn for resistance to corn earworm (Lepidoptera: Noctuidae), fall armyworm (Lepidoptera: Noctuidae), and southwestern corn borer (Lepidoptera: Crambidae) in a laboratory bioassay. *Journal of Agricultural Entomology, 15*(2), 105–112.

Williams, W. P., Sagers, J. B., Hanten, J. A., Davis, F. M., & Buckley, P. M. (1997). Transgenic corn evaluated for resistance to fall armyworm and southwestern corn borer. *Crop Science, 37*(3), 957–962. https://doi.org/10.2135/cropsci1997.0011183X003700030042x

Wu, C., Fan, Y., Zhang, C., Oliva, N., & Datta, S. K. (1997). Transgenic fertile japonica rice plants expressing a modified *cryIA (b)* gene resistant to yellow stem borer. *Plant Cell Reports, 17*, 129–132. https://doi.org/10.1007/s002990050365

Wu, E., Shi, M., Su, H., & Yang, Y. (2012). Effect of transgenic cottons on *Sylepta derogata* and its parasitoid *Apanteles derogatae*. *Chinese Journal of Biological Control, 28*(2), 205–211.

Wu, G., Cui, H., Ye, G., Xia, Y., Sardana, R., Cheng, X., Li, Y., Altosaar, I., & Shu, Q. (2002). Inheritance and expression of the *cry1Ab* gene in Bt (*Bacillus thuringiensis*) transgenic rice. *Theoretical and Applied Genetics, 104*, 727–734. https://doi.org/10.1007/s001220100689

Wünn, J., Klöti, A., Burkhardt, P. K., Biswas, G. C. G., Launis, K., Iglesias, V. A., & Potrykus, I. (1996). Transgenic indica rice breeding line IR58 expressing a synthetic *cryIA (b)*

gene from *Bacillus thuringiensis* provides effective insect pest control. *Bio/technology*, *14*(2), 171–176. https://doi.org/10.1038/nbt0296-171

Xiao, Y., & Wu, K. (2019). Recent progress on the interaction between insects and *Bacillus thuringiensis* crops. *Philosophical Transactions of the Royal Society B*, *374*, 20180316. http://dx.doi.org/10.1098/rstb.2018.0316

Xie, M., Zhang, Y. J., Peng, D. L., Wu, G., Xu, P., Zhao, J. J., & Zhang, Z. R. (2016). Field studies show no significant effect of a Cry1Ab/Ac producing transgenic cotton on the fungal community structure in rhizosphere soil. *European Journal of Soil Biology*, *73*, 69–76. https://doi.org/10.1016/j.ejsobi.2016.01.006

Xu, C., Cheng, J., Lin, H., Lin, C., Gao, J., & Shen, Z. (2018). Characterization of transgenic rice expressing fusion protein Cry1Ab/Vip3A for insect resistance. *Scientific Reports*, *8*(1), 15788. https://doi.org/10.1038/s41598-018-34104-4

Yadav, S., Singh, K., Divekar, P. A., & Karkute, S. G. (2023). Cry2Aa delta-endotoxin confers strong resistance against brinjal fruit and shoot borer in transgenic brinjal (*Solanum melongena* L.) plants. *Pakistan Journal of Zoology*. https://dx.doi.org/10.17582/journal.pjz/20230516110558

Yang, F., Kerns, D. L., Brown, S., Kurtz, R., Dennehy, T., Braxton, B., Head, G., & Huang, F. (2016). Performance and cross-crop resistance of Cry1F-maize selected *Spodoptera frugiperda* on transgenic Bt cotton: Implications for resistance management. *Scientific Reports*, *6*(1), 28059. https://doi.org/10.1038/srep28059

Yang, Y. Y., Liu, Y., Cao, F., Chen, X., Cheng, L., Romeis, J., Li, Y., & Peng, Y. (2014b). Consumption of Bt rice pollen containing Cry1C or Cry2A protein poses a low to negligible risk to the silkworm *Bombyx mori* (Lepidoptera: Bombyxidae). *PLoS One*, *9*(7), e102302. https://doi.org/10.1371/journal.pone.0102302

Yang, Y. Y., Mei, F., Zhang, W., Shen, Z., & Fang, J. (2014a). Creation of Bt rice expressing a fusion protein of Cry1Ac and Cry1I-like using a green tissue-specific promoter. *Journal of Economic Entomology*, *107*(4), 1674–1679. https://doi.org/10.1603/EC13497.

Yaoyu, B., Mingxing, J., & Jiaan, C. (2006). Impacts of environmental factors on degradation of Cry1Ab insecticidal protein in leaves of Bt rice. *Scientia Agricultura Sinica*, *39*(4), 721–727.

Yates-Stewart, A. D., Yorke, B. T., Willse, A., Fridley, J., & Head, G. P. (2023). Using sentinel plots to monitor for changes in thrips susceptibility to MON 88702 cotton containing the Cry51Aa2.834_16 Bt Protein. *Insects*, *14*(6), 497. https://doi.org/10.3390/insects14060497

Ye, R., Huang, H., Yang, Z., Chen, T., Liu, L., Li, X., Chen, H., & Lin, Y. (2009). Development of insect-resistant transgenic rice with Cry1C*-free endosperm. *Pest Management Science*, *65*(9), 1015–1020. https://doi.org/10.1002/ps.1788

Yu, H. L., Li, Y. H., Li, X. J., Romeis, J., & Wu, K. M. (2013). Expression of Cry1Ac in transgenic Bt soybean lines and their efficiency in controlling lepidopteran pests. *Pest Management Science*, *69*(12), 1326–1333. https://doi.org/10.1002/ps.3508

Zakon, H. H. (2016). Electric fields of flowers stimulate the sensory hairs of bumble bees. *Proceedings of the National Academy of Sciences*, *113*(26), 7020–7021. https://doi.org/10.1073/pnas.160742611

Zhang, G. F., Wan, F. H., Liu, W. X., & Guo, J. Y. (2006). Early instar response to plant-delivered Bt-toxin in a herbivore (*Spodoptera litura*) and a predator (*Propylaea japonica*). *Crop Protection*, *25*(6), 527–533. https://doi.org/10.1016/j.cropro.2005.08.008

Zhang, M. J., Feng, M. C., Xiao, L. J., Song, X. Y., Yang, W. D., & Ding, G. W. (2015). Impact of water content and temperature on the degradation of Cry1Ac protein in leaves and buds of Bt cotton in the soil. *PLoS One*, *10*(1), e115240. https://doi.org/10.1371/journal.pone.0115240

Zhang, M. J., Meichen, F., Lujie, X., Xiaoyan, S., Guangwei, D., & Wude, Y. (2019). Persistence of Cry1Ac protein from transgenic Bt cotton cultivation and residue returning in fields and its effect on functional diversity of soil microbial communities. *Pedosphere*, *29*(1), 114–122. https://doi.org/10.1016/S1002-0160(17)60475-2

Zhang, Y., Liu, Y., Ren, Y., Liu, Y., Liang, G., Song, F., Bai, S., Wang, J., & Wang, G. (2013). Overexpression of a novel *Cry1Ie* gene confers resistance to Cry1Ac-resistant cotton bollworm in transgenic lines of maize. *Plant Cell, Tissue and Organ Culture (PCTOC)*, *115*, 151–158. https://doi.org/10.1007/s11240-013-0348-5

Zhao, Y., Ma, Y., Niu, L., Ma, W., Mannakkara, A., Chen, L., & Lei, C. (2013). Bt cotton expressing Cry1Ac/Cry2Ab or Cry1Ac/epsps does not harm the predator *Propylaea japonica* through its prey *Aphis gossypii*. *Agriculture, Ecosystems & Environment*, *179*, 163–167. https://doi.org/10.1016/j.agee.2013.08.005

Zhaolei, L., Naishun, B., Xueping, C., Jun, C., Manqiu, X., Zhiping, S., Ming, N., & Changming, F. (2018). Soil incubation studies with Cry1Ac protein indicate no adverse effect of Bt crops on soil microbial communities. *Ecotoxicology and Environmental Safety*, *152*, 33–41. https://doi.org/10.1016/j.ecoenv.2017.12.054

13 Insects' Resistance to Entomopathogenic Bacteria

ABSTRACT

The role of entomopathogenic bacteria (EPB) in modern-day pest management is enormous. Still, their applicability is highly threatened by the evolution of resistance in insects. To date, a number of laboratory and field-evolved cases of resistance have been reported against two major EPBs, namely, *Bacillus thuringiensis* (Bt) and *Lysinibacillus spharicus* (Ls). Evolution of resistance also imposes a great threat to transgenic Bt crop technology. Certain mechanisms, including decreased toxin binding, enhanced immune response, imbalanced proteolytic processing of toxins, and intestinal regeneration, contribute to the development of resistance. Among them, an alteration in the toxin receptor that prevents toxin binding is the key mechanism. This situation demands urgent requirement of appropriate resistance management strategies, like resistance monitoring, crop rotation, refuge planting, gene pyramiding, toxin synergy, use of modified toxins, and other novel approaches, to delay the resistance evolution in insects. Thus, we can protect the applicability of EPBs from facing a complete downfall.

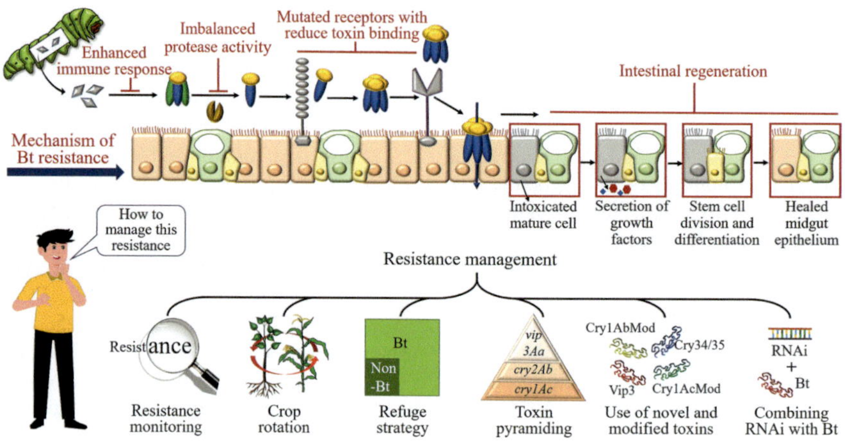

13.1 INTRODUCTION

Undoubtedly, the role of entomopathogenic bacteria (EPBs) in modern-day integrated pest management (IPM) programmes is enormous (Ali et al., 2022; Ruiu, 2015). For the past 50–60 years, different EPB-based bioinsecticides have been utilized worldwide to control a wide range of economically important insect pests. Apart from that, through genetic engineering (GE), a number of agricultural and horticultural crops carrying toxin genes from *Bacillus thuringiensis* Berliner (Bt) have been developed, controlling different insect pests in an efficient and environmentally friendly way (Babu et al., 2003; Gassmann & Reisig, 2023; Kumar et al., 2008). Thus, EPBs become an eco-friendly alternative to chemical insecticides by reducing the usage of these toxic chemicals (Azizoglu et al., 2020).

Because of the co-existence and co-evolution of these EPBs with different insects for millions of years, people started to assume that insects would never be able to develop resistance against these EPBs or their insecticidal toxins. Furthermore, it was assumed that these EPBs might either modify their produced toxins to suit any insect defence mechanism or develop a variety of toxins with particular activities, allowing them to adapt to pest defences. However, both adaptability and co-evolution are relevant factors. Insects are more likely to acquire resistance to transgenic plants and various EPB-based biopesticides if we tamper with nature by implementing various dubious management methods that include the intensive and excessive utilization of these EPBs and their toxins. One could argue that the evolutionary pressure from EPBs and their toxins has pushed insect pests to develop defence mechanisms that can withstand the bacterial attack (Ibrahim et al., 2010). To date, a number of insect species have developed resistance against different EPB-based biopesticides as well as against different transgenic plants under both laboratory and field conditions (Tabashnik et al., 2003, 2009, 2023a, 2023b). In this chapter, we discuss the resistance issues in insects against the two major EPBs, that is, Bt and *Lysinibacillus spharicus* (Meyer and Neide) (Ls).

DOI: 10.1201/9781003426882-13

13.2 INSECTS' RESISTANCE TO *BACILLUS THURINGIENSIS* (BT)

The continued success of using Bt in crop protection is highly threatened due to the onset of resistance in insects against Bt toxins. These insects not only become immune to Bt-based bioinsecticides but also ruin Bt transgenic crops. Several hundreds of research papers have already been published and continue to be published on laboratory and field-evolved resistance in insects against these Bt toxins. The potentiality of insects to develop resistance to these toxins increases due to repeated exposure and is a cause of great concern in the widespread cultivation of Bt crops. Such resistance can leave Bt, one of the most environmentally benign insecticides today, useless (Morin et al., 2003).

The evolution of resistance refers to a genetically determined reduction in the susceptibility of a population to a toxin (Tabashnik, 1994). Usually, insect populations frequently exhibit inherent genetic diversity that influences their sensitivity to toxins, with certain alleles making them susceptible while others render them resistant (Tabashnik et al., 2009). Before being exposed to a Bt toxin, insect populations rarely have resistance-inducing genes. Empirical estimates put the frequency of these genes at close to one in a thousand (Bentur et al., 2000; Bernardi et al., 2015; Burd et al., 2003; Downes et al., 2009; Gould et al., 1997; Huang et al., 2009; Mahon et al., 2007). Field-evolved resistance occurs when a field population of insects is exposed to a Bt toxin that increases the resistance-conferring allele's frequency in subsequent generations. Hence, field-evolved resistance is not signified by the inherent low susceptibility of an insect species to a toxin (Tabashnik et al., 2009). Usually, this resistance is assessed by laboratory bioassays by calculating the resistance ratio, which is measured by dividing the LC_{50} (lethal concentration of toxin causing 50% mortality) of a field-derived strain with the LC_{50} of a susceptible strain (Tabashnik, 1994; Tabashnik et al., 2008).

McGaughey (1985) was one of the earliest ones to report that the Indian meal moth, *Plodia interpunctella* (Hübner), which is an important stored grain pest, was able to develop resistance to Bt formulations within a few generations. Later on, the diamondback moth (DBM), *Plutella xylostella* (Linnaeus), became the first insect with field-evolved resistance to Bt (Tabashnik et al., 1990, 1994). Thereafter, several reports on the development of resistance in different insect pests against Bt have also been published. Among them, the most infamous is the DBM, as it develops a high level of resistance against Bt in the field, particularly in nations like China, Japan, the Philippines, Malaysia, and North America, where Bt-based products are widely utilized (Gujar & Mohan, 2003). Not only insects of field crops but also household pests like mosquitoes become less susceptible to mosquitocidal strains of *Bt* subsp. *israelensis* (Bti) (Wirth, 2010; Wirth et al., 2012). Apart from that, the development of high levels of resistance to one Cry toxin in an insect might lead to a significant increase in resistance to other Cry toxins (Gould et al., 1992; Niu et al., 2016; Tabashnik et al., 1994), making multiple toxins useless.

13.2.1 Global Status of Insects' Resistance to Bt Transgenic Crops

The role of transgenic Bt crops in modern-day pest management programmes is enormous. In 2019, millions of farmers across 27 nations on six continents cultivated about 109 mi ha (million hectares) of Bt crops, including brinjal, corn, cotton, soybeans, and sugarcane (ISAAA, 2019). However, the development of resistance in target pest populations against different Bt toxins is continuously threatening the sustainability of this technology and reducing the benefits of Bt crops (Gould, 1998; Tabashnik et al., 2009). To date, a number of insect pests have already developed varying levels of resistance against Bt crops. A number of research papers have been published on the field-evolved resistance in insects against Bt crops. Usually, the major portion of a pest population contains susceptible strains to the Bt trait at the initial stage. Still, the population may have some portion of naturally occurring genetically resistant individuals. Selection pressure exerted on the pest is responsible for how quickly a population develops genetic resistance. *Selection pressure* refers to those factors that are responsible for changing a population genetically. For example, a field with a 100% Bt crop would likely eradicate all the susceptible individuals and leave only the resistant individuals that will produce resistant offspring in subsequent generations. This is an example of very high selection pressure, and such things must be avoided. If it happens, it will cause resistant individuals to dominate the pest population, indicating the resistance has set in. This will ultimately lead to a similar prevalence of insect pests in the Bt crop fields as they are in the non-Bt crop fields.

For example, global monitoring data involving six lepidopteran pests revealed the occurrence of resistance in some populations of *Helicoverpa zea* (Boddie) within the first decade of Bt crops (Tabashnik et al., 2008). According to the data, 14 strains of *H. zea*, collected from Bt cotton fields between 2003 and 2006, showed >100 resistance ratios, including two strains having a resistance ratio of >1,000. In contrast, prior to the commercialization of Bt cotton, the highest resistance ratio for *H. zea* was 1.2. It is evident from the data on *H. zea* resistance both before and after commercialization of Bt cotton that some field populations of *H. zea* developed resistance to Cry1Ac as a result of rapid exposure to the toxin (Tabashnik et al., 2008).

One of the notable examples of failure of Bt crops to resist the target pests due to resistance development is bollworm resistance to Bt cotton in India. The scenario of Bt cotton in India is among the most intensively monitored scenarios regarding the proliferation and effects of genetically modified (GM) crops. As per the report of ISAAA (2019), 94% of cotton grown in India is Bt cotton, covering about 12 mi ha of area and making a positively significant long-term contribution in the country (Kranthi & Stone, 2020).

However, the development of resistance in cotton pests against Bt cotton becomes a serious threat to the cotton growers of the country. In November 2009, scientists from Monsanto India Limited reported the first case of pink bollworm (PBW), *Pectinophora gossypiella* (Saunders), resistance to first-generation Bt cotton (producing only Cry1Ac toxin) under field conditions from four districts of Gujarat, India (Bagla, 2010; Dhurua & Gujar, 2011). By 2009, the widespread adoption of Bt plants facilitated the swift proliferation of resistant PBW, and by 2015, it became a 'nightmare' and dominated the headlines (Buradikatti, 2015). As a consequence, the cotton growers started spending more on chemical pesticides than before the introduction of Bt cotton (Kranthi & Stone, 2020). Apart from India, cases of bollworm resistance to first-generation Bt cotton were also reported from Australia, China, Spain, and the USA (Tabashnik et al., 2008). To tackle this situation, Monsanto had introduced second-generation Bt cotton that produces two Bt toxins.

Tabashnik and Carrière (2017, 2019) have categorized insect resistance to Bt crops into three categories, namely, practical resistance, early warning of resistance, and no decrease in susceptibility. Among these three categories, two are the subdivisions of *field-evolved resistance*, defined as a genetically determined reduction in an insect population's susceptibility to a Bt toxin resulting from natural selection under the field condition (Tabashnik & Carrière, 2017, 2019). Of them, practical resistance indicates that over 50% of individuals within an insect population exhibit resistance, resulting in diminished field efficacy of the Bt crop and thus indicating its practical implication in pest management (Tabashnik et al., 2014). In their very recent publications, Tabashnik et al. (2023a, 2023b) have summarized the global resistance monitoring data by reviewing 73 cases of resistance monitoring data published from the initial 25 years of Bt crops, encompassing 24 species of insect pests (including 22 moths and 2 beetles) across 12 nations on six continents. Here they have classified all the 73 published cases among these three categories and reviewed the evolutionary principles underlying resistance management, clarified some common misconceptions, compared field outcomes to evolutionary principles–based predictions, and concluded with a summary and outlook for the future. Among these 73 cases described, 26 are cases of practical resistance (36%), 17 are cases of early warning of resistance (23%), and 30 cases are of no decline in susceptibility (41%). Each case describes the resistance or susceptibility of one species of insect pest to one Bt toxin expressed in one or more Bt crops in one country.

13.2.2 Mechanism of Bt Resistance in Insects

Prolonged exposure of an insect to Bt toxins induces physiological alterations and enhances the immunological response in the insect. As outlined in Chapter 4, Cry toxins exhibit toxicity through a series of steps, beginning with an activation phase in the host's midgut, followed by toxin binding to receptors and subsequent insertion into the enterocyte membrane, leading to pore formation, enterocyte osmotic death, and ultimately, the collapse of the intestinal barrier. This collapse permits gut bacteria to infiltrate the haemocoel, resulting in septicaemia and, eventually, insect death. Alterations in any of these sequential steps in the mode of action (MoA) of Bt insecticidal proteins (IPs) stimulate the resistance development in insects against the insecticidal activity of Bt. Possible scenarios that can explain the loss in sensitivity in insects to Bt toxins are described hereunder.

13.2.2.1 Decreased Toxin Binding

As mentioned previously, binding of Cry toxins with receptors is an important step in the MoA of Cry toxins. That is why changes in the receptor binding are considered the most frequently occurring resistance mechanism against Cry toxins (Figure 13.1c). Specifically, for most Bt-resistant lepidopterans, resistance correlates with reduced binding of at least one Cry toxin in the Bt formulation (Ferré & Van Rie, 2002). This type of resistance is responsible for the highest levels of resistance to specific Bt formulations (de Bortoli & Jurat-Fuentes, 2019). In one of the earliest studies, Van Rie et al. (1990) found a 50-fold reduction in the binding affinity of a Bt IP to the membrane receptor in a laboratory-selected resistant strain of *P. interpunctella*. Later on, Ferré et al. (1991) have compared the binding characteristics of Cry1Ab toxin between a laboratory strain and a field population of *P. xylostella*. The field population showed >200-fold resistance to Cry1Ab as compared to the laboratory strain due to the lack of binding of the protein to the brush border membrane of the midgut epithelial cells.

The binding ability of Bt toxins to receptors may be hampered due to a reduction in the number of toxin-binding sites or due to mutations that hamper the ability of receptors to bind to the Cry toxins (Ferré et al., 1991; Gahan et al., 2001; Loseva et al., 2002). From saturation-binding assays, it was found that the brush border membrane vesicles (BBMV) from the resistant strain of Colorado potato beetle (CPB), *Leptinotarsa decemlineata* Say, bound ~60% less Cry3A toxin than the BBMV from the sensitive strain. However, homologous competition inhibition of ^{125}I-Cry3A binding to BBMV revealed no difference in the binding affinity between the resistant and the sensitive strains (Loseva et al., 2002). The laboratory-selected Cry1Ac-resistant line YHD2 of *H. virescens* showed that a single mutation can cause 40–80% resistance to Cry1Ac, and it was linked to a retrotransposon insertion in the cadherin (CADR) gene (Gahan et al., 2001). Such mutation or disruption of the CADR gene also causes reduced binding and resistance to Cry1Ac in field-derived and laboratory-selected *P. gossypiella* (Morin et al., 2003; Wang et al., 2018, 2022), *H. armigera* (Xu et al., 2005; Yang et al., 2007), and *H. zea* (Fritz et al., 2020). There is a vital role of amino acids Leu^{1425} and Phe^{1429} in the interaction between CADR and Cry1Ac toxin. If there is any replacement of these amino acids with charged amino acids, the toxin will not be able to bind to CADR, which may lead to Cry1Ac resistance (Jurat-Fuentes & Adang, 2004). CADR mislocalization may happen on the surface of the midgut epithelium due to single-point mutation, leading to higher resistance to Cry1Ac in the bollworm complex of cotton (Xia et al., 2017). Mutation of ATP binding cassette (ABC) transporter proteins also causes reduction in Cry1Ac binding in *H. virescens* (Gahan et al., 2010) and *H. armigera* (Xiao et al., 2014), Cry1Ab binding in *Bombyx mori* (Linnaeus) (Tanaka et al., 2013), and Cry2Ab binding in *H. punctigera* and *H. armigera* (Tay et al., 2015). In addition, reduced levels of membrane-bound glycosylphosphatidylinositol (GPI)–anchored alkaline phosphatase (mALP) in the midgut are also associated with reduced binding and resistance to Cry1Ac in *H. virescens* (Jurat-Fuentes et al., 2011) and to Cry1Fa in *S. frugiperda* (Jakka et al., 2016). Absence or slicing of aminopeptidase N (APN), a putative receptor for Cry1C toxin, is also found in the resistant strains of *S. exigua* (Herrero et al., 2005) and *S. litura* (Rajagopal et al., 2002), respectively. Unlike the aforementioned cases, where resistance to Bt toxins was recessive or near recessive, a dominant resistance mechanism to Cry1Ac was also reported for field-derived *H. armigera*, in which resistance is associated with a dominant point mutation in a tetraspanin gene (Jin et al., 2018).

Apart from those, situations of cross-resistance occur when IPs share binding sites with the most potent Cry toxin in the Bt formulation, which is likely to be the selecting agent (de Bortoli & Jurat-Fuentes, 2019). There are also certain examples where selection with a Bt toxin led to cross-resistance to toxins with different binding sites than the selective one (Moar et al., 1995; Tabashnik & Carrière, 2020; Welch et al., 2015). In those cases, binding of toxins is not affected in resistant insects; rather, the insects develop resistance due to alteration of a common step in the toxin's MoA. Such mechanisms involve alteration in the proteolytic activity of the gut, toxin sequestration, intestinal regeneration, etc. (de Bortoli & Jurat-Fuentes, 2019). These resistance development mechanisms are mentioned in what follows. However, these mechanisms lead to lower levels of resistance and do not commonly occur in field condition (Ferré & Van Rie, 2002). That is why, in scenarios like Bt crops, where the insects are exposed to high levels of toxin, reduction in the toxin binding is the only way for an insect to develop resistance (Jakka et al., 2016; Ojha et al., 2014). In Table 13.1, we have listed the insects resistant to various Bt toxins, along with the receptors/enzymes associated with resistance development.

13.2.2.2 Enhanced Immune Response

An elevated immune response, characterized by modifications in mucosal surface activity leading to an upsurge in the release of pro-coagulants and protease enzymes, is also responsible for the development of resistance in insects against Bt toxins (Figure 13.1a). For instance, a 75 kDa toxin precipitating protein (TPP-75), characterized as a serine protease that resembles elastase, is produced in the gut juice of spruce budworm, *Choristoneura fumiferana* (Clemens). A contiguous region from Cry1Aa toxin interacts with this TPP-75 protein that causes the *Bt* subsp. *sotto* to precipitate protoxin. This precipitation leads to selective attenuation of the δ-endotoxin's activity prior to its binding with the potential receptors on susceptible cells. So for those larvae that do not display a receptor-based resistance, TPP-75 might be the potential resistance mechanism for them (Milne et al., 1998).

An elevation in the insect's immune response is also correlated with enhanced melanization. For example, the Cry toxin–tolerant larvae of *Ephestia kuehniella* Zeller showed a significantly higher melanization reaction than the susceptible ones (Rahman et al., 2004). A study involving both Bt-susceptible and Bt-resistant laboratory strains of *H. armigera* showed that the resistant strain had an augmented immune response characterized by enhanced melanization and the presence of a soluble toxin–binding glycoprotein in both its haemolymph and gut lumen. This systemic immune-induction mechanism of resistance is also transmitted from one generation to the next by a maternal effect (Ma et al., 2005). Apart from that, elevated levels of esterase were also found in Cry1Ac-resistant *H. armigera* larvae, as the esterase has the ability to bind with activated Cry1Ac and its protoxin and thus detoxify a considerable amount of Cry1Ac (Gunning et al., 2005). In addition, in the case of Egyptian cotton leafworm, *Spodoptera littoralis* (Boisduval), rapid hydrolysis of the Cry1C toxin was noticed in the resistant line compared to the sensitive one (Moussa et al., 2020).

13.2.2.3 Imbalanced Proteolytic Processing of Bt Toxins

During larval growth and development, the concentration of the gut juice proteases (that are involved in the activation of protoxin) in the alimentary canal has increased considerably. This upshot in concentration of the gut proteases results in the reduced sensitivity to Cry toxins (Figure 13.1b). For instance, in an experiment, a resistant strain of the CPB was subjected to function-based activity profiling using zymographic gels that contain gelatine as a substrate. The experiment showed the presence of specific proteolytic bands in the midgut extracts and BBMV of the resistant strain, which were not apparent in the sensitive strain. Furthermore, in contrast to the sensitive strain, the resistant one exhibited increased activity of the aminopeptidase linked to the membrane vesicles (Loseva et al., 2002). Thus, the resistance mechanism in the CPB is not only the result of a decrease in the binding ability of the toxins but also occurs due to the specific alterations in the activity of protease in the midgut. Unlike the CPB, considerably lower proteolytic activation of *Bt* subsp. *kurstaki* protoxins was noted in a resistant strain of *S. littoralis* (Ibrahim et al., 2010). Similarly, a decrease in the midgut protease activity reduced the activation of Cry toxins and showed resistance in *H. virescens* (Forcada et al., 1996) and *P. interpunctella* (Candas et al., 2003; Oppert et al., 1997). These indicate

TABLE 13.1
Insects Resistant to Bt Toxins and the Associated Receptors or Enzymes

Insect	Receptors/Enzymes	Resistance Mechanism	Bt Toxin	Reference
Chrysomela tremula Fabricius	ABCB1	Mutation	Cry3Aa	(Pauchet et al., 2016)
Chilo suppressalis (Walker)	Cadherin	Knockdown	Cry2A Cry1Ac	(Zhang et al., 2017a)
	ALP	Down-regulation	Cry1A Cry2Aa Cry1Ca	(Qiu et al., 2018)
Helicoverpa armigera (Hübner)	ABCA2	Mutations of three independent indel	Cry2Ab	(Tay et al., 2015)
	ALP	Digestion of *N*-glycosidase	Cry1Ac	(Ning et al., 2010)
	ABCC2	Inserting 73 bp in cDNA results in 6 bp deletion at the splicing site		(Xiao et al., 2014)
	Cadherin	Premature stop codon		(Xu et al., 2005)
		Point mutation causes mislocalization		(Xia et al., 2017)
	APN	Deletion mutation		(Zhang et al., 2009)
	Trypsin	Mutations in promoter region		(Liu et al., 2014)
	Protease	Improper protoxin processing due to altered protease profile		(Rajagopal et al., 2009)
	Tetraspanin	Mutation		(Jin et al., 2018)
Helicoverpa punctigera (Wallengren)	ABCA2	Deletion of 14 bp causes loss of tpm2 transporter motif in NBF2	Cry2Ab	(Tay et al., 2015)
	Cadherin	Mutation	Cry1Ac	(Walsh et al., 2018)
Helicoverpa zea (Boddie)	Cadherin	Mutation	Cry1Ac	(Fritz et al., 2020)
Heliothis virescens (Fabricius)	ALP	Down-regulation	Cry1Ac	(Jurat-Fuentes & Adang, 2004)
	ABCC2	Inactivating mutation		(Gahan et al., 2010)
	Cadherin	Retrotransposon-mediated insertion		(Gahan et al., 2001)
		Single-nucleotide mutation		(Xie et al., 2005)
Ostrinia furnacalis Guenée	ABCG1 APN	Down-regulation	Cry1Ab Cry1Ac	(Zhang et al., 2017b)
	Cadherin	Down-regulation and mutation	Cry1Ac	(Jin et al., 2014)
Ostrinia nubilalis (Hübner)	ABCC2	Mutation	Cry1Fa	(Coates & Siegfried, 2015)
	Cadherin	Premature termination codons/huge deletions	Cry1Ab	(Bel et al., 2009)
	APN	Down-regulation		(Coates et al., 2013)
	Aminopeptidase-P like gene	Mutation		(Khajuria et al., 2011)
Pectinophora gossypiella (Saunders)	Cadherin	Three mutant alleles in toxin-binding region	Cry1Ac	(Morin et al., 2003)
		207 bp deletion and transmembrane domain loss		(Wang et al., 2018)
		Premature stop codon or at least 99 bp deletion		(Fabrick et al., 2014)
		Insertion of intact CR1 retrotransposon		(Fabrick et al., 2011)
	ABCA2	Mutation	Cry2Ab	(Mathew et al., 2018)
Plutella xylostella (Linnaeus)	ABCG1 ALP	MAPK pathway–mediated down-regulation	Cry1Ac	(Guo et al., 2015)
	ABCC2	Mutation		(Baxter et al., 2011)
Spodoptera exigua (Hübner)	ALP	Knockdown	Cry2Aa	(Yuan et al., 2017)
	APN	Down-regulation	Cry1Ca Cry1Ac	(Herrero et al., 2005)
	ABCC2	Mutation	Cry1Ac	(Park et al., 2014)
Spodoptera frugiperda (J.E. Smith)	ALP	Down-regulation	Cry1Fa	(Jakka et al., 2016)
	ABCC2	Mutation	Cry1Fa	(Boaventura et al., 2020)
			Cry1F Cry1A.105	(Flagel et al., 2018)
Trichoplusia ni (Hübner)	APN	Down-regulation	Cry1Ac	(Tiewsiri & Wang, 2011)
	ABCA2	Mutation	Cry2Ab	(Yang et al., 2019)

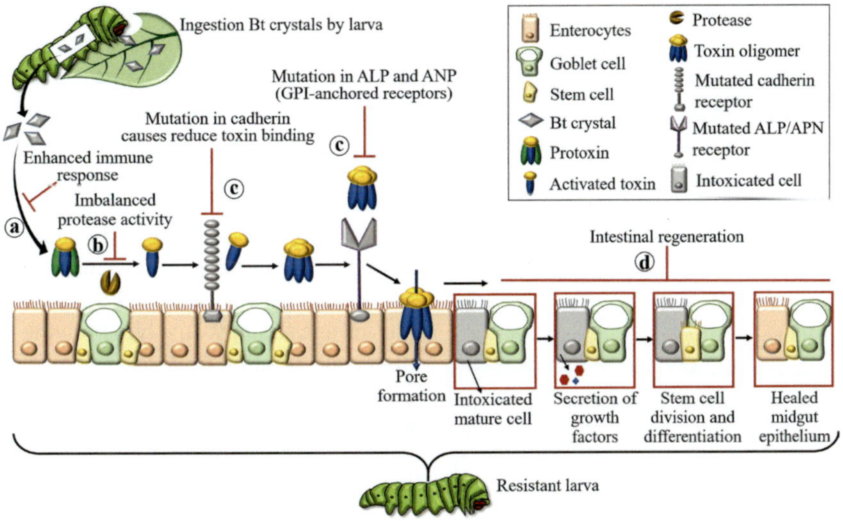

FIGURE 13.1 Possible resistance mechanisms in insects against Bt insecticidal proteins. There are four possible ways responsible for resistance development in insect larvae: (a) heightened immune response preventing protoxin from dissociating and being accessible; (b) erroneous production of protease or defect in protease activity interfering with protoxin/toxin activation; (c) decrease in binding sites or mutated receptors that block the proper toxin–receptor interaction; (d) healing of damaged midgut epithelial cells restricting the gut bacteria from entering the haemocoel.

that an imbalanced situation in the proteolytic processing of Bt toxins in the gut of insects is responsible for the resistance mechanism.

13.2.2.4 Intestinal Regeneration

There are very few cases where resistance to Cry toxins involves intestinal regeneration. To display their activity, Bt toxins require the intestinal barrier to collapse, allowing gut bacteria to enter the haemocoel and cause septicaemia. However, healing of this intestinal barrier prevents the toxin's activity. For instance, in a scanning electron microscope (SEM) study, Forcada et al. (1999) found significant damage on the midgut epithelium of both susceptible and resistant *H. virescens* in the initial hour after intoxication. However, unlike the susceptible strain, the midgut epithelium of the resistant one started healing as time passed and became fully recovered by 48 hours following intoxication. In another instance, Martinez-Ramirez et al. (1999) also assumed that the resistant larvae of *H. virescens* could repair (or substitute) the damaged cells more readily. In both cases, resistance was involved against Cry1Ac toxin. This enhanced midgut healing in resistant larvae may be due to increased production of mitogens or novel midgut differentiating factors (MDFs) or to variations in stem cell sensitivity to mitogens. Usually, lepidoptera larvae have four major cell types in their monolayer epithelium that include intestinal stem cells (ISCs), goblet cells (GCs), enterocytes or columnar cells (ECs), and enteroendocrine cells (EEs). Among them, the ISCs are the only cells that are capable of division, thus serving as the exclusive source of new cells for tissue healing and growth. When the gut epithelium faces various biotic and abiotic injuries, it stimulates proliferation and differentiation of ISCs in order to restore the integrity of the gut epithelium. When intoxication with Bt causes injury in midgut epithelial cells, in resistant insects, it might start producing MDFs that induce the ISCs to differentiate into mature midgut cell types and restore gut epithelial integrity (Figure 13.1d) (Castagnola & Jurat-Fuentes, 2016). Similarly, Loeb et al. (2001) also detected 1.5-times more number of midgut cells producing MDF1 peptide after treatment with Bt toxins. Thus, a mechanism like this prevents the resident gut bacteria from penetrating the haemocoel to cause septicaemia and may limit the efficiency of alternative entomopathogens that infect insects through oral routes.

These non-receptor-based resistance mechanisms lead to lower levels of resistance and do not commonly occur in the field condition. However, their occurrence poses a serious threat to the use of Bt-based pesticides and Bt crops since they impact steps common to all Cry toxins, resulting in cross-resistance to a wide spectrum of IPs of Bt. Figure 13.1 illustrates the potential phases involved in the Bt resistance mechanism in insects.

13.2.3 Resistance Management for Bt

With increasing evidence of insect resistance to Bt toxins, it is vital to adopt effective resistance management strategies to protect Bt's applicability. Resistance management for Bt refers to the reduction in the selection pressure while ensuring that the crop is properly protected. Actually, development of resistance is inevitable, as the growers always try to control the pests, and the pests are always working on developing strategies to overcome these control measures that ultimately lead to resistance issues. However, many strategies can be adopted to slow down the rate of resistance development, such as the following.

13.2.3.1 Insect Resistance Monitoring

The main purpose of monitoring insect resistance to Bt-based products and transgenic Bt crops is to detect the development of field-evolved resistance in insects early on, allowing us to take efficient, proactive management decisions before control breakdowns (Siegfried et al., 2007; Tabashnik et al., 2009). Therefore, documenting field-evolved resistance is not always an implication of occurrence of control problems in the field; rather, based on it, regulatory decisions on the continued use of Bt-based products or Bt crops can be taken by considering the relationship between field-evolved resistance and field control issues (Tabashnik et al., 2009). Early detection of field-evolved resistance requires sampling and testing of target insect pests surviving on or near Bt-treated or Bt transgenic crops (Tabashnik et al., 2008). Failure in sampling such insects favours underestimating of the frequency of resistance, which might delay identification and contradict the fundamental goal of resistance monitoring (Tabashnik et al., 2009). Usually, to detect the resistance development in insects, their susceptibility against Bt is determined. Susceptibility is measured by exposing the laboratory-reared progeny from a field population of insects to diets treated with Bt toxins or to Bt plant parts in laboratory bioassays (Tabashnik et al., 2009). Thus, after measuring the level of susceptibility, one can monitor the development of resistance in insects to Bt and thereby predict the next step to take.

13.2.3.2 Crop Rotation

Crop rotation is the practice of cultivating different crops in the same field in alternate years rather than the same crop in the same field for several years. It can be done in two ways. Firstly, rotation of the Bt hybrid with any non-host crop can hamper the growth and development of the pest. For example, if Bt maize and soybean are grown in rotation across years, larvae of western corn rootworm (WCR), *Diabrotica virgifera virgifera* LeConte, that are hatched out from the eggs laid in the maize field will not be able to complete their development on soybean, planted in the same field in the subsequent year. That is why Carrière et al. (2020a) noted about a 92% decline in the frequency of WCR problem fields when Bt maize is grown in rotation with soybean. Secondly, instead of cultivating Bt crops year after year, non-Bt hybrids of the same crop with rationale use of insecticides can be grown in rotation with Bt hybrids to delay the resistance development (Gassmann, 2021). This will cause less exposure of insects to Bt proteins and also conserve the susceptible insect population, thereby slowing down the process of resistance development. Moreover, Bt hybrids containing different toxins with varying MoAs can also be grown in rotation to control the target insects. This will not allow the pest to be in continuous touch with a particular toxin for a longer period, and thus, it will delay the resistance development. For example, if Bt crops producing Cry1A and Vip3A toxins are grown in rotation, then it will reduce the chances of resistance development in insects. Apart from that, if the insects even develop resistance to Cry1A toxin, Vip3A toxin in the subsequent year can control that Cry1A-resistant insect population.

13.2.3.3 Refuge Planting

With increasing concerns of potential evolution of insect resistance to Bt crops, the 'high dose plus refuge strategy' was mandated as a resistance management practice in various nations (Gould, 1998; Tabashnik, 2008). This 'high dose plus refuge strategy' involves growing Bt crops expressing a high concentration of Bt toxins and planting of refuges that act as a potential host for the pest attacking the Bt crop in parts of the Bt field or adjacent to the Bt field. These refuges can be a non-Bt hybrid of the same crop (e.g. non-Bt maize counterpart of Bt maize). Refuges can also be any other cultivated or wild non-Bt host crop for that pest (e.g. non-Bt maize as a refuge in a Bt cotton field). Sometimes, refuges might be another Bt crop that acts as a host for a subset of pests (e.g. Bt maize producing lepidopteran-specific toxins can be used as a refuge for root-feeding beetles that attack Bt maize producing coleopteran-specific toxins) (Tabashnik et al., 2023b). This refuge acts as a reservoir for the susceptible insect populations. If such a susceptible insect population co-exists with a resistant population, the susceptible insects will be able to mate with the resistant one and produce susceptible offspring. According to the results from modelling studies, the resistance may not evolve as quickly if

1. rare homozygous-resistant insects (RR) evolved in Bt crops mate with homozygous-susceptible insects (SS) present in refuges, and
2. the high concentration of Cry toxin in Bt crops kills the heterozygous offspring (RS) that come out from such mating.

(Bravo & Soberón, 2008)

Currently, in many countries, refuge planting with the purpose of resistance management is becoming a mandate, and in others, it occurs voluntarily (Carrière et al., 2020b). Apart from that, in countries like China, where refuge planting is not mandatory, unintentional occurrence of refuges (growing for reasons having no relation to resistance management) plays an important role in reducing insect resistance issues (Jin et al., 2015; Wan et al., 2017). Refuge planting has two essential criteria. Firstly, an efficient percentage of the total crop field needs to be planted with refuge crops so that it can ensure the generation of a sufficient number of susceptible insects. Secondly, the refuge area must be planted close to the Bt-cropped area that will ensure mating of the susceptible insects with resistant ones. The risk of resistance development is essential to be calculated beforehand to determine the size of the refuge area. Different approaches of refuge planting are discussed in the following.

Structured Refuges. In this type of refuge planting, a predetermined area of the field is devoted to refuge plants. This is the most common refuge planting strategy and has been adopted widely in the USA since 1996 (Carrière et al., 2016). Here, the refuges can be planted as blocks in the field, as brackets in either side of the Bt crops, as border rows surrounding the Bt crops, or as rows/stripes within the Bt field (Figure 13.2). Structured refuge has two key components: its size (as a proportion of the corresponding Bt field) and proximity of the refuge area to the Bt field (Reynolds & Martinez, 2021). This proportion of refuge-cropped area usually varies based on the type of Bt crop. For example, in the case of Bt maize, 20% of the total maize field needs to be planted with refuge plants. For Bt cotton, the proportion of structured refuge is increased to 50%. However, in Bt crops with pyramided toxins, the proportion of structured refuge can be reduced to only 5% of the total cropped area. Coming to the proximity of the refuge area, for Bt maize it needs to be within only 1/4 mile to 1/2 mile of Bt maize when borer pests are considered. However, in case of rootworms, the refuge area needs to be immediately adjacent to the Bt maize field (Bessin, n.d.).

Seed Blends or Seed Mixes or Refuge-in-the-Bag. One of the shortcomings of structured refuge is that its efficiency to delay the evolution of resistance to Bt depends on farmers' compliance with refuge requirements, which can be difficult and costly (Carroll et al., 2012). The fact that farmers occasionally fail to comply with refuge regulations raises the possibility that resistance evolution may be more likely than previously believed (Jaffe, 2009). In such cases where farmers are not growing refuges, seed blends or seed mixes or refuge-in-the-bag strategy can be adopted as an alternative approach that involves mixing Bt seeds with non-Bt seeds in the same seed bag, assuring refuges at the time of seeding (Carroll et al., 2012). Starting in 2010, this concept has been adopted in the USA to manage the resistance evolution to pyramided maize (Carrière et al., 2016). Seed mixes allow the transfer of the responsibility of implementing refuge strategy from farmers to the company associated with the supply of the seeds. Use of a seed mix guarantees that a sufficient number of suitable refuge plants is grown in each Bt field while also distributing refuge plants reasonably consistently inside the Bt fields, minimizing the likelihood of mating among Bt-resistant adults when compared to a separate block refuge (Carroll et al., 2012). In Figure 13.3, a pictorial representation on how the refuge strategy can delay insect resistance through structured refuge and seed mixes is presented.

However, seed mixes also have some drawbacks. According to mathematical models, seed mixes involving single or pyramided Bt crops can significantly accelerate the evolution of resistance as compared to the refuge blocks because of the extensive movements of larvae between Bt and refuge plants (Heuberger et al., 2011; Ives et al., 2011). Thus, it increases the exposure of susceptible insects to Bt toxins and reduces their survival or increases the dominance of resistant and heterozygous insects thereby. However, laboratory and greenhouse experiments suggest that the dominance of resistance in seed mixes of single-toxin crops occurs specifically in those pests having low levels of susceptibility to Bt toxins (Carrière et al., 2016). For instance, Brévault et al. (2015) recorded a significantly higher dominance of resistance in *H. zea* having low susceptibility to Cry1Ac cotton in seed mixes as compared to refuge blocks due to the higher survival of heterozygotes than susceptible individuals. In contrast, experiments with *P. gossypiella* and *P. xylostella* with high inherent Cry1Ac susceptibility suggested no dominance of resistance in seed mixes (Heuberger et al., 2014; Shelton et al., 2000). Apart from larval movement between Bt and non-Bt plants, pollen-mediated gene flow can also hasten the resistance evolution in seed mixes as compared to structured refuges, specifically for insects feeding on kernels. Such gene flow between Bt and non-Bt plants yields a mosaic of Bt and non-Bt kernels in seed mixes that kills the susceptible larvae in refuge plants and thus increases the dominance of resistance (Yang et al., 2014). So to sustain the efficiency of refuge strategy in Bt crops against pests having low

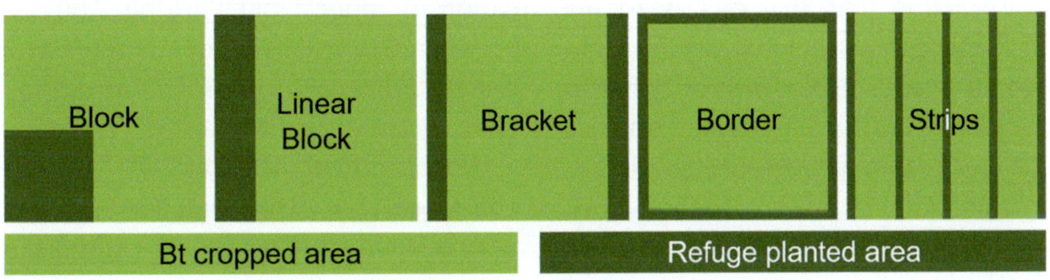

FIGURE 13.2 Different ways of arranging structured refuge.

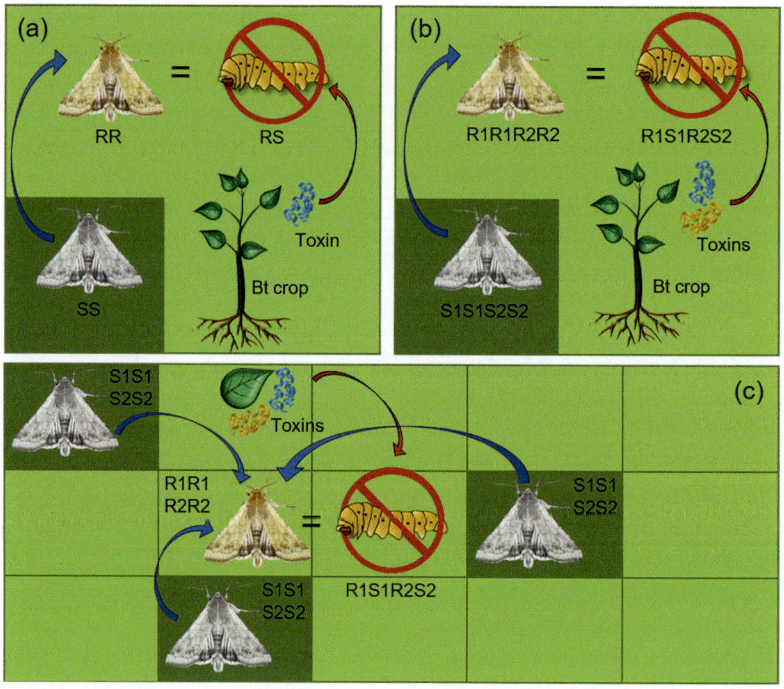

FIGURE 13.3 Pictorial representation on how the refuge strategy can delay insect resistance to (a) single-toxin-expressing Bt crops, (b) pyramided Bt crops with structured refuges, and (c) seed blends of pyramided Bt and non-Bt crops. *Note:* SS in (a) and S1S1S2S2 in (b, c) are susceptible homozygous moths; RR in (a) and R1R1R2R2 in (b, c) are resistant homozygous moths; RS in (a) and R1S1R2S2 in (b, c) are heterozygous offspring.

inherent susceptibility to Bt toxins, it is necessary to increase the proportion of refuge coupled with other pest management strategies.

Natural Refuge. *Natural refuge* implies allowing the growth of wild hosts such as weeds or other cultivated crops of that particular insect pest, which can harbour the susceptible insects. Such a type of refuge is effective against the pests that feed on multiple host plants and does not specialize solely on Bt crops (Reynolds & Martinez, 2021).

13.2.3.4 Crops Expressing Novel Bt Toxins

There are several Bt strains capable of producing other types of insecticidal toxins having no sequence similarity with Cry toxins. These toxins include Mtx-like-Cry, Bin-like-Cry, and Vip toxins that have also been expressed in transgenic crops. These novel toxins do not share any midgut receptor with any Cry toxin and work in a completely different mode of action than that of Cry toxins, perhaps enabling them to manage insects that exhibit resistance to Cry toxins (Bhatti et al., 2023; Bravo & Soberón, 2008). For instance, Cry34/35 are binary toxins requiring two components for toxicity and expressed successfully in transgenic maize to control *D. virgifera* (Hitchon et al., 2015). Similarly, transgenic plants expressing proteins from Vip3 group can efficiently control lepidopteran pests with a broader range of toxicity as compared to individual Cry toxins (Adamczyk & Mahaffey, 2008; Bhatti et al., 2023).

13.2.3.5 Pyramiding Toxins

It was Gould (1986) who had first pointed out that the pyramided cultivars with susceptible refuges can offer the highest levels of relative durability. In the case of Bt crop technology, the term 'pyramid' is applicable when many toxins with distinct MoA (binding to various receptor) are expressed in the same plant (Bravo & Soberón, 2008). In general, the concentration of each component toxin alone is sufficient to provide effective control and is expressed in all plant tissues throughout the larval feeding period for continuous exposure (Storer et al., 2012). When multiple toxins are expressed in a single plant, the possibility of developing resistance in insects gets reduced by several folds due to the requirement of multiple mutations to resist the activity of all the expressed toxins (Zhao et al., 2003). Here, the primary criteria of selecting the toxins involved in pyramid is their different MoA with binding to different receptors, because a heightened resistance to Cry toxins is an indication of modest but varying degrees of cross-resistance between Cry toxins, which is more pronounced among the closely related Cry toxins than the more distantly related ones (Ferré & Van Rie, 2002; Tabashnik, 1994).

For instance, Bollgard II® is a double-gene-stacked cotton variety expressing Cry1Ac and Cry2Ab proteins, where both proteins used to interact with different midgut receptors. This pyramided cotton variety was introduced to replace Bollgard® cotton (expressing only Cry1Ac protein), against which different insects became resistant

(Chitkowski et al., 2003). Further progress was made by introducing Bollgard III®, a triple-gene-stacked cotton with an additional Vip3a protein that has no similarity with Cry toxins (Whitehouse et al., 2014). Similarly, in 2009, maize hybrids (YieldGard®) expressing two structurally dissimilar Cry proteins (Cry2Ab2 + Cry1A.105) with differences in their way of midgut binding were released (US EPA, 2010). Similarly, transgenic maize with three dissimilar lepidopteran active Cry genes (*Cry1A.1.05*, *Cry2Ab*, and *Cry1F*) and three dissimilar coleopteran active Cry genes (*Cry34Ab/Cry35Ab* and *Cry3Bb*) were also developed (Bravo & Soberón, 2008). These pyramided crops significantly reduce the evolution of resistance in target pests compared to single-toxin crops.

Results from different mathematical models have shown that, under certain situations, crops expressing pyramided toxins with very small refuges can significantly delay the onset of resistance than the single-toxin-expressing crops and provide hundreds of generations of durability (Gould et al., 2006; Ives et al., 2011; Onstad & Meinke, 2010; Roush, 1998). However, plants expressing both single and pyramided toxins are vulnerable to resistance development when they meet conditions deviated from the conditions assumed in these models (Carrière et al., 2016; Storer et al., 2012). For example, extensive exposure of pests with inherently low susceptibility to either of the Bt toxins in the pyramid prior to the adoption of pyramided plants can significantly reduce the benefits of pyramid technology (Brévault et al., 2013). Concurrent use of Bt plants expressing single and two toxins accelerates the selection for resistance to pyramided Bt plants more swiftly than the sole use of two-gene plants (Zhao et al., 2005). Furthermore, development of cross-resistance to Cry toxins with different receptor binding is also possible (Gahan et al., 2005; Jurat-Fuentes et al., 2003). These issues suggest that gene pyramiding may not be a 'one-size-fits-all' method to mitigate insect resistance to Bt crop.

13.2.3.6 Modified Bt Toxins

To address the shortcomings of pyramided Bt toxins, integration of modified Bt proteins with enhanced potency and specificity is an efficient solution to overcome the resistance issues. These modifications enhance the toxins' stability, binding affinity to midgut receptors, oligomerization, and membrane perforation capabilities (Nagaraj et al., 2024). For instance, the binding of Cry1A to cadherin receptor promotes the proteolytic cleavage of α-1 helix of the toxin, resulting in toxin oligomerization and pore formation (Gómez et al., 2002). However, modified forms of Cry1Ab and Cry1Ac (i.e. Cry1AbMod and Cry1AcMod, respectively), lacking in α-1 helix, are capable of killing the insects by forming oligomers and pores in the absence of cadherin (Soberón et al., 2007). So in cases where the mutation of cadherin receptor is responsible for resistance, these modified toxins can act satisfactorily. Even in the case of resistant *P. xylostella* and *O. nubilalis* strains, where resistance was not linked to cadherin mutations, these modified toxins showed >350-fold higher potency relative to native toxins (Tabashnik et al., 2011). Similarly, engineering of Cry2Ab toxin, using site-specific saturation mutation, exhibited enhanced oligomerization and pore formation activities against *P. xylostella* than the wild-type Cry2Ab (Fu et al., 2024). Furthermore, eCry3.1Ab, a Cry3A variant, showed novel activity against WCR by forming a chymotrypsin/cathepsin G protease recognition site (Walters et al., 2008). Likewise, generation of novel Cry3Bb1.11098 protein through engineering of *Cry3Bb1* gene with substitution of five amino acids increased the natural protein's activity against WCR (Domínguez-Arrizabalaga et al., 2020). Therefore, a combination of these modified toxins and gene pyramiding could be an efficient way to delay the resistance development in pests.

13.2.3.7 Domain Exchange

Another strategy of generating toxins with enhanced efficacy and unique targets is to exchange the domains from different Cry proteins. For example, construction of recombinant toxin through the exchange of pore-forming domain I of *Cry1Ac* and domains II and III of *Cry9Aa* gene provided around five-fold higher toxicity to *H. armigera* than the native Cry1Ac (Shah et al., 2017). Swapping of domain III of Cry3Aa with the same domain from Cry1Ab resulted in induced activity against WCR larvae (Walters et al., 2010). Development of chimeric toxins through domain swapping or domain rearrangement restricts the insects from developing resistance against Cry proteins. Thus, transgenic cotton expressing novel *Cry1AcF* (developed through domain exchange) and *Cry2Aa* (developed through codon alteration) genes provided 80–100% mortality of *H. armigera* larvae and reduced the possibility of resistance development thereby (Karthik et al., 2021).

13.2.3.8 Fusion Protein

Generally speaking, a *fusion protein* is the one that is genetically engineered by incorporating the components of several insecticidal proteins to increase its efficacy against target pests. These fusion proteins with enhanced insecticidal properties are formed by combining segments or domains of different Bt proteins (Nagaraj et al., 2024). For example, transgenic cotton plants expressing a novel *Cry2AX1* gene that was created by combining the sequences from *Cry2Aa* and *Cry2Ac* genes showed effectiveness against *H. armigera* (Sakthi et al., 2015). Similarly, transgenic rice plants containing a fusion protein of Cry1Ab and Vip3A proteins showed a high level of resistance against Asiatic rice borer, *Chilo suppressalis* (Walker), and rice leaf folder, *Cnaphalocrocis medinalis* (Guenée) (Xu et al., 2018).

13.2.3.9 RNA Interference (RNAi) Technology

As a novel approach to target insect pest management, plant host–mediated 'RNA interference' (RNAi) technology has great potential in controlling coleopteran and lepidopteran pests. RNAi is a new strategy and emerged as a potent reverse genetics tool that uses double-stranded RNA (dsRNA) to silence a target gene at the cellular level, thereby suppressing the target gene's expression. This mechanism is likely to come under the next generation of insect-resistant transgenic plants. Using this technique, the host plant becomes able to express the dsRNAs that directly target the functional genes in insects. Body of phytophagous insects absorbs those highly expressed dsRNAs and then induces the system RNAi response (Zhang, 2017c). The efficiency of this technique highly relies on the choice of ideal target insect, target gene(s) for silencing, vector construct design, and the method for dsRNA delivery (Nagaraj et al., 2024).

In one of the earliest instances, this RNAi technology was employed to specifically target the cytochrome P450 gene (*CYP6AE14*) that detoxifies the cotton metabolite gossypol in *H. armigera* (Mao et al., 2007). So this strategy has a practical implication in enhancing the susceptibility of insects to insecticides (Ullah et al., 2023; Lv et al., 2023). Unlike Bt, RNAi targets a specific insect pest and thus protects the non-target organisms (Li et al., 2015). Another advantage of RNAi over Bt is that its application can provide control over a broader range of insect pests, specifically for sap-sucking insects, against which transgenic Bt crops have failed (Dong et al., 2022; Li et al., 2015). That is why this strategy is crucial in situations where the widespread adoption of Bt crops lets sucking pests attain a major status (Zhang et al., 2023). Although this technique opens up new opportunities for eco-friendly insect pest management practices in farming, it still has some limitations in insect resistance management that include target gene mutation and overexpression, apprehensions over off-target consequences, activation of alternate pathways, etc. (Nagaraj et al., 2024).

Mechanism of RNAi. The principle behind RNAi technology is that there is a protein responsible for a specific function in insects, and that protein is synthesized from a specific messenger RNA (mRNA) through translation. So if it is possible to degrade that mRNA, the insect will not be able to produce that protein. That will ultimately result in the inability of that insect to play the function, leading to the death of the target insect. The steps involved in mRNA degradation are thus (Figure 13.4):

- The gene responsible for production of the specific protein in insects is identified.
- The target gene is transferred into the plant cell through the *Agrobacterium* vector.
- Both the sense and antisense strands of the double-stranded DNA (dsDNA) undergo transcription and produce sense and antisense RNA. As the transcripts are complementary, they undergo base pairing and form dsRNA.
- In the presence of the corresponding single-stranded RNA (ssRNA) of dsRNA, dicer enzymes (RNAse III type enzymes) get activated and bind with the dsRNA, breaking it into a number of small fragments (double-stranded), called small interfering RNA (siRNA). These siRNAs contain 20–25 nucleotides.
- When insects feed on plant, this siRNA enters the insect body and binds with the Argonaute protein. This siRNA then unzips, and only one strand of it remains bound with the Argonaute protein, which is called the guide strand. The guide strand of siRNA and Argonaute protein, along with other proteins, form a complex called the RNA-induced silencing complex (RISC).
- siRNA directs the RISC to bind with specific mRNA. Binding is precise due to the base pairing between siRNA and target mRNA.

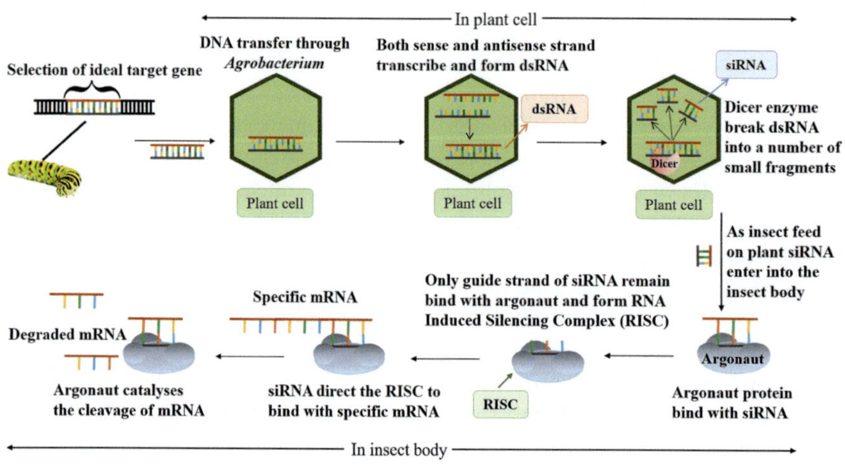

FIGURE 13.4 Steps involved in RNAi-mediated gene silencing.

siRNA has perfect complementarity to its target site.

- Once siRNA binds with the target mRNA, Argonaute protein catalyzes the cleavage of the mRNA and breaks it down.
- This degraded mRNA is unable to translate and does not produce the protein. As a result, the insect dies due to the absence of that specific function. Thus, RNAi interferes with the target gene's expression and achieves insecticidal property.

Pyramiding RNAi with Bt Genes. Transgenic crops with Bt pyramiding, which is one of the most widely adopted strategies by farmers in recent days, are under a serious threat due to the occurrence of cross-resistance in insects against multiple Bt toxins. This condition requires the implementation of alternate management strategies, wherein RNAi might be optimally integrated with Bt toxins, providing an efficient and durable resistance against multiple insect pests. The goal of RNAi is achieved by restricting the expression of target genes that encode proteins essential for the pests to develop resistance, but not for other organisms. On the basis of a large number of RNAi target screenings, the key genes that encode proteins responsible for controlling the juvenile hormone (JH) metabolism in insects were selected as the most suitable targets (Li et al., 2022). Thus, incorporation of RNAi traits that interfere with the JH metabolism with Bt in transgenic cotton showed high larval mortality in the Bt-resistant strain of *H. armigera* (Ni et al., 2017). This RNAi cotton successfully expresses the dsRNA cassettes that significantly decrease the mRNA levels of the target gene within the tested insect pests, resulting in the significant reduction in the JH synthesis. In addition, analysis of multi-year insect resistance data revealed that the RNAi cotton is able to delay the resistance development in cotton bollworm (Zhang, 2017c). In another study, transgenic maize containing two coleopteran-specific Bt toxins (Cry3Bb and Cry34/35Ab) and a dsRNA transcript having a 240–base pair fragment of WCR gene encoding DvSnf7 protein depicted that the addition of DvSnf7 can reduce root injury under high WCR densities and prolong the durability of Bt toxins (Head et al., 2017). Likewise, transgenic crops with multiple gene pyramiding and silencing (MGPS) approach coupled with refuge strategy also proved to be effective in controlling *S. frugiperda* (Ren et al., 2019; Zafar et al., 2020). Based on these observations, it can be stated that the resistance in insects to single-strategy transgenic cotton can easily be overcome by using the RNAi + Bt pyramiding strategy.

13.2.3.10 Other Resistance Management Strategies

In addition to the aforementioned approaches of resistance management to Bt, other eco-friendly and emerging pest management tactics, like release of sterile insects, genome editing, use of other microbial and botanical pesticides, protease inhibitors, lectins, etc., can be adopted along with Bt to delay the resistance evolution in insects against Bt (Nagaraj et al., 2024).

13.3 INSECTS' RESISTANCE TO *LYSINIBACILLUS SPHARICUS* (LS)

After Bt, another EPB that earned much attention in commercial applications is *Lysinibacillus sphaericus* (Ls), which is a dipteran-specific EPB and mostly utilized in mosquito control programmes, specifically against *Culex* mosquitoes (described in Chapter 5). The larvicidal activity of this bacterium depends on the binary (Bin) toxin, which is a heteromorphic protein containing two subunits (BinA and BinB) and synthesized during stage III of sporulation (Berry, 2012). Similar to Bt, the applicability of Ls is also threatened due to the evolution of resistance in mosquitoes against its toxins. Specifically, in subtropical and tropical countries, environmental conditions favour the proliferation of mosquitoes around the year that requires continuous application of Ls-based pesticides, imposing more selection pressure on these populations thereby (Silva Filha et al., 2014). In addition, higher persistency of this bacterium in mosquito breeding sites (Karch et al., 1988; Nicolas et al., 1987) also ensures prolonged larval exposure to its insecticidal crystals. Such higher selection pressure due to uncontrolled usage of Ls ultimately triggers the resistance development in mosquitoes.

Resistance against Ls was reported under both laboratory and field conditions. *Culex pipiens* Linnaeus populations have shown varying levels of resistance against Ls. Initially, cases of resistance have been reported in field populations of *C. pipiens* complex from France (Sinègre et al., 1994), India (Rao et al., 1995), Brazil (Silva-Filha et al., 1995), and China (Yuan et al., 2000). A high level of resistance (>10,000-fold) was reported in a field-selected *C. pipiens* colony from France due to one major sex-linked gene (Nielsen-LeRoux et al., 1997). The first resistant laboratory mosquito colony (i.e. GEO) was established from field samples of *Cx. pipiens* from California, having a resistance ratio of about 100,000 after 12 generations of imposing selection pressure (Wirth et al., 2000a). Rapid resistance has also been reported in *Culex quinquefasciatus* Say against Ls with a resistance ration of 35- to 150,000-fold and 10- to 10,000-fold under laboratory and field conditions, respectively, against different highly toxic strains of this bacterium (Pei et al., 2002; Yuan et al., 2003). It was also assumed that resistance to certain Ls strains may cause more or less cross-resistance to other strains producing the same group of toxins (Nielsen-LeRoux et al., 2001; Yuan et al., 2003). This makes most of the Ls strains ineffective against resistant *Culex* larvae. However, some strains of Ls that remain toxic to the resistant larval population led to the discovery of different other toxins (Nielsen-LeRoux et al., 2001; Pei et al., 2002). Furthermore, significant reduction in fertility and fecundity of Ls resistant larvae (Oliveira et al., 2003; Rodcharoen & Mulla, 1997) and recessive inheritance of resistance genes (Nielsen-LeRoux et al., 1997; Oliveira et al., 2004; Wirth et al., 2000a) make resistance management strategies for Ls viable under field conditions (Mulla et al., 2003).

13.3.1 Mechanism of Ls Resistance in Mosquitoes

It becomes relatively easy for mosquitoes to develop high levels of resistance against Ls strains producing only one spore-associated toxin (Bin) (Nielsen-LeRoux et al., 1995). In addition, there is involvement of only a single-type midgut receptor with which this heteromorphic Bin toxin interacts via its BinB component (Silva-Filha et al., 1999). So any failure in this binding step of the BinB component to its specific receptor facilitates the evolution of resistance in mosquito larvae under continuous selection pressure (Oliveira et al., 2004). That is why most of the reported cases of insect resistance against this bacterium involve any changes in the targeted gut receptor of the toxin that prevent localization of the toxin to host cell membranes (Berry, 2012). There is involvement of a variety of mutations for resistance mechanisms. For example, occurrence of mutation causing premature termination of translation of the target receptor protein resulted in truncation of its GPI-anchoring site that prevents the processing of this protein and eliminates its GPI anchorage from the plasma membrane, even though the toxin binding site and receptor's enzymatic activity remain unaffected. This truncated protein is incapable of functioning as a Bin toxin receptor due to its incompleteness as a membrane-bound protein (Darboux et al., 2002; Guo et al., 2013; Romão et al., 2006). Resistance may also emerge due to the insertion of a transposable element (like DNA) into the midgut toxin receptor's coding sequence. This induces a novel mRNA splicing event that unmasks cryptic donor and acceptor sites within the host gene. Thus, the creation of a new intron expressing an altered membrane protein makes it unable to interact with the toxin (Darboux et al., 2007). So any alteration in receptor binding has a crucial role in resistance development in mosquitoes. Apart from the changes in receptor binding, Oliveira et al. (2009) proposed that secretions of apocrine from gut cells of mosquitoes may also act as a defence mechanism against Ls because of more intense secretion of apocrine in the most resistant species.

13.3.2 Resistance Management for Ls

Resistance management for Ls requires an integrated vector-control approach. The major goal of this approach is to reduce the population density of disease-causing insect vectors by adopting different eco-friendly methods in a cost-effective way (WHO, 2004). By utilizing multiple management strategies, it is possible to lower down the selection pressure.

One of the most important approaches is the utilization of Bti or other mosquitocidal Bt strains, like *Bt* serovar. *jegathesan* and *Bt* serovar. *medellin*, that are able to kill Bin-resistant mosquito populations (Silva Filha et al., 2014). Development recombinant strains that express both Bin toxin of Ls and Cry/Cyt toxins from different Bt strains, with a synergistic action between the toxins, proved to have enhanced toxicity against Ls-resistant colonies of *Culex* mosquitoes (Poncet et al., 1994, 1997; Thiéry et al., 1998). Similarly, a mixture of Cyt toxins from Bti with Bin toxins from Ls showed activity against Bin-resistant *Culex* population (Wirth et al., 2000b, 2001, 2004, 2005), as well as against the larvae of *Aedes aegypti* (Linnaeus) (Wirth et al., 2000c), which shows poor susceptibility to Ls binary toxins. Apart from that, recombinant Bti producing an association of Bt and Bin toxins also showed improved toxicity against Bin resistant *Culex* strains (Park et al., 2003, 2005). Moreover, incorporation of Ls and Bt toxins in other aquatic microorganisms has also been attempted to achieve a combined expression and found to be suitable delivery in mosquito larval habitats (Yap et al., 1994). Thus, it is clear that synergy between Bt and Ls (Bin) toxins can efficiently delay the resistance development. However, an important consideration is that the introduction of genes from Bt to Ls strains has been often characterized by the instability and/or low expression of the respective foreign proteins (Federici et al., 2010).

Another approach is the use of alternative Ls toxins. As most of the Ls strains possess Bin toxins, the occurrence of cross-resistance is highly possible even if their Bin toxins come under different classes. That's why Ls strains producing other groups of toxins, like Cry48Aa/Cry49Aa and Mtx toxins, etc., can be employed to overcome Bin resistance (Silva Filha et al., 2014). For example, low amounts of Mtx1 and Mtx2 when used in combination with Bin toxin can significantly synergize the Bin toxin's activity against resistant *Culex* mosquitoes (Wirth et al., 2007). Similarly, Cry48Aa/Cry49Aa combination usually targets midgut molecules that are completely different from Bin receptors, resulting in its higher activity against Bin-resistant *Culex* larvae (de Melo et al., 2009). So use of biolarvicides with such multi-toxin traits could be an efficient solution in managing Ls resistance in mosquitoes. In addition to the two approaches mentioned, integration of other biocontrol agents, such as aquatic predators (aquatic insects, fishes), entomopathogenic fungi, and nematodes, in combination with Ls, is also a viable option to reduce the larval exposure to Ls toxins (Lacey, 2007). Use of insect growth regulators and spinosyns with a narrow toxicity spectrum also showed promising results when applied in combination with other biolarvicides (Silva Filha et al., 2014).

13.4 CONCLUSION

The utilization of microbial pesticides based on EPBs is increasing due to the demand for organic products and their advantages over synthetic pesticides. That is why, for many years, Bt- and Ls-based products and transgenic Bt crops have been successfully employed in managing different insect pests. However, with time, insects also developed multiple strategies to overcome the toxins produced by these EPBs, for which the adoption of unscientific pest management tactics is mainly responsible. This situation necessitates the implementation of alternate techniques to safeguard the applicability of EPBs. This literature provides substantial evidence that resistance to commercial EPBs targets the breakdown of toxin binding. Accordingly, the utilization of combinations of toxins targeting various receptors substantially mitigates the risk of elevated resistance levels. For Bt crops, refuge planting and crop rotation have a huge role to conserve the susceptible insect population. These are all simple strategies that can easily be employed in farmers' fields. However, it requires government interventions to mandate these strategies. Several innovative novel strategies are also emerging to address this problem, necessitating further research to mitigate insect resistance to EPBs. Another thing that needs to be kept in mind is that these strategies minimize but do not completely eradicate the risk of resistance evolution in insects. That is why proper use of these strategies is essential.

REFERENCES

Adamczyk Jr., J. J., & Mahaffey, J. S. (2008). Efficacy of Vip3A and Cry1Ab transgenic traits in cotton against various lepidopteran pests. *Florida Entomologist*, *91*(4), 570–575. https://doi.org/10.1653/0015-4040-91.4.570

Ali, S., Aqueel, M. A., Saeed, M. F., Shakeel, Q., Raheel, M., & Ullah, M. I. (2022). Utilization of entomopathogenic bacteria for modern insect pest management. In S. D. Mandal, G. Ramkumar, S. Karthi, & F. Jin (Eds.), *New and future development in biopesticide research: Biotechnological exploration* (pp. 93–113). Springer. https://doi.org/10.1007/978-981-16-3989-0_3

Azizoglu, U., Jouzani, G. S., Yilmaz, N., Baz, E., & Ozkok, D. (2020). Genetically modified entomopathogenic bacteria, recent developments, benefits and impacts: A review. *Science of the Total Environment*, *734*, 139169. https://doi.org/10.1016/j.scitotenv.2020.139169

Babu, R. M., Sajeena, A., Seetharaman, K., & Reddy, M. S. (2003). Advances in genetically engineered (transgenic) plants in pest management—An overview. *Crop Protection*, *22*(9), 1071–1086. https://doi.org/10.1016/S0261-2194(03)00142-X

Bagla, P. (2010). Hardy cotton-munching pests are latest blow to GM crops. *Science*, *327*(5972), 1439. https://doi.org/10.1126/science.327.5972.1439

Baxter, S. W., Badenes-Pérez, F. R., Morrison, A., Vogel, H., Crickmore, N., Kain, W., Wang, P., Heckel, D. G., & Jiggins, C. D. (2011). Parallel evolution of *Bacillus thuringiensis* toxin resistance in lepidoptera. *Genetics*, *189*(2), 675–679. https://doi.org/10.1534/GENETICS.111.130971

Bel, Y., Siqueira, H. A. A., Siegfried, B. D., Ferré, J., & Escriche, B. (2009). Variability in the cadherin gene in an *Ostrinia nubilalis* strain selected for Cry1Ab resistance. *Insect Biochemistry and Molecular Biology*, *39*(3), 218–223. https://doi.org/10.1016/J.IBMB.2008.11.005

Bentur, J. S., Andow, D. A., Cohen, M. B., Romena, A. M., & Gould, F. (2000). Frequency of alleles conferring resistance to a *Bacillus thuringiensis* toxin in a Philippine population of *Scirpophaga incertulas* (Lepidoptera: Pyralidae). *Journal of Economic Entomology*, *93*(5), 1515–1521. https://doi.org/10.1603/0022-0493-93.5.1515

Bernardi, O., Bernardi, D., Ribeiro, R. S., Okuma, D. M., Salmeron, E., Fatoretto, J., Medeiros, F. C. L., Burd, T., & Omoto, C. (2015). Frequency of resistance to Vip3Aa20 toxin from *Bacillus thuringiensis* in *Spodoptera frugiperda* (Lepidoptera: Noctuidae) populations in Brazil. *Crop Protection*, *76*, 7–14. https://doi.org/10.1016/j.cropro.2015.06.006

Berry, C. (2012). The bacterium, *Lysinibacillus sphaericus*, as an insect pathogen. *Journal of Invertebrate Pathology*, *109*(1), 1–10. https://doi.org/10.1016/j.jip.2011.11.008

Bessin, R. (n.d.). *Bt corn refuges for corn borer management.* Department of Entomology: Martin-Gatton College of Agriculture, Food and Environment. https://entomology.ca.uky.edu/ef

Bhatti, M. U., Tabassum, B., Berry, C., Khan, A., Qaisar, U., Ali, E., Khalid, R., Farooq, A. M., Tariq, M., & Ayaz, H. (2023). Transgenic maize inbred lines expressing high levels of *Bacillus thuringiensis* vegetative insecticidal protein (Vip3Aa86) offer effective control of maize stem borer (*Chilo partellus*). *Plant Cell, Tissue and Organ Culture (PCTOC)*, *153*(2), 417–427. https://doi.org/10.1007/s11240-023-02483-w

Boaventura, D., Ulrich, J., Lueke, B., Bolzan, A., Okuma, D., Gutbrod, O., Geibel, S., Zeng, Q., Dourado, P. M., Martinelli, S., & Flagel, L. (2020). Molecular characterization of Cry1F resistance in fall armyworm, *Spodoptera frugiperda* from Brazil. *Insect Biochemistry and Molecular Biology*, *116*, 103280. https://doi.org/10.1016/j.ibmb.2019.103280

Bravo, A., & Soberón, M. (2008). How to cope with insect resistance to Bt toxins? *Trends in Biotechnology*, *26*(10), 573–579. https://doi.org/10.1016/j.tibtech.2008.06.005

Brévault, T., Heuberger, S., Zhang, M., Ellers-Kirk, C., Ni, X., Masson, L., Li, X., Tabashnik, B. E., & Carrière, Y. (2013). Potential shortfall of pyramided transgenic cotton for insect resistance management. *Proceedings of the National Academy of Sciences*, *110*(15), 5806–5811. https://doi.org/10.1073/pnas.1216719110

Brévault, T., Tabashnik, B. E., & Carrière, Y. (2015). A seed mixture increases dominance of resistance to Bt cotton in *Helicoverpa zea*. *Scientific Reports*, *5*(1), 9807. https://doi.org/10.1038/srep09807

Buradikatti, K. (2015, December 5). Pink bollworm a nightmare for Bt cotton growers. *The Hindu.* https://www.thehindu.com/news/national/karnataka/pink-bollworm-a-nightmare-for-bt-cotton-growers/article7950687.ece

Burd, A. D., Gould, F., Bradley, J. R., Van Duyn, J. W., & Moar, W. J. (2003). Estimated frequency of nonrecessive Bt resistance genes in bollworm, *Helicoverpa zea* (Boddie) (Lepidoptera: Noctuidae) in eastern North Carolina. *Journal of Economic Entomology*, *96*(1), 137–142. https://doi.org/10.1093/jee/96.1.137

Candas, M., Loseva, O., Oppert, B., Kosaraju, P., & Bulla, L. A. (2003). Insect resistance to *Bacillus thuringiensis*: Alterations in the Indian meal moth larval gut proteome. *Molecular & Cellular Proteomics: MCP*, *2*(1), 19–28. https://doi.org/10.1074/MCP.M200069-MCP200

Carrière, Y., Brown, Z., Aglasan, S., Dutilleul, P., Carroll, M., Head, G., Tabashnik, B. E., Jørgensen, P. S., & Carroll, S. P. (2020a). Crop rotation mitigates impacts of corn rootworm resistance to transgenic Bt corn. *Proceedings of the National Academy of Sciences*, *117*(31), 18385–18392. https://doi.org/10.1073/pnas.2003604117

Carrière, Y., Brown, Z. S., Downes, S. J., Gujar, G., Epstein, G., Omoto, C., Storer, N. P., Mota-Sanchez, D., Søgaard Jørgensen, P., & Carroll, S. P. (2020b). Governing evolution: A socioecological comparison of resistance management for insecticidal transgenic Bt crops among four countries. *Ambio*, *49*, 1–16. https://doi.org/10.1007/s13280-019-01167-0

Carrière, Y., Fabrick, J. A., & Tabashnik, B. E. (2016). Advances in managing pest resistance to Bt crops: Pyramids and seed mixtures. In A. R. Horowitz, & I. Ishaaya (Eds.), *Advances in insect control and resistance management* (pp. 263–286). Springer. https://doi.org/10.1007/978-3-319-31800-4_14

Carroll, M. W., Head, G., & Caprio, M. (2012). When and where a seed mix refuge makes sense for managing insect resistance to Bt plants. *Crop Protection*, *38*, 74–79. https://doi.org/10.1016/j.cropro.2012.02.015

Castagnola, A., & Jurat-Fuentes, J. L. (2016). Intestinal regeneration as an insect resistance mechanism to entomopathogenic bacteria. *Current Opinion in Insect Science*, *15*, 104–110. https://doi.org/10.1016/j.cois.2016.04.008

Chitkowski, R. L., Turnipseed, S. G., Sullivan, M. J., & Bridges Jr., W. C. (2003). Field and laboratory evaluations of transgenic cottons expressing one or two *Bacillus thuringiensis* var. *kurstaki* Berliner proteins for management of noctuid (Lepidoptera) pests. *Journal of Economic Entomology*, *96*(3), 755–762. https://doi.org/10.1093/jee/96.3.755

Coates, B. S., & Siegfried, B. D. (2015). Linkage of an ABCC transporter to a single QTL that controls *Ostrinia nubilalis* larval resistance to the *Bacillus thuringiensis* Cry1Fa toxin. *Insect Biochemistry and Molecular Biology*, *63*, 86–96. https://doi.org/10.1016/J.IBMB.2015.06.003

Coates, B. S., Sumerford, D. V., Siegfried, B. D., Hellmich, R. L., & Abel, C. A. (2013). Unlinked genetic loci control the reduced transcription of aminopeptidase N 1 and 3 in the European corn borer and determine tolerance to *Bacillus thuringiensis* Cry1Ab toxin. *Insect Biochemistry and Molecular Biology*, *43*(12), 1152–1160. https://doi.org/10.1016/J.IBMB.2013.09.003

Darboux, I., Charles, J. F., Pauchet, Y., Warot, S., & Pauron, D. (2007). Transposon-mediated resistance to *Bacillus sphaericus* in a field-evolved population of *Culex pipiens* (Diptera: Culicidae). *Cellular Microbiology*, *9*(8), 2022–2029. https://doi.org/10.1111/j.1462-5822.2007.00934.x

Darboux, I., Pauchet, Y., Castella, C., Silva-Filha, M. H., Nielsen-LeRoux, C., Charles, J. F., & Pauron, D. (2002). Loss of the membrane anchor of the target receptor is a mechanism of bioinsecticide resistance. *Proceedings of the National Academy of Sciences*, *99*(9), 5830–5835. https://doi.org/10.1073/pnas.09261539

de Bortoli, C. P., & Jurat-Fuentes, J. L. (2019). Mechanisms of resistance to commercially relevant entomopathogenic bacteria. *Current Opinion in Insect Science*, *33*, 56–62. https://doi.org/10.1016/j.cois.2019.03.007

de Melo, J. V., Jones, G. W., Berry, C., Vasconcelos, R. H. T., de Oliveira, C. M. F., Furtado, A. F., Peixoto, C. A., & Silva-Filha, M. H. N. L. (2009). Cytopathological effects of *Bacillus sphaericus* Cry48Aa/Cry49Aa toxin on binary toxin-susceptible and-resistant *Culex quinquefasciatus* larvae. *Applied and Environmental Microbiology*, *75*(14), 4782–4789. https://doi.org/10.1128/AEM.00811-09

Dhurua, S., & Gujar, G. T. (2011). Field-evolved resistance to Bt toxin Cry1Ac in the pink bollworm, *Pectinophora gossypiella* (Saunders) (Lepidoptera: Gelechiidae), from India. *Pest Management Science*, *67*(8), 898–903. https://doi.org/10.1002/ps.2127

Domínguez-Arrizabalaga, M., Villanueva, M., Escriche, B., Ancín-Azpilicueta, C., & Caballero, P. (2020). Insecticidal activity of *Bacillus thuringiensis* proteins against coleopteran pests. *Toxins*, *12*(7), 430. https://doi.org/10.3390/toxins12070430

Dong, Y., Wu, M., Zhang, Q., Fu, J., Loiacono, F. V., Yang, Y., Wang, Z., Li, S., Chang, L., Bock, R., & Zhang, J. (2022). Control of a sap-sucking insect pest by plastid-mediated RNA interference. *Molecular Plant*, *15*(7), 1176–1191. https://doi.org/10.1016/j.molp.2022.05.008

Downes, S., Parker, T. L., & Mahon, R. J. (2009). Frequency of alleles conferring resistance to the *Bacillus thuringiensis* toxins Cry1Ac and Cry2Ab in Australian populations of *Helicoverpa punctigera* (Lepidoptera: Noctuidae) from 2002 to 2006. *Journal of Economic Entomology*, *102*(2), 733–742. https://doi.org/10.1603/029.102.0234

Fabrick, J. A., Mathew, L. G., Tabashnik, B. E., & Li, X. (2011). Insertion of an intact CR1 retrotransposon in a cadherin gene linked with Bt resistance in the pink bollworm, *Pectinophora gossypiella*. *Insect Molecular Biology*, *20*(5), 651–665. https://doi.org/10.1111/J.1365-2583.2011.01095.X

Fabrick, J. A., Ponnuraj, J., Singh, A., Tanwar, R. K., Unnithan, G. C., Yelich, A. J., Li, X., Carrière, Y., & Tabashnik, B. E. (2014). Alternative splicing and highly variable cadherin transcripts associated with field-evolved resistance of pink bollworm to bt cotton in India. *PLoS One*, *9*, e97900. https://doi.org/10.1371/JOURNAL.PONE.0097900

Federici, B. A., Park, H. W., & Bideshi, D. K. (2010). Overview of the basic biology of *Bacillus thuringiensis* with emphasis on genetic engineering of bacterial larvicides for mosquito control. *Open Toxinology Journal*, *3*(2), 83–100.

Ferré, J., Real, M. D., Van Rie, J., Jansens, S., & Peferoen, M. (1991). Resistance to the *Bacillus thuringiensis* bioinsecticide in a field population of *Plutella xylostella* is due to a change in a midgut membrane receptor. *Proceedings of the National Academy of Sciences*, *88*(12), 5119–5123. https://doi.org/10.1073/PNAS.88.12.5119

Ferré, J., & Van Rie, J. (2002). Biochemistry and genetics of insect resistance to *Bacillus thuringiensis*. *Annual Review of Entomology*, *47*, 501–533. https://doi.org/10.1146/ANNUREV.ENTO.47.091201.145234

Flagel, L., Lee, Y. W., Wanjugi, H., Swarup, S., Brown, A., Wang, J., Kraft, E., Greenplate, J., Simmons, J., Adams, N., & Wang, Y. (2018). Mutational disruption of the ABCC2 gene in fall armyworm, *Spodoptera frugiperda*, confers resistance to the Cry1Fa and Cry1A. 105 insecticidal proteins. *Scientific Reports*, *8*(1), 7255. https://doi.org/10.1038/s41598-018-25491-9

Forcada, C., Alcácer, E., Garcerá, M. D., & Martínez, R. (1996). Differences in the midgut proteolytic activity of two *Heliothis virescens* strains, one susceptible and one resistant to *Bacillus thuringiensis* toxins. *Archives of Insect Biochemistry and Physiology*, *31*(3), 257–272. https://doi.org/10.1002/(SICI)1520-6327(1996)31:3<257::AID-ARCH2>3.0.CO;2-V

Forcada, C., Alcácer, E., Garcerá, M. D., Tato, A., & Martínez, R. (1999). Resistance to *Bacillus thuringiensis* Cry1Ac toxin in three strains of *Heliothis virescens*: Proteolytic and SEM study of the larval midgut. *Archives of Insect*

Biochemistry and Physiology, *42*(1), 51–63. https://doi.org/10.1002/(SICI)1520-6327(199909)42:1<51::AID-ARCH6>3.0.CO;2-6

Fritz, M. L., Nunziata, S. O., Guo, R., Tabashnik, B. E., & Carrière, Y. (2020). Mutations in a novel cadherin gene associated with Bt resistance in *Helicoverpa zea*. *G3: Genes, Genomes, Genetics*, *10*(5), 1563–1574. https://doi.org/10.1534/g3.120.401053

Fu, B. W., Xu, L., Zheng, M. X., Shi, Y., & Zhu, Y. J. (2024). Engineering of *Bacillus thuringiensis* Cry2Ab toxin for improved insecticidal activity. *AMB Express*, *14*(1), 15. https://doi.org/10.1186/s13568-024-01669-5

Gahan, L. J., Gould, F., & Heckel, D. G. (2001). Identification of a gene associated with Bt resistance in *Heliothis virescens*. *Science*, *293*(5531), 857–860. https://doi.org/10.1126/SCIENCE.1060949

Gahan, L. J., Ma, Y. T., MacgregorCoble, M. L., Gould, F., Moar, W. J., & Heckel, D. G. (2005). Genetic basis of resistance to Cry1Ac and Cry2Aa in *Heliothis virescens* (Lepidoptera: Noctuidae). *Journal of Economic Entomology*, *98*(4), 1357–1368. https://doi.org/10.1603/0022-0493-98.4.1357

Gahan, L. J., Pauchet, Y., Vogel, H., & Heckel, D. G. (2010). An ABC transporter mutation is correlated with insect resistance to *Bacillus thuringiensis* Cry1Ac toxin. *PLoS Genetics*, *6*(12), e1001248. https://doi.org/10.1371/journal.pgen.1001248

Gassmann, A. J. (2021). Resistance to Bt maize by western corn rootworm: Effects of pest biology, the pest–crop interaction and the agricultural landscape on resistance. *Insects*, *12*(2), 136. https://doi.org/10.3390/insects12020136

Gassmann, A. J., & Reisig, D. D. (2023). Management of insect pests with Bt crops in the United States. *Annual Review of Entomology*, *68*(1), 31–49. https://doi.org/10.1146/annurev-ento-120220-105502

Gómez, I., Sánchez, J., Miranda, R., Bravo, A., & Soberón, M. (2002). Cadherin-like receptor binding facilitates proteolytic cleavage of helix α-1 in domain I and oligomer pre-pore formation of *Bacillus thuringiensis* Cry1Ab toxin. *FEBS Letters*, *513*(2–3), 242–246. https://doi.org/10.1016/S0014-5793(02)02321-9

Gould, F. (1986). Simulation models for predicting durability of insect-resistant germ plasm: Hessian fly (Diptera: Cecidomyiidae)-resistant winter wheat. *Environmental Entomology*, *15*(1), 11–23. https://doi.org/10.1093/ee/15.1.11

Gould, F. (1998). Sustainability of transgenic insecticidal cultivars: Integrating pest genetics and ecology. *Annual Review of Entomology*, *43*(1), 701–726. https://doi.org/10.1146/annurev.ento.43.1.701

Gould, F., Anderson, A., Jones, A., Sumerford, D., Heckel, D. G., Lopez, J., Micinski, S., Leonard, R., & Laster, M. (1997). Initial frequency of alleles for resistance to *Bacillus thuringiensis* toxins in field populations of *Heliothis virescens*. *Proceedings of the National Academy of Sciences*, *94*(8), 3519–3523. https://doi.org/10.1073/pnas.94.8.3519

Gould, F., Cohen, M. B., Bentur, J. S., Kennedy, G. G., & VanDuyn, J. (2006). Impact of small fitness costs on pest adaptation to crop varieties with multiple toxins: A heuristic model. *Journal of Economic Entomology*, *99*(6), 2091–2099. https://doi.org/10.1093/jee/99.6.2091

Gould, F., Martinez-Ramirez, A., Anderson, A., Ferre, J., Silva, F. J., & Moar, W. J. (1992). Broad-spectrum resistance to *Bacillus thuringiensis* toxins in *Heliothis virescens*. *Proceedings of the National Academy of Sciences*, *89*(17), 7986–7990. https://doi.org/10.1073/pnas.89.17.7986

Gujar, G. T., & Mohan, M. (2003). Diamondback moth resistance to *Bacillus thuringiensis* and its toxins: An Indian experience. In O. Koul, G. S. Dhaliwal, S. S. Marwaha, & J. K. Arora (Eds.), *Biopesticides and pest management* (pp. 96–112). Campus Books International.

Gunning, R. V., Dang, H. T., Kemp, F. C., Nicholson, I. C., & Moores, G. D. (2005). New resistance mechanism in *Helicoverpa armigera* threatens transgenic crops expressing *Bacillus thuringiensis* Cry1Ac toxin. *Applied and Environmental Microbiology*, *71*(5), 2558–2563. https://doi.org/10.1128/AEM.71.5.2558-2563.2005

Guo, Q. Y., Cai, Q. X., Yan, J. P., Hu, X. M., Zheng, D. S., & Yuan, Z. M. (2013). Single nucleotide deletion of *cqm1* gene results in the development of resistance to *Bacillus sphaericus* in *Culex quinquefasciatus*. *Journal of Insect Physiology*, *59*(9), 967–973. https://doi.org/10.1016/j.jinsphys.2013.07.002

Guo, Z., Kang, S., Chen, D., Wu, Q., Wang, S., Xie, W., Zhu, X., Baxter, S. W., Zhou, X., Jurat-Fuentes, J. L., & Zhang, Y. (2015). MAPK signaling pathway alters expression of midgut ALP and ABCC genes and causes resistance to *Bacillus thuringiensis* Cry1Ac toxin in diamondback moth. *PLoS Genetics*, *11*(4). https://doi.org/10.1371/JOURNAL.PGEN.1005124

Head, G. P., Carroll, M. W., Evans, S. P., Rule, D. M., Willse, A. R., Clark, T. L., Storer, N. P., Flannagan, R. D., Samuel, L. W., & Meinke, L. J. (2017). Evaluation of SmartStax and SmartStax PRO maize against western corn rootworm and northern corn rootworm: Efficacy and resistance management. *Pest Management Science*, *73*(9), 1883–1899. https://doi.org/10.1002/ps.4554

Herrero, S., Gechev, T., Bakker, P. L., Moar, W. J., & de Maagd, R. A. (2005). *Bacillus thuringiensis* Cry1Ca-resistant *Spodoptera exigua* lacks expression of one of four Aminopeptidase N genes. *BMC Genomics*, *6*, 96. https://doi.org/10.1186/1471-2164-6-96

Heuberger, S., Crowder, D. W., Brévault, T., Tabashnik, B. E., & Carrière, Y. (2011). Modeling the effects of plant-to-plant gene flow, larval behavior, and refuge size on pest resistance to Bt cotton. *Environmental Entomology*, *40*(2), 484–495. https://doi.org/10.1603/EN10247

Heuberger, S., Ellers-Kirk, C., Yafuso, C., Gassmann, A. J., Tabashnik, B. E., Dennehy, T. J., & Carrière, Y. (2014). Effects of refuge contamination by transgenes on Bt resistance in pink bollworm (Lepidoptera: Gelechiidae). *Journal of Economic Entomology*, *101*(2), 504–514. https://doi.org/10.1093/jee/101.2.504

Hitchon, A. J., Smith, J. L., French, B. W., & Schaafsma, A. W. (2015). Impact of the Bt corn proteins Cry34/35Ab1 and Cry3Bb1, alone or pyramided, on western corn rootworm (Coleoptera: Chrysomelidae) beetle emergence in the field. *Journal of Economic Entomology*, *108*(4), 1986–1993. https://doi.org/10.1093/jee/tov125

Huang, F., Parker, R., Leonard, R., Yong, Y., & Liu, J. (2009). Frequency of resistance alleles to *Bacillus thuringiensis*-corn in Texas populations of the sugarcane borer, *Diatraea saccharalis* (F.) (Lepidoptera: Crambidae). *Crop Protection*, *28*(2), 174–180. https://doi.org/10.1016/j.cropro.2008.10.002

Ibrahim, M. A., Griko, N., Junker, M., & Bulla, L. A. (2010). *Bacillus thuringiensis*: A genomics and proteomics perspective. *Bioengineered Bugs*, *1*(1), 31–50. https://doi.org/10.4161/bbug.1.1.10519

ISAAA. (2019). *Global status of commercialized of biotech/GM crops in 2019: Biotech crops drive socio-economic development and sustainable environment in the new frontier* (Brief 55). International Service for the Acquisition of Agri-biotech Applications.

Ives, A. R., Glaum, P. R., Ziebarth, N. L., & Andow, D. A. (2011). The evolution of resistance to two-toxin pyramid transgenic crops. *Ecological Applications*, *21*(2), 503–515. https://doi.org/10.1890/09-1869.1

Jaffe, G. (2009). *Complacency on the farm: Significant non-compliance with EPA's refuge requirements threatens the future effectiveness of genetically engineered pest-protected corn*. Center for Science in the Public Interest.

Jakka, S. R. K., Gong, L., Hasler, J., Banerjee, R., Sheets, J. J., Narva, K., Blanco, C. A., & Jurat-Fuentes, J. L. (2016). Field-evolved mode 1 resistance of the fall armyworm to transgenic Cry1Fa expressing corn associated with reduced Cry1Fa toxin binding and midgut alkaline phosphatase expression. *Applied and Environmental Microbiology*, *82*(4), 1023–1034. https://doi.org/10.1128/AEM.02871-15

Jin, L., Wang, J., Guan, F., Zhang, J., Yu, S., Liu, S., Xue, Y., Li, L., Wu, S., Wang, X., Yang, Y., Abdelgaffar, H., Jurat-Fuentes, J. L., Tabashnik, B. E., & Wu, Y. (2018). Dominant point mutation in a tetraspanin gene associated with field-evolved resistance of cotton bollworm to transgenic Bt cotton. *Proceedings of the National Academy of Sciences*, *115*(46), 11760–11765. https://doi.org/10.1073/pnas.1812138115

Jin, L., Zhang, H., Lu, Y., Yang, Y., Wu, K., Tabashnik, B. E., & Wu, Y. (2015). Large-scale test of the natural refuge strategy for delaying insect resistance to transgenic Bt crops. *Nature Biotechnology*, *33*(2), 169–174. https://doi.org/10.1038/nbt.3100

Jin, T., Chang, X., Gatehouse, A. M. R., Wang, Z., Edwards, M. G., & He, K. (2014). Downregulation and mutation of a Cadherin gene associated with Cry1Ac resistance in the Asian Corn Borer, *Ostrinia furnacalis* (Guenée). *Toxins*, *6*(9), 2676–2693. https://doi.org/10.3390/TOXINS6092676

Jurat-Fuentes, J. L., & Adang, M. J. (2004). Characterization of a Cry1Ac-receptor alkaline phosphatase in susceptible and resistant *Heliothis virescens* larvae. *European Journal of Biochemistry*, *271*(15), 3127–3135. https://doi.org/10.1111/J.1432-1033.2004.04238.X

Jurat-Fuentes, J. L., Gould, F. L., & Adang, M. J. (2003). Dual resistance to *Bacillus thuringiensis* Cry1Ac and Cry2Aa toxins in *Heliothis virescens* suggests multiple mechanisms of resistance. *Applied and Environmental Microbiology*, *69*(10), 5898–5906. https://doi.org/10.1128/AEM.69.10.5898-5906.2003

Jurat-Fuentes, J. L., Karumbaiah, L., Jakka, S. R. K., Ning, C., Liu, C., Wu, K., Jackson, J., Gould, F., Blanco, C., Portilla, M., Perera, O., & Adang, M. (2011). Reduced levels of membrane-bound alkaline phosphatase are common to lepidopteran strains resistant to Cry toxins from *Bacillus thuringiensis*. *PLoS One*, *6*(3), e17606. https://doi.org/10.1371/journal.pone.0017606

Karch, S., Monteny, N., & Coz, J. (1988). Persistence of *Bacillus sphaericus* in a mosquito breeding site 4 years after its introduction for microbial control. *Comptes Rendus de l'Académie des Sciences, III (Sciences de la Vie)*, *307*(5), 289–292.

Karthik, K., Negi, J., Rathinam, M., Saini, N., & Sreevathsa, R. (2021). Exploitation of novel Bt ICPs for the management of *Helicoverpa armigera* (Hübner) in cotton (*Gossypium hirsutum* L.): A transgenic approach. *Frontiers in Microbiology*, *12*, 661212. https://doi.org/10.3389/fmicb.2021.661212

Khajuria, C., Buschman, L. L., Chen, M. S., Siegfried, B. D., & Zhu, K. Y. (2011). Identification of a novel aminopeptidase P-like gene (*OnAPP*) possibly involved in Bt toxicity and resistance in a major corn pest (*Ostrinia nubilalis*). *PloS One*, *6*(8), e23983. https://doi.org/10.1371/JOURNAL.PONE.0023983

Kranthi, K. R., & Stone, G. D. (2020). Long-term impacts of Bt cotton in India. *Nature Plants*, *6*(3), 188–196. https://doi.org/10.1038/s41477-020-0615-5

Kumar, S., Chandra, A., & Pandey, K. C. (2008). *Bacillus thuringiensis* (Bt) transgenic crop: An environment friendly insect-pest management strategy. *Journal of Environmental Biology*, *29*(5), 641–653.

Lacey, L. A. (2007). *Bacillus thuringiensis* serovariety *israelensis* and *Bacillus sphaericus* for mosquito control. *Journal of the American Mosquito Control Association*, *23*(sp2), 133–163. https://doi.org/10.2987/8756-971X(2007)23[133:BTSIAB]2.0.CO;2

Li, H., Guan, R., Guo, H., & Miao, X. (2015). New insights into an RNAi approach for plant defence against piercing-sucking and stem-borer insect pests. *Plant, Cell & Environment*, *38*(11), 2277–2285. https://doi.org/10.1111/pce.12546

Li, Y. Y., Chen, J. J., Liu, M. Y., He, W. W., Reynolds, J. A., Wang, Y. N., Wang, M. Q., & Zhang, L. S. (2022). Enhanced degradation of juvenile hormone promotes reproductive diapause in the predatory ladybeetle *Coccinella septempunctata*. *Frontiers in Physiology*, *13*, 877153. https://doi.org/10.3389/fphys.2022.877153

Liu, C., Xiao, Y., Li, X., Oppert, B., Tabashnik, B. E., & Wu, K. (2014). Cis-mediated down-regulation of a trypsin gene associated with Bt resistance in cotton bollworm. *Scientific Reports*, *4*, 7219. https://doi.org/10.1038/SREP07219

Loeb, M. J., Martin, P. A., Hakim, R. S., Goto, S., & Takeda, M. (2001). Regeneration of cultured midgut cells after exposure to sublethal doses of toxin from two strains of *Bacillus thuringiensis*. *Journal of Insect Physiology*, *47*(6), 599–606. https://doi.org/10.1016/S0022-1910(00)00150-5

Loseva, O., Ibrahim, M., Candas, M., Koller, C. N., Bauer, L. S., & Bulla, L. A. (2002). Changes in protease activity and Cry3Aa toxin binding in the Colorado potato beetle: Implications for insect resistance to *Bacillus thuringiensis* toxins. *Insect Biochemistry and Molecular Biology*, *32*(5), 567–577. https://doi.org/10.1016/S0965-1748(01)00137-0

Lv, H., Li, X., Li, J., Yu, C., Zeng, Q., Ning, G., Wan, H., Li, j., Ma, K., & He, S. (2023). Overcoming resistance in insect pest with a nanoparticle-mediated dsRNA and insecticide co-delivery system. *Chemical Engineering Journal*, *475*, 146239. https://doi.org/10.1016/j.cej.2023.146239

Ma, G., Roberts, H., Sarjan, M., Featherstone, N., Lahnstein, J., Akhurst, R., & Schmidt, O. (2005). Is the mature endotoxin Cry1Ac from *Bacillus thuringiensis* inactivated by a coagulation reaction in the gut lumen of resistant *Helicoverpa armigera* larvae? *Insect Biochemistry and Molecular Biology*, *35*(7), 729–739. https://doi.org/10.1016/j.ibmb.2005.02.011

Mahon, R. J., Olsen, K. M., Downes, S., & Addison, S. (2007). Frequency of alleles conferring resistance to the Bt toxins Cry1Ac and Cry2Ab in Australian populations of *Helicoverpa armigera* (Lepidoptera: Noctuidae). *Journal of Economic Entomology*, *100*(6), 1844–1853. https://doi.org/10.1093/jee/100.6.1844

Mao, Y. B., Cai, W. J., Wang, J. W., Hong, G. J., Tao, X. Y., Wang, L. J., Huang, Y. P., & Chen, X. Y. (2007). Silencing a cotton bollworm P450 monooxygenase gene by plant-mediated RNAi impairs larval tolerance of gossypol. *Nature Biotechnology*, *25*(11), 1307–1313. https://doi.org/10.1038/nbt1352

Martinez-Ramirez, A. C., Gould, F., & Ferre, J. (1999). Histopathological effects and growth reduction in a susceptible and a resistant strain of *Heliothis virescens* (Lepidoptera: Noctuidae) caused by sublethal doses of pure Cry1A crystal proteins from *Bacillus thuringiensis*.

Biocontrol Science and Technology, *9*(2), 239–246. https://doi.org/10.1080/09583159929811

Mathew, L. G., Ponnuraj, J., Mallappa, B., Chowdary, L. R., Zhang, J., Tay, W. T., Walsh, T. K., Gordon, K. H., Heckel, D. G., Downes, S., & Carrière, Y. (2018). ABC transporter mis-splicing associated with resistance to Bt toxin Cry2Ab in laboratory-and field-selected pink bollworm. *Scientific Reports*, *8*(1), 13531. https://doi.org/10.1038/s41598-018-31840-5

McGaughey, W. H. (1985). Insect resistance to the biological insecticide *Bacillus thuringiensis*. *Science*, *229*(4709), 193–195. https://doi.org/10.1126/science.229.4709.1

Milne, R., Wright, T., Kaplan, H., & Dean, D. (1998). Spruce budworm elastase precipitates *Bacillus thuringiensis* delta-endotoxin by specifically recognizing the C-terminal region. *Insect Biochemistry and Molecular Biology*, *28*(12), 1013–1023. https://doi.org/10.1016/S0965-1748(98)00090-3

Moar, W. J., Pusztai-Carey, M., Van Faassen, H. E. N. K., Bosch, D., Frutos, R., Rang, C., Luo, K., & Adang, M. J. (1995). Development of *Bacillus thuringiensis* CryIC resistance by *Spodoptera exigua* (Hubner) (Lepidoptera: Noctuidae). *Applied and Environmental Microbiology*, *61*(6), 2086–2092. https://doi.org/10.1128/aem.61.6.2086-2092.1995

Morin, S., Biggs, R. W., Sisterson, M. S., Shriver, L., Ellers-Kirk, C., Higginson, D., Holley, D., Gahan, L. J., Heckel, D. G., Carrière, Y., Dennehy, T. J., Brown, J. K., & Tabashnik, B. E. (2003). Three cadherin alleles associated with resistance to *Bacillus thuringiensis* in pink bollworm. *Proceedings of the National Academy of Sciences*, *100*(9), 5004–5009. https://doi.org/10.1073/PNAS.0831036100

Moussa, S., Biaomy, F., Aiad, K., Khalil, H., & Abd El-latif, A. O. (2020). *Bacillus thuringiensis* Cry1C resistance development and its processing pattern in Egyptian cotton leaf worm: *Spodoptera littoralis* (Boisd.) (Lepidoptera: Noctuidae). *Egyptian Journal of Biological Pest Control*, *30*, 36. https://doi.org/10.1186/s41938-020-00237-w

Mulla, M. S., Thavara, U., Tawatsin, A., Chomposri, J., & Su, T. (2003). Emergence of resistance and resistance management in field populations of tropical *Culex quinquefasciatus* to the microbial control agent *Bacillus sphaericus*. *Journal of the American Mosquito Control Association*, *19*(1), 39–46.

Nagaraj, S., Rajasekaran, R., Palaniappan, J., Rangasamy, S., Narayanasamy, C., & Narayanan, M. B. (2024). Emerging technological developments to address pest resistance in Bt cotton. *Journal of Cotton Research*, *7*, 30. https://doi.org/10.1186/s42397-024-00192-z

Ni, M., Ma, W., Wang, X., Gao, M., Dai, Y., Wei, X., Zhang, L., Peng, Y., Chen, S., Ding, L., Tian, Y., Li, J., Wang, H., Wang, X., Xu, G., Guo, W., Yang, Y., Wu, Y., Heuberger, S., Tabashnik, B. E., Zhang, T., & Zhu, Z. (2017). Next-generation transgenic cotton: Pyramiding RNAi and Bt counters insect resistance. *Plant Biotechnology Journal*, *15*(9), 1204–1213. https://doi.org/10.1111/pbi.12709

Nicolas, L., Dossou-Yovo, J., & Hougard, J. M. (1987). Persistence and recycling of *Bacillus sphaericus* 2362 spores in *Culex quinquefasciatus* breeding sites in West Africa. *Applied Microbiology and Biotechnology*, *25*, 341–345. https://doi.org/10.1007/BF00252544

Nielsen-LeRoux, C., Charles, J. F., Thiéry, I., & Georghiou, G. P. (1995). Resistance in a laboratory population of *Culex quinquefasciatus* (Diptera: Culicidae) to *Bacillus sphaericus* binary toxin is due to a change in the receptor on midgut brush-border membranes. *European Journal of Biochemistry*, *228*(1), 206–210. https://doi.org/10.1111/j.1432-1033.1995.0206o.x

Nielsen-LeRoux, C., Pasquier, F., Charles, J. F., SinÈGre, G., Gaven, B., & Pasteur, N. (1997). Resistance to *Bacillus sphaericus* involves different mechanisms in *Culex pipiens* (Diptera: Culicidae) larvae. *Journal of Medical Entomology*, *34*(3), 321–327. https://doi.org/10.1093/jmedent/34.3.321

Nielsen-LeRoux, C., Rao, D. R., Murphy, J. R., Carron, A., Mani, T. R., Hamon, S., & Mulla, M. S. (2001). Various levels of cross-resistance to *Bacillus sphaericus* strains in *Culex pipiens* (Diptera: Culicidae) colonies resistant to *B. sphaericus* strain 2362. *Applied and Environmental Microbiology*, *67*(11), 5049–5054. https://doi.org/10.1128/AEM.67.11.5049-5054.2001

Ning, C., Wu, K., Liu, C., Gao, Y., Jurat-Fuentes, J. L., & Gao, X. (2010). Characterization of a Cry1Ac toxin-binding alkaline phosphatase in the midgut from *Helicoverpa armigera* (Hübner) larvae. *Journal of Insect Physiology*, *56*(6), 666–672. https://doi.org/10.1016/J.JINSPHYS.2010.02.003

Niu, Y., Head, G. P., Price, P. A., & Huang, F. (2016). Performance of Cry1A. 105-selected fall armyworm (Lepidoptera: Noctuidae) on transgenic maize plants containing single or pyramided Bt genes. *Crop Protection*, *88*, 79–87. https://doi.org/10.1016/j.cropro.2016.06.005

Ojha, A., Sree, K. S., Sachdev, B., Rashmi, M. A., Ravi, K. C., Suresh, P. J., Mohan, K. J., & Bhatnagar, R. K. (2014). Analysis of resistance to Cry1Ac in field-collected pink bollworm, *Pectinophora gossypiella* (Lepidoptera: Gelechiidae), populations. *GM Crops & Food*, *5*(4), 280–286. https://doi.org/10.4161/21645698.2014.947800

Oliveira, C. D., Tadei, W. P., & Abdalla, F. C. (2009). Occurrence of apocrine secretion in the larval gut epithelial cells of *Aedes aegypti* L., *Anopheles albitarsis* Lynch-Arrıbálzaga and *Culex quinquefasciatus* Say (Diptera: Culicidae): A defense strategy against infection by *Bacillus sphaericus* Neide? *Neotropical Entomology*, *38*, 624–631. https://doi.org/10.1590/S1519-566X2009000500010

Oliveira, C. M., Beltran, J. E., Silva-Filha, M. H., & Regis, L. (2003). Biological fitness of a *Culex quinquefasciatus* population and its resistance to *Bacillus sphaericus*. *Journal of the American Mosquito Control Association*, *19*(2), 125–129.

Oliveira, C. M., Silva-Filha, M. H., Nielsen-Leroux, C., Pei, G., Yuan, Z., & Regis, L. (2004). Inheritance and mechanism of resistance to *Bacillus sphaericus* in *Culex quinquefasciatus* (Diptera: Culicidae) from China and Brazil. *Journal of Medical Entomology*, *41*(1), 58–64. https://doi.org/10.1603/0022-2585-41.1.58

Onstad, D. W., & Meinke, L. J. (2010). Modeling evolution of *Diabrotica virgifera virgifera* (Coleoptera: Chrysomelidae) to transgenic corn with two insecticidal traits. *Journal of Economic Entomology*, *103*(3), 849–860. https://doi.org/10.1603/EC09199

Oppert, B., Kramer, K. J., Beeman, R. W., Johnson, D., & McGaughey, W. H. (1997). Proteinase-mediated insect resistance to *Bacillus thuringiensis* toxins. *Journal of Biological Chemistry*, *272*(38), 23473–23476. https://doi.org/10.1074/jbc.272.38.23473

Park, H. W., Bideshi, D. K., & Federici, B. A. (2003). Recombinant strain of *Bacillus thuringiensis* producing Cyt1A, Cry11B, and the *Bacillus sphaericus* binary toxin. *Applied and Environmental Microbiology*, *69*(2), 1331–1334. https://doi.org/10.1128/AEM.69.2.1331-1334.2003

Park, H. W., Bideshi, D. K., Wirth, M. C., Johnson, J. J., Walton, W. E., & Federici, B. A. (2005). Recombinant larvicidal bacteria with markedly improved efficacy against *Culex* vectors of West Nile virus. *American Journal of Tropical Medicine and Hygiene*, *72*(6), 732–738. https://doi.org/10.4269/ajtmh.2005.72.732

Park, Y., González-Martínez, R. M., Navarro-Cerrillo, G., Chakroun, M., Kim, Y., Ziarsolo, P., Blanca, J., Cañizares, J., Ferré, J., & Herrero, S. (2014). ABCC transporters mediate insect resistance to multiple Bt toxins revealed by bulk segregant analysis. *BMC Biology*, *12*, 46. https://doi.org/10.1186/1741-7007-12-46

Pauchet, Y., Bretschneider, A., Augustin, S., & Heckel, D. G. (2016). A P-glycoprotein is linked to resistance to the *Bacillus thuringiensis* Cry3Aa toxin in a leaf beetle. *Toxins*, *8*(12), 362. https://doi.org/10.3390/toxins8120362

Pei, G., Oliveira, C. M., Yuan, Z., Nielsen-LeRoux, C., Silva-Filha, M. H., Yan, J., & Regis, L. (2002). A strain of *Bacillus sphaericus* causes slower development of resistance in *Culex quinquefasciatus*. *Applied and Environmental Microbiology*, *68*(6), 3003–3009. https://doi.org/10.1128/AEM.68.6.3003-3009.2002

Poncet, S., Bernard, C., Dervyn, E., Cayley, J., Klier, A., & Rapoport, G. (1997). Improvement of *Bacillus sphaericus* toxicity against dipteran larvae by integration, via homologous recombination, of the Cry11A toxin gene from *Bacillus thuringiensis* subsp. *israelensis*. *Applied and Environmental Microbiology*, *63*(11), 4413–4420. https://doi.org/10.1128/aem.63.11.4413-4420.1997

Poncet, S., Delecluse, A., Anello, G., Klier, A., & Rapoport, G. (1994). Transfer and expression of the *cryIVB* and *cryIVD* genes of *Bacillus thuringiensis* subsp. *israelensis* in *Bacillus sphaericus* 2297. *FEMS Microbiology Letters*, *117*(1), 91–95. https://doi.org/10.1111/j.1574-6968.1994.tb06747.x

Qiu, L., Wang, P., Wu, T., Li, B., Wang, X., Lei, C., Lin, Y., Zhao, J., & Ma, W. (2018). Downregulation of *Chilo suppressalis* alkaline phosphatase genes associated with resistance to three transgenic *Bacillus thuringiensis* rice lines. *Insect Molecular Biology*, *27*(1), 83–89. https://doi.org/10.1111/IMB.12349

Rahman, M. M., Roberts, H. L., Sarjan, M., Asgari, S., & Schmidt, O. (2004). Induction and transmission of *Bacillus thuringiensis* tolerance in the flour moth *Ephestia kuehniella*. *Proceedings of the National Academy of Sciences*, *101*(9), 2696–2699. https://doi.org/10.1073/pnas.0306669101

Rajagopal, R., Arora, N., Sivakumar, S., Rao, N. G. V., Nimbalkar, S. A., & Bhatnagar, R. K. (2009). Resistance of *Helicoverpa armigera* to Cry1Ac toxin from *Bacillus thuringiensis* is due to improper processing of the protoxin. *The Biochemical Journal*, *419*(2), 309–316. https://doi.org/10.1042/BJ20081152

Rajagopal, R., Sivakumar, S., Agrawal, N., Malhotra, P., & Bhatnagar, R. K. (2002). Silencing of midgut aminopeptidase N of *Spodoptera litura* by double-stranded RNA establishes its role as *Bacillus thuringiensis* toxin receptor. *Journal of Biological Chemistry*, *277*(49), 46849–46851. http://doi.org/10.1074/jbc.C200523200

Rao, D. R., Mani, T. R., Rajendran, R., Joseph, A. S., Gajanana, A., & Reuben, R. (1995). Development of a high level of resistance to *Bacillus sphaericus* in a field population of *Culex quinquefasciatus* from Kochi, India. *Journal of the American Mosquito Control Association-Mosquito News*, *11*(1), 1–5.

Ren, M., Zafar, M. M., Mo, H., Yang, Z., & Li, F. (2019). Fighting against fall armyworm by using Multiple Genes Pyramiding and Silencing (MGPS) technology. *Science China Life Sciences*, *62*(12), 1703–1706. https://doi.org/10.1007/s11427-019-1586-7

Reynolds, A., & Martinez, J. (2021, December 1). *Insect resistance management for Bt plant-incorporated protectants*. US EPA. https://www.epa.gov/regulation-biotechnology-under-tsca-and-fifra/insect-resistance-management-bt-plant-incorporated

Rodcharoen, J., & Mulla, M. S. (1997). Biological fitness of *Culex quinquefasciatus* (Diptera: Culicidae) susceptible and resistant to *Bacillus sphaericus*. *Journal of Medical Entomology*, *34*(1), 5–10. https://doi.org/10.1093/jmedent/34.1.5

Romão, T. P., de Melo Chalegre, K. D., Key, S., Ayres, C. F. J., de Oliveira, C. M. F., de-Melo-Neto, O. P., & Silva-Filha, M. H. N. L. (2006). A second independent resistance mechanism to *Bacillus sphaericus* binary toxin targets its α-glucosidase receptor in *Culex quinquefasciatus*. *The FEBS Journal*, *273*(7), 1556–1568. https://doi.org/10.1111/j.1742–4658.2006.05177.x

Roush, R. T. (1998). Two–toxin strategies for management of insecticidal transgenic crops: Can pyramiding succeed where pesticide mixtures have not? *Philosophical Transactions of the Royal Society of London. Series B: Biological Sciences*, *353*(1376), 1777–1786. https://doi.org/10.1098/rstb.1998.0330

Ruiu, L. (2015). Insect pathogenic bacteria in integrated pest management. *Insects*, *6*(2), 352–367. https://doi.org/10.3390/insects6020352

Sakthi, A. R., Naveenkumar, A., Deepikha, P. S., Balakrishnan, N., Kumar, K. K., Devi, E. K., Balasubramani, V., Arul, L., Singh, P. K., Sudhakar, D., Udayasuriyan, V., & Balasubramanian, P. (2015). Expression and inheritance of chimeric *cry2AX1* gene in transgenic cotton plants generated through somatic embryogenesis. *In Vitro Cellular & Developmental Biology-Plant*, *51*, 379–389. https://doi.org/10.1007/s11627-015-9695-8

Shah, J. V., Yadav, R., & Ingle, S. S. (2017). Engineered Cry1Ac-Cry9Aa hybrid *Bacillus thuringiensis* delta-endotoxin with improved insecticidal activity against *Helicoverpa armigera*. *Archives of Microbiology*, *199*, 1069–1075. https://doi.org/10.1007/s00203-017-1407-9

Shelton, A. M., Tang, J. D., Roush, R. T., Metz, T. D., & Earle, E. D. (2000). Field tests on managing resistance to Bt-engineered plants. *Nature Biotechnology*, *18*(3), 339–342. https://doi.org/10.1038/73804

Siegfried, B. D., Spencer, T., Crespo, A. L., Storer, N. P., Head, G. P., Owens, E. D., & Guyer, D. (2007). Ten years of Bt resistance monitoring in the European corn borer: What we know, what we don't know, and what we can do better. *American Entomologist*, *53*(4), 208–214. https://doi.org/10.1093/ae/53.4.208

Silva-Filha, M. H. N. L., Berry, C., & Regis, L. (2014). *Lysinibacillus sphaericus*: Toxins and mode of action, applications for mosquito control and resistance management. In T. Dhadialla, & S. Gill (Eds.), *Advances in insect physiology: Vol. 47. Insects midgut and insecticidal proteins* (pp. 89–176). Elsevier. https://doi.org/10.1016/B978-0-12-800197-4.00003-8

Silva-Filha, M. H. N. L., Nielsen-LeRoux, C., & Charles, J. F. (1999). Identification of the receptor for *Bacillus sphaericus* crystal toxin in the brush border membrane of the mosquito *Culex pipiens* (Diptera: Culicidae). *Insect Biochemistry and Molecular Biology*, *29*(8), 711–721. https://doi.org/10.1016/S0965-1748(99)00047-8

Silva-Filha, M. H. N. L., Regis, L., Nielsen-Leroux, C., & Charles, J. F. (1995). Low-level resistance to *Bacillus sphaericus* in a field-treated population of *Culex quinquefasciatus* (Diptera: Culicidae). *Journal of Economic Entomology*, *88*(3), 525–530. https://doi.org/10.1093/jee/88.3.525

Sinègre, G., Babinot, M., Quermel, J. M., & Gaven, B. (1994). First field occurrence of *Culex pipiens* resistance to *Bacillus sphaericus* in southern France. *Proceedings, 8th European Meeting of Society for Vector Ecology* (pp. 5–8). Society for Vector Ecology.

Soberón, M., Pardo-López, L., López, I., Gómez, I., Tabashnik, B. E., & Bravo, A. (2007). Engineering modified Bt toxins to counter insect resistance. *Science*, *318*(5856), 1640–1642. https://doi.org/10.1126/science.1146453

Storer, N. P., Thompson, G. D., & Head, G. P. (2012). Application of pyramided traits against Lepidoptera in insect resistance management for Bt crops. *GM Crops & Food*, *3*(3), 154–162. https://doi.org/10.4161/gmcr.20945

Tabashnik, B. E. (1994). Evolution of resistance to *Bacillus thuringiensis*. *Annual Review Entomology*, *39*, 47–79. https://doi.org/10.1146/annurev.en.39.010194.000403

Tabashnik, B. E. (2008). Delaying insect resistance to transgenic crops. *Proceedings of the National Academy of Sciences*, *105*(49), 19029–19030. https://doi.org/10.1073/pnas.0810763106

Tabashnik, B. E., & Carrière, Y. (2017). Surge in insect resistance to transgenic crops and prospects for sustainability. *Nature Biotechnology*, *35*(10), 926–935. https://doi.org/10.1038/nbt.3974

Tabashnik, B. E., & Carrière, Y. (2019). Global patterns of resistance to Bt crops highlighting pink bollworm in the United States, China, and India. *Journal of Economic Entomology*, *112*(6), 2513–2523. https://doi.org/10.1093/jee/toz173

Tabashnik, B. E., & Carrière, Y. (2020). Evaluating cross-resistance between Vip and Cry toxins of *Bacillus thuringiensis*. *Journal of Economic Entomology*, *113*(2), 553–561. https://doi.org/10.1093/jee/toz308

Tabashnik, B. E., Carrière, Y., Dennehy, T. J., Morin, S., Sisterson, M. S., Roush, R. T., Shelton, A. M., & Zhao, J. Z. (2003). Insect resistance to transgenic Bt crops: Lessons from the laboratory and field. *Journal of Economic Entomology*, *96*(4), 1031–1038. https://doi.org/10.1093/jee/96.4.1031

Tabashnik, B. E., Carrière, Y., Wu, Y., & Fabrick, J. A. (2023a). Global perspectives on field-evolved resistance to transgenic Bt crops: A special collection. *Journal of Economic Entomology*, *116*(2), 269–274. https://doi.org/10.1093/jee/toad054

Tabashnik, B. E., Cushing, N. L., Finson, N., & Johnson, M. W. (1990). Field development of resistance to *Bacillus thuringiensis* in diamondback moth (Lepidoptera: Plutellidae). *Journal of Economic Entomology*, *83*(5), 1671–1676. https://doi.org/10.1093/jee/83.5.1671

Tabashnik, B. E., Fabrick, J. A., & Carrière, Y. (2023b). Global patterns of insect resistance to transgenic Bt crops: The first 25 years. *Journal of Economic Entomology*, *116*(2), 297–309. https://doi.org/10.1093/jee/toac183

Tabashnik, B. E., Finson, N., Johnson, M. W., & Heckel, D. G. (1994). Cross-resistance to *Bacillus thuringiensis* toxin CryIF in the diamondback moth (*Plutella xylostella*). *Applied and Environmental Microbiology*, *60*(12), 4627–4629. https://doi.org/10.1128/AEM.60.12.4627-4629.1994

Tabashnik, B. E., Gassmann, A. J., Crowder, D. W., & Carrière, Y. (2008). Insect resistance to Bt crops: Evidence versus theory. *Nature Biotechnology*, *26*(2), 199–202. https://doi.org/10.1038/nbt1382

Tabashnik, B. E., Huang, F., Ghimire, M. N., Leonard, B. R., Siegfried, B. D., Rangasamy, M., Yang, Y., Wu, Y., Gahan, L. J., Heckel, D. G., Bravo, A., & Soberón, M. (2011). Efficacy of genetically modified Bt toxins against insects with different genetic mechanisms of resistance. *Nature Biotechnology*, *29*(12), 1128–1131. https://doi.org/10.1038/nbt.1988

Tabashnik, B. E., Mota-Sanchez, D., Whalon, M. E., Hollingworth, R. M., & Carrière, Y. (2014). Defining terms for proactive management of resistance to Bt crops and pesticides. *Journal of Economic Entomology*, *107*(2), 496–507. https://doi.org/10.1603/EC13458

Tabashnik, B. E., Van Rensburg, J. B. J., & Carrière, Y. (2009). Field-evolved insect resistance to Bt crops: Definition, theory, and data. *Journal of Economic Entomology*, *102*(6), 2011–2025. https://doi.org/10.1603/029.102.0601

Tanaka, S., Miyamoto, K., Noda, H., Jurat-Fuentes, J. L., Yoshizawa, Y., Endo, H., & Sato, R. (2013). The ATP-binding cassette transporter subfamily C member 2 in *Bombyx mori* larvae is a functional receptor for Cry toxins from *Bacillus thuringiensis*. *The FEBS Journal*, *280*(8), 1782–1794. https://doi.org/10.1111/febs.12200

Tay, W. T., Mahon, R. J., Heckel, D. G., Walsh, T. K., Downes, S., James, W. J., Lee, S. F., Reineke, A., Williams, A. K., & Gordon, K. H. J. (2015). Insect resistance to *Bacillus thuringiensis* toxin Cry2Ab is conferred by mutations in an ABC transporter subfamily A protein. *PLoS Genetics*, *11*(11), e1005534. https://doi.org/10.1371/journal.pgen.1005534

Thiéry, I., Hamon, S., Delécluse, A., & Orduz, S. (1998). The introduction into *Bacillus sphaericus* of the *Bacillus thuringiensis* subsp. *medellin cyt1Ab1* gene results in higher susceptibility of resistant mosquito larva populations to *B. sphaericus*. *Applied and Environmental Microbiology*, *64*(10), 3910–3916. https://doi.org/10.1128/AEM.64.10.3910–3916.1998

Tiewsiri, K., & Wang, P. (2011). Differential alteration of two aminopeptidases N associated with resistance to *Bacillus thuringiensis* toxin Cry1Ac in cabbage looper. *Proceedings of the National Academy of Sciences*, *108*(34), 14037–14042. https://doi.org/10.1073/PNAS.1102555108

Ullah, F., Gul, H., Tariq, K., Hafeez, M., Desneux, N., & Song, D. (2023). Silencing of cytochrome P450 genes CYP6CY14 and CYP6DC1 in *Aphis gossypii* by RNA interference enhances susceptibility to clothianidin. *Entomologia Generalis*, *43*(3), 669–678. https://doi.org/10.1127/entomologia/2023/2002

US EPA. (2010). *Biopesticide registration action document: Bacillus thuringiensis Cry1A.105 and Cry2Ab2 insecticidal proteins and the genetic material necessary for their production in corn*. https://www3.epa.gov/pesticides/chem_search/reg_actions/pip/mon-89034-brad.pdf

Van Rie, J., McGaughey, W. H., Johnson, D. E., Barnett, B. D., & Van Mellaert, H. (1990). Mechanism of insect resistance to the microbial insecticide *Bacillus thuringiensis*. *Science*, *247*(4938), 72–74. https://doi.org/10.1126/science.2294593

Walsh, T., James, B., Chakroun, M., Ferré, J., & Downes, S. (2018). Isolating, characterising and identifying a Cry1Ac resistance mutation in field populations of *Helicoverpa punctigera*. *Scientific Reports*, *8*, 2626. https://doi.org/10.1038/s41598-018-21012-w

Walters, F. S., deFontes, C. M., Hart, H., Warren, G. W., & Chen, J. S. (2010). Lepidopteran-active variable-region sequence imparts coleopteran activity in eCry3.1Ab, an engineered *Bacillus thuringiensis* hybrid insecticidal protein. *Applied and Environmental Microbiology*, *76*(10), 3082–3088. https://doi.org/10.1128/AEM.00155-10

Walters, F. S., Stacy, C. M., Lee, M. K., Palekar, N., & Chen, J. S. (2008). An engineered chymotrypsin/cathepsin G site in domain I renders *Bacillus thuringiensis* Cry3A active against western corn rootworm larvae. *Applied and Environmental Microbiology*, *74*(2), 367–374. https://doi.org/10.1128/AEM.02165-07

Wan, P., Xu, D., Cong, S., Jiang, Y., Huang, Y., Wang, J., Wu, H., Wang, L., Wu, K., Carrière, Y., Mathias, A., Li, X., & Tabashnik, B. E. (2017). Hybridizing transgenic Bt cotton with non-Bt cotton counters resistance in pink bollworm. *Proceedings of the National Academy of Sciences*, *114*(21), 5413–5418. https://doi.org/10.1073/pnas.1700396114

Wang, L., Ma, Y., Wan, P., Liu, K., Xiao, Y., Wang, J., Cong, S., Xu, D., Wu, K., Fabrick, J. A., Li, X., & Tabashnik, B. E. (2018). Resistance to *Bacillus thuringiensis* linked with a cadherin transmembrane mutation affecting cellular trafficking in pink bollworm from China. *Insect Biochemistry and Molecular Biology*, *94*, 28–35. https://doi.org/10.1016/J.IBMB.2018.01.004

Wang, L., Xu, D., Huang, Y., Zhou, H., Liu, W., Cong, S., Wang, J., Li, W., & Wan, P. (2022). Mutation in the cadherin gene is a key factor for pink bollworm resistance to Bt cotton in China. *Toxins*, *14*(1), 23. https://doi.org/10.3390/toxins14010023

Welch, K. L., Unnithan, G. C., Degain, B. A., Wei, J., Zhang, J., Li, X., Tabashnik, B. E., & Carrière, Y. (2015). Cross-resistance to toxins used in pyramided Bt crops and resistance to Bt sprays in *Helicoverpa zea*. *Journal of Invertebrate Pathology*, *132*, 149–156. https://doi.org/10.1016/j.jip.2015.10.003

Whitehouse, M. E. A., Wilson, L. J., Davies, A. P., Cross, D., Goldsmith, P., Thompson, A., Harden, S., & Baker, G. (2014). Target and nontarget effects of novel "triple-stacked" Bt-transgenic cotton 1: Canopy arthropod communities. *Environmental Entomology*, *43*(1), 218–241. https://doi.org/10.1603/EN13167

WHO. (2004). *Global strategic framework for integrated vector management*. (WHO/CDS/CPE/PVC/2004.10).

Wirth, M. C. (2010). Mosquito resistance to bacterial larvicidal toxins. *The Open Toxinology Journal*, *3*(1), 126–140. https://doi.org/10.2174/1875414701003010126

Wirth, M. C., Delécluse, A., & Walton, W. E. (2001). Cyt1Ab1 and Cyt2Ba1 from *Bacillus thuringiensis* subsp. *medellin* and *B. thuringiensis* subsp. *israelensis* synergize *Bacillus sphaericus* against *Aedes aegypti* and resistant *Culex quinquefasciatus* (Diptera: Culicidae). *Applied and Environmental Microbiology*, *67*(7), 3280–3284. https://doi.org/10.1128/AEM.67.7.3280–3284.2001

Wirth, M. C., Federici, B. A., & Walton, W. E. (2000c). Cyt1A from *Bacillus thuringiensis* synergizes activity of *Bacillus sphaericus* against *Aedes aegypti* (Diptera: Culicidae). *Applied and Environmental Microbiology*, *66*(3), 1093–1097. https://doi.org/10.1128/AEM.66.3.1093-1097.2000

Wirth, M. C., Georghiou, G. P., Malik, J. I., & Abro, G. H. (2000a). Laboratory selection for resistance to *Bacillus sphaericus* in *Culex quinquefasciatus* (Diptera: Culicidae) from California, USA. *Journal of Medical Entomology*, *37*(4), 534–540. https://doi.org/10.1603/0022-2585-37.4.534

Wirth, M. C., Jiannino, J. A., Federici, B. A., & Walton, W. E. (2004). Synergy between toxins of *Bacillus thuringiensis* subsp. *israelensis* and *Bacillus sphaericus*. *Journal of Medical Entomology*, *41*(5), 935–941. https://doi.org/10.1603/0022-2585-41.5.935

Wirth, M. C., Jiannino, J. A., Federici, B. A., & Walton, W. E. (2005). Evolution of resistance toward *Bacillus sphaericus* or a mixture of *B. sphaericus*+ Cyt1A from *Bacillus thuringiensis*, in the mosquito, *Culex quinquefasciatus* (Diptera: Culicidae). *Journal of Invertebrate Pathology*, *88*(2), 154–162. https://doi.org/10.1016/j.jip.2005.01.003

Wirth, M. C., Walton, W. E., & Federici, B. A. (2000). Cyt1A from *Bacillus thuringiensis* restores toxicity of *Bacillus sphaericus* against resistant *Culex quinquefasciatus* (Diptera: Culicidae). *Journal of Medical Entomology*, *37*(3), 401–407. https://doi.org/10.1093/jmedent/37.3.401

Wirth, M. C., Walton, W. E., & Federici, B. A. (2012). Inheritance, stability, and dominance of Cry resistance in Culex quinquefasciatus (Diptera: Culicidae) selected with the three Cry toxins of *Bacillus thuringiensis* subsp. *israelensis*. *Journal of Medical Entomology*, *49*(4), 886–894. https://doi.org/10.1603/ME11192

Wirth, M. C., Yang, Y., Walton, W. E., Federici, B. A., & Berry, C. (2007). Mtx toxins synergize *Bacillus sphaericus* and Cry11Aa against susceptible and insecticide-resistant *Culex quinquefasciatus* larvae. *Applied and Environmental Microbiology*, *73*(19), 6066–6071. https://doi.org/10.1128/AEM.00654-07

Xia, Y., Dai, Q., Hu, R., Pacheco, S., Yang, Y., Liang, G., Sobero, M., Bravo, A., Liu, K., & Wu, K. (2017). A single point mutation resulting in cadherin mislocalization underpins resistance against *Bacillus thuringiensis* toxin in cotton bollworm. *The Journal of Biological Chemistry*, *292*(7), 2933–2943. https://doi.org/10.1074/JBC.M116.768671

Xiao, Y., Zhang, T., Liu, C., Heckel, D. G., Li, X., Tabashnik, B. E., & Wu, K. (2014). Mis-splicing of the ABCC2 gene linked with Bt toxin resistance in *Helicoverpa armigera*. *Scientific Reports*, *4*, 6184. https://doi.org/10.1038/srep06184

Xie, R., Zhuang, M., Ross, L. S., Gomez, I., Oltean, D. I., Bravo, A., Soberon, M., & Gill, S. S. (2005). Single amino acid mutations in the cadherin receptor from *Heliothis virescens* affect its toxin binding ability to Cry1A toxins. *The Journal of Biological Chemistry*, *280*(9), 8416–8425. https://doi.org/10.1074/JBC.M408403200

Xu, C., Cheng, J., Lin, H., Lin, C., Gao, J., & Shen, Z. (2018). Characterization of transgenic rice expressing fusion protein Cry1Ab/Vip3A for insect resistance. *Scientific Reports*, *8*(1), 15788. https://doi.org/10.1038/s41598-018-34104-4

Xu, X., Yu, L., & Wu, Y. (2005). Disruption of a cadherin gene associated with resistance to Cry1Ac δ-endotoxin of *Bacillus thuringiensis* in *Helicoverpa armigera*. *Applied and Environmental Microbiology*, *71*(2), 948–954. https://doi.org/10.1128/AEM.71.2.948-954.2005

Yang, F., Kerns, D. L., Head, G. P., Leonard, B. R., Levy, R., Niu, Y., & Huang, F. (2014). A challenge for the seed mixture refuge strategy in Bt maize: Impact of cross-pollination on an ear-feeding pest, corn earworm. *PLoS One*, *9*(11), e112962. https://doi.org/10.1371/journal.pone.0112962

Yang, X., Chen, W., Song, X., Ma, X., Cotto-Rivera, R. O., Kain, W., Chu, H., Chen, Y. R., Fei, Z., & Wang, P. (2019). Mutation of ABC transporter ABCA2 confers resistance to Bt toxin Cry2Ab in *Trichoplusia ni*. *Insect Biochemistry and*

Molecular Biology, *112*, 103209. https://doi.org/10.1016/j.ibmb.2019.103209

Yang, Y., Chen, H., Wu, Y., Yang, Y., & Wu, S. (2007). Mutated cadherin alleles from a field population of *Helicoverpa armigera* confer resistance to *Bacillus thuringiensis* toxin Cry1Ac. *Applied and Environmental Microbiology*, *73*(21), 6939–6944. https://doi.org/10.1128/AEM.01703-07

Yap, W. H., Thanabalu, T., & Porter, A. G. (1994). Expression of mosquitocidal toxin genes in a gas-vacuolated strain of *Ancylobacter aquaticus*. *Applied and Environmental Microbiology*, *60*(11), 4199–4202. https://doi.org/10.1128/aem.60.11.4199-4202.1994

Yuan, X., Zhao, M., Wei, J., Zhang, W., Wang, B., Myint Khaing, M., & Liang, G. (2017). New insights on the role of alkaline phosphatase 2 from *Spodoptera exigua* (Hübner) in the action mechanism of Bt toxin Cry2Aa. *Journal of Insect Physiology*, *98*, 101–107. https://doi.org/10.1016/J.JINSPHYS.2016.12.004

Yuan, Z. M., Pei, G. F., Regis, L., Nielsen-Leroux, C., & Cai, Q. X. (2003). Cross-resistance between strains of *Bacillus sphaericus* but not *B. thuringiensis israelensis* in colonies of the mosquito *Culex quinquefasciatus*. *Medical and Veterinary Entomology*, *17*(3), 251–256. https://doi.org/10.1046/j.1365-2915.2003.00429.x

Yuan, Z. M., Zhang, Y., Cai, Q., & Liu, E. Y. (2000). High-level field resistance to *Bacillus sphaericus* C3–41 in *Culex quinquefasciatus* from southern China. *Biocontrol Science and Technology*, *10*(1), 41–49. https://doi.org/10.1080/09583150029378

Zafar, M. M., Razzaq, A., Farooq, M. A., Rehman, A., Firdous, H., Shakeel, A., Mo, H., & Ren, M. (2020). Insect resistance management in *Bacillus thuringiensis* cotton by MGPS (multiple genes pyramiding and silencing). *Journal of Cotton Research*, *3*, 33. https://doi.org/10.1186/s42397-020-00074-0

Zhang, L., Wei, Y., Wei, L., Liu, X., & Liu, N. (2023). Effects of transgenic cotton lines expressing *dsAgCYP6CY3-P1* on the growth and detoxification ability of *Aphis gossypii* Glover. *Pest Management Science*, *79*(1), 481–488. https://doi.org/10.1002/ps.7220

Zhang, N. (2017c, April 10). *How to counter rapidly increasing pest resistance to transgenic crops producing single bacillus thuringiensis (Bt) toxins?* Chinese Academy of Sciences.

Zhang, S., Cheng, H., Gao, Y., Wang, G., Liang, G., & Wu, K. (2009). Mutation of an aminopeptidase N gene is associated with *Helicoverpa armigera* resistance to *Bacillus thuringiensis* Cry1Ac toxin. *Insect Biochemistry and Molecular Biology*, *39*(7), 421–429. https://doi.org/10.1016/J.IBMB.2009.04.003

Zhang, T., Coates, B. S., Wang, Y., Wang, Y., Bai, S., Wang, Z., & He, K. (2017b). Down-regulation of aminopeptidase N and ABC transporter subfamily G transcripts in Cry1Ab and Cry1Ac resistant Asian corn borer, *Ostrinia furnacalis* (Lepidoptera: Crambidae). *International Journal of Biological Sciences*, *13*(7), 835. https://doi.org/10.7150/ijbs.18868

Zhang, Z., Teng, X., Ma, W., & Li, F. (2017a). Knockdown of two Cadherin genes confers resistance to Cry2A and Cry1C in *Chilo suppressalis*. *Scientific Reports*, *7*, 5992. https://doi.org/10.1038/S41598-017-05110-9

Zhao, J. Z., Cao, J., Collins, H. L., Bates, S. L., Roush, R. T., Earle, E. D., & Shelton, A. M. (2005). Concurrent use of transgenic plants expressing a single and two *Bacillus thuringiensis* genes speeds insect adaptation to pyramided plants. *Proceedings of the National Academy of Sciences*, *102*(24), 8426–8430. https://doi.org/10.1073/pnas.0409324102

Zhao, J. Z., Cao, J., Li, Y., Collins, H. L., Roush, R. T., Earle, E. D., & Shelton, A. M. (2003). Transgenic plants expressing two *Bacillus thuringiensis* toxins delay insect resistance evolution. *Nature Biotechnology*, *21*(12), 1493–1497. https://doi.org/10.1038/nbt907

14 Selection Criteria and Legislation Procedure for Commercial Production of EPB

ABSTRACT

The selection of the most efficient entomopathogenic bacteria is very important from a commercial as well as scientific point of view. Several points should be taken care of in that respect, most importantly host specificity, efficiency, mortality factor, mass production capability, environmental safety, and effect of beneficial organisms. From an industrial point of view, there is a big role of its research and development sector. The legislation process goes on simultaneously with the industrial research. After the selection of the entomopathogenic strain, one should go for patenting of that particular strain before commercialization. Side by side, in the laboratory, the mass production strategy is to be finalized along with the fixation of the proper dose. The formulation is to be registered, and then it can be available for marketing. It is a long process from selection to commercialization, and proper evaluation and inspection at every step are highly required.

14.1 INTRODUCTION

The first step of starting any business is the preparation of a business plan. In a similar way, before starting the commercial production of entomopathogenic bacteria (EPB), the blueprint should be strongly built. The selected EPB should be competent enough to show better management, along with safety to non-targets and ease of mass culturing. After selecting a particular EPB, the protocol should be standardized for isolation, confirmation, mass production, and post-production strategies. All these steps generally come under the trade secrets of individual companies; still, the general flow of research and development is discussed in this chapter. After standardizing all protocols, the industry must legally register their product to be commercialized. The legislation procedure differs from country to country. This chapter aims to give a comprehensive overview of all the background procedures, starting from selection criteria, standardization, and finally legislation, which must be done before starting with mass production of EPB. It is dedicated to all the entrepreneurs who are investing or intend to invest their capital in this sector.

14.2 SELECTION CRITERIA FOR MASS PRODUCTION

The decision regarding the selection of any EPB is very crucial, as the success of the whole industry depends on these bacterial agents. Burges (1981) outlined three approaches for identifying and developing new microbial pest control agents (MPCAs), including (1) the serendipity approach, where a promising agent is discovered in a specific pest and then further developed to combat that pest, as in the instance of *Paenibacillus popilliae* Dutky against the Japanese beetle, *Popillia japonica* Newman; (2) the agent-specific approach, where a recognized agent is produced against as many pests as feasible, like in the case of *Bacillus thuringiensis* Berliner

DOI: 10.1201/9781003426882-14

(Bt) for the management of diverse lepidopteran pests; and finally, (3) the goal-oriented approach, wherein an MPCA is identified for the management of a certain pest. Burges claims that although this approach has had the least success, it has the potential to be improved for integrated pest management programmes that have gaps against specific pests that are challenging to control, and broad-spectrum insecticides are no longer being utilized.

Two approaches were put out by Ravensberg (2011) for the commercial development of biocontrol agents: (1) The problem-solving approach, where a biological control agent (BCA) can, in theory, be searched for and developed following appropriate selection criteria for the management of certain pests that cannot be controlled sufficiently in the current integrated pest management (IPM) programme and seek for a new solution. IPM programmes for greenhouse crops apply this approach of problem-solving. For this approach, we hope the previous chapters of this book will help the researchers. (2) Then the opportunity approach, where an opportunity arises when a novel BCA comes to light by chance and someone realizes, based on knowledge and experience, that a new product may be developed for which a market might exist. In commercial biocontrol of pests, this strategy is rare. For every approach, we should go for commercial production of such EPB based on different criteria to achieve the desired performance, described in what follows.

14.2.1 Host Specificity

As compared to entomopathogenic fungi, bacteria are more host-specific, but not as much as viruses (Deka et al., 2021). Host specificity in entomopathogens varies at the strain level and cannot be generalized by taxonomic group; some fungal strains (e.g., *Beauveria bassiana*) exhibit specificity comparable to or even narrower than certain bacterial strains. This moderate host specificity gives bacteria added advantages to excel among biopesticides, and maybe this is among the multiple reasons the market of biopesticides is reigned by EPB. A vast host range of any microbial agent increases the probability of being pathogenic on non-target organisms as well. Contrary to that, a very specific host range makes the agent only useful for controlling that very particular pest, not other similar ones. Most of the EPBs like Bt, *Lysinibacillus sphaericus* (Meyer & Neide) (Ls) or *Yersinia entomophaga* Hurst et al. have their effects on several insect orders, whereas *P. popillae* or *Serratia entomophila* Grimont et al. show specificity to a single host only (Couch et al., 2023). The EPB must be selected for commercial production based on their host range and assessed properly for host specificity and off-target effects. In order to determine the level of susceptibility of a pest, the first step in selecting one effective biological control agent is to screen the agent in a vast range of insects (Sabbahi et al., 2022).

14.2.2 Effect on Humans and the Environment

A major care has to be taken for the effect on non-targets. EPB with potential risk of off-target attack along with human effect or crop loss should be avoided for commercial production as such. They can be genetically engineered to mutate their side effects, and those need much research and development (R&D) before being considered for mass multiplication. Although genetic engineering of entomopathogenic bacteria (EPB) to reduce off-target effects has been explored in research, it is not widely adopted in commercial production due to stringent biosafety regulations and limited public acceptance. Several criteria have been published by the United States Environmental Protection Agency (USEPA) for assessing the ecotoxicity of biopesticides to non-targets, such as aquatic and terrestrial species (EPA, 1996a, 1996b). They are mostly similar to that of tests required for approval of chemical pesticides. The majority of studies that revealed any negative impact brought on by commercial EPB strains asserted that the adverse effect was justified by physical contact or by a high concentration of the agent(s) (Oliveira-Filho & Grisolia, 2022). Especially for Cry toxins, it has been experimentally proven in mice that they degrade in the gastrointestinal tract of mammals so mostly do not cause any harm when ingested orally (Mezzomo et al., 2013).

The safety of the applicator and crop workers is the primary issue in registration. In the case of bacteria, this is crucial since many of them produce metabolites, and many of the ones employed as biopesticides are closely linked to harmful bacteria, such as the bacteria belonging to the genus *Bacillus* and *Serratia* (Ravensberg, 2011).

14.2.3 Mortality Factors

It is obvious that a high mortality rate is a highly desired trait. The larval stage of susceptible pests is the target of bacterial biopesticides. However, larvae that survive exposure frequently exhibit fitness costs in later stages, such as the formation of smaller pupae that do not develop or show decreased fecundity as adults, as well as higher susceptibility to further exposures (Couch et al., 2023). In general, susceptibility to these biopesticides declines with larval development. Even though some bacteria, such as *Wolbachia*, may move vertically within the egg (Hurst et al., 1999), in the egg stage, *Spiroplasma* can kill male embryos (Harumoto & Lemaitre, 2018). The pathogenicity and virulence of the infection induce mortality, which is dose-dependent and reliant on the quality of the supplied inoculum. Numerous testing techniques, frequently bioassays, are reported in literature for a wide range of pathogens and target pests. The LD_{50}, LC_{50}, and LT_{50} values are used to compare virulence between species and strains. The LC_{50} is the concentration of an entomopathogen that must be present for a specific amount of time to eliminate 50% of the insect population. The LD_{50}, on the other hand, is the dose necessary to kill 50% of a population. Aside from this figure, the LT_{50}, or the time taken to kill 50% of a population at a particular dosage or concentration of a pathogen, is indicative of a pathogen's 'speed of kill' and is useful in the selection process. By evaluating various doses, a dose–mortality curve should be prepared, and the slope of the curve provides helpful data for comparing isolates. A steeper dose mortality curve indicates lower variability in host susceptibility, while a flatter curve may reflect higher variability or lower dose-dependence; virulence must be evaluated in conjunction with LC_{50}, LT_{50}, and other bioassay metrics. Obviously, the smaller the dose, the more likely it is that a product may be manufactured inexpensively. The screening procedure must take into account abiotic elements, like temperature, relative humidity (RH), light (UV), and rain. It is important to consider a wide range, and strains with a considerable tolerance for unfavourable circumstances would be preferable. Another relevant mortality factor is the persistence of the inoculum after spraying, which is always expected to be higher (Ravensberg, 2011).

14.2.4 Capability of Mass Production

To employ a pathogen as a biopesticide, it must have infective propagules that may be sprayed to a crop. Spore-producing EPBs are relatively easy to cultivate and formulate. For all the putative commercial entomopathogens, it is important to figure out how easy or difficult it is to mass-multiply them. This is especially true when it comes to the propagation of infectious propagules (Ravensberg, 2011). For bacteria which can be easily fermented, like Bt, their mass production is very easy (Couch et al., 2023). For fastidious bacteria, it is very difficult to propagate them in vitro and requires in vivo condition. Though in vivo production of spores was successful in the case of *P. popillae*, it has been observed that in vivo products are more efficient than in vitro products (Redmond & Potter, 1995). There are several bacteria which are unable to grow in artificial medium. These situations highly increase the production cost and decrease the probability of selection of bacterium for commercial production.

14.2.5 Efficiency

Having an extended shelf life is quite desired in addition to simplicity of production. For instance, the ability to produce spores in a bacterium affects the selection process, since spores are stable and may be kept for a long period (Ravensberg, 2011). Important factors that should be considered while choosing the strain include the capacity for mass production with significant yields, various plausible formulations, and ease of application by farmers (Sabbahi et al., 2022).

Burges (1981) recognized all the necessary conditions for the development of a microbial pest control agent. He realized that 'market size and competitiveness are vital', and considering biopesticide market demand, especially of EPB, entrepreneurs need not be bothered about in this aspect. He also mentioned some other important commercial criteria, like production cost and registration cost, to be crucial determining factors.

14.3 RESEARCH AND DEVELOPMENT OF MASS PRODUCTION STRATEGIES

After determining the bacterial agent with whom to proceed for commercial production, a major role is played by R&D. That is why in almost all agriculture-based companies, there is a dedicated section for R&D. Starting from selection of the agent, preparing standard protocol for mass multiplication, and post-production handling strategies, to quality maintenance and improvement, at every step, R&D plays a very important role. Accredited laboratories and private sector partners conduct the essential eco and mammalian toxicity evaluations and quality assurance for the commercialization of biopesticides. These partners also carry out the formulation and field effectiveness trials. In order to scale up the actual application of laboratory findings and promote the large-scale commercial development of microbial biopesticides, collaboration between industry and academia is very much crucial. More research is needed to determine the efficacy of EPB against target insect pests in various agricultural systems (Sabbahi et al., 2022).

14.3.1 Isolation of EPB

We have seen in the earlier chapters that different EPBs have different ranges of habitat and diverse growth requirements, so inevitably, their isolation processes are different from each other. Though enotomopathogenicity and other associated traits desired for commercialization are mostly specific to strain levels and the isolation and especially confirmation should be specific to the strain itself, here we try to discuss in general the isolation procedure of important EPB genera.

14.3.1.1 *Bacillus* sp.

Bacterial species belonging to class *Bacilli* are often isolated from soils, insects, and water samples. Some species, like Bt, develop heat-resistant endospores that facilitate differentiation of the targeted bacteria. The following is the standard procedure for bacterial isolation:

1. Isolation may be carried out using soils (2–4 g in 10 mL sterilized water), insects (0.2–0.4 g/mL sterilized water), or water samples (after concentration using a 0.22 μm filter).
2. Warm the samples in an 80°C water bath for 10 minutes, then quickly cool the tubes on ice. This eliminates vegetative cells and non-spore-producing bacteria while enriching *Bacillus* sp.
3. Carry out serial dilutions, typically at 10^{-2} and 10^{-3}, and cultivate the inoculums on sporulation media, such as minimal basal salt (MBS) medium, or common medium, like nutrient agar. Subculturing should be continued until pure cultures are achieved.

4. Identify bacteria using several biochemical assays and 16S rDNA sequencing.

(Fisher & Garczynski, 2012)

Other media, such as polymyxin pyruvate (to inhibit Gram-negative organisms), egg yolk, mannitol, and bromothymol blue agar (PEMBA), can be used to isolate Bt and *B. cereus* Frankland & Frankland. A liquid enrichment broth (5 g peptone, 5 g meat extract, 10 g KNO_3, made up to 1,000 mL with distilled water) might be used as an alternative. Microscopically, Bt isolates can be recognized by the presence of parasporal bodies, or crystals, in the sporangium. *B. cereus* colonies are somewhat rhizoid, have a distinctive turquoise blue hue, and are typically surrounded by an egg yolk precipitate of a similar tinge (Koppenhöfer et al., 2012).

14.3.1.2 *Paenibacillus* sp.

P. popilliae and *P. lentimorbus* (Dutky) are obligatory pathogens of scarabs (Coleoptera), since their proliferation and sporulation depend on the host. They exist as endospores in soils. These bacteria can be isolated from haemolymph, and the methods vary, depending on the bacterial species.

1. Apply 0.5% (v/v) sodium hypochlorite (NaOCl) to the surface of the grub larvae (*Coleoptera*) to disinfect it.
2. Use a sterilized needle to pinch the corpse, then collect the dripping blood in sterilized water.
3. Culture the diluted drops on Mueller–Hinton broth, yeast extract, potassium phosphate, glucose, and pyruvate (MYPGP) agar or St. Julian medium (J-Medium). Note: Use of 0.1% (w/v) tryptone solution is advised during bacterial dilutions to promote the germination of vegetative cells.

Spores should be heated for 15 minutes in a 1 M calcium chloride solution (pH 7.0) at 60°C, then suspend them in the haemolymph of the cabbage looper *Trichoplusia ni* Hübner (Lepidoptera: Noctuidae) and tyrosine at an alkaline pH. A French press can be used to exert pressure after heating the spores at 75°C for 30 minutes in order to enhance germination (Koppenhöfer et al., 2012).

Alternately, Milner presented a different technique that makes use of *P. popilliae* var. *rhopaea*'s poor germination (Milner, 1977):

1. Create suspensions of soil by mixing 2 g of soil with 20 mL of sterilized water.
2. Prepare a germination medium with 0.5% yeast extract and 0.1% glucose.
3. Bring the pH level to 6.5.
4. Mix germination media (1:50) into the soil suspension.
5. Apply a series of 20-minute heat shocks at 70°C seven times every hour.
6. Spread the aliquot over J-Medium, and conduct anaerobic incubation for 7 h at 28°C.

Stahly et al. (1992) proposed a new approach for quantifying spores that takes use of *P. popilliae* resistance to vancomycin. This technique involves plating soil suspensions on MYPGP agar with 0.015% (w/v) vancomycin. This approach should only be used with caution since not all *P. popilliae* strains are vancomycin-resistant. Furthermore, fungal contamination can be prevented by adding 0.01% (w/v) cycloheximide and incubating for 3 weeks at 30°C.

14.3.1.3 *Serratia* spp.

Serratia spp. are usually taken out of soils, and some of them may also be isolated from insect carcasses since they are saprophytes. O'Callaghan and Jackson presented a technique based on a selective agar medium to promote the development of insect-pathogenic *Serratia* spp., such *S. entomophila*, *S. proteamaculans* (Paine and Stansfield), and *S. marcescens* Bizio.

1. Soil inoculums or haemolymph from infected larvae can be used to isolate *Serratia* on Caprylate thallous agar (CTA).
2. Culturing is performed by pulling and separating the cadavers' anterior ends. After that, the intestinal contents are cultivated on CTA plates.
3. Red colonies are produced by *Serratia marcescens*. Cream-colored bacterial colonies produced on CTA can subsequently be moved onto other selective medium for *Serratia* spp. identification.
4. The presence of *Serratia* spp. is shown by the development of a halo on a Deoxyribonuclease (DNase)-Toluidine Blue agar after 24 hours of incubation at 30°C. *S. proteamaculans* is then confirmed by the development of blue or green colonies on adonitol agar. The presence of *S. entomophila* is suggested by the development of yellow colonies on adonitol agar, which is verified by the growth on itaconate agar at 30°C after 96 hours. Additional molecular techniques that focus on certain DNA sequences can separate pathogenic strains from non-pathogenic ones.

(O'Callaghan & Jackson, 1993; Sharma et al., 2021)

14.3.1.4 EPN-Symbiotic Bacteria

Entomopathogenic nematode–symbiotic bacteria, like *Photorhabdus* and *Xenorhabdus*, are isolated by dropping haemolymph of infected insect (mostly *Galleria mellonella* Linnaeus, which is used as trap to collect IJs of EPN like *Heterorhabditis* or *Steinernema* from soil) onto a nutrient bromothymol agar (NBTA) containing triphenyltetrazolium chloride and bromothymol blue. Incubate the streaked plate at 28°C, and continuously subculture in NBTA. This process is repeated until uniform colonies are obtained (Godjo et al., 2018).

14.3.1.5 *Yersinia entomophaga*

Yersinia entomophaga is isolated by cultivating the haemolymph of infected *Costelytra zealandica* (White) (Coleoptera: Scarabaeidae) grass-grub larvae onto Luria-Bertani (LB) agar, resulting in a colony of a cream hue (Hurst et al., 2011).

14.3.1.6 *Pseudomonas entomophila*

Isolating *P. entomophila* Mulet et al. from *Drosophila* can be done in LB media with the appropriate antibiotics (100 µg/mL rifampicin and 600 µg/mL carbenicilin). The bacterial culture can be maintained on LB media (Vodovar et al., 2005). From soil the bacterium can be isolated following the standard procedure of isolating *Pseudomonads* in general. Three ten-fold dilution series of soil suspension is spread on modified King's B agar and modified King's A agar. Plates were incubated at 28°C for 48 hours. Subsequently, the plates can be evaluated under UV light in order to detect fluorescent pigment production. Single colonies representative of phenotypical type are streaked onto petri dishes containing King's B agar; subsequently, they should be characterized physiologically, biochemically, and genetically (Villamizar et al., 2020).

14.3.1.7 *Chromobacterium*

Bacterial isolates from insects belonging to *Chromobacterium* display violet pigmentation when cultivated on L-agar or nutritional agar due to the presence of violacein. Bacteria can be isolated by plating dilutions of soil and water samples on nutrient agar with saline solution (0.9% NaCl) and incubating at 37°C for 48 hours. Individual colonies of violet-coloured bacteria should be picked up, and the pure culture should be isolated by streaking in nutrient agar. Every 4 weeks, the bacterial isolates are sub-cultured on nutrient agar at 4°C (Sharma et al., 2021; Vishnu & Palaniswamy, 2016).

14.3.2 Confirmation of the Identity of the Isolated EPB

The aim of identification is to match all the characteristics of a pure culture isolate with a well-defined taxon (Bochner, 2009). For bacteria, this has been accomplished for so long via a range of biochemical tests. To date, these tests are quick and pocket-friendly methods for identification of bacteria. But not a single test is confirmatory of bacterial identity. It might go beyond 20 tests for confirmation of a bacteria (Altheide, 2019). While biochemical tests offer preliminary insights into bacterial metabolism and physiology, definitive identification increasingly relies on molecular techniques such as 16S rDNA sequencing, multilocus sequence typing (MLST), or whole-genome analysis, particularly for strain-level confirmation of EPBs. These tests are based on their ability of metabolic utilization of different sources and detected by the colour changes due to the bacterial utilization of nutrients and their subsequent growth. Though the biochemical utilization behaviour of bacteria might vary with their adaptation for some specific cases, some specific biochemical tests are listed in Figure 14.1, along with the results against respective EPBs, represented by colours as the representation of test result, except for 'V' cells, indicating variable results, and white cells with black border indicate no data.

Some culture-independent approaches have also been used to detect and quantify EPB. As entomopathogenicity, in most cases, is strain-specific, there should be specific methods for the identification of particular EPB strains, and biochemical tests might need to be more sophisticated in this regard. For instance, creating primers for *Bt* serovar *israelensis* and testing them using soil samples, or amplifying the region of 16S ribosomal DNA from the bacteria *P. entomophila* using a duplex polymerase chain reaction (PCR) and further validating the method in *P. entomophila*-infected *Drosophila melanogaster* Meigen (Sharma et al., 2021).

14.3.3 Protocol Fixing and Standardization for Mass Production

For commercial purpose, there should be a standard protocol which can be followed by some semi-skilled labourers to carry forward the mass production. These protocols vary from bacteria to bacteria, and also from company to company. Thus, it comes under trade secret. All the steps, including inoculum preparation, seed suspension, fermentation, filtration, and condensation, finally adding adjuvants to prepare formulation, quality testing, and then packaging, should be well standardized beforehand so that the company may perform smoothly and with highest efficiency.

14.3.3.1 Culture Media

Culture media and provided growth conditions are the main regulators of growth of EPB. Selected nutrients significantly impact the bacterial growth rate, production capacity, and ultimate infectiousness. Nutrients that need to be given include trace elements, as well as a carbon source like glucose or molasses and a nitrogen source like yeast extract or soybean meal. Nutrients chosen for culture are determined by the organism's needs, availability, and cost. However, the media utilized only makes up a minor portion of the operating expenses for utilities, maintenance, and equipment (Vimala Devi, 2018). The bacteria-specific requirements of culture media have been discussed in detail in Chapter 15.

14.3.3.2 Fermentation Technique

Bacillus thuringiensis (Bt) is the predominant EPB used in large-scale production, while *Lysinibacillus sphaericus* (Ls) is primarily restricted to mosquito control and produced on a relatively smaller commercial scale. The two main EPB

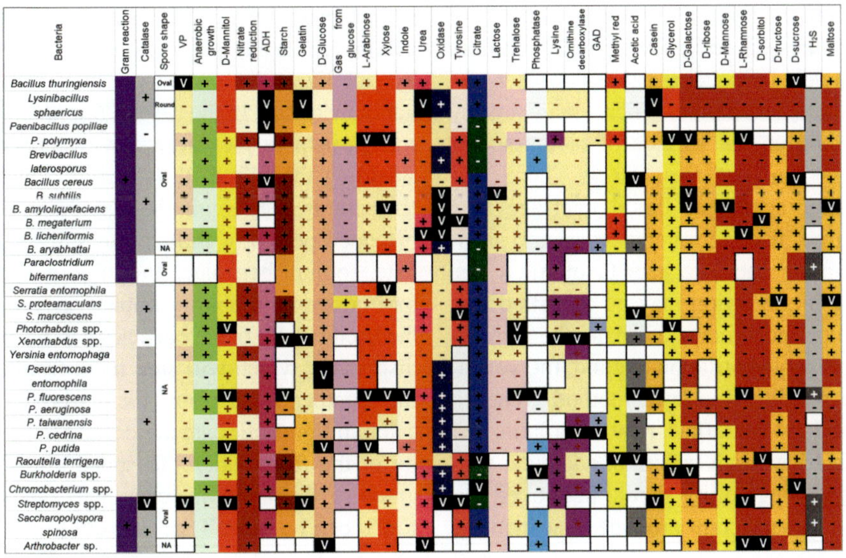

FIGURE 14.1 Biochemical characterization of various entomopathogenic bacteria. *Abbreviations: VP* = Voges–Proskauer test; *ADH*= arginine dihydrolase; *GAD* = glutamic acid decarboxylase; H_2S = hydrogen sulphide.

***Source*: Abo-Bakr et al. (2020); Adibi et al. (2017); Afrin and Bhuiyan (2023); Allwin et al. (2007); Ameer and Kadium (2021); Al-Saadi et al. (2013); Ali et al. (2022); Aryal (2022a (2022b (2022c (2022d); Asolkar et al. (2014); Atta (2015); Bator et al. (2020); Behrendt et al. (2009); Boemare and Akhurst (1988 (2015); Chantarasiri et al. (2017); Cordova-Kreylos et al. (2013); Dever and Lopez (2016); Dewi et al. (2022); Dieppois et al. (2015); Elarabi et al. (2020); Fisher and Garczynski (2012); Funke et al. (1996); Gate (2009); Ghosh et al. (2017); Gil et al. (1986); Glare et al. (2017); Gorashi et al. (2014); Grimont et al. (1978 (1982 (1988); Hamdi et al. (2013); Hazir et al. (2004); He et al. (2007); Hoffmann and Altenbuchner (2015); Hurst et al. (2011); Hurst and Glare (2018); Islam et al. (2014); Kämpfer et al. (2013); Kutsuna et al. (2019); Lee et al. (1986); Lee et al. (2017); Logan and De Vos (2015); López and Alippi (2009); Machado et al. (2023); Mathur and Kumar (2013); Martin and Travers (1989); Maymon et al. (2015); Mertz and Yao (1990); Moffett and Colwell (1968); Molina et al. (2019); Mulet et al. (2012); Nagai et al. (1996); Nepali et al. (2018); Ngalimat et al. (2021); Nikel et al. (2015); Pagare et al. (2018); Pastor et al. (2014); Peel et al. (1999); Pettersson et al. (1999); Rajashekhar et al. (2017); Regnum Prokaryotae (n.d. a, b); Reimer et al. (2022); Rhodes (1959); Rubikas et al. (1987); Saha et al. (2005); Saikia et al. (2011); Sangeetha et al. (2016); Sankar et al. (2016); Shahbaz-Mohammadi and Omidinia (2011); Sheikhi et al. (2012); Shivaji et al. (2009); Sivendra et al. (1975); Song et al (2011); Sura and Hiremath (2019); Suryawanshi et al. (2014); Taddei et al. (2006); Te Giffel et al. (1997); UK Health Security Agency (n.d.); VetBact (2023); von der Weid et al. (2000); Wang et al. (2010); Wohlers et al. (2021); Wordofa and Kristensen (2018); Yamanaka et al. (1992); Yu et al. (2011); Zhang et al. (2013).**

mass-produced, that is, Bt and Ls, are fermentable, and thus, their mass production strategy is much easier as compared to those of the non-fermentable ones. The two most prevalent forms of fermentation are submerged and solid; however, there are numerous more. While submerged fermentation is commonly used for EPBs like Bt, solid-state fermentation remains a viable and scalable option for producing spore-forming entomopathogens, especially under resource-limited conditions.

Submerged or deep-tank fermentation, as the name indicates, is the growth of microorganisms in a completely liquid medium. This technique was developed by the Pacific Yeast Product Company in 1956 (Glare & O'Callaghan, 2000). There are benefits to submerged fermentation, including high automation degree, high density, and high control accuracy. Due to these benefits, liquid-state submerged fermentation technology is used in Bt-industrialized fermentation production. The key factors of liquid-submerged fermentation include the strain, culture medium, and culture conditions; postprocessing technology; and safeguarding the fermentation process against phage or other microorganism invasions. There are two types of submerged fermentation: fed-batch and batch (Li et al., 2017). In batch fermentation, all the medium components are placed in the closed reactor at the start of cultivation and thus is a discontinuous process. A fed-batch culture is the modified version of batch fermentation in which nutrients are provided systematically in a semi-closed system, resulting in a semi-continuous reaction. The internal condition remains constant in fed-batch system, unlike the previous one. Batch fermentation systems are comparatively easier to manage and less prone to contamination than continuous systems, but strict aseptic conditions are still essential to prevent microbial invasions that compromise yield and product safety. Fed-batch cultures can be performed in small fermenters, and there is more control over microbial growth and desired product (Magar, 2021). At current times, large-scale manufacturers primarily use fed-batch submerged fermentation because batch fermentation allows the inhibitory effects of substrates and metabolites on cell growth and endotoxin synthesis, while negatively influencing oxygen transfer.

Solid-state fermentation cultivates microorganisms by using nutrients adsorbed on the surface of biomasses or nutrients given by the particles themselves. These biomasses can offer a significant gas–liquid interface in a comparatively small area, meeting water, oxygen, and nutritional requirements for the aerobic growth of microbes. Traditional solid-state fermentation may be classified into four categories based on the anticipated scale: the terrace-type method, the large-pool ventilation method, the vessel-box method, and the net-dish thin-layer method. When it comes to selecting solid raw materials, new, affordable raw materials (carriers) that have rich nutrition, adequate ventilation, and a strong wetting ability should be looked for. Additionally, appropriate surfactant products that increase fermentation level and toxicity potency while reducing wetting time should be chosen. The two types of carriers that are frequently utilized are inorganic and organic biomasses. Biomass products, such as peanut meal, rice bran, wheat bran, and yellow bean cake powder, can be used as carriers and sources of nutrients. Expanded perlite and fine sand are examples of inorganic carriers that function well in ventilation but require more nutrients. The primary control over the air permeability of the culture medium is accomplished by modification of the moisture content. The culture medium can be kept suitably loose by having an appropriate initial moisture content. Furthermore, a certain distance between the particles facilitates oxygen transport and aids in the bacteria's absorption of nutrients from the culture medium, supporting the development and multiplication of the target bacteria. Increased water content can cause the culture medium to clump together and lose porosity, which will hinder the transport of oxygen. Low water content, on the other hand, can limit the degree of culture medium expansion and water activity, which will prevent the growth of bacteria (Li et al., 2017).

14.3.3.3 Post-Processing Techniques

Adjuvants are components that are added to plant protection products (like microbial insecticides) or used on their own to increase their efficacy. The four major types of adjuvants utilized in the manufacturing of microbial pesticides are carriers, nutritional adjuvants, protective agents, and surfactants. Wetting agents, which decrease the interfacial tension between liquid and solid to increase liquid adhesion to solid surfaces, include sodium lauryl sulphate, stretch powder, polyethylene glycol, calcium lignosulfonate, sodium carboxymethyl cellulose, sodium dodecylbenzene sulfonate, sodium tripolyphosphate, lignin, sodium sulfonate, lignin sulphur salt, and fatty amine polyoxyethylene ether. The effect of a wetting agent on the water activity of microbiological pesticides should be addressed during storage and transportation. Important adjuvants for wettable powders include stabilizers, such as calcium carbonate, potassium dihydrogen phosphate, sodium carboxymethyl cellulose, and methylcellulose. Droplet sizes of the mixture can be regulated, and drifting during spraying is prevented by stabilizers such as magnesium aluminium silicate and xanthogen glue. In the case of *Bacillus*, spores play a crucial role in helping the bacteria withstand harsh conditions by shielding them from the deterioration brought on by high temperatures. To retain the suspension concentrations for an extended period of time, preservatives such potassium sorbate, sodium benzoate, and ethyl p-hydroxybenzoate must have a strong inhibitory impact on mould, yeast, and aerobic bacteria. Pesticide particles can be made smaller using a dispersion, which will stop them from settling and clumping together. Dispersants like carboxymethyl cellulose and calcium lignosulfonate are crucial for preserving the stability of the formulation. To speed up the breakdown of the granules, disintegrating chemicals such as NaCl, $AlCl_3$, and $CaCl_2$ can be used. The viscosity of the formulation can be changed by adding xanthan gum. Film-forming chemicals are essential in the formulation because they can join and wrap uniformly around the exterior of the seeds. These are polymer composite materials with good water and air permeability, such as gum arabic, PVA series, and PEG macromolecule series. To avoid misuse during storage, colourants like red rhododendron are also used (Lin et al., 2023).

Water-dispersible granules, wettable powders or pellets, and liquid base formulations are the most common forms of microbial biopesticides. Formulations are chosen according to the type of available dispensing equipment, end-user preferences, and ease of use. It is frequently possible to generate a single microbial biopesticide in all four formulations; the specific choice will depend on the needs of the field. The dependability and residual activity of microbial plant protection agents might be improved by nanoformulations and microencapsulation technologies, which could lead to an increase in their field use. Depending on the type of target pest and the mode of action, microbial biopesticides are commonly delivered as foliar sprays, root dips, soil additives, seed treatments, or a feasible mix of multiple approaches. Foliar treatment is spraying microbial biopesticides directly into the leaves to decrease visible pests, followed by treating the plant throughout the season to keep pest numbers below the economic threshold level. The ease of use of foliar sprays is one of the key drivers of expansion in the microbial biopesticide sector (Sabbahi et al., 2022).

In order to satisfy the demands of production and management, a set of technical standards as quality standardization are developed following scientific study, field testing, and inquiry. To meet the demands of industrial production, product standardization is established. The goal of quality standardization for EPB products involves two components. The first factor is the standardization of industrial products, or the products produced by the factories under quality control. The second facet pertains to worldwide standardization, including products made by various nations, manufacturing methodologies, and equivalent subspecies or strains of biopesticides.

Research and development investment in microbial biopesticide firms varies widely; while small enterprises in developing countries may invest under 1%, established firms often allocate significantly higher percentages depending on regional policy support and market competitiveness. As a result, there are few environmentally acceptable products available, waste disposal is not guaranteed to be efficient, applied product structures are not logically formed, and products have minimal impact on invasive, frequent, and refractory pests as well as nematodes and underground and subterranean pests. Currently, EPB businesses are dedicated to promoting formulations that lead to effective and secure products, such as dust-free, water-based, and release-controlled forms, as well as the creation of affordable and environmentally friendly packaging materials. They should also rigorously regulate the use of toxic and hazardous solvents and adjuvants (toluene, xylene, etc.), and they should give special attention to the creation of pesticide intermediates and unique adjuvants (Li et al., 2017).

The packaging is also a very crucial point, as shelf life highly depends on packaging quality. Liquid formulations can be packed in HDPE bottles, cans, or drums (Venkatesh, 2020), whereas wettable powder formulations can be packed in LDPE bags (NVBDCP, 2024).

14.4 LEGISLATION PROCESS

Before commercialization, legislation plays a great role, and the procedure goes parallelly with R&D. Two major roles played by legislative norms during mass production of EPB are patent filing for the selected strain for commercialization and registration of the final formulation, along with production procedure and safety. One industrialist can escape from the first step by searching in the list of latest filed patents and having contract with the patent owner, because in that case, independent researchers explore some strain for entomopathogenicity and do patenting for the same. The US Patent Application Publication 2012/0100236 A1 is about emerging EPB for insect pests (Ruiu et al., 2013).

The three main things to consider while registering a formulation are the environment, human safety, and product reliability. The process of obtaining commercial authorizations for microbial biopesticides is extensive and involves testing to ascertain the mechanism of action, host range, and effectiveness of the products, in addition to toxicological and environmental assessments. For local firms, toxicological tests might occasionally be too expensive and complicated. The effectiveness and labelling of products for microbial plant protection must also meet the requirements set forth by the authorities for safe handling. It might be challenging for industrialists to achieve these requirements.

Countries have different laws and processes developed for registering EPBs. For example, the differences in procedures between the EU and member states (MS) levels, a lack of flexibility, and a significant degree of actor heterogeneity make the EU registration process for EPBs more difficult. As a result, the average time for EU registration is 1.6 years (43%) more than that of US registration.

Undoubtedly, the globalization of some biopesticide registration procedures would promote their commercialization on a global scale. The most notable efforts towards this harmonization have been made thus far by the European and Mediterranean Plant Protection Organization, the Organization for Economic Cooperation and Development, and the International Organization for Biological Control, although their progress has been inconsistent. The present trend of lowering the usage of chemical pesticides and streamlining the low-risk chemical approval process might make EPBs more readily available and mass-produced internationally (Sabbahi et al., 2022).

14.5 CONCLUSION

From the discovery of EPB to its practical application in the field, there are several steps that should be followed by one willing entrepreneur. This chapter tries to give them a clearer view of the pathway to accomplish their business endeavour. This chapter also tries to focus on the role of research and development throughout this lengthy journey and tries to influence researchers of it plausible scope for their employment and work satisfaction. We have also tried to give a brief overview of the legislative procedure before proceeding to mass production, which would help industrialists chalk out their plan of work for finalization of their intended commercial venture.

REFERENCES

Abo-Bakr, A., Fahmy, E. M., Badawy, F., Abd El-latif, A. O., & Moussa, S. (2020). Isolation and characterization of the local entomopathogenic bacterium, *Bacillus thuringiensis* isolates from different Egyptian soils. *Egyptian Journal of Biological Pest Control*, *30*, 54. https://doi.org/10.1186/s41938-020-00250-z

Adibi, A., Rees, E. R., McCarley, S., Sica, V. P., & Oberlies, N. H. (2017). Characterization and isolation of peptide metabolites of an antifungal bacterial isolate identified as *Bacillus amyloliquefaciens* subspecies *plantarum* strain FZB42. *Journal of Microbiology, Biotechnology and Food Sciences*, *6*(6), 1309–1313. https://doi.org/10.15414/jmbfs.2017.6.6.1309-1313

Afrin, S., & Bhuiyan, M. N. I. (2023). Antagonistic activity of *Bacillus amyloliquefaciens* subsp. *amyloliquefaciens* against multidrug resistant *Serratia rubidaea*. *Current Research in Microbial Sciences*, *5*, 100206. https://doi.org/10.1016/j.crmicr.2023.100206

Ali, N. S., Huang, F., Qin, W., & Yang, T. C. (2022). Identification and characterization of a new *Serratia proteamaculans* strain that naturally produces significant amount of extracellular laccase. *Frontiers in Microbiology*, *13*, 878360. https://doi.org/10.3389/FMICB.2022.878360

Allwin, L., Kennedy, J. S., & Radhakrishnan, V. (2007). Characterization of different geographical strains of *Bacillus thuringiensis* from Tamil Nadu. *Research Journal of Agri culture and Biological Sciences*, *3*(5), 362–366.

Al-Saadi, A., Hameed, N. M., & Mohammad Jaralla, E. (2013). Isolation and identification of *Streptomyces* from different sample of soils. *Journal of Biology and Medical Sciences*, *1*, 31–36.

Altheide, S. T. (2019). Biochemical and culture-based approaches to identification in the diagnostic microbiology laboratory. *American Society for Clinical Laboratory Science*, *32*(4), 166–175. https://doi.org/10.29074/ASCLS.2019001875

Ameer, M. A. A., & Kadium, S. W. (2021). Isolating and diagnosing *Burkholderia* spp. from soil and soil surrounding onion roots in Najaf Governorate. *Annals of the Romanian Society for Cell Biology*, *25*(6), 14826 14840.

Aryal, S. (2022a, August 9). *Biochemical test and identification of bacillus subtilis*. Microbiology Info.com. Retrieved October 10, 2023, from https://microbiologyinfo.com/biochemical-test-and-identification-of-bacillus-subtilis

Aryal, S. (2022b, August 9). *Biochemical test and identification of serratia marcescens*. Microbiology Info.com. Retrieved October 10, 2023, from https://microbiologyinfo.com/biochemical-test-and-identification-of-serratia-marcescens

Aryal, S. (2022c, April 4). *Biochemical test of bacillus cereus*. Microbe Notes. Retrieved October 10, 2023, from https://microbenotes.com/biochemical-test-of-bacillus-cereus

Aryal, S. (2022d, May 23). *Biochemical test of pseudomonas aeruginosa*. Microbe Notes. Retrieved October 10, 2023, from https://microbenotes.com/biochemical-test-of-pseudomonas-aeruginosa

Asolkar, R., Koivunen, M., Marrone, P., Cordova-krevlos, A. L., & Huang, H. (2014). *Isolated bacterial strain of the genus Burkholderia and pesticidal metabolites therefrom* (Patent No. US 8,822, 193 B2). U.S. Patent and Trademark Office. https://patents.google.com/patent/WO2014149241A1/en

Atta, H. M. (2015). Biochemical studies on antibiotic production from *Streptomyces* sp.: Taxonomy, fermentation, isolation and biological properties. *Journal of Saudi Chemical Society*, *19*(1), 12–22. https://doi.org/10.1016/j.jscs.2011.12.011

Bator, I., Wittgens, A., Rosenau, F., Tiso, T., & Blank, L. M. (2020). Comparison of three xylose pathways in *Pseudomonas putida* KT2440 for the synthesis of valuable products. *Frontiers in Bioengineering and Biotechnology*, *7*, 480. https://doi.org/10.3389/fbioe.2019.00480

Behrendt, U., Schumann, P., Meyer, J. M., & Ulrich, A. (2009). *Pseudomonas cedrina* subsp. *fulgida* subsp. nov., a fluorescent bacterium isolated from the phyllosphere of grasses; emended description of *Pseudomonas cedrina* and description of *Pseudomonas cedrina* subsp. *cedrina* subsp. nov. *International Journal of Systematic and Evolutionary Microbiology*, *59*(6), 1331–1335. https://doi.org/10.1099/ijs.0.005025-0

Bochner, B. R. (2009). Global phenotypic characterization of bacteria. *FEMS Microbiology Reviews*, *33*(1), 191–205. https://doi.org/10.1111/J.1574-6976.2008.00149.X

Boemare, N. E., & Akhurst, R. J. (1988). Biochemical and physiological characterization of colony form variants in *Xenorhabdus* spp. (Enterobacteriaceae). *Microbiology*, *134*(3), 751–761. https://doi.org/10.1099/00221287-134-3-751

Boemare, N. E., & Akhurst, R. J. (2015). Photorhabdus. In M. E. Trujillo, S. Dedysh, P. DeVos, B. Hedlund, P. Kämpfer, F. A. Rainey, & W. B. Whitman (Eds.), *Bergey's manual of systematics of archaea and bacteria* (pp. 1–15). Wiley. https://doi.org/10.1002/9781118960608.gbm01159

Burges, H. D. (1981). Strategy for the microbial control of pests in 1980 and beyond. In H. D. Burges (Ed.), *Microbial control of pests and plant diseases 1970–1980* (pp. 797–836). Academic Press.

Chantarasiri, A., Boontanom, P., & Nuiplot, N. O. (2017). Isolation and characterization of *Lysinibacillus sphaericus* BR2308 from coastal wetland in Thailand for the biodegradation of lignin. *AACL Bioflux*, *10*(2), 200–209.

Cordova-Kreylos, A. L., Fernandez, L. E., Koivunen, M., Yang, A., Flor-Weiler, L., & Marrone, P. G. (2013). Isolation and characterization of *Burkholderia rinojensis* sp. nov., a non-*Burkholderia cepacia* complex soil bacterium with insecticidal and miticidal activities. *Applied and Environmental Microbiology*, *79*(24), 7669–7678. https://doi.org/10.1128/AEM.02365-13

Couch, T. L., Jackson, T. A., & Jurat-Fuentes, J. L. (2023). Commercial production of entomopathogenic bacteria. In J. A. Morales-Ramos, M. G. Rojas, & D. Shapiro-Ilan (Eds.), *Mass production of beneficial organisms* (2nd ed., pp. 359–373). Academic Press. https://doi.org/https://doi.org/10.1016/B978-0-12-822106-8.00018-X

Deka, B., Baruah, C., & Babu, A. (2021). Entomopathogenic microorganisms: Their role in insect pest management. *Egyptian Journal of Biological Pest Control*, *31*(1), 121. https://doi.org/10.1186/s41938-021-00466-7

Dever, C., & Lopez, J. (2016, January 29). *Brevibacillus laterosporus*. Microbe Wiki. Retrieved October 9, 2023, from https://microbewiki.kenyon.edu/index.php/Courtney_Dever_and_Jennifer_Lopez

Dewi, N. D. K., Hidayati, E., Sarkono, S., Prasedya, E. S., & Suryadi, B. F. (2022). Isolation of entomopathogenic *Lysinibacillus sphaericus* from sewage at some housing complex in Mataram city and evaluation of its toxicity against *Aedes aegypti* larvae in laboratory. *Biotropika: Journal of Tropical Biology*, *10*(1), 55–61. https://doi.org/10.21776/ub.biotropika.2022.010.01.07

Dieppois, G., Opota, O., Lalucat, J., & Lemaitre, B. (2015). *Pseudomonas entomophila*: A versatile bacterium with entomopathogenic properties. In J. L. Ramos, J. Goldberg, & A. Filloux (Eds.), *Pseudomonas* (pp. 25–49). Springer. https://doi.org/10.1007/978-94-017-9555-5_2

Elarabi, N. I., Abdelhadi, A. A., Ahmed, R. H., Saleh, I., Arif, I. A., Osman, G., & Ahmed, D. S. (2020). *Bacillus aryabhattai* FACU: A promising bacterial strain capable of manipulate the glyphosate herbicide residues. *Saudi Journal of Biological Sciences*, *27*(9), 2207–2214. https://doi.org/10.1016/J.SJBS.2020.06.050

EPA. (1996a). *Microbial pesticide test guidelines: Freshwater aquatic invertebrate testing, tier I* (Report no. OPPTS 885.4240). United States Environmental Protection Agency.

EPA. (1996b). *Microbial pesticide test guidelines: Background—Mammalian toxicity/pathogenicity/infectivity* (Report no. OPPTS 885.3000). United States Environmental Protection Agency.

Fisher, T. W., & Garczynski, S. F. (2012). Isolation, culture, preservation, and identification of entomopathogenic bacteria of the *Bacilli*. In L. A. Lacey (Ed.), *Manual of techniques in invertebrate pathology* (2nd ed., pp. 75–99). Academic Press. https://doi.org/10.1016/B978-0-12-386899-2.00003-8

Funke, G., Hutson, R. A., Bernard, K. A., Pfyffer, G. E., Wauters, G., & Collins, M. D. (1996). Isolation of *Arthrobacter* spp. from clinical specimens and description of *Arthrobacter cumminsii* sp. nov. and *Arthrobacter woluwensis* sp. nov. *Journal of Clinical Microbiology*, *34*(10), 2356–2363. https://doi.org/10.1128/jcm.34.10.2356-2363.1996

Gate, V. L. (2009). *Characterization of pseudomonas fluorescens and siderophore mediated antibiosis against soil borne fungal pathogens* (Unpublished doctoral dissertation, Dr. Panjabrao Deshmukh Krishi Vidyapeeth, Akola).

Ghosh, T. S., Chatterjee, S., Azmi, S. A., Mazumdar, A., & Dangar, T. K. (2017). Virulence assay and role of *Bacillus thuringiensis* TS110 as biocontrol agent against the larval stages of rice leaffolder *Cnaphalocrocis medinalis*. *Journal of Parasitic Diseases*, *41*(2), 491–495. https://doi.org/10.1007/s12639-016-0835-9

Gil, M. C., Rosa, M. D. L., Mosso, M. A., & Arribas, M. G. (1986). Numerical taxonomy of *Bacillus* isolated from orally administered drugs. *Journal of Applied Microbiology*, *61*(4), 347–356. https://doi.org/10.1111/j.1365-2672.1986.tb04296.x

Glare, T. R., Jurat-Fuentes, J.-L., & O'Callaghan, M. (2017). Basic and applied research. In L. A. Lacey (Ed.), *Microbial control of insect and mite pests* (pp. 47–67). Elsevier. https://doi.org/10.1016/B978-0-12-803527-6.00004-4

Glare, T. R., & O'Callaghan, M. (2000). *Bacillus thuringiensis: Biology, ecology, and safety*. Wiley.

Godjo, A., Afouda, L., Baimey, H., Decraemer, W., & Willems, A. (2018). Molecular diversity of *Photorhabdus* and *Xenorhabdus* bacteria, symbionts of *Heterorhabditis* and *Steinernema* nematodes retrieved from soil in Benin. *Archives of Microbiology*, *200*(4), 589–601. https://doi.org/10.1007/s00203-017-1470-2

Gorashi, N. E., Tripathi, M., Kalia, V., & Gujar, G. T. (2014). Identification and characterization of the Sudanese *Bacillus thuringiensis* and related bacterial strains for their efficacy against *Helicoverpa armigera* and *Tribolium castaneum*. *Indian Journal of Experimental Biology*, *52*(6), 637–649.

Grimont, P. A., Grimont, F., & Irino, K. (1982). Biochemical characterization of *Serratia liquefaciens sensu stricto, Serratia proteamaculans*, and *Serratia grimesii* sp. nov. *Current Microbiology*, *7*, 69–74. https://doi.org/10.1007/BF01568416

Grimont, P. A., Grimont, F., & Starr, M. P. (1978). *Serratia proteamaculans* (paine and atansfield) comb. nov., a senior subjective synonym of *Serratia liquefaciens* (grimes and hennerty) Bascomb et al. *International Journal of Systematic and Evolutionary Microbiology*, *28*(4), 503–510. https://doi.org/10.1099/00207713-28-4-503

Grimont, P. A., Jackson, T. A., Ageron, E., & Noonan, M. J. (1988). *Serratia entomophila* sp. nov. associated with amber disease in the New Zealand grass grub *Costelytra zealandica*. *International Journal of Systematic and Evolutionary Microbiology*, *38*(1), 1–6. https://doi.org/10.1099/00207713-38-1-1

Hamdi, C., Essanaa, J., Sansonno, L., Crotti, E., Abdi, K., Barbouche, N., Balloi, A., Gonella, E., Alma, A., Daffonchio, D., & Boudabous, A. (2013). Genetic and biochemical diversity of *Paenibacillus larvae* isolated from Tunisian infected honey bee broods. *BioMed Research International*, *2013*, 479893. https://doi.org/10.1155/2013/479893

Harumoto, T., & Lemaitre, B. (2018). Male-killing toxin in a bacterial symbiont of *Drosophila*. *Nature*, *557*(7704), 252–255. https://doi.org/10.1038/s41586-018-0086-2

Hazir, S., Stackebrandt, E., Lang, E., Schumann, P., Ehlers, R. U., & Keskin, N. (2004). Two new subspecies of *Photorhabdus luminescens*, isolated from *Heterorhabditis bacteriophora* (Nematoda: Heterorhabditidae): *Photorhabdus luminescens* subsp. *kayaii* subsp. nov. and *Photorhabdus luminescens* subsp. *thracensis* subsp. nov. *Systematic and Applied Microbiology*, *27*(1), 36–42. https://doi.org/10.1078/0723-2020-00255

He, Z., Kisla, D., Zhang, L., Yuan, C., Green-Church, K. B., & Yousef, A. E. (2007). Isolation and identification of a *Paenibacillus polymyxa* strain that coproduces a novel lantibiotic and polymyxin. *Applied and Environmental Microbiology*, *73*(1), 168–178. https://doi.org/10.1128/AEM.02023-06

Hoffmann, J., & Altenbuchner, J. (2015). Functional characterization of the mannitol promoter of *Pseudomonas fluorescens* DSM 50106 and its application for a mannitol-inducible expression system for *Pseudomonas putida* KT2440. *PLoS One*, *10*(7), e0133248. https://doi.org/10.1371/journal.pone.0133248

Hurst, G. D. D., Jigging, F. M., Von Der Schulenburg, J. H. G., Bertrand, D., West, S. A., Goriacheva, I. I., Zakharov, I. A., Werren, J. H., Stouthamer, R., & Majerus, M. E. N. (1999). Malekilling *Wolbachia* in two species of insect. *Proceedings of the Royal Society of London. Series B: Biological Sciences*, *266*(1420), 735–740. https://doi.org/10.1098/RSPB.1999.0698

Hurst, M. R. H., Becher, S. A., Young, S. D., Nelson, T. L., & Glare, T. R. (2011). *Yersinia entomophaga* sp. nov., isolated from the New Zealand grass grub *Costelytra zealandica*. *International Journal of Systematic and Evolutionary Microbiology*, *61*(4), 844–849. https://doi.org/10.1099/IJS.0.024406-0

Hurst, M. R. H., & Glare, T. R. (2018). *Insecticidal Yersinia entomophaga compositions and uses thereof* (Patent No. US 10, 039, 286 B2). U.S. Patent and Trademark Office. https://patents.google.com/patent/US10039286B2/en

Islam, M. S., Aktar, M. B., Rahman, M. M., & Uddin, K. M. M. (2014). Isolation and characterization of *Streptomyces* spp collected from Bangladeshi soils on the basis of morphological and biochemical studies. *International Journal of Current Microbiology and Applied Science*, *3*(11), 734–742.

Kämpfer, P., Martin, K., & Glaeser, S. P. (2013). *Lysinibacillus contaminans* sp. nov., isolated from surface water. *International Journal of Systematic and Evolutionary Microbiology*, *63*(9), 3148–3153. https://doi.org/10.1099/ijs.0.049593-0

Koppenhöfer, A. M., Jackson, T. A., & Klein, M. G. (2012). Bacteria for use against soil-inhabiting insects. In L. A. Lacey (Ed.), *Manual of techniques in invertebrate pathology* (2nd ed., pp. 129–149). Elsevier. https://doi.org/10.1016/B978-0-12-386899-2.00005-1

Kutsuna, R., Miyoshi-Akiyama, T., Mori, K., Hayashi, M., Tomida, J., Morita, Y., Tanaka, K., & Kawamura, Y. (2019). Description of *Paraclostridium bifermentans* subsp. *muricolitidis* subsp. nov., emended description of *Paraclostridium bifermentans* (Sasi Jyothsna et al., 2016), and creation of *Paraclostridium bifermentans* subsp. *bifermentans* subsp. nov. *Microbiology and Immunology*, *63*(1), 1–10. https://doi.org/10.1111/1348-0421.12663

Lee, A., Cheng, K. C., & Liu, J. R. (2017). Isolation and characterization of a *Bacillus amyloliquefaciens* strain with zearalenone removal ability and its probiotic potential. *PLoS One*, *12*(8), e0182220. https://doi.org/10.1371/journal.pone.0182220

Lee, H. H., Park, M. Y., & Lee, C. W. (1986). Biochemical characterization *of Bacillus thuringiensis*, 23 serovars. *Microbiology and Biotechnology Letters*, *14*(2), 205–208.

Li, L., Chen, Z., & Yu, Z. (2017). Mass production, application and market development of *Bacillus thuringiensis* biopesticides in China. In L. M. Fiuza, R. Polanczyk, & N. Crickmore (Eds.), *Bacillus thuringiensis and Lysinibacillus sphaericus* (pp. 185–212). Springer. https://doi.org/10.1007/978-3-319-56678-8_12

Lin, F., Mao, Y., Zhao, F., Idris, A. L., Liu, Q., Zou, S., Guan, X., & Huang, T. (2023). Towards sustainable green adjuvants for microbial pesticides: Recent progress, upcoming challenges, and future perspectives. *Microorganisms*, *11*(2), 364. https://doi.org/10.3390/microorganisms11020364

Logan, N. A., & De Vos, P. (2015). Bacillus. In *Bergey's manual of systematics of archaea and bacteria* (pp. 1–163). Wiley. https://doi.org/10.1002/9781118960608.gbm00530

López, A. C., & Alippi, A. M. (2009). Diversity of *Bacillus megaterium* isolates cultured from honeys. *LWT-Food Science and Technology*, *42*(1), 212–219. https://doi.org/10.1016/j.lwt.2008.05.001

Machado, R. A. R., Bhat, A. H., Castaneda-Alvarez, C., Askary, T. H., Půža, V., Pagès, S., & Abolafia, J. (2023). *Xenorhabdus aichiensis* sp. nov., *Xenorhabdus anantnagensis* sp. nov., and *Xenorhabdus yunnanensis* sp. nov., isolated from *Steinernema* Entomopathogenic nematodes. *Current Microbiology*, *80*, 300. https://doi.org/10.1007/s00284-023-03373-2

Magar, S. T. (2021, September 28). *Batch vs fed-batch vs continuous culture—20 key differences*. Microbe Notes. Retrieved October 9, 2023, from https://microbenotes.com/batch-vs-fed-batch-vs-continuous-culture

Martin, P. A., & Travers, R. S. (1989). Worldwide abundance and distribution of *Bacillus thuringiensis* isolates. *Applied and Environmental Microbiology*, *55*(10), 2437–2442. https://doi.org/10.1128/aem.55.10.2437-2442.1989

Mathur, N., & Kumar, A. (2013). Decolorization of methyl red by an isolated *Pseudomonas putida* strain MR1. *African Journal of Microbiology Research*, *7*(12), 983–989. https://doi.org/10.5897/AJMR2013.2537

Maymon, M., Martínez-Hidalgo, P., Tran, S. S., Ice, T., Craemer, K., Anbarchian, T., Sung, T., Hwang, L. H., Chou, M., Fujishige, N. A., Villella, W., Ventosa, J., Sikorski, J., Sanders, E. R., Faull, K. F., & Hirsch, A. M. (2015). Mining the phytomicrobiome to understand how bacterial coinoculations enhance plant growth. *Frontiers in Plant Science*, *6*, 784. https://doi.org/10.3389/FPLS.2015.00784

Mertz, F. P., & Yao, R. C. (1990). *Saccharopolyspora spinosa* sp. nov. isolated from soil collected in a sugar mill rum still. *International Journal of Systematic and Evolutionary Microbiology*, *40*(1), 34–39. https://doi.org/10.1099/00207713-40-1-34

Mezzomo, B. P., Miranda-Vilela, A. L., Freire, I. de S., Barbosa, L. C. P., Portilho, F. A., Lacava, Z. G. M., & Grisolia, C. K. (2013). Hematotoxicity of *Bacillus thuringiensis* as spore-crystal strains Cry1Aa, Cry1Ab, Cry1Ac or Cry2Aa in Swiss albino mice. *Journal of Hematology & Thromboembolic Diseases*, *1*(1), 1000104. https://doi.org/10.4172/2329-8790.1000104

Milner, R. J. (1977). A method for isolating milky disease, *Bacillus popilliae* var. *rhopaea*, spores from the soil. *Journal of Invertebrate Pathology*, *30*(3), 283–287. https://doi.org/10.1016/0022-2011(77)90135-5

Moffett, M. L., & Colwell, R. R. (1968). Adansonian analysis of the Rhizobiaceae. *Journal of General Microbiology*, *51*(2), 245–266. https://doi.org/10.1099/00221287-51-2-245

Molina, L., Rosa, R. L., Nogales, J., & Rojo, F. (2019). *Pseudomonas putida* KT2440 metabolism undergoes sequential modifications during exponential growth in a complete medium as compounds are gradually consumed. *Environmental Microbiology*, *21*(7), 2375. https://doi.org/10.1111/1462-2920.14622

Mulet, M., Gomila, M., Lemaitre, B., Lalucat, J., & García-Valdés, E. (2012). Taxonomic characterisation of *Pseudomonas* strain L48 and formal proposal of *Pseudomonas entomophila* sp. nov. *Systematic and Applied Microbiology*, *35*(3), 145–149. https://doi.org/10.1016/j.syapm.2011.12.003

Nagai, K., Ohta, S., Zenda, H., Matsumoto, H., & Makino, M. (1996). Biochemical characterization of a *Pseudomonas fluorescens* strain isolated from a benzalkonium chloride solution. *Biological and Pharmaceutical Bulletin*, *19*(6), 873–875. https://doi.org/10.1248/bpb.19.873

Nepali, B., Bhattarai, S., & Shrestha, J. (2018). Identification of *Pseudomonas fluorescens* using different biochemical tests. *International Journal of Applied Biology*, *2*(2), 27–32. https://doi.org/10.20956/ijab.v2i2.5260

Ngalimat, M. S., Yahaya, R. S. R., Baharudin, M. M. A. A., Yaminudin, S. M., Karim, M., Ahmad, S. A., & Sabri, S. (2021). A review on the biotechnological applications of the operational group *Bacillus amyloliquefaciens*. *Microorganisms*, *9*(3), 614. https://doi.org/10.3390/microorganisms9030614

Nikel, P. I., Romero-Campero, F. J., Zeidman, J. A., Goñi-Moreno, Á., & de Lorenzo, V. (2015). The glycerol-dependent metabolic persistence of *Pseudomonas putida* KT2440 reflects the regulatory logic of the GlpR repressor. *MBio*, *6*(2), e00340–15. https://doi.org/10.1128/mbio.00340-15

NVBDCP. (2024). *Technical specification of (Bti 5% WP) as larvicide|DTE, national vector borne disease control programme* (pp. 1–5). Government of Madhya Pradesh Supply.

O'Callaghan, M., & Jackson, T. A. (1993). Isolation and enumeration of *Serratia entomophila*—A bacterial pathogen of the New Zealand grass grub, *Costelytra zealandica*. *Journal of Applied Bacteriology*, *75*(4), 307–314. https://doi.org/10.1111/j.1365-2672.1993.tb02781.x

Oliveira-Filho, E. C., & Grisolia, C. K. (2022). The ecotoxicology of microbial insecticides and their toxins in genetically modified crops: An overview. *International Journal of Environmental Research and Public Health*, *19*(24), 16495. https://doi.org/10.3390/ijerph192416495

Pagare, K. A., Khot, G. G., & Mohite, P. B. (2018). Isolation and identification of *Bacillus thuringiensis*. *Journal of Pharmacognosy and Phytochemistry*, *7*(6), 1015–1017. http://www.ncbi.nlm.nih.gov/pubmed/710921

Pastor, N., Rosas, S., Luna, V., & Rovera, M. (2014). Inoculation with *Pseudomonas putida* PCI2, a phosphate solubilizing rhizobacterium, stimulates the growth of tomato plants. *Symbiosis*, *62*(3), 157–167. https://doi.org/10.1007/S13199-014-0281-3

Peel, M. M., Alfredson, D. A., Gerrard, J. G., Davis, J. M., Robson, J. M., McDougall, R. J., Scullie, B. L., & Akhurst, R. J. (1999). Isolation, identification, and molecular characterization of strains of *Photorhabdus luminescens* from infected humans in Australia. *Journal of Clinical Microbiology*, *37*(11), 3647–3653. https://doi.org/10.1128/JCM.37.11.3647-3653.1999

Pettersson, B., Rippere, K. E., Yousten, A. A., & Priest, F. G. (1999). Transfer of *Bacillus lentimorbus* and *Bacillus popilliae* to the genus *Paenibacillus* with emended descriptions of *Paenibacillus lentimorbus* comb. nov. and *Paenibacillus popilliae* comb. nov. *International Journal of Systematic Bacteriology*, *49*(2), 531–540. https://doi.org/10.1099/00207713-49-2-531

Rajashekhar, M., Shahanaz, E., & Vinay, K. (2017). Biochemical and molecular characterization of *Bacillus* spp. isolated from insects. *Journal of Entomology and Zoology Studies*, *5*(5), 581–588.

Ravensberg, W. J. (2011). Selection of a microbial pest control agent. In W. J. Ravensberg (Ed.), *A roadmap to the successful development and commercialization of microbial pest control products for control of arthropods* (Vol. 10, pp. 23–57). Springer. https://doi.org/10.1007/978-94-007-0437-4

Redmond, C. T., & Potter, D. A. (1995). Lack of efficacy of in vivo—And putatively in vitro-produced *Bacillus popilliae* against field populations of Japanese beetle (Coleoptera:

Scarabaeidae) grubs in Kentucky. *Journal of Economic Entomology*, *88*(4), 846–854. https://doi.org/10.1093/JEE/88.4.846

Regnum Prokaryotae. (n.d.a). *ABIS encyclopedia*: *Bacillus megaterium*. https://www.tgw1916.net/Bacillus/megaterium.html

Regnum Prokaryotae. (n.d.b). *ABIS encyclopedia*: *Raoultella terrigena*. https://www.tgw1916.net/Enterobacteria/raoultella_terrigena.html

Reimer, L. C., Sardà Carbasse, J., Koblitz, J., Ebeling, C., Podstawka, A., & Overmann, J. (2022). Bac Dive in 2022: The knowledge base for standardized bacterial and archaeal data. *Nucleic Acids Research*, *50*(D1), D741–D746. https://doi.org/10.1093/nar/gkab961

Rhodes, M. E. (1959). The characterization of *Pseudomonas fluorescens*. *Microbiology*, *21*(1), 221–263. https://doi.org/10.1099/00221287-21-1-221

Rubikas, J., Androsiuniene, D., Chestukhina, G., Smirnova, T., Kapitonova, O., & Stepanov, V. (1987). Crystal protein formed by *Bacillus subtilis* cells. *Journal of Bacteriology*, *169*(11), 5258–5262. https://doi.org/10.1128/JB.169.11.5258-5262.1987

Ruiu, L., Satta, A., & Floris, I. (2013). Emerging entomopathogenic bacteria for insect pest management. *Bulletin of Insectology*, *66*(2), 181–186.

Sabbahi, R., Hock, V., Azzaoui, K., Saoiabi, S., & Hammouti, B. (2022). A global perspective of entomopathogens as microbial biocontrol agents of insect pests. *Journal of Agriculture and Food Research*, *10*, 100376. https://doi.org/10.1016/J.JAFR.2022.100376

Saha, P., Mondal, A. K., Mayilraj, S., Krishnamurthi, S., Bhattacharya, A., & Chakrabarti, T. (2005). *Paenibacillus assamensis* sp. *nov*., a novel bacterium isolated from a warm spring in Assam, India. *International Journal of Systematic and Evolutionary Microbiology*, *55*(6), 2577–2581. https://doi.org/10.1099/ijs.0.63846-0

Saikia, R., Gogoi, D. K., Mazumder, S., Yadav, A., Sarma, R. K., Bora, T. C., & Gogoi, B. K. (2011). *Brevibacillus laterosporus* strain BPM3, a potential biocontrol agent isolated from a natural hot water spring of Assam, India. *Microbiological Research*, *166*(3), 216–225. https://doi.org/10.1016/j.micres.2010.03.002

Sangeetha, B. G., Jayaprakas, C. A., Siji, J. V., Rajitha, M., Shyni, B., & Mohandas, C. (2016). Molecular characterization and amplified ribosomal DNA restriction analysis of entomopathogenic bacteria associated with *Rhabditis* (*Oscheius*) spp. *3 Biotech*, *6*, 32. https://doi.org/10.1007/s13205-015-0326-1

Sankar, B. S., Kishore, S., Saurav, D., & Madhumita, B. (2016). Structural and functional analysis of glutamate decarboxylase system in *Bacillus aryabhattai*. *Research Journal of Biotechnology*, *11*(11), 1–11.

Shahbaz-Mohammadi, H., & Omidinia, E. (2011). Screening and characterization of proline dehydrogenase flavoenzyme producing *Pseudomonas entomophila*. *Iranian Journal of Microbiology*, *3*(4), 201.

Sharma, L., Bohra, N., Rajput, V. D., Roberto Quiroz-Figueroa, F., Singh, R. K., & Marques, G. (2021). Advances in entomopathogen isolation: A case of bacteria and fungi. *Microorganisms*, *9*(1), 16. https://doi.org/10.3390/microorganisms9010016

Sheikhi, F., Roayaei Ardakani, M., Enayatizamir, N., & Rodriguez-Couto, S. (2012). The determination of assay for laccase of *Bacillus subtilis* WPI with two classes of chemical compounds as substrates. *Indian Journal of Microbiology*, *52*, 701–707. https://doi.org/10.1007/s12088-012-0298-3

Shivaji, S., Chaturvedi, P., Begum, Z., Pindi, P. K., Manorama, R., Padmanaban, D. A., Shouche, Y. S., Pawar, S., Vaishampayan, P., Dutt, C. B. S., Datta, G. N., Manchanda, R. K., Rao, U. R., Bhargava, P. M., & Narlikar, J. V. (2009). *Janibacter hoylei* sp. nov., *Bacillus isronensis* sp. nov. and *Bacillus aryabhattai* sp. nov., isolated from cryotubes used for collecting air from the upper atmosphere. *International Journal of Systematic and Evolutionary Microbiology*, *59*(12), 2977–2986. https://doi.org/10.1099/ijs.0.002527-0

Sivendra, R., Lo, H. S., & Lim, K. T. (1975). Identification of *Chromobacterium violaceum*: Pigmented and non pigmented strains. *Journal of General Microbiology*, *90*(1), 21–31. https://doi.org/10.1099/00221287-90-1-21

Song, Z., Liu, K., Lu, C., Yu, J., Ju, R., & Liu, X. (2011). Isolation and characterization of a potential biocontrol *Brevibacillus laterosporus*. *African Journal of Microbiology Research*, *5*(18), 2675–2681.

Stahly, D. P., Takefman, D. M., Livasy, C. A., & Dingman, D. W. (1992). Selective medium for quantitation of *Bacillus popilliae* in soil and in commercial spore powders. *Applied and Environmental Microbiology*, *58*(2), 740–743. https://doi.org/10.1128/AEM.58.2.740-743.1992

Sura, N. K., & Hiremath, L. (2019). Isolation of *Bacillus megaterium* and its commercial importance. *International Journal of ChemTech Research*, *12*(4), 30–36. https://doi.org/10.20902/ijctr.2019.120405

Suryawanshi, R. K., Patil, C. D., Borase, H. P., Salunke, B. K., & Patil, S. V. (2014). Studies on production and biological potential of prodigiosin by *Serratia marcescens*. *Applied Biochemistry and Biotechnology*, *173*, 1209–1221. https://doi.org/10.1007/s12010-014-0921-3

Taddei, A., Rodríguez, M. J., Márquez-Vilchez, E., & Castelli, C. (2006). Isolation and identification of *Streptomyces* spp. from Venezuelan soils: Morphological and biochemical studies. I. *Microbiological Research*, *161*(3), 222–231. https://doi.org/10.1016/j.micres.2005.08.004

TeGiffel, M. C., Beumer, R. R., Klijn, N., Wagendorp, A., & Rombouts, F. M. (1997). Discrimination between *Bacillus cereus* and *Bacillus thuringiensis* using specific DNA probes based on variable regions of 16S rRNA. *FEMS Microbiology Letters*, *146*(1), 47–51. https://doi.org/10.1111/j.1574-6968.1997.tb10169.x

UK Health Security Agency. (n.d.). *Bacteria collection: Pseudo monas putida additional information*. https://www.culturecollections.org.uk/nop/product/pseudomonas-putida-2

Venkatesh. (2020). *VectoGuardTM*. Agri Life. https://agrilife.in/products/microbial_vectoguard.php

VetBact. (2023, March 8). *Bacillus lichenformis*. Swedish University of Agricultural Sciences—Faculty of Veterinary Medicine and Animal Science. Retrieved October 11, 2023, from https://www.vetbact.org/?artid=12

Villamizar, S., Ferro, J. A., Caicedo, J. C., & Alves, L. M. C. (2020). Bactericidal effect of entomopathogenic bacterium *Pseudomonas entomophila* against *Xanthomonas citri* reduces citrus canker disease severity. *Frontiers in Microbiology*, *11*, 1431. https://doi.org/10.3389/fmicb.2020.01431

Vimala Devi, P. S. (2018). Mass production of *Bacillus thuringiensis* and *Nomuraea rileyi*. In C. Chattopadhyay, R. K. Tanwar, M. Sehgal, A. Birah, S. Bhagat, N. Ahmad, & N. Mehta (Eds.), *Handbook of integrated pest management* (pp. 353 357). Indian Council of Agricultural Research.

Vishnu, T. S., & Palaniswamy, M. (2016). Isolation and identification of *Chromobacterium* sp. from different ecosystems. *Asian Journal of Pharmaceutical and Clinical Research*, *9*, 253–257. https://doi.org/10.22159/ajpcr.2016.v9s3.14847

Vodovar, N., Vinals, M., Liehl, P., Basset, A., Degrouard, J., Spellman, P., Boccard, F., & Lemaitre, B. (2005). Drosophila host defense after oral infection by an entomopathogenic *Pseudomonas* species. *Proceedings of the National Academy of Sciences*, *102*(32), 11414–11419. https://doi.org/10.1073/PNAS.0502240102

von der Weid, I., Paiva, E., Nóbrega, A., van Elsas, J. D., & Seldin, L. (2000). Diversity of *Paenibacillus polymyxa* strains isolated from the rhizosphere of maize planted in Cerrado soil. *Research in Microbiology*, *151*(5), 369–381. https://doi.org/10.1016/S0923-2508(00)00160-1

Wang, L.-T., Tai, C.-J., Wu, Y.-C., Chen, Y.-B., Lee, F.-L., & Wang, S.-L. (2010). *Pseudomonas taiwanensis* sp. nov., isolated from soil. *International Journal of Systematic and Evolutionary Microbiology*, *60*, 2094–2098. https://doi.org/10.1099/ijs.0.014779-0

Wohlers, K., Wirtz, A., Reiter, A., Oldiges, M., Baumgart, M., & Bott, M. (2021). Metabolic engineering of *Pseudomonas putida* for production of the natural sweetener 5-ketofructose from fructose or sucrose by periplasmic oxidation with a heterologous fructose dehydrogenase. *Microbial Biotechnology*, *14*(6), 2592–2604. https://doi.org/10.1111/1751-7915.13913

Wordofa, G. G., & Kristensen, M. (2018). Tolerance and metabolic response of *Pseudomonas taiwanensis* VLB120 towards biomass hydrolysate-derived inhibitors. *Biotechnology for Biofuels*, *11*(1), 199. https://doi.org/10.1186/s13068-018-1192-y

Yamanaka, S., Hagiwara, A., Nishimura, Y., Tanabe, H., & Ishibashi, N. (1992). Biochemical and physiological characteristics of *Xenorhabdus* species, symbiotically associated with entomopathogenic nematodes including *Steinernema kushidai* and their pathogenicity against *Spodoptera litura* (Lepidoptera: Noctuidae). *Archives of Microbiology*, *158*(6), 387–393. https://doi.org/10.1007/BF00276297

Yu, V. C.-H., Yu, P. H., Ho, K., & Lee, F. W. (2011). Isolation and identification of a new tetrodotoxin-producing bacterial species, *Raoultella terrigena*, from Hong Kong marine puffer fish *Takifugu niphobles*. *Marine Drugs*, *9*(11), 2384–2396. https://doi.org/10.3390/md9112384

Zhang, Y., Li, S., Liu, L., & Wu, J. (2013). Acetoin production enhanced by manipulating carbon flux in a newly isolated *Bacillus amyloliquefaciens*. *Bioresource technology*, *130*, 256–260. https://doi.org/10.1016/j.biortech.2012.10.036

15 Mass Production and Commercial Formulation of Some Important Entomopathogenic Bacteria

ABSTRACT

The positive shift from chemical to organic farming and the increased demand for biopesticides have created an urgent requirement to uplift the production of entomopathogenic bacteria for the fulfilment of that market demand. There are several commercial biopesticide products available in the market that are made up of entomopathogenic bacteria. Most of them comprise *Bacillus thuringiensis*, followed by *Lysinibacillus sphaericus*. Liquid-state fermentation (LSF) is mostly used in in vitro production methods for industrial purposes, whereas solid-state fermentation (SSF) is performed in small-scale production units. *Paenibacillus popilliae* is obligatorily multiplied via in vivo method. Mass production strategies for most other entomopathogenic bacteria are mostly at the research level. Despite minor problems in production and commercialization, EPB products have a bright future.

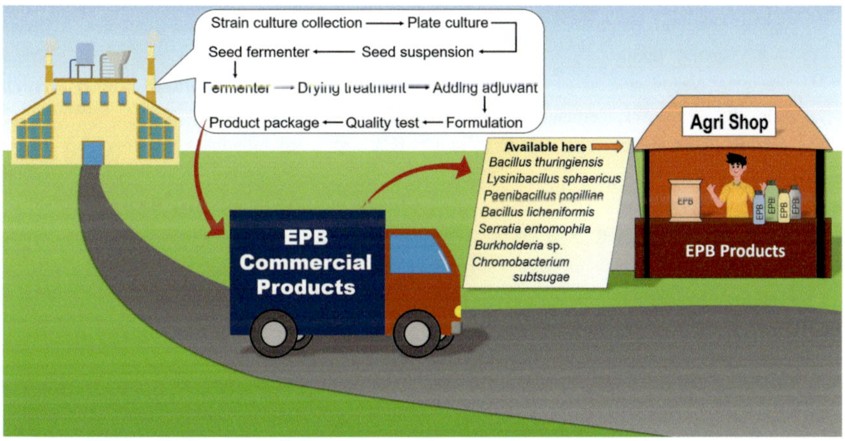

15.1 INTRODUCTION

To overcome the side effects of chemical pesticides, the world is now in search of strong organic alternatives. There are several biological agents that have pesticidal properties. Among them, entomopathogenic bacteria (EPB) play a major role. In the previous chapters, we have come to know the entomopathogenic nature of diverse groups of bacteria, but all do not qualify for commercial production. The first step in selecting an EPB for commercial production is to determine whether there is a commercial market for the bacterium. This decision is crucial because a huge amount of capital is required for scaling up, producing, and formulating an EPB on a commercial scale. Couch and Jurat-Fuentes (2014) defined *commercial-scale production* as deep tank fermentation in a fermenter with a minimum working volume of 30,000 L. This chapter deals with a detailed explanation of the commercial production strategies of major EPB and the present market status of the same.

15.2 COMMERCIAL PRODUCTS OF DIFFERENT ENTOMOPATHOGENIC BACTERIA

The need is more, and the supply is less—this is all about the present situation when it comes to looking for microbial biopesticides. But we have to admit that the last few decades have seen a considerable improvement in research on EPB which resulted in the production of a large number of commercial products, paving the path for their steady marketability at the cost of conventional pesticides. There is no example of commercial production of EPB lacking cell walls yet. This reflects the difficulty in their propagation and, in some cases, the potentiality of pathogenicity in vertebrates. Different bacterial species having commercially utilization as insecticides are presented in Table 15.1. Recently, attempts have been made to make biopesticides more efficient using nanoformulations. Sodium titanate–Bt came up as a new nanopesticide for cotton leafworm (Zaki et al., 2017).

DOI: 10.1201/9781003426882-15

TABLE 15.1
Entomopathogenic Bacteria: List of Commercial Products Based on EPB

Bacteria	Trade Names	Toxin Involved	Effective Against	Reference
Bacillus thuringiensis var. *kurstaki* (strain HD-1)	Bactospeine, Biobit, Dipel, Foray, Javelin, Thruicide, Vault	Cry1Aa, Cry1Ab, Cry1Ac, Cry2Aa, and Cry2Ab	Lepidoptera	(Agrichem, n.d.; Fernández-Chapa et al., 2019; Hazarika et al., 2005; Johnson & Yancy, 1997; Ladurner et al., 2008; Levin, 2009; Monterey, n.d.; Prasad, 2019; Ruiu, 2018; Sabbahi et al., 2022; Sanahuja et al., 2011; Seven Springs Farm Supply, n.d.)
Bt var. *kurstaki* (strain SA-11)	Delfin, Javelin-WG			
Bt var. *kurstaki* (strain SA-12)	Costar-WG, Deliver, Monterey Bt			
Bt var. *kurstaki* (strain Z-52)	Biolep			
Bt var. *kurstaki* (strain EG 2348)	Cordalene, Lepinox Plus, Rapax			
Bt var. *kurstaki* (strain EG 7841)	Crymax-WDG			
Bt var. *kurstaki* (serovar *kurstaki*) (Indigenous)	Halt			
Bt var. *galleriae*	Spicturin	Cry8Da	Lepidoptera	(Fernández-Chapa et al., 2019; Prasad, 2019)
Bt var. *tenebrionis*	Di Terra, Foll, M-one, M-Trak, Novodor, Trident	Cry3Aa	Coleoptera like Colorado potato beetle and adult elm beetles	(Fernández-Chapa et al., 2019; Khachatourians, 2019; Prasad, 2019; Ruiu, 2018)
Bt var. *israelensis*	Acrobe, Mosquito Beater WSP, Skeetal, Teknar, Vectobac, Vectobar	Cry4A, Cry4B, Cry11A, and Cyt1Aa	Mosquitoes and Blackflies	(Fernández-Chapa et al., 2019; Prasad, 2019; Ruiu, 2018; Sabbahi et al., 2022)
Bt var. *aizawai*	Able-WG, Agree-WP, Certan, Design, Florbac, XenTari	Cry1Aa, Cry1Ab, Cry1Ba, Cry1Ca, and Cry1Da	Armyworms, diamondback moth caterpillar	(Fernández-Chapa et al., 2019; Prasad, 2019; Ruiu, 2018; Sanahuja et al., 2011)
Lysinibacillus sphaericus	Larvicide, Spherimos, Spherifix, Vectolex and Vectomax	Bin	Mosquitoes and black flies	(Prasad, 2019; Ruiu, 2018; Silva Filha et al., 2014)
Paenibacillus popilliae	Doom, Milky Spore Powder	—	Japanese beetle	(Prasad, 2019; Sabbahi et al., 2022)
Bacillus licheniformis	Novozymes	—	*Aedis aegypti*	(Ruiu, 2015)
Serratia entomophila	Invade	—	Grass-grub	(Johnson et al., 2001)
Burkholderia spp.	Majestene, Venerate	—	Chewing and sucking insects, mites, nematodes	(Ruiu, 2018)
Chromobacterium subtsugae	Grandevo	—	Chewing and sucking insects and mites	(Ruiu, 2018)
Saccharopolyspora spinosa	Conserve, SpinTor, Success, Tracer™ 120	—	Insects	(Ruiu, 2018; Salgado & Sparks, 2005)

15.3 MASS PRODUCTION PROTOCOL OF MAJOR ENTOMOPATHOGENIC BACTERIA

The production protocol is always a trade secret and not disclosed to others, but new players of biocontrol industry hence need to reinvent the already-invented protocols. It has a good side, in that different pathways would come up with different time and input requirements and different efficiencies. Yet there is a need for a general pathway that describes all phases that will guide start-up enterprises through the development of new products in an optimal way (Ravensberg, 2011). The maximum toxin production can be achieved only by careful attention to the interaction of fermentation conditions, media, and the isolates involved, as no medium is best suited to all isolates (Vimala Devi, 2018).

15.3.1 *Bacillus thuringiensis*

Dating back to 1938, the first commercial product of *Bacillus thuringiensis* Berliner (Bt) was introduced in France, that is, Sporeine. Since then, there has been a continuous rise in the development of advanced products with high virulence and field persistence. Bt-based bioinsecticides are classified under fourth-generation insecticides (Fernández-Chapa et al., 2019).

15.3.1.1 Culturing on Media

Before mass production begins, several important steps need to be followed. A spore stock is prepared, and after a series of tests for purity, the spores are stored in freeze-dried ampoules or under liquid nitrogen. Observations are made by culturing the spore stock on agar plates and incubating these at 30°C for 5 days to prevent contaminating bacteria and other pollutants or bacteria phage (Couch & Jurat-Fuentes, 2014). Bt can be easily multiplied on Luria–Bertani broth (LB), nutrient agar, or T3 agar media. Pure culture of Bt is maintained on the agar slants while it is multiplied as broth cultures in flasks for testing against insects. For storage purposes, it is essential to store the pure culture of Bt in 10% glycerol at −80°C, and we can revive the cultures on Luria–Bertani agar (LA) slants when required. It is desirable to perform sub-culturing of Bt to two different slants at 15-day intervals to avoid chances of mutations in the cultures (Vimala Devi, 2018). Whenever new batches of the stock culture are to be made, the starting culture should pass through testing to ensure purity and insecticidal activity at levels equivalent to the parent culture (Couch & Jurat-Fuentes, 2014). For multiplication, first, we have to perform inoculation in the media under a laminar airflow chamber, then the flasks are plugged with sterile cotton and placed on a shaker at 30°C. Depending on the strain/isolate, the time taken for complete growth, sporulation, and lysis can exist up to 48–72 hours. The Bt spores and crystals are then recovered from broth through centrifugation, and the resultant pellet is dried and milled to get a powder to use in bioassays (Vimala Devi, 2018).

15.3.1.2 Growth Phases

After inoculation, there is an initial slow-growing lag phase, followed by a rapid-growing exponential phase, when intensive culture growth occurs. This phase persists for 16–18 hours. At the end of this exponential phase, sporulation is initiated, and the cells enter the stationary phase. Spores start appearing within the cells and parasporal inclusion bodies of the crystalline toxin. Sporulation is completed within 48–50 hours. This is followed by the lytic/death phase, where the cells undergo lysis, releasing spores and crystalline inclusions into the fermentation liquid. After 60–72 hours, almost 90–98% of spores and inclusion bodies are liberated. The change in pH during this process depends on the composition of the medium used. After sterilizing the fermentation medium, the pH should ideally be 6.8–7.2. During the exponential phase, acids are formed from the saccharides. As a consequence, pH will drop to 5.8–6.0 after 10–12 hours, it will rise again to about 7.5 after 25 hours, further rise is achieved to 8.0 after 30–35 hours, and ultimately this rise stops to 8.8 at 50–60 hours (Vimala Devi, 2018). During the fermentation period, the pH should be maintained between 6.8 and 7.2 by adding 1 N sodium hydroxide/1 N hydrochloric acid (Couch & Jurat-Fuentes, 2014).

15.3.1.3 Fermentation Techniques for Mass Production

After thorough investigation and examination of growth and sporulation phases at the laboratory level, we can go for mass production of Bt on a commercial range with various raw materials. The standard method for the production of microorganisms is fermentation. There are many types of fermentation; the two most common are submerged and solid. Media optimization is vital for mass production in consideration of both vegetative and generative phases (Ravensberg, 2011). Bt consumes sugars, often glucose, fructose, maltose, ribose, molasses, starch, dextrin, wheat flour, and inulin, and produces acid throughout the fermentation process (El-Bendary, 2006). Özkan et al. (2003) investigated several dietary and cultural factors influencing *Bt* var. *israelensis* (Bti) HD500's production of Diptera-specific endotoxins. They claimed that although glucose, starch, and molasses were suppressive, inulin, dextrin, maltose, lactose, sucrose, whey, and glycerol were stimulatory. As the sole nitrogen provider, Bt prefers amino acids, except cysteine and cystine, for growth, rather than inorganic nitrogen sources. Peptone is best for the organic source of nitrogen. Metal ions like K^+, Ca^{2+}, Mg^{2+}, Mn^{2+}, Zn^{2+}, Cu^{2+}, and Fe^{2+} are also required for better sporulation (El-Bendary, 2006).

The important factors for the optimum growth of Bt are pH, temperature, and oxygen. All these parameters can be controlled in fermenters. Hence, Bt is conventionally produced in liquid media in large-capacity industrial fermenters by employing the submerged or deep-tank fermentation method, which is, as the name implies, the growing of microorganisms in a system consisting of only liquid.

There are many advantages of using complete liquid systems, such as the ability to maintain constant temperature and pH, to pump large quantities of air into the system and disperse with the help of stirring impellers, and to generate reasonably homogeneous conditions to maximize the growth of microorganisms. At the beginning of a small-scale production unit, fermenters like seed fermenters of 20–40 L capacity which contain 10–25 L of the nutrient medium can be used. Around 1–3 L of culture are taken out from the shaker or from the laboratory fermenter to serve as the inoculum; this covers 1–3% of the total volume of the medium required for fermentation. These fermentation tanks are made up of stainless steel. An aeration ring is used, along with an agitator or an air outlet under a propeller, to aerate the culture (Vimala Devi, 2018). The air volume used for aeration should be between 0.1 and 1.0 volumes of air per minute per batch volume (Couch & Jurat-Fuentes, 2014). Unless foam-suppressing agents are used, the fermentation liquid will face intensive foaming at the beginning of fermentation and sporulation time (after about 24 hours). Silicone-based de-foaming agents can be used to mitigate this problem. Relatively, solid-state fermentation (SSF) is easy to develop and maintain on a small-scale basis. Scaling them up to the desired sizes is necessary for commercial production, but for SSF, there are numerous problems. Aeration becomes a major difficulty as the volume of the solid mass increases more rapidly than the available surface area. Substrates like wheat (*Triticum* sp.) bran and cotton

(*Gossypium* sp.) seed meal have been employed successfully for multiplication Bt through SSF (Vimala Devi, 2018).

All these processes do vary on the strategies of different companies, which are patented and highly secret. It is safe to assume that commercial companies exercise extreme care during these steps to preserve the insecticidal activity of the EPB (Couch & Jurat-Fuentes, 2014).

Bt var. *kurstaki* isolate can be mass-produced at a low cost using the principle of SSF. This can be successfully undertaken on a solid medium containing wheat bran (150 g), yeast extract (0.72 g), molasses (3.6 g), potassium salts (0.36 g of di-potassium hydrogen orthophosphate and potassium dihydrogen phosphate), and 250 mL distilled water. Autoclaving of the medium (pH 7.2–7.5) is done in polythene covers, and after cooling, it is dispensed into sterile plastic tubs aseptically. In the medium, the seed culture of the isolate (multiplied in nutrient broth) is added, mixed thoroughly, and closed in fully airtight condition with a polythene sheet. In the laminar airflow chamber, the tubs are incubated at 30°C temperature with intermittent aerations for 72 hours. After complete lysis, sterile distilled water is added to the fermented medium and filtered thoroughly using a muslin cloth. The filtrate is centrifuged to remove the water-soluble exotoxin, and the supernatant is discarded. The pellet is mixed with the filtration residue, shade-dried, powdered, and passed through a sieve to get the formulation (Vimala Devi, 2018).

In an experiment on the efficiency of the medium in the mass production of *Bt* var. *tolworthi*, Valicente et al. (2010) continued culturing the inoculum in three different media. Medium 1 was composed of Luria Bertani (LB) plus salts ($FeSO_4$, $ZnSO_4$, $MnSO_4$, $MgSO_4$) and 0.2% glucose; medium 2 contained 1.5% glucose, 0.5% soybean flour, plus salts; and medium 3 was prepared with liquid swine manure at 4% and 0.2% glucose. During the fermentation period, medium 1 showed the highest spore concentration, at 1.4×10^9 spores/mL after 96 hours of fermentation. The efficiency of the Bt produced in all three media was tested against *Spodoptera frugiperda* (J. E. Smith) first instar larvae, and results showed that each product killed above 60% of the population in the highest concentrations.

To make Bt production economically efficient, cheaper raw material is essential. In developing countries like India, Bt biopesticides are produced by utilizing several low-cost raw materials. Several agricultural by-products and waste materials, such as bird feathers and other residues from chicken slaughter house, dried animal blood, fish meal, citrus peels, wheat bran, cornmeal, seeds of dates, silkworm pupal skin, ground nut cake, cane molasses, cotton seed meal, soybean meal, fodder yeast, cheese whey, and corn steep liquor have been documented as alternate sources for Bt production culture medium (El-Bendary, 2006). All these raw materials are nutrient-rich (carbohydrate and proteins). Also, charcoal, plaster of Paris, and fly ash can be used as insecticide carriers. The utilization of fly ash in the formulation of biopesticide can contribute to its proper utilization, thus decreasing environmental pollution. Bti formulation based on fly ash has been evaluated against *Culex* mosquito in field conditions of the natural ecosystem and found efficient to reduce larval population, regardless of habitats (Tikar & Prakash, 2017).

One major aspect influencing the Bt field control impact is formulation. The categories of Bt formulations include emulsified oil, microcapsules, wettable powder, suspending agent, water-dispersible granules, and so on (Li et al., 2017). Lecithin, propionic acid–based adjuvants, siliconized adjuvants, oils, surfactants, and LI can be used as adjuvants of different Bt formulations (dos Santos et al., 2019, 2021). From the 1950s through the 1980s, plate-frame filter technology was used in liquid-state fermentation; nevertheless, it was limited to producing Bt-suspending agents with low-toxicity potency. Continuous centrifugal separation has been used since the 1990s to enhance the procedure and raise the toxicity potency. The toxicity potency of Bt products was markedly enhanced when considering the ultrafiltration condensation and spray-drying technology employed in the medical production process. Furthermore, it is possible to

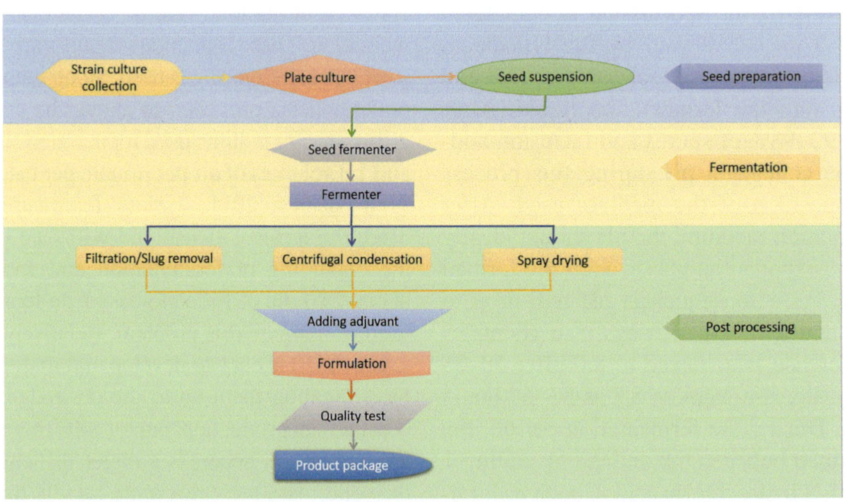

FIGURE 15.1 Mass production protocol of *Bacillus thuringiensis*.

create Bt wettable powder with a high ICP (insecticidal crystal protein) concentration. The powder can be crushed using airflow smashing technology to meet the criteria for manufacturing fineness and formulation processing.

The quality indicators of Bt products at least include (1) the subspecies or strains of fermentation production; effective ingredients, such as parasporal crystals, thuringiensins, spores; and main adjuvant ingredients, and (2) the expression of toxicity potency, such as the number of spores and the toxicity of parasporal crystal and thuringiensin. When a biological assay is performed to determine toxicity, the standard product should be used. Meanwhile, the insects that are used for testing should be specified (Li et al., 2017).

The packaging and labelling of any insecticide are very crucial things, so as for EPB. The packaging material should be good-quality HDPE containers for liquid formulations (Venkatesh, 2020) or LDPE bags for wettable powders (NVBDCP, 2024). There are several types of labelling, like primary labelling, secondary labelling, and big labelling. All the EPBs come under the green label. The label should contain information on the name of the material, name of the manufacturer, spore count, manufacturing date, expiry date, batch number, net quantity, nominal bacterial content (%w/w), minimum cautionary notice, etc. (DPPQS, n.d.). The workflow for the mass production of *Bacillus thuringiensis* bioproducts is depicted in Figure 15.1.

15.3.2 *Lysinibacillus sphaericus*

L. sphaericus (Meyer and Neide) (Ls) lacks relevant biochemical pathways, like the Embden Meyerhoff and hexose monophosphate pathways, thus cannot use sugars as fermentation metabolites. The significant carbon and energy sources used by several strains of Ls to grow were acetate, pyruvate, lactate, glutamate, succinate, histidine, and arginine (El-Bendary, 2006). Most Ls strains cannot use phenylalanine or proline but need arginine, glutamic acid, methionine, threonine, serine, alanine, and lysine for growth and spore germination, with arginine, methionine, and glutamic acid proving to be most effective in germination activation (Shevtsov et al., 1990). Several ions, like potassium ions, phosphate ions, Mn^{2+}, Mg^{2+}, Ca^{2+}, Zn^{2+}, Fe^{2+}, Ni^{2+}, and Cu^{2+}, are essential for the enhancement of toxin production (El-Bendary, 2006).

Except for this, mostly the process is similar to that of Bt. Strains 2362 or C3-41 are the basis of commercial Ls pesticides. Nutrient broth supplemented with 0.05% yeast extract is used for seed inoculum preparation. For special applications of the Ls granules in killing mosquitoes, it is generally packed into water-soluble pouches for application to containers and stormwater catch basins (Couch & Jurat-Fuentes, 2014).

15.3.3 *Paenibacillus popilliae*

As *P. papillae* Dutky does not produce spores in vitro, we have to rely on in vivo techniques only for the mass production of the bacterium. This technique has both its advantages and disadvantages. Start-ups can be started with low capital, as no high technical expertise is required, and the developmental costs are also low. It is easy to scale up when labour cost is low. Here only lies the disadvantage as well that labour is required in high numbers. Also, there is high risk involved for contamination (Ravensberg, 2011).

15.4 MARKET OVERVIEW

Before starting any commercial project, it is highly needed to analyze the market properly. It is regardless to mention that the market demand of EPB is always rising, yet the local and global scenario should be very clear, as a start-up businessman, whether after this lengthy process of R&D, fixed capital investment, processing of products, legal registration, and all much benefit is going to be achieved. Presently, more than 90% of microbial biopesticides on the market are derived from EPB, more specifically from Bt only (Kumar et al., 2021). So the current market situation of microbial bioagents as a whole and precisely for EPB is elucidated in what follows, as both the statistics are very much related to each other.

15.4.1 Market Overview of Microbial Bioagents

As per recent projections, the global biopesticides market is expected to grow at a CAGR of 14.7% from an estimated value of USD 4.3 billion in 2020 to reach USD 8.5 billion by 2025 (DPPQS, 2021). In the US market, around 600 products are available, while the European Union market is struggling to reach the number 100 (Acheuk et al., 2022; Marrone, 2023). The EU is thought to have fewer active biopesticide ingredients registered than the USA, India, Brazil, or China. By the early 2050s, it is anticipated that the usage of biopesticides would be comparable to that of synthetic pesticides; however, there are significant uncertainties concerning the acceptance rates, particularly in regions like Africa and Southeast Asia, which account for the majority of the flexibility in such forecasts (Kumar et al., 2021). As a major developing country of South Asia, we can see India. In India, there are only 14 types of biopesticides registered under the Insecticide Act 1968 at present. A total of 970 biopesticides companies are registered in India under the Central Insecticides Board and Registration Committee (CIBRC). This governing body screens potential biopesticides for biosafety (DPPQS, 2021). The cumulative consumption in the United States, Canada, and Mexico is approximately 45%, inclusive of all globally sold biopesticides, while European Union uses 20%. Globally, the amount of biopesticides is limited to only 4.5% of total pesticides production. In the USA, it is 6%, whereas in India, it is only 3%, according to Subhash Chandra Santra, professor in the Department of Environmental Science at Kalyani University, Kolkata (Padma, 2019). Worldwide, the use of biopesticides increases annually at a rate of 16%, which represents approximately 8% of the world pesticide trade (Fernández-Chapa et al., 2019). In India, the annual growth rate of biopesticide production is 2.5% (Padma, 2019). Although biopesticides contribute only 2.89% of the overall pesticide market in India, their demand is rising steadily in all parts of the country and the whole world.

The increased industrial interest in biopesticides is responsible for changing the global scenario featured with strategic partnerships, mergers, and acquisitions of companies. Significant examples of this trend are the acquisitions of the US-based biological company AgraQuest and the German company Prophyta GmbH by Bayer CropScience. Similarly, Becker Underwood Inc. was acquired by BASF, while Syngenta agreed to acquire the US-based biotechnology company Pasteuria Bioscience Inc. and the Swiss Devgen. On the other side, Monsanto and Novozymes have established a historical BioAg Alliance committed to discover and develop new microbial solutions for agriculture (Ruiu, 2015).

15.4.2 Market Overview of EPB

Among the 14 biopesticides registered under the Insecticide Act 1968, at present in India, three strains of Bt, namely, *israelensis*, *kurstaki*, and *galleriae*, and Ls are mass-produced for utilization as EPB (DPPQS, 2021). Around 78 EPB formulations, which are registered in India under the Central Insecticides Board and Registration Committee (CIBRC), are represented in Table 15.2. It is needless to

TABLE 15.2
Entomopathogenic Bacterial Formulations Available in India

Sl. No.	Bacterial Agent	Company	Formulation	Section
1.	Bt	M/s Sandoz (I) Ltd., Bombay	—	9(3B)
2.	Bt	M/s Lupin Agrochemical India Ltd., Bombay	—	9(3B)
3.	Ls	M/s Biotech International Ltd., New Delhi	Tech and 1.15%WP	9(3B)
4.	Bti	M/s Biotech International Ltd., New Delhi	Tech and WP	9(3B)
5.	Ls, Bt	M/s Tuticorin Alkali Chemical & Fertilizers Ltd., Madras	Tech	9(3B)
6.	Bt	M/s Lupin Agrochemicals (I) Ltd., Bombay From M/s Abbott Laboratories, USA & Australia	Formulation	9(3B)
7.	Bti	M/s Biotech International Ltd., New Delhi	Tech	9(3B)
8.	Ls	M/s Biotech International Ltd., New Delhi	Tech	9(3B)
9.	Btk HPWP	M/s HIL, New Delhi	—	9(3B)
10.	Btk HPWP	M/s Rallis India Ltd., Bangalore	—	9(3B)
11.	Bt (BIOBIT)	M/s Rallis India Ltd., Bangalore	Biobit II WP	9(3B)
12.	Btk HPWP	M/s Rallis India Ltd., Bangalore from M/s Nova Nordisk, Denmark	—	9(3B)
13.	Btk	M/s HIL.	—	9(3B)
14.	Btk	M/s Sandoz (i) Ltd.,	Delfin WG	9(3)
15.	Btg	M/s Tuticorin alkali chemicals & fertilizers Ltd.	Tech	9(3)
16.	Ls (strain 1543M)	M/s Tuticorin alkali chemicals & fertilizers Ltd.	Tech	9(3)
17.	Bti	M/s Biotech International Ltd.	Tech (I) 1.15% WP	9(3)
18.	Ls	M/s Biotech International Ltd.	Tech (I) 1.15% WP	9(3)
19.	Btk	M/s Wock lard (P) Ltd.,	150% WP	9(B)
20.	Btk	M/s Wock lard (P) Ltd.,	Tech.	9(3)
21.	Bt	M/s Wockhardt Ltd.	Tech	9(3)
22.	Bti	M/s HIL., New Delhi	—	9(3B)
23.	Btk	M/s Lupin Agro Chemical India Ltd., Pameli	For port	9(3)
24.	Bti	M/s Hoechest Schering AgrEvo, Mumbai	Suspension Formulation	9(3B)
25.	Bti	Biotech International Ltd.	—	9(3)
26.	Bti	M/s Biotech International Ltd., New Delhi	—	9(3) port
27.	Btk	M/s Ajay Biotech (India) Ltd. Pune	—	9(3b)
28.	Bti (strain VCRC B-17)	M/s Tuticorin Alkali Chemical & Fertilizer Ltd., Chennai	Tech, containing 20% of delta endotoxin	9(3B)
29.	Bti	M/s Aventis Crop Science India Ltd., Mumbai	Formulation for port	9(3B)
30.	Bti	M/s Tuticorin Alkali Chemicals & Fertilizers Ltd., Chennai	Tech	9(3)
31.	Bti	M/s tuticorin Alkali chem & fertilizers, Chennai	Bti (AS)	9(3)
32.	Bti	M/s Tuticorin Alkali Chemicals & Fertilisers Ltd., Chennai	—	9(3B)
33.	Btk	M/s Tuticorin Alkali Chem & Fertilizers, Chennai	Tech and 2.5% AS	9(3B)
34.	Bti	M/s Aventis Crop sci. Ltd.,	Vectobac 12 AS	9(3)
35.	Btk	Ajay Biotech Lited	7.5% WP	9(3B)
36.	Btk	Ajay Biotech Lited	7.5% WP	9(3B)

(Continued)

TABLE 15.2 (*Continued*)
Entomopathogenic Bacterial Formulations Available in India

Sl. No.	Bacterial Agent	Company	Formulation	Section
37.	Bti	M/s R.K. Biotech Pvt. Lited	5% WP	9(3B)
38.	Bti	M/s Wockhardt Life Science Ltd.	WP	9(3)
39.	Btk	M/s Biotech International Ltd., New Delhi for of changed name of the foreign manufacturer in their Registration Certificates of Btk & Bti formulation and Technical.	—	endorsement
40.	Btk	Directorate of Oilseeds Research, Hyderabad	0.5% WP	9(3B)
41.	Bti	M/s R.K. Biotech Pvt. Ltd.	5% AS	9(3)
42.	Bti	M/s Bacto Power India Pvt. Ltd	5 AS	9(3B)
43.	Btk	M/s Multiplex Bio-tech Pvt. Ltd., Bangalore	5% WP	9(3B)
44.	Btk	M/s Indore Biotech Inputs & Research P, Ltd.	0.5% WP	9(3B)
45.	Btk (Strain: DOR bt-1)	M/s Sri Biotech, Hyderabad	0.5% WP	9(3B)
46.	Btk	M/s Microplex Biotech & Agrochem Pvt. Ltd., Wardha	0.5% WP	9(3B)
47.	Btk (Strain DOR bt-1)	M/s Varsha Bioscience & Technology, Hyderabad	0.5%WP	9(3B)
48.	Btk	M/s Kan Biosys Pvt. Ltd., Pune	0.5% WP	9(3B)
49.	Btk	M/s Shree Shiva Bio-Tech.	0.5% WP	9(3B)
50.	Btk	M/s Neelagriva Biosciences Pvt. Ltd., Hyderabad	0.5% WP	9(3B)
51.	Btk (Strain: DOR bt-1)	M/s Shri Ram Solvent Extractions Pvt. Ltd., Jaspur	0.5% WP	9(3B)
52.	Btk (Strain: DOR bt-1)	M/s Surya Bio Products, Eluru (AP)	0.5% WP	9(3B)
53.	Btk (Strain: DOR Bt-1)	M/s. Neesa Agritech Pvt. Ltd., Ahmedabad	0.5% WP	9(3B)
54.	Btk (Strain: DOR Bt-1)	M/s. Gujarat Eco Microbial Technologies	0.5% WP	9(3B)
55.	Btk (Strain: DOR Bt-1)	M/s. Prathibha Biotech, Hyderabad	0.5% WP	9(3B)
56.	Btk	M/s Amit Biotech Pvt. Ltd., Kolkata	0.5% WP	9(3B)
57.	Btk	M/s Junna LifeSciences Pvt. Ltd., Hyderabad	0.5% WP	9(3B)
58.	Bti	M/s. Amit Biotech (P) Ltd., Kolkota	5% AS	9(3B)
59.	Bti	M/s. Neelgriva Bioscience	5.0% AS (potency 4100 ITU)	9(3B)
60.	Btk (strain DOR Bt-I)	M/s Bio-Control Laboratory, Varanasi	0.5% WP	9(3B)
61.	Btk (strain DOR Bt-1)	M/s Poshak Bio research (P) Ltd., Gujarat	0.5% WP	9(3B)
62.	Btk (strain DOR Bt-1)	M/s Agro Biotech Research Centre, Kottoyam	0.5% WP	9(3B)
63.	Btk (strain DOR Bt-1)	M/s Ruchi Biochemicals, Goregaon, Mumbai	0.5% WP	9(3B)
64.	Btk (strain DOR Bt-1)	M/s Hindustan Bioenergy Ltd., Lucknow	0.5% WP	9(3B)
65.	Btk (strain DOR Bt-1)	M/s Nitapol Industries, Kolkata	0.5% WP	9(3B)
66.	Btk (strain DOR Bt-1)	M/s Bharat Biocon Pvt. Ltd., New Delhi	0.5% WP	9(3B)
67.	Btk	M/s Viswa Mitra Bio Agro (P) Ltd., Guntur	0.5% WP	9(3B)
68.	Btk	M/s Devi Biotech Pvt. Ltd., Madurai	0.5% WP	9(3B)
69.	Btk (strain DOR-Bt-I)	M/s Kaveri Seed Company Ltd., Secunderabad	0.5% WP	9(3B)
70.	Btk	M/s Krishi Bio-Product & Research Pvt. Ltd., Indore	0.5% WP	9(3B)
71.	Btk	M/s Gujarat Chemicals & Fertilizers Trading Company	0.5% WP	9(3B)
72.	Btk	M/s Krishi Vikas Sahakari Samiti Ltd., Jaipur	0.5% WP	9(3B)
73.	Btk (accession no. NACC-B-01118)	M/s Chaitra Agri Organics, Mysore	0.5% WP	9(3B)
74.	Btk (strain DOR-Bt-1) (accession no. NACC-B-01118).	M/s Ajay Bio-Tech (India) Ltd., Pune (MS)	0.5% WP	9(3B)
75.	Btk (strain DOR-Bt-1) (accession no. NACC-B-01118).	M/s International Panaacea Ltd., New Delhi	0.5% WP	9(3B)
76.	Bti (strain: VCRC, B-17) (accession no. MTCC-5596).	M/s Kilpest India Ltd., Bhopal	5.0% AS	9(3B)
77.	Bti	M/s Fine Trap India, Yavatmal	5% AS	9(3B)
78.	Bti (strain designation ABIL (accession no. NACC-B-01318)	M/s Ajay Biotech (India) Ltd., Pune MS.	5% WP	9(3)

Note: Bt: *Bacillus thuringiensis*; Ls: *Lysinibacillus sphaericus*; Bti: *Bt* var. *israelensis*; Btk: *Bt* var. *kurstaki*; Btg: *Bt* var. *galleriae* (Source: DPPQS, 2021) mention that Bt is one of the most promising biopesticides, representing around 70% of the global biopesticide market (Tikar & Prakash, 2017).

Bt biopesticide market is dominated by Abbott Laboratories (Chicago, IL) (since the acquisition in 1995 of Novo-Nordisk's biopesticide business) and Novartis (created through the merger in 1996 of Ciba and Sandoz); these together account for >70% of global production. The other 30% is contributed approximately by 30 companies with more than 100 Bt products, most of them containing a single Bt toxin, but some may have combinations of up to 5 (Sanahuja et al., 2011).

15.4.3 Problems and Scope of EPB Market

The higher cost of registration of biopesticides and the long-drawn-out process deter companies from engaging in research and development on biological pesticides in India. Another limitation is the requirement to verify the microorganism for biosafety before registration, multiplication, and marketing of biopesticides. Many universities and research institutions that conduct the initial R&D on biopesticides are mostly unable to bear the additional costs required for extensive safety and allergic tests (Padma, 2019).

The key active ingredients used to produce biopesticides are imported from China, and other key manufacturing companies are in Europe and North America. During the crisis of COVID-19, there was low availability of raw materials with the progress of the cropping season. This created a scarcity of products in even established markets, such as the United States, Brazil, Spain, and Italy. Rather, this created a huge scope for utilization of diverse raw materials as local source for the production of biopesticides. And in post-COVID scenario, the dependency on different countries for import is rather reducing. Also, there was a decline in the availability of active chemical ingredients, which also led to a scarcity of chemical products in the market, ultimately encouraging the farmers to adopt an alternative solution for control of pests. This situation also boosted the demand for biopesticides with the increasing health consciousness and demand for healthy and organic food in the human society (Markets and Markets, 2023). As a result, there is need of huge revolution in research on biopesticides and EPB and discoveries are expected shortly.

15.5 CONCLUSION

The growing health hazards and resulting awareness are increasing the demands of eco-friendly approaches to pest management. Though R&D is showing EPB as a very promising approach, there is difference between experimental productions in lab conditions and mass-level commercial productions. The last one is really challenging, and it is very much needed by a wise businessman to thoroughly understand the protocol as well as the market situation to gain maximum profit. This chapter presents a crisp and clear picture of the current entrepreneurs along with the interested would-be ones to encourage them to join in this commercial forum and explore more of these environment-friendly and market-friendly pesticides.

REFERENCES

Acheuk, F., Basiouni, S., Shehata, A. A., Dick, K., Hajri, H., Lasram, S., Yilmaz, M., Emekci, M., Tsiamis, G., Spona-Friedl, M., May-Simera, H., Eisenreich, W., & Ntougias, S. (2022). Status and prospects of botanical biopesticides in Europe and Mediterranean countries. *Biomolecules, 12*(2), 311. https://doi.org/10.3390/BIOM12020311/S1

Agrichem. (n.d.). *Cordalene®*. Retrieved August 23, 2021, from https://agrichembio.com/en/producto/cordalene

Couch, T. L., & Jurat-Fuentes, J. L. (2014). Commercial production of entomopathogenic bacteria. In J. A. Morales-Ramos, M. G. Rojas, & D. I. Shapiro-Ilan (Eds.), *Mass production of beneficial organisms: Invertebrates and entomopathogens* (1st ed., pp. 415–435). Academic Press. https://doi.org/10.1016/B978-0-12-391453-8.00012-1

dos Santos, C. A. M., da Silva Santos, R. T., Della'Vechia, J. F., Griesang, F., Polanczyk, R. A., & Ferreira, M. da C. (2019). Effect of addition of adjuvants on physical and chemical characteristics of Bt bioinsecticide mixture. *Scientific Reports, 9*, 12525. https://doi.org/10.1038/s41598-019-48939-y

dos Santos, C. A. M., do Nascimento, J., Gonçalves, K. C., Smaniotto, G., de Freitas Zechin, L., da Costa Ferreira, M., & Polanczyk, R. A. (2021). Compatibility of Bt biopesticides and adjuvants for *Spodoptera frugiperda* control. *Scientific Reports, 11*, 5271. https://doi.org/10.1038/s41598-021-84871-w

DPPQS. (2021). *Bio-pesticide registrant*. Directorate of Plant Protection, Quarantine & Storage of Government of India. http://ppqs.gov.in/divisions/cib-rc/bio-pesticide-registrant

DPPQS. (n.d.). *Bacillus thuringiensis var. israelensis packaging*. DPPQS (pp. 1–13). https://ppqs.gov.in/sites/default/files/bacillus_thuringiensis_var.israelensis_5_as_vcrc-b-17_mtcc-5596_1.pdf

El-Bendary, M. A. (2006). *Bacillus thuringiensis* and *Bacillus sphaericus* biopesticides production. *Journal of Basic Microbiology, 46*(2), 158–170. https://doi.org/10.1002/jobm.200510585

Fernández-Chapa, D., Ramírez-Villalobos, J., & Galán-Wong, L. (2019). Toxic potential of *Bacillus thuringiensis*: An overview. In Y. Jia (Ed.), *Protecting rice grains in the post-genomic era* (pp. 1–22). IntechOpen. https://doi.org/10.5772/intechopen.85756

Hazarika, L. K., Bhattacharrya, B., Kalita, S., & Das, P. (2005). Bt as a biocide and its role in management of tea pests. *International Journal of Tea Science, 4*(3 & 4), 7–16.

Johnson, T. B., & Yancy, J. (1997). CRYMAX. *Proceedings of the Beltwide Cotton Conference, 1*, 54–55.

Johnson, V. W., Pearson, J., & Jackson, T. A. (2001). Formulation of *Serratia entomophila* for biological control of grass grub. *New Zealand Plant Protection, 54*, 125–127. https://doi.org/10.30843/NZPP.2001.54.3752

Khachatourians, G. G. (2019). Insecticides, microbial. In *Reference module in life sciences*. Elsevier. https://doi.org/10.1016/b978-0-12-809633-8.13066-3

Kumar, J., Ramlal, A., Mallick, D., & Mishra, V. (2021). An overview of some biopesticides and their importance in plant protection for commercial acceptance. *Plants, 10*(6), 1185. https://doi.org/10.3390/plants10061185

Ladurner, E., Benuzzi, M., & Franceschini, S. (2008). *Btk strain EG 2348: A valuable tool for the control of a wide range of Lepidopteran pests*. ABIM. https://www.abim.ch/fileadmin/abim/documents/presentations2008/session4/3_Ladurner_ABIM-2008.pdf

Levin, D. B. (2009). Human health effects resulting from exposure to *Bacillus thuringiensis* applied during insect control programmes. In A. E. Hajek, T. R. Glare, & M. O'Callaghan (Eds.), *Use of microbes for control and eradication of invasive arthropods* (pp. 291–303). Springer. https://doi.org/10.1007/978-1-4020-8560-4_16

Li, L., Chen, Z., & Yu, Z. (2017). Mass production, application and market development of *Bacillus thuringiensis* biopesticides in China. In L. M. Fiuza, R. Polanczyk, & N. Crickmore (Eds.), *Bacillus thuringiensis and Lysinibacillus sphaericus* (pp. 185–212). Springer. https://doi.org/10.1007/978-3-319-56678-8_12

Markets and Markets. (2023). *Biopesticides market statistics, growth dynamics and future trends | COVID-19 impact on biopesticides market*. Retrieved August 28, 2021, from https://www.marketsandmarkets.com/Market-Reports/biopesticides-267.html

Marrone, P. G. (2023). Status of the biopesticide market and prospects for new bioherbicides. *Pest Management Science*, *80*(1), 81–86. https://doi.org/10.1002/ps.7403

Monterey. (n.d.). *Monterey Bt. RTU (ready-to-use)*. Retrieved July 1, 2023, from https://www.montereylawngarden.com/product/monterey-b-t-rtu/?attachment_id=918&download_file=5b96aa340995e

NVBDCP. (2024). *Technical specification of (Bti 5% WP) as larvicide|DTE, national vector borne disease control programme* (pp. 1–5). Government of Madhya Pradesh Supply. https://mpphscl.mp.gov.in/Files/PDF/107fc729-2218-47ef-9475-33ca9162b29c_1_ItemListwithSpecificationT-590.pdf

Özkan, M., Dilek, F. B., Yetis, Ü., & Özcengiz, G. (2003). Nutritional and cultural parameters influencing antidipteran delta-endotoxin production. *Research in Microbiology*, *154*(1), 49–53. https://doi.org/10.1016/S0923-2508(02)00006-2

Padma, T. V. (2019, July 9). *What is preventing the widespread adoption of biopesticides in India?* Mongabay. Retrieved July 1, 2023, from https://india.mongabay.com/2019/07/what-is-preventing-the-widespread-adoption-of-biopesticides-in-india/

Prasad, T. V. (2019). Biological control. In *Handbook of entomology* (4th ed., pp. 178–234). New Vishal's Publication.

Ravensberg, W. J. (2011). Selection of a microbial pest control agent. In W. J. Ravensberg (Ed.), *A roadmap to the successful development and commercialization of microbial pest control products for control of arthropods* (Vol. 10, pp. 23–57). Springer. https://doi.org/10.1007/978-94-007-0437-4_2

Ruiu, L. (2015). Insect pathogenic bacteria in integrated pest management. *Insects*, *6*(2), 352–367. https://doi.org/10.3390/insects6020352

Ruiu, L. (2018). Microbial biopesticides in agroecosystems. *Agronomy*, *8*(11), 235. https://doi.org/10.3390/agronomy8110235

Sabbahi, R., Hock, V., Azzaoui, K., Saoiabi, S., & Hammouti, B. (2022). A global perspective of entomopathogens as microbial biocontrol agents of insect pests. *Journal of Agriculture and Food Research*, *10*, 100376. https://doi.org/10.1016/J.JAFR.2022.100376

Salgado, V. L., & Sparks, T. C. (2005). The spinosyns: Chemistry, biochemistry, mode of action, and resistance. In L. I. Gilbert, & S. S. Gill (Eds.), *Insect control: Biological and synthetic agents* (pp. 137–173). Academic Press. https://doi.org/10.1016/B0-44-451924-6/00078-8

Sanahuja, G., Banakar, R., Twyman, R. M., Capell, T., & Christou, P. (2011). *Bacillus thuringiensis*: A century of research, development and commercial applications. *Plant Biotechnology Journal*, *9*(3), 283–300. https://doi.org/10.1111/J.1467-7652.2011.00595.X

Seven Springs Farm Supply. (n.d.). *Deliver BT-1 lb bag*. Retrieved August 23, 2021, from https://www.7springsfarm.com/products/deliver-bt-1-lb-bag

Shevtsov, V. V., Petrova, T. M., Khovrychev, M. P., Letunova, E. V., & Shevtsov, V. V. (1990). Faktory rosta i prorastaniia spor razlichnykh shtammov *Bacillus sphaericus* [Growth and spore germination factors of various strains of *Bacillus sphaericus*]. *Mikrobiologiia*, *59*(3), 453–459.

Silva Filha, M. H. N. L., Berry, C., & Regis, L. (2014). *Lysinibacillus sphaericus*: Toxins and mode of action, applications for mosquito control and resistance management. In T. S. Dhadialla, & S. S. Gill (Eds.), *Advances in insect physiology* (Vol. 47, pp. 89–176). Academic Press. https://doi.org/10.1016/B978-0-12-800197-4.00003-8

Tikar, S., & Prakash, S. (2017). Fly ash-based *Bacillus thuringiensis israelensis* formulation: An ecofriendly approach. *Indian Journal of Medical Research*, *146*(6), 680. https://doi.org/10.4103/ijmr.IJMR_1679_16

Valicente, F. H., De Souza Tuelher, E., Isabella, M., Leite, S., Freire, F. L., & Vieira, C. M. (2010). Production of *Bacillus thuringiensis* biopesticide using commercial lab medium and agricultural by-products as nutirent sources. *Revista Brasileira de Milho e Sorgo*, *9*(1), 1–11. https://doi.org/10.18512/1980-6477/rbms.v9n01p1-11

Venkatesh. (2020). *VectoGuardTM*. Agri Life. https://agrilife.in/products/microbial_vectoguard.php

Vimala Devi, P. S. (2018). Mass production of *Bacillus thuringiensis* and *Nomuraea rileyi*. In C. Chattopadhyay, R. K. Tanwar, M. Sehgal, A. Birah, S. Bhagat, N. Ahmad, & N. Mehta (Eds.), *Handbook of integrated pest management* (pp. 353–357). Indian Council of Agricultural Research.

Zaki, A. M., Zaki, A. H., Farghali, A. A., & Abdel-Rahim, E. F. (2017). Sodium titanate-Bacillus as a new nanopesticide for cotton leaf-worm. *Journal of Pure and Applied Microbiology*, *11*(2), 725–732. https://doi.org/10.22207/JPAM.11.2.11

16 Ecological and Economical Analysis of Entomopathogenic Bacteria

ABSTRACT

Entomopathogenic bacteria (EPB) play a noteworthy role in the promotion of ecosystem services. The utilization of EPB has the potential to regulate insect pests, thereby leading to a rise in crop productivity and a decrease in the reliance on chemical pesticides. Provisioning services include the potential enhancement of the output of food and other resources. In addition, EPB can improve ecosystem-supporting services, which include soil formation and nutrient cycling, by breaking down insect corpses and releasing nutrients into the soil thereafter. By reducing the use of chemical pesticides which have been shown to negatively impact the environment and human health, EPB has the potential to improve cultural services as well. In cost–benefit analysis, it has been found that the application of EPB and other biocontrol agents can result in almost double income over investment. In developing countries like India, there is a constant rise in the demand for biocontrol products, which ensures a better scope. Around 78 commercial formulations of entomopathogenic bacteria are registered in India. The incorporation of biocontrol agents (EPB as a component) in several IPM programmes gave phenomenal success when implemented but is now facing challenges that require further research.

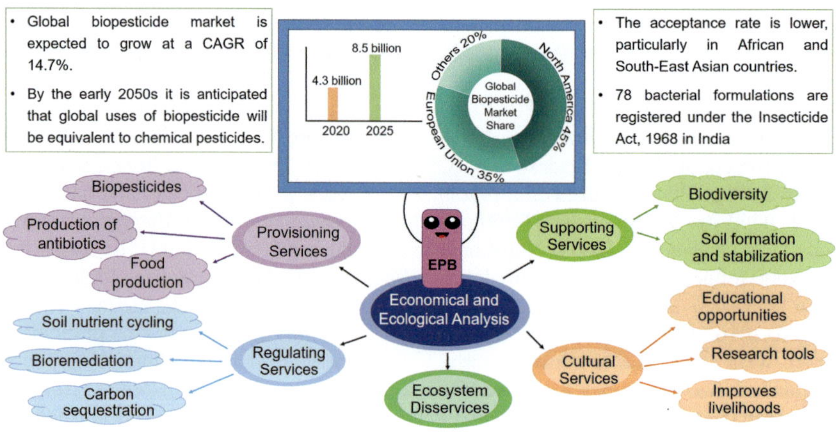

16.1 INTRODUCTION

The earlier chapters have enlightened us regarding the need for entomopathogenic bacteria, their history of discovery to date, different EPBs, their selection criteria for commercialization, and other market details, but the final decision always depends on cost efficiency. How much benefit we get against the cost incurred is very vital from an investor's point of view. This chapter will discuss the benefits and costs from both ecological and economical perspectives. Compared to other modules, EPB, singly or along with other biocontrol agents, serves the ecosystem with several types of services and gives a margin of profit to the farmers. Also, certain disservices cannot be neglected. Similarly, while scrutinizing the market status of EPB, we can see a lot of scopes; still, it is not without thorns. The services and benefits of EPBs, and the associated problems, are discussed in the chapter, along with successful case studies of EPB utilization in pest management.

16.2 ECOSYSTEM SERVICES FROM ENTOMOPATHOGENIC BACTERIA

The potential of EPB to regulate pest populations in different agroecosystems has been widely acknowledged. The application of EPB has resulted in enhanced crop productivity and reduced reliance on chemical pesticides (Sharma et al., 2019). The benefits of EPB extend beyond pest control and encompass a range of ecosystem services, such as provisioning, regulating, supporting, and cultural services. The aforementioned services hold significant importance in preserving the sustainability of agroecosystems and augmenting their ability to adapt to alterations in the environment. In contemporary times, the notion of ecosystem services has garnered considerable academic and research interest, as it underscores the critical role of ecological systems in furnishing advantages to human civilization. Ecosystem services refer to the advantages that ecosystems offer to both humans and other living organisms. The

DOI: 10.1201/9781003426882-16

classification of ecosystem services comprises four main categories, namely, provisioning services, regulating services, supporting services, and cultural services (Reid et al., 2005). The category of provisioning services encompasses the generation of sustenance, potable water, and additional commodities (Power, 2010). Ecosystem services that involve the regulation of climate, purification of water, and control of pests are commonly referred to as *regulating services* (Plaas et al., 2019). The provision of nutrient cycling and soil formation are among the supporting services (Kassam et al., 2022). *Cultural services* encompass a range of activities that provide individuals with opportunities for leisure, learning, and spiritual fulfilment (Baumber et al., 2019). *Ecosystem disservices* refer to the adverse effects that result from the processes and functions of an ecosystem (Daily, 1997).

The augmentation of food and other resources production can potentially enhance the provisioning of services. Entomopathogenic bacteria have the potential to regulate insect populations, thereby enhancing regulating services, such as pest control, pollination, and disease regulation (Azizoglu et al., 2020). Furthermore, EPB has the potential to facilitate the provision of supporting ecosystem services, such as nutrient cycling and soil formation, through the process of decomposing insect carcasses and subsequently releasing nutrients into the soil. EPB has the potential to make a significant contribution to cultural services through the reduction of chemical pesticide usage, which has been known to have adverse effects on both human health and the environment. In general, it is evident that entomopathogenic bacteria possess the potential to exert a substantial influence on the delivery of ecosystem services, encompassing both advantageous and disadvantageous effects. Although significant research has been conducted on the potential of entomopathogenic nematodes as biopesticides and biological control agents, their potential contribution to soil nutrient cycling and carbon sequestration warrants further investigation (Sharma et al., 2019). Nonetheless, akin to any technological advancement, the utilization of entomopathogenic bacteria also entails plausible hazards and apprehensions. The aforementioned factors encompass the potential for inadvertent impacts on non-targeted organisms, the emergence of resistance, and the peril of accidental discharge into the ecosystem. The aforementioned services hold significance in preserving the sustainability of agroecosystems and augmenting their ability to adapt to alterations in the environment.

Enhancing the output of food and other resources can potentially facilitate the provisioning of services (Sharma et al., 2019). Furthermore, EPB has the potential to enhance ecosystem services, such as nutrient cycling and soil formation, through the process of decomposing insect carcasses and subsequently releasing nutrients into the soil. The provision of these services is often a key component of social welfare programmes and public policy initiatives aimed at promoting social and economic development.

16.2.1 Provisioning Services by Entomopathogenic Bacteria

The term 'provisioning services' pertains to the advantages that individuals obtain from ecosystems in the form of commodities and merchandise. The enhancement of crop productivity and quality through the regulation of pest populations is a potential contribution of EPB to provisioning services (Vandenberghe et al., 2021). Biopesticides that are based on EPB have the potential to effectively manage a diverse array of pests, such as lepidopteran larvae, coleopteran larvae, dipteran larvae, and a variety of soil-dwelling pests. The outcome of this phenomenon can be an increase in agricultural productivity and enhanced produce standards, thereby contributing to the amelioration of food insecurity and augmenting the revenue of farmers (Ali et al., 2022).

16.2.2 Regulating Services by Entomopathogenic Bacteria

Regulating services pertain to the advantages that ecosystems offer by regulating crucial ecological processes and functions. The implementation of an ecosystem-based approach can potentially serve as a means of regulating services through the management of pest populations, the mitigation of chemical pesticide usage, and the preservation of natural ecosystem equilibria (da Silva et al., 2020). Biopesticides based on EPB have the potential to mitigate the adverse effects of chemical pesticides on non-target organisms and promote the biodiversity of agroecosystems (Rao & Jurat-Fuentes, 2020). Moreover, the implementation of EPB has the potential to regulate nutrient cycling and carbon sequestration through the facilitation of beneficial soil microorganisms and the improvement of soil fertility.

16.2.3 Supporting Services by Entomopathogenic Bacteria

Supporting services are defined as the advantages that ecosystems offer by facilitating crucial ecological processes and functions. The utilization of EPB has the potential to bolster support services through the improvement of soil health and fertility, facilitation of plant growth and development, and optimization of soil water-holding capacity (Sharma et al., 2019). The utilization of EPB has the potential to aid in the stabilization of soils, thereby mitigating the occurrence of erosion and minimizing the likelihood of soil degradation.

16.2.4 Cultural Services by Entomopathogenic Bacteria

Cultural services pertain to intangible advantages that ecosystems offer, encompassing cultural, spiritual, and aesthetic values. The promotion of sustainable and eco-friendly pest control methods, the enhancement of agroecosystem biodiversity, and the preservation of traditional ecological knowledge are potential ways in which EPB can contribute to cultural services (Noureldeen et al., 2022). EPB may indirectly contribute to cultural services by supporting eco-friendly agriculture and health-conscious practices (Sajnaga & Kazimierczak, 2020).

16.3 ECOSYSTEM DISSERVICES BY ENTOMOPATHOGENIC BACTERIA

The term *ecosystem disservice* pertains to the adverse effects that ecosystems may have on the welfare of human beings. The implementation of EPB has the potential to result in ecosystem disservices as it may cause harm to non-target organisms, such as beneficial insects and soil-dwelling organisms. Moreover, biopesticides based on EPB have the potential to cause inadvertent discharge, which may result in the pollution of non-target environments (Sandhi & Reddy, 2019; Santhana Bharathi et al., 2019). The details of different ecosystem services and disservices of EPB are represented with examples in Table 16.1.

16.4 ECONOMICAL ANALYSIS OF ENTOMOPATHOGENIC BACTERIA

Before starting any commercial project, it is highly needed to analyze the market properly. It is regardless to mention that the market demand for entomopathogenic bacteria is always rising, yet the local and global scenario should be very clear as a start-up businessman; after the lengthy process of R&D, fixed capital investment, processing of products, legal registration, and all, how much benefit is going to be achieved? Presently, more than 90% of microbial biopesticides on the market are derived from entomopathogenic bacteria, more specifically from *Bacillus thuringiensis* or Bt only (Kumar et al., 2021). So the current market situation of microbial bioagents as a whole, and precisely for entomopathogenic bacteria, is elucidated in what follows, as both the statistics are very much related to each other.

16.4.1 Market Overview of Microbial Bioagents

As per recent projections, the global biopesticides market is expected to grow at a CAGR of 14.7% from an estimated value of USD 4.3 billion in 2020 to reach USD 8.5 billion by 2025 (DPPQS, 2021). In the US market, around 600 products are available, while the European Union market is struggling to reach the number 100 (Acheuk et al., 2022; Marrone, 2023). The EU is thought to have fewer active biopesticide ingredients registered than the USA, India, Brazil, or China. By the early 2050s, it is anticipated that the usage of biopesticides would be comparable to that of synthetic pesticides; however, there are significant uncertainties concerning the acceptance rates, particularly in regions like Africa and Southeast Asia, which account for the majority of the flexibility in such forecasts (Kumar et al., 2021). As a major developing country of South Asia, we can see India. In India, there are only 14 types of biopesticides registered under the Insecticide Act 1968 at present. A total of 970 biopesticides companies are registered in India under the Central Insecticides Board and Registration Committee (CIBRC). This governing body screens potential biopesticides for biosafety (DPPQS, 2021). The cumulative consumption in the United States, Canada, and Mexico is approximately 45%, inclusive of all globally sold biopesticides, while European Union uses 20%. Globally, the amount of biopesticides is limited to only 4.5% of total pesticides production. In the USA, it is 6%, whereas in India, it is only 3%, according to Subhash Chandra Santra, professor in the Department of Environmental Science at Kalyani University, Kolkata (Padma, 2019). Worldwide, the use of biopesticides increases annually at a rate of 16%, which represents approximately 8% of the world pesticide trade (Fernández-Chapa et al., 2019). In India, the annual growth rate of biopesticide production is 2.5% (Padma, 2019). Although biopesticides contribute only 2.89% of the overall pesticide market in India, their demand is rising steadily in all parts of the country and the whole world.

The increased industrial interest in biopesticides is responsible for changing the global scenario featured with strategic partnerships, mergers, and acquisitions of companies. Significant examples of this trend are the acquisitions of the US-based biological company AgraQuest and the German company Prophyta GmbH by Bayer CropScience. Similarly, Becker Underwood Inc. was acquired by BASF, while Syngenta agreed to acquire the US-based biotechnology company Pasteuria Bioscience Inc. and the Swiss Devgen. On the other side, Monsanto and Novozymes have established a historical BioAg Alliance committed to discover and develop new microbial solutions for agriculture (Ruiu, 2015).

The market for biological pest control is still developing, and there is limited large-scale demand compared to conventional pesticides. This can lead to higher per-unit production costs, though costs may decrease as demand grows and manufacturing processes improve. Let us take India as a developing country to understand the present scenario. As the major biopesticides consumed in the country as well as globally for pest management are EPB products (Tikar & Prakash, 2017) and we do not have accurate statistical data for the latter, we can compare the biopesticide scenario to the EPB market. In Figure 16.1, we can see a clear uprising trendline of the demand for biopesticides. Still, compared to that, the trendline of the consumption of biopesticides is almost parallel over the years (DPPQS, 2024a, 2024b). This is a good indication that there is a need for a greater number of biopesticide-producing farms for more quantity and quality production, which would be very helpful in bringing down the cost of biopesticides and making them accessible to a greater number of small and marginal farmers of the country.

TABLE 16.1
Ecosystem Services and Disservices from EPB with Suitable Examples

Ecosystem Services	Service Type	Description	Some Examples	References
Provisioning services	Biopesticides	The utilization of entomopathogenic bacteria as a substitute for chemical pesticides presents a viable and environmentally conscious approach to managing pests.	*Bacillus thuringiensis* (Bt) *Bacillus subtilis* *Pseudomonas fluorescens* *Serratia marcescens* *Xenorhabdus* spp.	(Rao & Jurat-Fuentes, 2020)
	Production of antibiotics	Certain types of entomopathogenic bacteria, such as *Streptomyces*, have the ability to synthesize antibiotics that possess therapeutic potential for the treatment of bacterial infections in both human and animal populations.	*Streptomyces* spp. *Bacillus subtilis* *Pseudomonas fluorescens* *Serratia marcescens* *Xenorhabdus nematophila* *Photorhabdus luminescens* *Paenibacillus polymyxa* *Lysobacter* spp.	(Zakalyukina et al., 2022)
	Food production	EPB controls harmful pests, which indirectly helps in improving crop yields.	*Bacillus thuringiensis* (Bt) *Bacillus popilliae* (Bp) *Bacillus sphaericus* (Bs) *Xenorhabdus* spp. *Serratia entomophila* *Pseudomonas fluorescens* *Paenibacillus* spp.	(Ali et al., 2022)
	Biological control of pests	In order to manage pests that harm crops and plants, entomopathogenic bacteria can be used as biopesticides, which eliminates the need for chemical pesticides.	*Bacillus thuringiensis* (Bt) *Paenibacillus popilliae* *Lysinibacillus sphaericus* *Xenorhabdus* spp. *Serratia entomophila* *Pseudomonas fluorescens*	(Wang & Zhao, 2022)
Regulating services	Soil nutrient cycling	Certain entomopathogenic bacteria, such as *Bacillus thuringiensis*, have the potential to facilitate the breakdown of organic matter within soil, thereby liberating essential nutrients that are required for plant development.	*Photorhabdus luminescens* *Xenorhabdus* spp. *Bacillus thuringiensis* (Bt) *Paenibacillus* spp. *Pseudomonas fluorescens* *Serratia entomophila*	(Sharma et al., 2019)
	Bioremediation	The utilization of entomopathogenic bacteria has been proposed as a potential strategy for the bioremediation of polluted environments, including soil, water, and air, with the aim of mitigating the negative impacts of environmental contamination.	*Photorhabdus luminescens* *Xenorhabdus* spp. *Bacillus thuringiensis* (Bt) *Paenibacillus* spp. *Pseudomonas fluorescens* *Serratia entomophila*	(Vyatchina et al., 2021)
	Disease control	The utilization of entomopathogenic bacteria (*Xenorhabdus* and *Photorhabdus*) has the potential to regulate the transmission of diseases, such as malaria, through the eradication of disease-carrying insects, such as mosquitoes.	*Xenorhabdus* sp. *Photorhabdus* sp.	(da Silva et al., 2020)
	Soil formation and stabilization	Certain types of entomopathogenic bacteria have the ability to facilitate the formation and stabilization of soil, thereby mitigating the potential for erosion and preserving soil integrity.	*Bacillus thuringiensis* (Bt) *Pseudomonas fluorescens*	

(Continued)

TABLE 16.1 (*Continued*)
Ecosystem Services and Disservices from EPB with Suitable Examples

Ecosystem Services	Service Type	Description	Some Examples	References
Supporting services	Biodiversity	Entomopathogenic bacteria play a significant role in the biodiversity of ecosystems, offering crucial ecological services and serving as a source of nourishment for other living entities.	*Bacillus thuringiensis* (Bt) *Pseudomonas fluorescens*	(Azizoglu et al., 2020)
	Pollination	Certain types of entomopathogenic bacteria have the potential to augment the population of pollinators within a given region, thereby enhancing the process of pollination and ultimately resulting in amplified agricultural productivity. Most of the EPBs are non-destructive for bees, like synthetic insecticides; hence, it indirectly supports pollination.	*Bacillus thuringiensis* (Bt) *Pseudomonas fluorescens*	(Azizoglu et al., 2020)
	Research tools	The utilization of entomopathogenic bacteria as model organisms in scientific inquiry pertaining to genetics, ecology, and evolution has yielded valuable insights into the functioning of biological systems.	*Photorhabdus luminescens* *Xenorhabdus* spp. *Bacillus thuringiensis* (Bt) *Paenibacillus* spp. *Pseudomonas fluorescens* *Serratia entomophila*	(Sajnaga & Kazimierczak, 2020)
Cultural services	Education	The utilization of entomopathogenic bacteria as pedagogical instruments in academic institutions, such as schools and universities, can serve as an effective means of imparting knowledge on subjects such as microbiology, ecology, and environmental science to students.	*Photorhabdus luminescens* *Xenorhabdus* spp. *Bacillus thuringiensis* (Bt) *Paenibacillus* spp. *Pseudomonas fluorescens* *Serratia entomophila*	(Noureldeen et al., 2022)
	Improves livelihoods	EPB is a comparably low-cost method to control certain pests, which saves money and ultimately reflects in improving livelihoods.	*Photorhabdus luminescens* *Xenorhabdus* spp. *Bacillus thuringiensis* (Bt) *Paenibacillus* spp. *Pseudomonas fluorescens* *Serratia entomophila*	(Thakur et al., 2020)
	Impact on non-target organisms	The potential of entomopathogenic bacteria to cause harm to non-target organisms, including beneficial insects and soil microbes, has been documented.	*Bacillus thuringiensis* (Bt) *Photorhabdus luminescens* *Xenorhabdus* spp. *Serratia entomophila*	(Sandhi & Reddy, 2019)
Ecosystem disservices	Resistance development	The excessive utilization of entomopathogenic bacteria may result in the emergence of resistance in pests, thereby posing a challenge to their management in subsequent periods; though it is not proven, there is huge possibility of resistance development.		(Santhana Bharathi et al., 2019)
	Disruption of natural ecosystems	The incorporation of entomopathogenic bacteria has the potential to cause perturbations in ecological systems, resulting in inadvertent outcomes and probable ecological disequilibria.		(Chin et al., 2021)

(*Continued*)

TABLE 16.1 (*Continued*)
Ecosystem Services and Disservices from EPB with Suitable Examples

Ecosystem Services	Service Type	Description	Some Examples	References
	Potential for human illness	Certain entomopathogenic bacteria, specifically those classified under the genus *Clostridium*, possess the ability to induce sickness in humans upon ingestion or inhalation.	*Proteus mirabilis* *Bacillus cereus* *Bacillus anthracis* *Yersinia pestis* *Clostridium botulinum*	(Ebani & Mancianti, 2021)
	Risk of accidental release	Unintended outcomes may arise from the inadvertent dissemination of entomopathogenic bacteria into the ecosystem, including the proliferation of novel illnesses or the pollution of water reservoirs.		

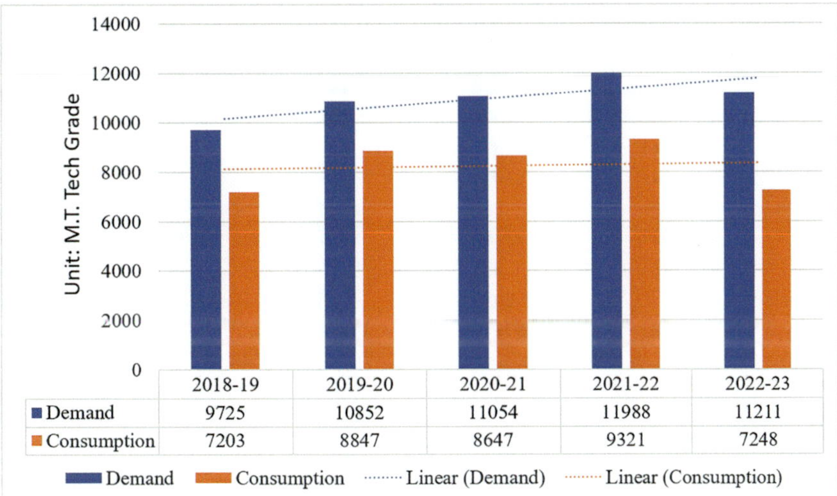

FIGURE 16.1 Indian scenario of demand and consumption of biopesticides over the last 5 years (2018–2019 to 2022–2023).

***Source*: https://ppqs.gov.in/statistical-database.**

A very positive scenario has arisen regarding the use of biopesticides over the last 5 years in the country (Table 16.2) (DPPQS, 2024b, 2024c). There is a positive change of 0.62% in the year 2022–2023 as compared to 2028–2019 in the use of biopesticides, which indicates a constant market for biopesticides, where there is a significant drop in the use of chemical pesticides with a negative shift of 12.07% across India. It is a very positive sign from both the ecological perspective and economic perspective of EPB. The best adaptation of biopesticides has been observed in Rajasthan, where there is more than 1,000% increase in the use of biopesticides along with more than 18% decrease in the use of chemical pesticides over the last 5 years. There is also a good level of increase in the use of biopesticides observed in states like Telangana, Andhra Pradesh, Jharkhand, and Uttarakhand. Also, there are few states, like Himachal Pradesh and Karnataka, where people are focusing more on chemical pesticides than biopesticides and consumption of biopesticides has decreased over the years. Several states like Maharashtra, Orissa, Punjab, and Jammu and Kashmir showed an overall reduction in pesticide use.

16.4.2 Market Overview of Entomopathogenic Bacteria

It is needless to mention that Bt is one of the most promising biopesticides, representing around 70% of the global biopesticide market (Tikar & Prakash, 2017). Bt biopesticide market is dominated by Abbott Laboratories (Chicago, IL) (since the acquisition in 1995 of Novo-Nordisk's biopesticide business) and Novartis (created through the merger in 1996 of Ciba and Sandoz); these together account for >70% of global production. The other 30% is contributed approximately by 30 companies with more than 100 Bt products, most of them containing a single Bt toxin, but some may have combinations of up to 5 (Sanahuja et al., 2011).

TABLE 16.2
State-Wise Consumption of Biopesticides and Chemical Pesticides in India

		2018–2019		2022–2023		Percentage Change	
Sl. No.	States/UTs	Biopesticide	Chemical Pesticide	Biopesticide	Chemical Pesticide	Biopesticide	Chemical Pesticide
1	Andhra Pradesh	10	1689	51	2001	414.77	18.45
2	Bihar	350	850	360	995	2.86	17.06
3	Chhattisgarh	505	1770	740	1775	46.53	0.28
4	Goa	6	25	5	34	−11.67	34.05
5	Gujarat	306	1608	548	1750	79.08	8.83
6	Haryana	410	4015	440	4066	7.32	1.27
7	Himachal Pradesh	2	322	0.37	470	−75.48	45.89
8	Jharkhand	41	646	93	450	124.48	−30.38
9	Karnataka	544	1524	454	1669	−16.54	9.51
10	Kerala	862	995	628	504	−27.13	−49.33
11	Madhya Pradesh	322	540	375	598	16.46	10.74
12	Maharashtra	1164	11746	15	6814	−98.74	−41.99
13	Orissa	310	1609	81	1348	−73.87	−16.22
14	Punjab	246	5543	172	5130	−30.08	−7.45
15	Rajasthan	15	2290	170	1865	1033.33	−18.56
16	Tamil Nadu	500	1901	907	1952	81.40	2.68
17	Telangana	84	4894	522	4920	521.43	0.53
18	Uttar Pradesh	47	11049	53	11824	12.66	7.01
19	Uttarakhand	52	195	120	111	132.38	−42.95
20	West Bengal	997	3190	1234	3321	23.77	4.10
22	Assam	234	256	272	455	16.60	77.61
23	Jammu and Kashmir	2	2459	0.98	384	−38.75	−84.39
Grand Total		**7203**	**59670**	**7248**	**52466**	**0.62**	**−12.07**

Note: Unit: MT formulation.
Source: DPPQS (2024b, 2024c).

16.4.3 Cost–Benefit Assessment of Entomopathogenic Bacteria in Pest Management

The cost–benefit assessment of every production strategy is very crucial before its implementation, as this economic perspective is the main and sometimes only concern of the farmers. They want to ensure the enhancement of their income while shifting from their conventional practices to newer ones. So the utilization of entomopathogenic bacteria in pest management should be discussed in this direction as well so that they gain more acceptability among the farmers, the consumers of pesticides, which would eventually create increased demands and would encourage industrialists and entrepreneurs to start investment in production and marketing of EPB. And this is equally applicable to other entomopathogens as well.

In the adopted village of Ashta, Maharashtra, the IPM biointensive module comprising *Bacillus thuringiensis* var *kurstaki* Bulla et al. as a component had almost double benefit than the cost (Singh et al., 2005). In an experiment on lepidopterous defoliating pest management of soybean in Manipur, it has been found that treatments of *Beauveria bassiana* (Balsamo) (500 mL/ha) + Bt commercial (500 mL/ha) (1:8.21) and *Nomuraea rileyi* (Farlow) (500 mL/ha) + Bt commercial (500 mL/ha) (1:8.14) have the highest incremental cost–benefit ratio over several other combinations, along with Chlorantraniliprole 18.5% SC as standard check (Lalsangzuali et al., 2023).

16.5 CASE STUDIES ON THE USAGE OF ENTOMOPATHOGENIC BACTERIA IN PEST MANAGEMENT

Entomopathogenic bacteria have been successfully incorporated into integrated pest management (IPM) programmes across various regions and crop systems. A major success has been found in cotton cultivation in the United States in 1995 and India in 2002 (Abbas, 2018). Genetically engineered Bt cotton is used in these regions alongside monitoring practices, crop rotation, and natural predators to control pest populations, which gives good control over pests like cotton bollworm (*Helicoverpa armigera* [Hübner]) and pink bollworm (*Pectinophora gossypiella* [Saunders]). In 1997–1998, there was a dedicated IPM programme in the Ashta village of Maharashtra. The biointensive module comprising spray of *Bacillus thuringiensis* var. *kurstaki* (Dipel 8L) at 1 kg/ha along with Aphidin at 4 mL/L of water against sucking pests, release of *Trichogramma chilonis* Ishii (1.5 lakh/ha.), use of pheromone traps (10/ha) for monitoring, and spray of 5% neem seed kernel extract (NSKE) and sulphur 80 WP at 25 g/10 Lof water. This module led to the minimization of the bollworm complex in the village as compared to last year, 1996, when chemical pesticides failed to stop *H. armigera* epidemics. However, the resistance outbreak is forcing farmers to spend more on pesticides today than before the introduction of Bt. So this aspect should be seriously taken care of (Kranthi & Stone, 2020).

16.6 PROBLEMS AND SCOPE OF EPB MARKET

The higher cost of registration of biopesticides and the long-drawn-out process deter companies from engaging in research and development on biological pesticides in India. Another limitation is the requirement to verify the microorganism for biosafety before registration, multiplication, and marketing of biopesticides. Many universities and research institutions that conduct the initial R&D on biopesticides are mostly unable to bear the additional costs required for extensive safety and allergic tests (Padma, 2019).

The key active ingredients used to produce biopesticides are imported from China, and other key manufacturing companies are in Europe and North America. During the crisis of COVID-19, there was low availability of raw materials with the progress of the cropping season. This created a scarcity of products even in established markets, such as the United States, Brazil, Spain, and Italy. Rather, this created a huge scope for the utilization of diverse raw materials as local sources for the production of biopesticides. And in the post-COVID scenario, the dependency on different countries for import is rather reducing. Also, there was a decline in the availability of active chemical ingredients, which also led to a scarcity of chemical products in the market, ultimately encouraging the farmers to adopt an alternative solution for the control of pests. This situation also boosted the demand for biopesticides with the increasing health consciousness and demand for healthy and organic food in human society (Markets and Markets, 2023). As a result, there is a need for a huge revolution in research on biopesticides and EPB, and discoveries are expected shortly.

16.7 CONCLUSION

From the preceding discussion, we can understand the ecological and economical values of entomopathogenic bacteria, making them potent pest management agents. Though involved in different types of ecosystem services like provisioning services, regulating services, supporting services, and cultural services, it has also some ecosystem disservices as well. There is an increasing demand for biopesticides, which signifies a good possibility in the flourishment of the EPB market; still, it maintains a constant level of consumption without any uprise. Despite regulatory and cost-related constraints, the EPB market has witnessed a steady rise in demand, driven by increasing resistance to chemical pesticides, regulatory incentives, and organic farming expansion. There are several success stories via incorporation of EPB as well as biocontrol agents in IPM, but issues of resistance development against major EPBs like Bt have also come up. This can be tackled by more research and coming up with new agents; several potent yet-to-be-commercialized EPBs are discussed in the previous chapters with this intention only. The purpose of the present chapter is to influence researchers and industrialists to explore opportunities in ecology and economy-friendly agent for insect pest management for a better future.

REFERENCES

Abbas, M. S. T. (2018). Genetically engineered (modified) crops (*Bacillus thuringiensis* crops) and the world controversy on their safety. *Egyptian Journal of Biological Pest Control*, *28*(1), 52. https://doi.org/10.1186/s41938-018-0051-2

Acheuk, F., Basiouni, S., Shehata, A. A., Dick, K., Hajri, H., Lasram, S., Yilmaz, M., Emekci, M., Tsiamis, G., Spona-Friedl, M., May-Simera, H., Eisenreich, W., & Ntougias, S. (2022). Status and prospects of botanical biopesticides in Europe and Mediterranean countries. *Biomolecules*, *12*(2), 311. https://doi.org/10.3390/BIOM12020311/S1

Ali, S., Aqueel, M. A., Saeed, M. F., Shakeel, Q., Raheel, M., & Ullah, M. I. (2022). Utilization of entomopathogenic bacteria for modern insect pest management. In S. D. Mandal, G. Ramkumar, S. Karthi, & F. Jin (Eds.), *New and future development in biopesticide research: Biotechnological exploration* (pp. 93–113). Springer Nature. https://doi.org/10.1007/978-981-16-3989-0_3

Azizoglu, U., Jouzani, G. S., Yilmaz, N., Baz, E., & Ozkok, D. (2020). Genetically modified entomopathogenic bacteria, recent developments, benefits and impacts: A review. *Science of the Total Environment*, *734*, 139169. https://doi.org/10.1016/j.scitotenv.2020.139169

Baumber, A., Metternicht, G., Cross, R., Ruoso, L. E., Cowie, A. L., & Waters, C. (2019). Promoting co-benefits of carbon farming in Oceania: Applying and adapting approaches and metrics from existing market-based schemes. *Ecosystem Services*, *39*, 100982. https://doi.org/10.1016/j.ecoser.2019.100982

Chin, K. L., H'ng, P. S., Lee, C. L., Wong, W. Z., Go, W. Z., Khoo, P. S., Luqman, A. C., & Ashaari, Z. (2021). Application strategies by selective medium treated with entomopathogenic bacteria *Serratia marcescens* and *Pseudomonas aeruginosa* as potential biocontrol against *Coptotermes curvignathus*. *Royal Society Open Science*, *8*(4), rsos.201311. https://doi.org/10.1098/rsos.201311

Daily, G. C. (1997). Introduction: What are ecosystem services. In G. C. Daily (Ed.), *Nature's services: Societal dependence on natural ecosystems* (pp. 1–10). Island Press.

da Silva, W. J., Pilz-Júnior, H. L., Heermann, R., & da Silva, O. S. (2020). The great potential of entomopathogenic bacteria *Xenorhabdus* and *Photorhabdus* for mosquito control: A review. *Parasites & Vectors*, *13*, 376. https://doi.org/10.1186/s13071-020-04236-6

DPPQS. (2021). *Bio-pesticide registrant*. Directorate of Plant Protection, Quarantine & Storage of Government of India. http://ppqs.gov.in/divisions/cib-rc/bio-pesticide-registrant

DPPQS. (2024a). *Estimated demand of bio-pesticides in states/UTs*. Directorate of Plant Protection, Quarantine & Storage. https://ppqs.gov.in/statistical-database

DPPQS. (2024b). *Statewise consumption of bio-pesticides*. Directorate of Plant Protection, Quarantine & Storage. https://ppqs.gov.in/statistical-database

DPPQS. (2024c). *Statewise consumption of chemical pesticides*. Directorate of Plant Protection, Quarantine & Storage. https://ppqs.gov.in/statistical-database

Ebani, V. V., & Mancianti, F. (2021). Entomopathogenic fungi and bacteria in a veterinary perspective. *Biology*, *10*(6), 479. https://doi.org/10.3390/biology10060479

Fernández-Chapa, D., Ramírez-Villalobos, J., & Galán-Wong, L. (2019). Toxic potential of *Bacillus thuringiensis*: An overview. In Y. Jia (Ed.), *Protecting rice grains in the post-genomic era* (pp. 1–22). IntechOpen. https://doi.org/10.5772/intechopen.85756

Kassam, A., Friedrich, T., & Derpsch, R. (2022). Successful experiences and lessons from conservation agriculture worldwide. *Agronomy*, *12*(4), 769. https://doi.org/10.3390/agronomy12040769

Kranthi, K. R., & Stone, G. D. (2020). Long-term impacts of Bt cotton in India. *Nature Plants*, *6*(3), 188–196. https://doi.org/10.1038/s41477-020-0615-5

Kumar, J., Ramlal, A., Mallick, D., & Mishra, V. (2021). An overview of some biopesticides and their importance in plant protection for commercial acceptance. *Plants*, *10*(6), 1185. https://doi.org/10.3390/plants10061185

Lalsangzuali, C., Karam, N., Devi, H. N., Devi, T. S., Sophia, L., Singh, D., & Anando, N. (2023). Incremental cost benefit ratio of different combination of microbial insecticides for management of lepidopterous defoliating pests of soybean in Manipur. *The Pharma Innovation Journal*, *12*(8), 1958–1961.

Markets and Markets. (2023). *Biopesticides market statistics, growth dynamics and future trends | COVID-19 impact on biopesticides market*. Retrieved August 28, 2021, from https://www.marketsandmarkets.com/Market-Reports/biopesticides-267.html

Marrone, P. G. (2023). Status of the biopesticide market and prospects for new bioherbicides. *Pest Management Science*, *80*(1), 81–86. https://doi.org/10.1002/ps.7403

Noureldeen, A., Kumar, U., Asad, M., Darwish, H., Alharthi, S., Fawzy, M. A., Al-Barty, A. M., Alotaibi, S. S., Fallatah, A., & Alghamdi, A. (2022). Aphicidal activity of five plant extracts applied singly or in combination with entomopathogenic bacteria, *Xenorhabdus budapestensis* against rose aphid, *Macrosiphum rosae* (Hemiptera: Aphididae). *Journal of King Saud University-Science*, *34*(8), 102306. https://doi.org/10.1016/j.jksus.2022.102306

Padma, T. V. (2019, July 9). *What is preventing the widespread adoption of biopesticides in India?* Mongabay. Retrieved July 1, 2023, from https://india.mongabay.com/2019/07/what-is-preventing-the-widespread-adoption-of-biopesticides-in-india/

Plaas, E., Meyer-Wolfarth, F., Banse, M., Bengtsson, J., Bergmann, H., Faber, J., Potthoff, M., Runge, T., Schrader, S., & Taylor, A. (2019). Towards valuation of biodiversity in agricultural soils: A case for earthworms. *Ecological Economics*, *159*, 291–300. https://doi.org/10.1016/j.ecolecon.2019.02.003

Power, A. G. (2010). Ecosystem services and agriculture: Tradeoffs and synergies. *Philosophical Transactions of the Royal Society B: Biological Sciences*, *365*(1554), 2959–2971. https://doi.org/10.1098/rstb.2010.0143

Rao, T., & Jurat-Fuentes, J. L. (2020). Advances in the use of entomopathogenic bacteria/Microbial Control Agents (MCAs) as biopesticides in suppressing crop insect pests. In N. Birch, & T. Glare (Eds.), *Biopesticides for sustainable agriculture* (pp. 99–134). Burleigh Dodds Science Publishing. https://doi.org/10.19103/AS.2020.0073.06

Reid, W. V., Mooney, H. A., Cropper, A., Capistrano, D., Carpenter, S. R., Chopra, K., Dasgupta, P., Dietz, T., Duraiappah, A. K., Hassan, R., Kasperson, R., Leemans, R., May, R. M., McMichael, A. J., Pingali, P., Samper, C., Scholes, R., Watson, R. T., Zakri, A. H., Shidong, Z., Ash, N. J., Bennett, E., Kumar, P., Lee, M. J., Raudsepp-Hearne, C., Simons, H., Thonell, J., & Zurek, M. B. (2005). *Ecosystems and human well-being—Synthesis: A report of the millennium ecosystem assessment*. Island Press.

Ruiu, L. (2015). Insect pathogenic bacteria in integrated pest management. *Insects*, *6*(2), 352–367. https://doi.org/10.3390/insects6020352

Sajnaga, E., & Kazimierczak, W. (2020). Evolution and taxonomy of nematode-associated entomopathogenic bacteria of the genera *Xenorhabdus* and *Photorhabdus*: An overview. *Symbiosis*, *80*(1), 1–13. https://doi.org/10.1007/s13199-019-00660-0

Sanahuja, G., Banakar, R., Twyman, R. M., Capell, T., & Christou, P. (2011). *Bacillus thuringiensis*: A century of research, development and commercial applications. *Plant Biotechnology Journal*, *9*(3), 283–300. https://doi.org/10.1111/J.1467-7652.2011.00595.X

Sandhi, R. K., & Reddy, G. V. P. (2019). Effects of entomopathogenic nematodes and symbiotic bacteria on non-target arthropods. In M. Khan, & W. Ahmad (Eds.), *Microbes for sustainable insect pest management: An eco-friendly approach* (1st ed., pp. 247–273). Springer. https://doi.org/10.1007/978-3-030-23045-6_9

Santhana Bharathi, N., Deepa, A., & Chithra Pand Sujatha, K. (2019). The pathogenic potential of Steinernema sp. against the dengue virus vector *Aedes aegypti*. *Journal of Entomology and Zoology Studies*, *7*(5), 682–687.

Sharma, L., Bohra, N., Singh, R. K., & Marques, G. (2019). Potential of entomopathogenic bacteria and fungi. In M. Khan, & W. Ahmad (Eds.), *Microbes for sustainable insect pest management: An eco-friendly approach* (pp. 115–149). Springer. https://doi.org/10.1007/978-3-030-23045-6_4

Singh, A., Sharma, O. P., Lavekar, R. C., Bambawale, O., Murthy, K. S., & Dhandapani, A. (2005). IPM technology for rainfed cotton. *National Centre For Integrated Pest Management*. https://doi.org/10.13140/2.1.1451.6488

Thakur, N., Kaur, S., Tomar, P., Thakur, S., & Yadav, A. N. (2020). Microbial biopesticides: Current status and advancement for sustainable agriculture and environment. In A. A. Rastegar, A. N. Yadav, & N. Yadav (Eds.), *New and future developments in microbial biotechnology and bioengineering* (pp. 243–282). Elsevier. https://doi.org/10.1016/B978-0-12-820526-6.00016-6

Tikar, S., & Prakash, S. (2017). Fly ash-based *Bacillus thuringiensis israelensis* formulation: An ecofriendly approach. *Indian Journal of Medical Research*, *146*(6), 680. https://doi.org/10.4103/ijmr.IJMR_1679_16

Vandenberghe, L. P. S., Pandey, A., Carvalho, J. C., Letti, L. A. J., Woiciechowski, A. L., Karp, S. G., Thomaz-Soccol, V., Martínez-Burgos, W. J., Penha, R. O., & Herrmann, L. W.

(2021). Solid-state fermentation technology and innovation for the production of agricultural and animal feed bioproducts. *Systems Microbiology and Biomanufacturing*, *1*, 142–165. https://doi.org/10.1007/s43393-020-00015-7

Vyatchina, O. F., Karavaeva, E. A., Tolstoy, M. Y., Atanova, M. V., & Kovalenko, S. A. (2021). Influence of humates on the growth of microorganisms. *Sustainable and Efficient Use of Energy, Water and Natural Resources*, 180.

Wang, L., & Zhao, Y. (2022). Transcriptome analysis of *Callosobruchus chinensis*: Insight into the biological control using entomopathogenic bacteria, *Bacillus thuringiensis*. *Food Science and Technology*, *42*, e26122. https://doi.org/10.1590/fst.26122

Zakalyukina, Y. V., Pavlov, N. A., Lukianov, D. A., Marina, V. I., Belozerova, O. A., Tashlitsky, V. N., Guglya, E. B., Osterman, I. A., & Biryukov, M. V. (2022). A new albomycin-producing strain of *Streptomyces globisporus* subsp. *globisporus* may provide protection for ants *Messor structor*. *Insects*, *13*(11), 1042. https://doi.org/10.3390/insects13111042

17 Possible Way Forward and Policy Perspective of the Utilization of Entomopathogenic Bacteria as an Alternative to GM Crops and Insecticides

ABSTRACT

Entomopathogenic bacteria (EPB) for insect pest management offer various benefits in comparison to chemical insecticides. These advantages include their comparably more selectivity towards specific insect targets than conventional synthetic chemical insecticides, reduced environmental footprint, and prolonged persistence in the ecosystem. Although entomopathogenic bacteria are eco-friendly and selective, their persistence in field environments is often limited by UV degradation, temperature fluctuations, and microbial competition. The mechanism of their operation entails the generation of toxins, enzymes, and additional secondary metabolites that can elicit diverse symptoms in the insect host. Certain parameters are important and taken into consideration in the effective management of pests through EPB. Correct identification and characterization of bacteria are important, which determine the success of EPB. Virulent strain selection and genetic manipulation, along with innovative formulations, such as nanoparticle- and microencapsulation-based formulations, give added advantage to the entomopathogenic bacteria. Although genetic manipulation of EPBs holds promise for enhancing virulence or stability, its application is limited commercially due to regulatory restrictions and public resistance to genetically modified microorganisms. EPB is additionally utilized in antibiotics, in antimicrobial production, as well as in bioremediation. Certain entomopathogenic *Bacillus* species, such as *B. subtilis* and *B. amyloliquefaciens*, produce antimicrobial compounds; however, this is not a universal feature across all EPBs.

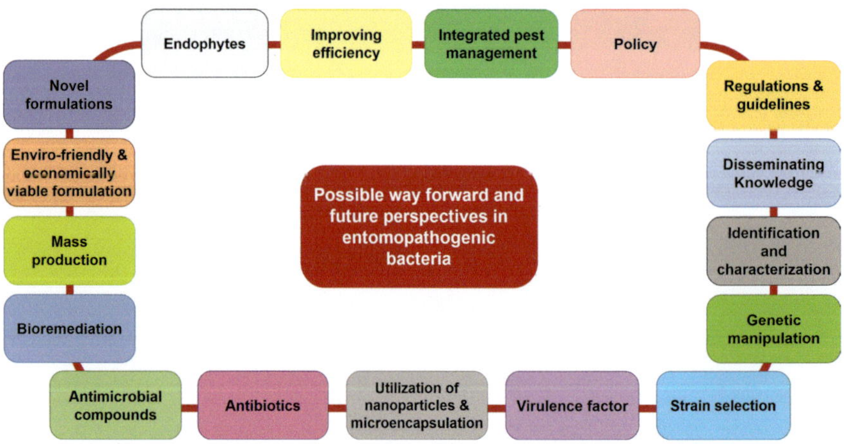

17.1 INTRODUCTION

Entomopathogenic bacteria (EPB) refer to a group of microorganisms that have the ability to infect and induce mortality in a diverse array of insect species, especially crop insect pests (Glare et al., 2017). They have been utilized for the biological management of insect pests for numerous decades, and their efficacy and safety have been firmly established (Tozlu et al., 2022). In recent years, due to more insecticide contamination and pollution cases, there has been a renewed interest in the potential of EPB as the demand for sustainable and eco-friendly pest control methods continues to rise (Sajnaga & Kazimierczak, 2020). The present discourse aims to examine the contemporary state of research on EPB and its potential for future advancements.

Entomopathogenic bacteria for insect pest management offer various benefits in comparison to chemical insecticides. These advantages include their comparably

DOI: 10.1201/9781003426882-17

more selectivity towards specific insect targets than conventional synthetic chemical insecticides, reduced environmental footprint, and prolonged persistence in the ecosystem. The mechanism of their operation entails the generation of toxins, enzymes, and additional secondary metabolites that can elicit diverse symptoms in the insect host.

17.2 APPLICATION OF ENTOMOPATHOGENIC BACTERIA

Bacillus thuringiensis is widely recognized as the most prominent example of an entomopathogenic bacterium and has been extensively employed for safeguarding crops on a global scale (Lacey et al., 2001). *Xenorhabdus* and *Photorhabdus* have been the subject of research due to their prospective utilization in the biological management of subterranean insect pests (da Silva et al., 2020). The limited efficacy of EPB as biopesticides against certain insect pests, particularly those with tough or waxy cuticles, represents a significant challenge in their development. An additional obstacle pertains to the elevated expenses associated with the production, quality control, shelf life, and formulation processes, thereby impeding their potential for commercialization (Kusakabe et al., 2022). In order to address these obstacles, scholars are investigating diverse approaches to enhance the efficiency and economic viability of biopesticides based on entomopathogenic bacteria.

An approach that can be taken is to enhance the management of entomopathogenic bacteria to the intended insect host. Scientists are currently investigating diverse techniques, including the utilization of nanoparticles and microencapsulation, to amplify the insect cuticle's penetration and endurance of EPB (Damalas & Koutroubas, 2018). An alternative strategy involves augmenting the pathogenicity of EPB through genetic manipulation (Azizoglu et al., 2020) or the process of strain selection (de Bortoli & Jurat-Fuentes, 2019). Researchers have identified several factors that can enhance the effectiveness of EPB, including insecticidal proteins and chitinases (Hirschmann et al., 2017).

Apart from their utilization as biopesticides, EPBs possess various other potential applications. An area that shows promise is the utilization of EPB as a potential source for innovative antibiotics (Cai et al., 2017). EPB is capable of synthesizing a diverse array of antimicrobial compounds that exhibit inhibitory effects against pathogenic bacteria and fungi (De Mandal et al., 2021). Scholars are currently investigating the potential of antibiotics derived from EPB as a viable substitute for traditional antibiotics, which are experiencing a growing resistance.

Entomopathogenic bacteria may have potential applications in the field of bioremediation (Jamal & Ahmad, 2022). They exhibit the capability to deteriorate a diverse array of organic contaminants, encompassing pesticides, herbicides, and industrial chemicals. EPB-based bioremediation is being investigated by scholars as a viable and economical approach to remediate polluted soils and water bodies (Ahsan & Shimizu, 2021).

The new dimension of EPB as a plant endophyte has also been explored, where *Bacillus thuringiensis* has successfully colonized inside the plant as an endophyte through seed treatment and spraying. This work could replace the tedious procedure of developing genetically modified crops, and thereby, without altering the plant gene, pests can be controlled through environmentally sustainable ways (Pola et al., 2022). It could also save huge costs involved in GM crop production. The use of other entomopathogenic bacteria as endophytes further needs to be explored.

The potential for future advancements in EPB research appears to be optimistic, with a number of research domains expected to garner heightened focus in the forthcoming years. An area of focus pertains to the identification and characterization of newly discovered EPB strains that exhibit enhanced effectiveness and safety (Alotaibi et al., 2021). Scientists are currently investigating diverse origins, including microbiomes present in soil and insect gut, with the aim of identifying novel strains of EPB.

An additional research domain pertains to the formulation of environment-friendly and economically viable techniques for the manufacturing of biopesticides that are based on EPB (Sharma et al., 2019). Researchers are exploring various mass production methods, such as solid-substrate fermentation and *in vivo* production, to optimize the mass production of EPB (Jackson, 2017). In addition, researchers are developing new formulation methods, such as wettable powders, granules, and sprays, to improve the distribution and perseverance of EPB on the target insect (Couch et al., 2023). Scholars are currently investigating diverse techniques to enhance the efficiency of EPB manufacturing. Furthermore, scholars are currently devising novel techniques for formulating EPBs, such as granulation and spraying, with the aim of enhancing its efficacy and durability against the intended insect species (Korany et al., 2019). The promising area of research lies in the inclusion of entomopathogenic bacteria as an integral component in the development of integrated pest management (IPM) strategies. Scholars are currently investigating the viability of using EPB-derived biopesticides as a component of integrated pest management strategies for diverse crops and insect species (Ali et al., 2022).

In general, it can be inferred that entomopathogenic bacteria exhibit significant potential in the realm of biological pest management. These substances exhibit eco-friendliness and negligible adverse effects on non-target organisms, rendering them a viable and sustainable substitute for chemical pesticides (da Silva et al., 2020). The progress in biotechnology is expected to facilitate the enhancement of bacteria as biopesticides, encompassing their efficacy, applicability, and compatibility with other pest control methodologies. In order to fully harness the capabilities of entomopathogenic bacteria, it is imperative to persist in conducting research on their ecology, genetics, and molecular biology (Chakroun et al., 2016). Further research is required to comprehensively comprehend the mechanisms of infection and the genetic underpinnings of virulence, alongside the formulation of techniques for large-scale production and distribution. Furthermore, there is a necessity for the ongoing advancement of strains that exhibit specificity towards targeted pests and can endure diverse environmental circumstances (Soumia et al., 2020).

Furthermore, it is imperative to guarantee the conscientious utilization of entomopathogenic bacteria. The implementation of suitable regulations and guidelines is imperative to mitigate the excessive utilization of biopesticides and to guarantee their safe and efficacious application (Ruiu, 2015). Disseminating knowledge regarding the advantages of biological control and the appropriate application of biopesticides to both farmers and the general populace is of paramount significance (Sajnaga & Kazimierczak, 2020).

17.3 ROLE OF ENTOMOPATHOGENS IN ACHIEVING SUSTAINABLE DEVELOPMENT GOALS (SDGs)

Notwithstanding these apprehensions, entomopathogenic bacteria possess the capacity to contribute significantly towards attaining the United Nations' Sustainable Development Goals, specifically in the domains of ensuring food security, conserving biodiversity, and mitigating climate change. The attainment of the Sustainable Development Goals (SDGs) poses a significant societal challenge, given the persistent strain that human activities exert on ecosystems. The Sustainable Development Goals were introduced by the United Nations in 2015 as a means of tackling worldwide issues, including but not limited to poverty, inequality, climate change, and environmental degradation (Sachs, 2012). Through a comprehensive investigation and comprehension of the ecosystem services and plausible hazards linked with these microorganisms, we can persist in exploiting their potential while alleviating any adverse consequences. Hence, it is imperative to discern tactics that can bolster ecosystem services and aid in accomplishing the Sustainable Development Goals (Wood et al., 2018).

Entomopathogenic bacteria have the potential to make a positive impact towards the attainment of various Sustainable Development Goals, such as SDG 2: Zero Hunger; SDG 3: Good Health and Well-Being; and SDG 12: Responsible Consumption and Production. The implementation of EPB has the potential to address the issue of food insecurity by enhancing agricultural productivity and diminishing the reliance on synthetic pesticides, thereby augmenting the accessibility and cost-effectiveness of food. Employing environmentally preferable practices, such as

TABLE 17.1

Potential Capacity of Entomopathogenic Bacteria in Achieving UN Sustainable Development Goals

Sustainable Development Goals	Potential Contributions of Entomopathogenic Bacteria	References
SDG 1: No Poverty	The utilization of entomopathogenic bacteria has the potential to mitigate poverty by enhancing agricultural productivity and augmenting food safety.	(Marwa et al., 2020)
SDG 2: Zero Hunger	The utilization of entomopathogenic bacteria has the potential to mitigate the deleterious effects of pests on crops, thereby resulting in a boost in agricultural yield and an enhancement of food security.	(Nawaz et al., 2022)
SDG 3: Good Health and Well-Being	Entomopathogenic bacteria possess the ability to regulate the proliferation of disease-transmitting insects, including mosquitoes, and can synthesize antibiotics that have potential therapeutic applications for bacterial infections in both human and animal populations.	(Gnambani et al., 2020)
SDG 6: Clean Water and Sanitation	Certain strains of entomopathogenic bacteria possess the ability to decompose organic pollutants present in soil and water, thereby playing a crucial role in the preservation of aquatic ecosystems.	(Azizoglu et al., 2020)
SDG 12: Sustainable Consumption and Production	The utilization of entomopathogenic bacteria as biopesticides has the potential to mitigate the adverse ecological and health consequences linked to synthetic pesticides.	(Rao & Jurat-Fuentes, 2020)
SDG 14: Life below Water	The utilization of entomopathogenic bacteria has the potential to aid in the preservation of aquatic ecosystems through the breakdown of organic pollutants present in both soil and water.	(Kebede et al., 2022)
SDG 15: Life on Land	The utilization of entomopathogenic bacteria has the potential to enhance the sustainability of agricultural practices through the mitigation of pest-related crop damage and the enhancement of soil fertility. In addition, they have the potential to mitigate the adverse effects of chemical pesticides on non-target organisms and aid in the preservation of land-based ecosystems through the enhancement of biodiversity.	(Reinbacher et al., 2021)

FIGURE 17.1 Potential contributions of EPB in achieving UN SDGs.

reducing the usage of chemical pesticides, can potentially enhance human health and well-being by mitigating the adverse effects of such substances. Implementing EPB can potentially foster responsible consumption and production practices by mitigating agriculture's environmental footprint. Table 17.1 and Figure 17.1 outline the possible impact of EPB on the attainment of the United Nations' Sustainable Development Goals.

Policies should encourage the use of EPBs as biopesticides to reduce reliance on synthetic chemicals and promote environmental sustainability (Bravo et al., 2011). Developing standardized regulations for producing, commercializing, and applying EPB is critical. Governments must ensure these biological agents are safe for non-target organisms, humans, and the environment. Policies must fund R&D to identify new EPB, optimize their production, and improve their efficacy. An integrated approach that applies ecological and social concepts and the principle of interaction of living beings could help achieve the SDGs (FAO, 2021). This includes studying potential resistance in target pests to prevent reduced effectiveness. Policy measures should focus on training programmes for farmers on the effective use of EPBs. Subsidies or financial incentives can encourage adoption, especially in regions where chemical pesticides dominate (Lacey & Kaya, 2007). Awareness campaigns are necessary to educate the public and farmers about the benefits and safety of EPB. Misconceptions about biological control agents can hinder their adoption. As pest outbreaks often cross borders, policies should foster international partnerships for shared R&D and harmonized regulations. Initiatives like the International Plant Protection Convention (IPPC) can incorporate EPB into pest management guidelines. Policies should encourage private sector participation to scale up the production of EPBs. This includes intellectual property protections and incentives for biotech companies to develop innovative products (Schnepf et al., 1998).

The utilization of EPB has been shown to effectively manage insect pests and regulate their populations, leading to a notable enhancement in food productivity. The current state of knowledge and future research directions in this field are also highlighted, with the ultimate goal of achieving the SDGs.

Entomopathogenic bacteria refer to a cluster of microorganisms that have the ability to invade and eliminate insects. Entomopathogenic bacteria are recognized as significant biological agents that help control insect populations by acting as natural enemies of insects (Rajamani & Negi, 2021). Entomopathogenic bacteria have been employed as a means of biological control to manage insect pests in various fields, such as agriculture, forestry, and public health (Sethi et al., 2022). Nevertheless, the precise function of EPBs in advancing ecosystem services and their contribution towards accomplishing the Sustainable Development Goals remain inadequately comprehended.

17.4 CONCLUSION

To conclude, entomopathogenic bacteria exhibit potential as a viable mechanism in the array of biological control methods employed to combat insect pests. It also exhibits promising prospective in the realm of pest management, provided that research efforts are sustained and their application is conducted in a responsible manner (Lacey et al., 2015). The utilization of non-chemical pesticides presents a promising and viable option that can effectively mitigate the detrimental effects of insect pests on agriculture and the environment while also promoting sustainability (Alotaibi et al., 2021).

REFERENCES

Ahsan, N., & Shimizu, M. (2021). *Lysinibacillus* species: Their potential as effective bioremediation, biostimulant, and biocontrol agents. *Reviews in Agricultural Science*, *9*, 103–116. https://doi.org/10.7831/ras.9.0_103

Ali, S., Aqueel, M. A., Saeed, M. F., Shakeel, Q., Raheel, M., & Ullah, M. I. (2022). Utilization of entomopathogenic bacteria for modern insect pest management. In S. D. Mandal, G. Ramkumar, S. Karthi, & F. Jin (Eds.), *New and future development in biopesticide research: Biotechnological exploration* (pp. 93–113). Springer Nature. https://doi.org/10.1007/978-981-16-3989-0_3

Alotaibi, S. S., Darwish, H., Alharthi, S., Alghamdi, A., Noureldeen, A., Fallatah, A. M., Fodor, A., Al-Barty, A., Albogami, B., & Baazeem, A. (2021). Control potentials of three entomopathogenic bacterial isolates for the carob moth, *Ectomyelois ceratoniae* (Lepidoptera: Pyralidae) in pomegranates. *Agriculture*, *11*(12), 1256. https://doi.org/10.3390/agriculture11121256

Azizoglu, U., Jouzani, G. S., Yilmaz, N., Baz, E., & Ozkok, D. (2020). Genetically modified entomopathogenic bacteria, recent developments, benefits and impacts: A review. *Science of the Total Environment*, *734*, 139169. https://doi.org/10.1016/j.scitotenv.2020.139169

Bravo, A., Likitvivatanavong, S., Gill, S. S., & Soberón, M. (2011). *Bacillus thuringiensis*: A story of a successful bioinsecticide. *Insect Biochemistry and Molecular Biology*, *41*(7), 423–431. https://doi.org/10.1016/j.ibmb.2011.02.006

Cai, X., Challinor, V. L., Zhao, L., Reimer, D., Adihou, H., Grün, P., Kaiser, M., & Bode, H. B. (2017). Biosynthesis of the antibiotic nematophin and its elongated derivatives in entomopathogenic bacteria. *Organic Letters*, *19*(4), 806–809.

Chakroun, M., Banyuls, N., Bel, Y., Escriche, B., & Ferré, J. (2016). Bacterial vegetative insecticidal proteins (Vip) from entomopathogenic bacteria. *Microbiology and Molecular Biology Reviews*, *80*(2), 329–350. https://doi.org/10.1128/mmbr.00060-15

Couch, T. L., Jackson, T. A., & Jurat-Fuentes, J. L. (2023). Commercial production of entomopathogenic bacteria. In J. A. Morales-Ramos, M. G. Rojas, & D. Shapiro-Ilan (Eds.), *Mass production of beneficial organisms* (2nd ed., pp. 359–373). Academic Press. https://doi.org/10.1016/B978-0-12-822106-8.00018-X

Damalas, C. A., & Koutroubas, S. D. (2018). Current status and recent developments in biopesticide use. *Agriculture*, *8*(1), 13. https://doi.org/10.3390/agriculture8010013

Da Silva, W. J., Pilz-Júnior, H. L., Heermann, R., & da Silva, O. S. (2020). The great potential of entomopathogenic bacteria *Xenorhabdus* and *Photorhabdus* for mosquito control: A review. *Parasites & Vectors*, *13*, 376. https://doi.org/10.1186/s13071-020-04236-6

De Bortoli, C. P., & Jurat-Fuentes, J. L. (2019). Mechanisms of resistance to commercially relevant entomopathogenic bacteria. *Current Opinion in Insect Science*, *33*, 56–62. https://doi.org/10.1016/j.cois.2019.03.007

De Mandal, S., Panda, A. K., Murugan, C., Xu, X., Senthil Kumar, N., & Jin, F. (2021). Antimicrobial peptides: Novel source and biological function with a special focus on entomopathogenic nematode/bacterium symbiotic complex. *Frontiers in Microbiology*, *12*, 555022. https://doi.org/10.3389/fmicb.2021.555022

FAO. (2021). Integrated pest management. Retrieved from FAO IPM Guidelines.

Glare, T. R., Jurat-Fuentes, J. L., & O'callaghan, M. (2017). Basic and applied research: Entomopathogenic bacteria. In L. A. Lacey (Ed.), *Microbial control of insect and mite pests* (pp. 47–67). Academic press. https://doi.org/10.1016/B978-0-12-803527-6.00004-4

Gnambani, E. J., Bilgo, E., Sanou, A., Dabiré, R. K., & Diabaté, A. (2020). Infection of highly insecticide-resistant malaria vector *Anopheles coluzzii* with entomopathogenic bacteria *Chromobacterium violaceum* reduces its survival, blood feeding propensity and fecundity. *Malaria Journal*, *19*(1), 352. https://doi.org/10.1186/s12936-020-03420-4

Hirschmann, M., Grundmann, F., & Bode, H. B. (2017). Identification and occurrence of the hydroxamate siderophores aerobactin, putrebactin, avaroferrin and ochrobactin C as virulence factors from entomopathogenic bacteria. *Environmental Microbiology*, *19*(10), 4080–4090. https://doi.org/10.1111/1462-2920.13845

Jackson, T. A. (2017). Entomopathogenic bacteria: Mass production, formulation, and quality control. In L. A. Lacey (Ed.), *Microbial control of insect and mite pests* (pp. 125–139). Academic Press. https://doi.org/10.1016/B978-0-12-803527-6.00008-1

Jamal, Q. M. S., & Ahmad, V. (2022). *Lysinibacilli*: A biological factories intended for bio-insecticidal, bio-control, and bioremediation activities. *Journal of Fungi*, *8*(12), 1288. https://doi.org/10.3390/jof8121288

Kebede, G., Abera, S., Haregu, S., Yeshitila, A., & Palanivel, H. (2022). Genetically modified bacteria for alleviating agrochemical impact on the environment. In M. Naeem, J. F. J. Bremont, A. A. Ansari, & S. S. Gill (Eds.), *Agrochemicals in soil and environment* (pp. 565–583). Springer Nature. https://doi.org/10.1007/978-981-16-9310-6_24

Korany, S. M., Mansour, A. N., El-Hendawy, H. H., Kobisi, A. N. A., & Aly, H. H. (2019). Entomopathogenic efficacy of the chitinolytic bacteria: *Aeromonas hydrophila* isolated from Siwa Oasis, Egypt. *Egyptian Journal of Biological Pest Control*, *29*, 16. https://doi.org/10.1186/s41938-019-0116-x

Kusakabe, A., Wang, C., Xu, Y., Molnár, I., & Stock, S. P. (2022). Selective toxicity of secondary metabolites from the entomopathogenic bacterium *Photorhabdus luminescens sonorensis* against selected plant parasitic nematodes of the Tylenchina suborder. *Microbiology Spectrum*, *10*(1), e02577–21. https://doi.org/10.1128/spectrum.02577-21

Lacey, L. A., Frutos, R., Kaya, H. K., & Vail, P. (2001). Insect pathogens as biological control agents: Do they have a future? *Biological Control*, *21*(3), 230–248. https://doi.org/10.1006/bcon.2001.0938

Lacey, L. A., Grzywacz, D., Shapiro-Ilan, D. I., Frutos, R., Brownbridge, M., & Goettel, M. S. (2015). Insect pathogens as biological control agents: Back to the future. *Journal of Invertebrate Pathology*, *132*, 1–41. https://doi.org/10.1016/j.jip.2015.07.009

Lacey, L. A., & Kaya, H. K. (2007). *Field manual of techniques in invertebrate pathology: Application and evaluation of pathogens for control of insects and other invertebrate pests* (2nd ed.). Springer. https://doi.org/10.1007/978-1-4020-5933-9

Marwa, A. M., Abdel-Aziz, A. K., Abdel-Hafez, F. H., & Zahia, K. M. (2020). Effect of sodium carbonate and mint oil as additives on the potency of the entomopathogenic bacteria, *Bacillus thuringiensis* against the cotton leaf worm *Spodoptera littoralis* (Lepidoptera: Noctuidae). *Egyptian Journal of Plant Protection Research Institute*, *3*(2), 654–661.

Nawaz, A., Gogi, M. D., Sufyan, M., Arshad, M., Ali, Q., Tayyib, M., Shahid, M. R., Razzaq, F., & Zhou, W. (2022). Advances in nanobiotechnology as sustainable solutions for insect pest management. In A. S. Tanda (Ed.), *Advances in integrated pest management technology: Innovative and applied aspects* (pp. 421–455). Springer. https://doi.org/10.1007/978-3-030-94949-5_16

Pola, S., Kesharwani, A. K., Singh, J., Singh, D., & Kalia, V. K. (2022). Endophytic ability of indigenous *Bacillus thuringiensis* strain VKK-BB2: New horizons for the development of novel insect pest-resistant crops. *Egyptian Journal of Biological Pest Control*, *32*, 8. https://doi.org/10.1186/s41938-022-00512-y

Rajamani, M., & Negi, A. (2021). Biopesticides for pest management. In V. Venkatramanan, S. Shah, & R. Prasad (Eds.), *Sustainable bioeconomy: Pathways to sustainable development goals* (pp. 239–266). Springer. https://doi.org/10.1007/978-981-15-7321-7_11

Rao, T., & Jurat-Fuentes, J. L. (2020). Advances in the use of entomopathogenic bacteria/Microbial Control Agents (MCAs) as biopesticides in suppressing crop insect pests. In N. Birch, & T. Glare (Eds.), *Biopesticides for sustainable agriculture* (pp. 99–134). Burleigh Dodds Science Publishing. https://doi.org/10.19103/AS.2020.0073.06

Reinbacher, L., Bacher, S., Praprotnik, E., & Grabenweger, G. (2021). Standard non-target tests for risk assessment of plant protection products are unsuitable for entomopathogenic fungi—a proposal for a new protocol. *Journal of Soils and Sediments*, *21*(6), 2357–2368. https://doi.org/10.1007/s11368-021-02919-w

Ruiu, L. (2015). Insect pathogenic bacteria in integrated pest management. *Insects*, *6*(2), 352–367. https://doi.org/10.3390/insects6020352

Sachs, J. D. (2012). From millennium development goals to sustainable development goals. *The Lancet*, *379*(9832), 2206–2211. https://doi.org/10.1016/S0140-6736(12)60685-0

Sajnaga, E., & Kazimierczak, W. (2020). Evolution and taxonomy of nematode-associated entomopathogenic bacteria of the genera *Xenorhabdus* and *Photorhabdus*: An overview. *Symbiosis*, *80*(1), 1–13. https://doi.org/10.1007/s13199-019-00660-0

Schnepf, E., Crickmore, N., Rie, J. V., Lereclus, D., Baum, J., Feitelson, J., Zeigler, D. R., & Dean, D. H. (1998). *Bacillus thuringiensis* and its pesticidal crystal proteins. *Microbiology and Molecular Biology Reviews*, *62*(3), 775. https://www.ncbi.nlm.nih.gov/pmc/articles/PMC98934/

Sethi, S., Dhakad, S., & Arora, S. (2022). The use of biopesticides for sustainable farming: Way forward toward Sustainable Development Goals (SDGs). In S. Arora, A. Kumar, S. Ogita, & Y.-Y. Yau (Eds.), *Biotechnological innovations for environmental bioremediation* (pp. 571–596). Springer Nature. https://doi.org/10.1007/978-981-16-9001-3_24

Sharma, L., Bohra, N., Singh, R. K., & Marques, G. (2019). Potential of entomopathogenic bacteria and fungi. In M. Khan, & W. Ahmad (Eds.), *Microbes for sustainable insect pest management: An eco-friendly approach* (pp. 115–149). Springer. https://doi.org/10.1007/978-3-030-23045-6_4

Soumia, P. S., Krishna, R., Jaiswal, D. K., Verma, J. P., Yadav, J., & Singh, M. (2020). Entomopathogenic microbes for sustainable crop protection: Future perspectives. In A. N. Yadav, J. Singh, C. Singh, & N. Yadav (Eds.), *Current trends in microbial biotechnology for sustainable agriculture* (pp. 469–497). Springer. https://doi.org/10.1007/978-981-15-6949-4_19

Tozlu, E., Tozlu, G., & Kotan, R. (2022). Biocontrol of *Cymbalophora rivularis* (Menetries) (Lepidoptera: Erebidae) larvae by entomopathogenic bacteria and fungi. *Egyptian Journal of Biological Pest Control*, *32*(1), 66. https://doi.org/10.1186/s41938-022-00565-z

Wood, S. L. R., Jones, S. K., Johnson, J. A., Brauman, K. A., Chaplin-Kramer, R., Fremier, A., Girvetz, E., Gordon, L. J., Kappel, C. V., Mandle, L., Mulligan, M., O'Farrell, P., Smith, W. K., Willemen, L., Zhang, W., & DeClerck, F. A. (2018). Distilling the role of ecosystem services in the sustainable development goals. *Ecosystem Services*, *29*(2018), 70–82. https://doi.org/10.1016/j.ecoser.2017.10.010

Index

Note: Page numbers in **bold** indicate a table on the corresponding page

D

E

Q

R

S

T

U

V

W

X

Y

Z

V

W

X

Y

Z

For Product Safety Concerns and Information please contact our
EU representative GPSR@taylorandfrancis.com Taylor & Francis
Verlag GmbH, Kaufingerstraße 24, 80331 München, Germany